'Maps move, and this Handbook assembles a variety of vantage points to witness such movements: textual, sensorial and the more-than-representational, cinematic and the virtual, resistive and mundane, grounded and atmospheric, monumental and ephemeral. Careful to not recuperate mapmaking but make it more responsible, more resonating, this collection bends, without breaking, the reverberative potential of the drawn line. It leaves mapmaking practices more curious, more open, more vibrational, without the privilege of an ahistorical treatment.'

Matthew W. Wilson, *Professor of Geography, University of Kentucky, USA*

'Tania Rossetto and Laura Lo Presti have compiled a state-of-the-art collection of commentaries on the many ways in which the humanities and cartography are joined at the hip. Bringing together an international and interdisciplinary cast of writers on the cutting edge of geohumanistic enquiry they show how the seemingly instrumental rationalities of the map have always been, and always should be, richly discursive endeavours embedded in strategies of domination and resistance. This is a must-read collection for scholars across the humanities interested in the role of cartography in human meaning-making.'

Tim Cresswell, *Ogilvie Professor of Geography, University of Edinburgh, UK*

'Mapping remains an extraordinarily diverse and generative technique for mediating the world. Committed to theoretical and methodological pluralism, this outstanding collection explores its technologies, politics and consequences through a rich range of case studies drawn from across "cartographic culture", both historical and contemporary.'

Gillian Rose, *Professor of Human Geography, University of Oxford, UK*

THE ROUTLEDGE HANDBOOK OF CARTOGRAPHIC HUMANITIES

The Routledge Handbook of Cartographic Humanities offers a vibrant exploration of the intersection and convergence between map studies and the humanities through the multifaceted traditions and inclinations from different disciplinary, geographical and cultural contexts.

With 42 chapters from leading scholars, this book provides an intellectual infrastructure to navigate core theories, critical concepts, phenomenologies and ecologies of mapping, while also providing insights into exciting new directions for future scholarship. It is organised into seven parts:

- Part 1 moves from the depths of the humans-maps relation to the posthuman dimension, from antiquity to the future of humanity, presenting a multidisciplinary perspective that bridges chronological distances, introspective instances and social engagements.
- Part 2 draws on ancient, archaeological, historical and literary sources, to consider the materialities and textures embedded in such texts. Fictional and non-fictional cartographies are explored, including layers of time, mobile historical phenomena, unmappable terrain features and even animal perspectives.
- Part 3 examines maps and mappings from a medial perspective, offering theoretical insight into cartographic mediality as well as studies of its intermedial relations with other media.
- Part 4 explores how a cultural cartographic perspective can be productive in researching the digital as a human experience, considering the development of a cultural attentiveness to a wide range of map-related phenomena that interweave human subjectivities and nonhuman entities in a digital ecology.
- Part 5 addresses a range of issues and urgencies that have been, and still are, at the centre of critical cartographic thinking, from politics, inequalities and discrimination.
- Part 6 considers the growing amount of literature and creative experimentation that involve mapping in practices of eliciting individual life histories, collective identities and self-accounts.
- Part 7 examines the variety of ways in which we can think of maps in the public realm.

This innovative and expansive Handbook will appeal to those in the fields of geography, art, philosophy, media and visual studies, anthropology, history, digital humanities and cultural studies as well as industry professionals.

Tania Rossetto is Associate Professor of Cultural Geography at the University of Padua, Italy.

Laura Lo Presti is Junior Assistant Professor of Geography at the University of Padua, Italy.

THE ROUTLEDGE HANDBOOK OF CARTOGRAPHIC HUMANITIES

Edited by Tania Rossetto and Laura Lo Presti

LONDON AND NEW YORK

Designed cover image: *Beneath the Lines*, artwork by Fabio Roncato in collaboration with the Museum of Geography of the University of Padua. Photo by Giada Peterle. Courtesy of the authors.

Artwork realised under the frame of the creative project *MAPPArti*, curated by Giada Peterle and Giovanni Donadelli, and exhibited at the Museum of Geography of the University of Padua, Sala della Musica from June, 30 to September, 8, 2023.

First published 2024
by Routledge
4 Park Square, Milton Park, Abingdon, Oxon OX14 4RN

and by Routledge
605 Third Avenue, New York, NY 10158

Routledge is an imprint of the Taylor & Francis Group, an informa business

British Library Cataloguing-in-Publication Data
A catalogue record for this book is available from the British Library

Library of Congress Cataloging-in-Publication Data
Names: Rossetto, Tania, editor. | Lo Presti, Laura, editor.
Title: The Routledge handbook of cartographic humanities / edited by Tania Rossetto and Laura Lo Presti.
Other titles: Handbook of cartographic humanities
Description: Abingdon, Oxon ; New York, NY : Routledge, 2024. | Includes bibliographical references and index.
Identifiers: LCCN 2023055432 (print) | LCCN 2023055433 (ebook) | ISBN 9781032355931 (hardback) | ISBN 9781032355948 (paperback) | ISBN 9781003327578 (ebook)
Subjects: LCSH: Cartography—History. | Humanities—History.
Classification: LCC GA108.7 .R68 2024 (print) | LCC GA108.7 (ebook) | DDC 526.09—dc23/eng/20240214
LC record available at https://lccn.loc.gov/2023055432
LC ebook record available at https://lccn.loc.gov/2023055433

ISBN: 978-1-032-35593-1 (hbk)
ISBN: 978-1-032-35594-8 (pbk)
ISBN: 978-1-003-32757-8 (ebk)

DOI: 10.4324/9781003327578

Typeset in Sabon
by Apex CoVantage, LLC

CONTENTS

FIGURES

TABLES

CONTRIBUTORS

Derek H. Alderman, University of Tennessee (USA), is a cultural-historical geographer specialising in public memory, race, civil rights, heritage tourism and critical approaches to mobility, place naming and mapping—often in the context of the African-American Freedom Struggle. He is (co)author of over 165 articles, chapters, and other essays.

Javier Arce-Nazario, University of North Carolina at Chapel Hill (USA), is an associate professor of geography. His work in South America and the Caribbean looks for new landscape histories and ways of mapping to increase the impact non-academic stakeholders have on landscape change, and the positive impact that landscape change research has on their communities.

Giorgio Avezzù, University of Bergamo (Italy), is an assistant professor of film and media studies. He is the author of two books dedicated to the relations between audiovisual content and geography; the most recent one discusses the geographies of audiovisual consumption in Italy.

Laura Bliss is a writer, editor and reporter. On staff at Bloomberg CityLab, she covers cities and the environment and is the founder of MapLab, a newsletter about maps. She was the host of Bedrock, USA, a podcast about extremism in local government, which was a finalist for a Webby Award and an Online Journalism Award. She was 2022–2023 Knight Science Journalism Fellow at MIT.

Barbara Brayshay, Royal Holloway University of London (UK), is a researcher in the Department of Geography specialising in the oral history of the UK Environmental Movement. She is also a director of LivingMaps Network, a group of radical cartographers, activists and artists with an interest in counter-mapping for social change.

Martin Brückner, University of Delaware (USA), is the director of the Winterthur Program in American Material Culture. He is the author of two award-winning books on geographic literacy and the social life of maps; his published work addresses the material and spatial imagination in early American literary and visual culture.

Tadas Bugnevicius, Columbia University (USA), is a film and media scholar working on theories of modernity, post-WW2 French and Francophone cinema and Eastern European postsocialism. He is currently writing a book manuscript that reassesses the politics and aesthetics of the French Tradition of Quality (1941–1959).

Sally Bushell, Lancaster University (UK), is a professor of Romantic and Victorian Literature. Her research is concerned with literary spatiality and the mapping of texts in a range of ways (across process; empirically; digitally). She is also interested in digital and spatial projects for the mapping of literature.

Sébastien Caquard, Concordia University (Montreal/Tiohtià:ke, Canada), is a human cartographer working at the intersection between mapping, technologies, oral history and memories. He has led the development of Atlascine, an open-source platform dedicated to map stories in depth, to listen to them and to reflect on cartographic processes and practices.

Valentina Carraro, University of Amsterdam (the Netherlands), is an assistant professor at the Department of Human Geography, Planning and International Development Studies. Her research and teaching sit at the intersection of critical cartography, digital and political geography, with a focus on how digital technologies and practices reconfigure political imaginaries and relations.

David Chandler, University of Westminster (UK), is a professor of international relations. He edits the journal *Anthropocenes: Human, Inhuman, Posthuman*, and his recent books include *Resilience in the Anthropocene: Governance and Politics at the End of the World* (2020) and *Ontopolitics in the Anthropocene: An Introduction to Mapping, Sensing and Hacking* (2018).

Christina E. Dando, University of Nebraska Omaha (USA), investigates the intersection of landscape, media and gender. She is the author of *Women and Cartography in the Progressive Era* (2019), an exploration of American women's utilisation of mapping to advance causes that were important to them, including community improvement and suffrage.

Aldo de Moor is the founder of CommunitySense, a Dutch research consultancy specialising in community informatics. With a PhD from Tilburg University, he focuses on participatory community network mapping, collaborative sensemaking, collaboration ecosystems analysis and social-technical systems design, addressing complex societal issues through social innovation.

Veronica della Dora, Royal Holloway University of London (UK), is a professor of human geography. Her research interests and publications span historical and cultural geography and the history of cartography. Her current project explores representations of the life journey metaphor in western culture from classical antiquity through the age of the Anthropocene.

Janet Downie, University of North Carolina at Chapel Hill (USA), is a classicist specialising in Greek literature of the Roman Imperial period. Her current research focuses on ancient geographical writing, literary topographies and constructions of Greek identity in the post-classical Mediterranean.

Mike Duggan, King's College London (UK), is a cultural geographer interested in how maps and mapping technologies shape social life and cultural practices. He is the editor-in-chief of the *Livingmaps Review*, a bi-annual journal for radical and critical cartography.

Nancy Duxbury, University of Coimbra (Portugal), is a senior researcher at the Centre for Social Studies. Her research spans cultural mapping, creative tourism development and culture in local sustainability. She leads the Horizon Europe project 'IN SITU: Place-based Innovation of Cultural and Creative Industries in Non-urban Areas' (https://insituculture.eu/, 2022–2026).

Matthew H. Edney, University of Southern Maine (USA), is Osher Professor in the History of Cartography and the Director of the History of Cartography Project (UW—Madison). He currently researches the historical emergence and articulation of fundamental concepts such as 'cartography' and 'the map'. See mappingasprocess.net for a full bibliography of his works.

W. F. Garrett-Petts, Thompson Rivers University (Canada), is a professor of English. He is engaged in exploring questions of visual and verbal culture, cultural mapping and the artistic animation of small cities. He has presented and published over 170 papers and articles, and 18 books, including, most recently, *Artistic Approaches to Cultural Mapping* (2019).

Joe Gerlach, University of Bristol (UK), is a cultural geographer with research interests in non-representational theory, critical cartography, ethics, geophilosophy, Spinoza, and micropolitics and minor theory.

Tiago Luís Gil, University of Brasilia (Brazil), is an associate professor of history of the Americas in the Department of History and the coordinator of the Digital Atlas of Portuguese America. He is the author of many books, chapters and articles. His research interests include HGIS, digital history, economic history and databases.

Piraye Hacıgüzeller, University of Antwerp (Belgium), is trained as an archaeologist and engineer. She works with geodata, metadata and machine learning within heritage, humanities and archaeological research, and on the philosophy of archaeological thought. She co-edited *Re-Mapping Archaeology: Critical Perspectives, Alternative Mappings* (2019).

Severin Halder, kollektiv orangotango/University of Münster (Germany), is an activist and geographer driven by experiences with everyday resistance in Rio de Janeiro, Bogotá, Berlin, Chiloé a.o. while working within kollektiv orangotango, community gardens and academia. He is currently working on the evolutions of the Not-an-Atlas project and is curating a compost festival.

Stephen P. Hanna, University of Mary Washington (USA), is a cartographer and cultural geographer focused on commemorative spaces and heritage tourism. His work in applying and developing critical mapping approaches to the ways public history is practised in landscapes can be found in over 70 articles, book, book chapters, research reports, opinion essays and other publications.

Tom Harper is the British Library's lead curator of antiquarian mapping. He has worked on numerous public map exhibitions including *Magnificent Maps: Power Propaganda and Art* (2010–2011), *Lines in the Ice: Seeking the Northwest Passage* (2014), and *Maps & the 20th Century: Drawing the Line* (2016–2017).

Sam Hind, University of Manchester (UK), is a media scholar researching digital navigation, sensing and autonomous driving. He is a co-author and co-editor of books on playful mapping and the praxeology of data.

Emmanuelle Kayiganwa is a retired Montrealer of Rwandan origin who shared her life story in 2009 as part of the Montreal Life Stories project. She has collaborated in the Atlas of Rwandan Life Stories since 2017 and worked closely with Élise Olmedo since 2020 on the co-construction of a *Subjective Atlas* of her own story (2023).

Ferran Larroya, Unversitat de Barcelona (Spain), is a physicist working on his PhD at the UB Institute of Complex Systems. His research focuses on the study of human behaviour in the framework of Complex Systems Physics and especially on pedestrian mobility with data collected through citizen science practices in public experiments.

Salvatore Liccardo, University of Vienna (Austria), is a historian working on processes of identity formation in Late Antiquity, with a special focus on examples of lists of ethnonyms. His research also covers late and post-Roman geographical knowledge and imaginations, with specific emphasis on the analysis of the Tabula Peutingeriana.

Laura Lo Presti, University of Padua (Italy), is a cultural geographer and (co)author of over 40 articles, chapters, books and other essays that explore cartography through theoretical and methodological contaminations coming from visual studies and contemporary art, mobility and migration studies, and critical theory.

Chris Lukinbeal, University of Arizona (USA), is a professor of geography and founding director of the Geographic Information Systems Technology Programs. He is the past president of the Association of Pacific Coast Geographers, an associate editor for *Geohumanities*, with books on media's mapping impulse, mediated geographies, place and television, and the geography of cinema.

Tommaso Morawski, Bibliotheca Hertziana—Max Planck Institute for Art History of Rome (Italy), obtained his PhD in philosophy and history of philosophy at Sapienza Università di Roma in 2017. His major research foci include aesthetics, German Enlightenment (especially Kant), philosophy of cartography, space, imagination and media theory. He is member of the editorial board of the international journal *Pòlemos. Materiali di Filosofia e critica sociale.*

Julien Nègre, École Normale Supérieure de Lyon (France), is an associate professor of American Studies and a junior fellow of the Institut Universitaire de France (IUF). His research explores the place of cartographic documents in American literary history. He is the author of a book on Thoreau's maps and of articles on William Byrd, Cooper, Melville and Kerouac.

Taien Ng-Chan, York University (Canada), is a writer and media artist working with emergent technologies (such as VR/AR or locative sound) and experimental mapping processes. In addition to her creative and scholarly essays, she has written for stage, screen and radio and exhibited in film festivals, conference events and art galleries internationally.

Élise Olmedo, Concordia University (Canada), is a postdoctoral researcher in geography. Her research focuses on the development of the concept of sensibility mapping ('Cartographie sensible' in French) by building bridges between mapping practices and the worlds of sensibility, emotion and intimacy, and between research and creation.

Roger Paez, ELISAVA Barcelona School of Design and Engineering, UVic-UCC (Spain), is a PhD architect working at the intersection of design, architecture and the city, focusing on temporality, experimentation and social impact. He is author of several publications where he explores maps as design tools in the expanded field of spatial design.

Davide Papotti, University of Parma (Italy), is a cultural geographer whose work mainly focuses on literary geographies and geographical imageries. He studied at the University of Parma (BA, 1993), at the University of Virginia (MA in Italian literature in 1996), and at the University of Padua (PhD in geography in 2002).

Andrea Pase, University of Padua (Italy), is a historical and social geographer. His research interests concern, among others, the unfolding of modern state territoriality, with particular reference to the use of cartography in the process of establishing borders (in Europe and in colonial and post-colonial Africa).

Marie Patino is a graphics reporter at *Bloomberg News*, where she builds interactive data visualisations for the newsroom. She makes maps and writes about them for MapLab with Laura Bliss.

Davi Pereira Junior, Maranhão State University (Brazil), is a Quilombola intellectual, activist and anthropologist. His research interests include Black social movements in the Americas; Black community struggles for collective land rights; traditional peoples in Brazil; and the role of cartography for inequality and sustainability issues facing Indigenous and Black people.

Josep Perelló, Universitat de Barcelona (Spain), is a full professor and researcher at the UB Institute of Complex Systems. He is the research leader of OpenSystems, a pioneering research group in the field of citizen science conducting public experiments in collaboration with artists, designers, museums and cultural festivals.

Giada Peterle, University of Padua (Italy), is a cultural geographer working on graphic geographies, art-research collaborations, narrative geographies and mobilities. She authored the book *Comics as a Research Practice: Drawing Narrative Geographies Beyond the Frame* (2021) and is the director of the Museum of Geography of the University of Padua.

John Pickles, University of North Carolina at Chapel Hill (USA), is the D. W. Patterson distinguished professor of geography and international studies. His work focuses on cultural and social theory, the political economy of development, economic integration in post-socialist Europe, and the politics of mapping and border-making.

Amy E. Potter, Georgia Southern University (USA), is a cultural geographer working in the area of tourism geographies, particularly the intersections of tourism, memory and race. She is the author of over 30 publications including co-editor of *Social Memory and Heritage Tourism Methodologies* (2015).

Claire Reddleman, University of Manchester (UK), teaches digital humanities and art history, with interests in digital cultural heritage, visual methods, mapping, contemporary art and 'ways of seeing' using new technologies. Claire's monograph *Cartographic Abstraction in Contemporary Art: Seeing With Maps* (2019) uses artworks to critique capitalist modes of abstraction.

Les Roberts, University of Liverpool (UK), is a reader in cultural and media studies in the School of the Arts and co-director of the Centre for Culture and Everyday Life. His most recent publication is the monograph *Posthuman Buddhism and the Digital Self: The Production of Dwellspace* (2023).

Piera Rossetto, Ca' Foscari University of Venice (Italy), is Rita Levi Montalcini assistant professor in Hebrew language and literature. She holds a PhD in social and historical anthropology and in languages and civilisations of Asia and North Africa. She is interested in Jewish post-colonial migrations from the MENA region.

Tania Rossetto, University of Padua (Italy), is a cultural geographer working on geographic epistemologies, geovisualities, cultural cartography and urban studies. She is the author of *Object-Oriented Cartography: Maps as Things* (2019), where she explores the possibilities of a speculative-realist map theory by bringing cartographic objects to the foreground.

Paul Schweizer, kollektiv orangotango/University of Halle (Germany), is a geographer and popular educator. As part of kollektiv orangotango, he co-conducts collective art interventions in public space. He co-edited *This Is Not an Atlas* and curates the Not-an-Atlas-platform. Currently he organises mapping processes in Europe and Latin America to facilitate a global dialogue of engaged cartographies.

Jörn Seemann, Ball State University (USA), is a cultural geographer who has been teaching cartography for more than 25 years. His research interests cover a wide range of topics from cultural perspectives in cartography, map history, geographic thought and visual narratives to qualitative methods, cultural landscapes and cultural geographies of Latin America.

Tim Shea, University of North Carolina at Chapel Hill (USA), is an assistant professor of classical archaeology specialising in the art, archaeology and topography of ancient Greece in the archaic and classical periods. He is currently exploring how Athenian immigrant communities expressed their identities through tombstones and burial plots.

Bjørn Sletto, University of Texas (USA), researches environmental and social justice, informality, and insurgent and decolonial planning in Latin America. His particular interests include knowledge co-production in planning processes and the role of citizen planners in producing just and sustainable urban landscapes.

Manuela Valtchanova, ELISAVA Barcelona School of Design and Engineering, UVic-UCC (Spain), is a practising architect and a researcher working between politics, space and intersubjectivity with a research interest in architecture of action. In her professional, academic and research career, she has developed different practices with a specific focus on temporary socio-spatial interventions and actions of spatial justice.

Toni Veneri, Wake Forest University (North Carolina, USA), is a literary scholar who conducts interdisciplinary research on early modern geographical and cartographical culture, exploring its environmental and artistic ramifications, especially in the Mediterranean. His publications include works on travel literature, maps and the imagination of the sea in Renaissance Venice.

Clancy Wilmott, University of California, Berkeley (USA), is an assistant professor of critical cartography, geovisualisation and design in the Berkeley Center for New Media and the Department of Geography. She researches cartography, colonialism, digitalities and spatial practice.

Bo Zhao, University of Washington, Seattle (USA), is a critical GIS scholar working on emerging GIS technologies and their social implications. He proposed a humanistic GIS agenda and urges geographers to examine deepfake geography, where he calls for more critical and ethical approaches to detecting and preventing spatial misinformation.

INTRODUCTION

Why Cartographic Humanities?

Tania Rossetto and Laura Lo Presti

Humanistic map studies: an expansive field

The humanities have been part of geographical knowledge and its expression for centuries, finding in maps and cartographic imaginations useful and intimate companions to reflect with, challenge and advance new spatial paradigms, methods and metaphors. After the more recent rise of the 'spatial turn' in the arts and humanities and the proliferation of digital technologies in several cultural domains, new research areas such as spatial digital humanities, geohumanities, deep mapping and map art, to name a few, have shown that the engagements of scholars and practitioners with cartography and mapping practices have expanded further, becoming increasingly diverse and highly mutable. In parallel to the growing fascination with cartography that arose within various humanistic fields, in the last 15 years, we have witnessed the emergence of 'map studies' as a transversal research area that is strongly affected by humanistic approaches and methodologies. This area intersects not only more established traditions such as the history of cartography and critical cartography but also the multifaceted realm of 'cultural cartography' (Cosgrove, 2008). The *Routledge Handbook of Cartographic Humanities* is precisely designed to explore the intersection and convergence between cultural map studies and the humanities, expressing multifaceted traditions and inclinations coming from different disciplinary, geographical and cultural contexts. Various humanistic understandings of maps have been published, yet often they have been presented as ancillary to the core cartographic content of a book or treated within a single humanistic domain (e.g. art and cartography; literature and cartography; media and cartography). Here, a broad humanistic gaze is adopted not just to show how cartography traverses distinct disciplinary fields of the humanities but also to suggest foundational nodes, shared instances, ongoing contradictions, recurring interrogations and present urgencies emerging within the developing global arena of humanistic map scholarship.

In inviting our contributors and assembling the handbook, we reflected on the point whether each singular nexus that currently entangles cartography with a humanistic field could be reframed under a broader but peculiar theoretical and practical space of encounter: the cartographic humanities. In parallel to the geohumanities (an umbrella term that embraces the growing interdisciplinary engagement between geography and arts and

DOI: 10.4324/9781003327578-1

humanities scholarship and practice), the 'cartohumanities' explore how cartographic knowledge is advanced across the humanities and how, vice versa, map scholarship contributes to the human sciences. Consequently, in our view, the cartographic humanities is an 'expansive' concept rather than a comprehensive, finite one. Working expansively rather than comprehensively means, in fact, that there is potential space and time for cartographic humanities to grow further and include always new mapping theories, methodologies and empirical works rather than defining the boundaries and centre of this new territory once and for all.

This handbook comes several years after first attempts to focus 'map studies' and collect foundational works on maps, such as *Rethinking Maps* (Dodge et al., 2009) and *The Map Reader* (Dodge et al., 2011), which were followed by the more recent *Routledge Handbook of Mapping and Cartography* (Kent and Vujakovic, 2018) and *Mapping Across Academia* (Brunn and Dodge, 2017) volumes. The *Routledge Handbook of Cartographic Humanities* now offers the possibility to trace and gauge, after more than a decade, the value and expansion of map studies, exploring the charming 'figure of the map' (Mitchell, 2008) in the humanities while giving visibility to a growing vibrant network of map researchers with diverse disciplinary backgrounds. Acknowledging the importance of merging cultural cartography with other academic fields, we aim to provide a kind of intellectual 'infrastructure' (Lo Presti, this volume) to navigate core theories, critical concepts, phenomenologies and ecologies of mapping, while also providing insights into exciting new directions for future scholarship on the broader frame of the cartographic humanities.

Being in dialogue with 'all the humanities'

Interviewed by John Fraser Hart at the University of Minnesota on 24 April 1972, about the meaning of the phrase 'humanistic geography' (distinct from 'human geography'), Yi-Fu Tuan responded that he meant 'an explicit recognition or ultimate interest in the human condition' and a consequent 'general interest with all the humanities'.[1] By associating geography with cartography, one could see the cartographic humanities simply as a byproduct of the now established geohumanities (Dear et al., 2011). Yet, as masterfully highlighted by della Dora in the opening chapter of this handbook, the phrase 'cartographic humanities' expresses a peculiar tension, one which can be found in the very act of 'humanising the map' (della Dora, this volume). In fact, while we are stressing how much cartography and mapping are today greatly valued within several cultural domains, we should not forget—as several of the chapters included in this handbook demonstrate—that maps have always been, and still are, contradictorily understood within the humanities.

One of the most common critical and sceptical considerations deriving from this contradictory view is that western mapping practices are inherently irreconcilable with the expression of human experiences and emotions. Expressive forms, such as writing, painting or photography, have been considered more apt ways of grasping the human condition: the map is detachment, whereas narration is involvement; the map is an act of measuring, whereas art has the power to reveal the nuances of human life; maps communicate spaces of homogeneity, whereas literary texts convey variations of human sentiments. Significantly, one of the early reactions of geographers when the idea of mapping literary works was advanced in the well-known *Atlas of the European Novel, 1800–1900* by Franco Moretti (1998) was to criticise this act as an attempt to apply a Cartesian grid to literary texts, thus neglecting the qualitative dimension of both textual and real places acknowledged by human geography (Rossetto, 2016: 261). Italian geographer Claudio Cerreti (1998), for instance, saw Moretti as influenced by spatial analysis rather than geography, with his denotative geometry of the *extensio* of spatial relations being far from a geography of the connoted *intensio* of places. Cartography, in summary, was seen as 'less human' than either geography or literature.

While the unmappability of literature is still a hotly debated issue, literary scholars have come to adopt nuanced, complex and productive approaches towards the relationships between cartography and map studies (see for instance Nègre, this volume; and Bushell, this volume). Yet it is not the case that when cartography is generally linked to the humanities, the association is most frequently presumed in technological terms, with reference to literary or historical geographical information systems (GIS) or other applications involving geolocated data. Significantly, when delineating the resurgence of some 'geographical primitives' within the newly established geohumanities, Hawkins (2020: 97–98) includes the idea of 'location' among such returned primitives. Attributed to 'that *cartographic* imperative that drove the discipline for centuries' (italics added), she sees the idea of location as now re-emerging in the context of the GIS-based digital humanities, 'which ensured that location is a crucial part of how the humanities are thinking about space and place' (Hawkins, 2020: 97–98). Yet this could lead one to think that cartography is nearly entirely inherent to forms of 'digital geohumanities' (Travis, 2020). Indeed, our delineation of the cartographic humanities obviously includes digital culture in a pervasive way (see in particular Part 4 in this volume), but it remains distant from a more tech-oriented application of cartography and GIScience, which is typically associated with the digital humanities. By being practised within a wide range of disciplines and fields—such as cultural geography, literary studies, history, philosophy, classical studies, archaeology, anthropology, film, media and visual studies, curatorial and art practice, architecture, ethnography, digital culture, political and post-colonial studies, among others—the cartographic humanities eludes a reductive affiliation to the digital humanities and shows the capacity to be in dialogue with 'all the humanities', as Tuan would say.

The productive contradictions of the cartohumanities

As we have seen, ingrained views on cartography (Edney, 2019) often bring with them a series of tensions, contradictions and binaries—the quantitative and the qualitative, data and feelings, the mathematical and the discursive, the abstract and the embodied, cartography and chorography, rationality and creativity, exactness and impression, to name a few—which are also reflected in several of the chapters of this handbook (see for instance Arce-Nazario et al., this volume; or Luckinbeal, this volume). As Crampton (2010: 56) reminded, while it was modern cartography that identified a scientific divide between the 'proper' and the 'transgressive' map, it was Harley (1989) who identified the (detrimental) tendency to apply binary thinking to maps and the delineation of cartography as a discourse of opposites (see also Kitchin et al., 2009: 2–4). The point of view of the cartographic humanities, to recall the expression used in a seminal article that largely contributed to opening up the perspectives of map studies 20 years ago (Del Casino and Hanna, 2005), is fundamentally to stay productively 'beyond the binaries'. The cartographic humanities, in fact, stems from the recognition that the complexity of the cartographic realm requires not only multidisciplinary investigations but also theoretical and methodological pluralism and inclusiveness. As Roberts (2012: 12) now famously put it, 'there is so much more to say about mapping than is often said in cartographic circles'. He noted that although the use of the cartographic lexicon by non-geo/cartographers

> has its problems and frustrations . . . at the same time the semantic ambiguity that has arguably dogged theoretical discourses in recent years presents us with challenges that can enliven and enrich, rather than inhibit, critical understandings of cultures of mapping.
>
> *(Roberts, 2012: 11)*

Indeed, this messiness is a clear symptom of what Lo Presti (2018) called 'an extroversion' which currently characterises cartography and mapping. The encounters between cartography and the humanities help in enlarging the horizons of map thinking, showing how maps are, and have always been, often simultaneously—rather than in oppositional ways—representations, practices, powerful devices, material objects, cognitive tools, embodied things, works of creativity, emplaced actions, living entities, technological machines, networks of feelings and much more. The recent multiplication and expansion of cartographic worlds, repertoires and phenomenologies that we encounter in our particularly 'rich cartographic culture', as Cosgrove (2008) suggested, have only made more palpable this need for a pluralistic, and humanistic, study of maps.

In this light, we see the cover of our handbook as particularly expressive. It portrays a work by artist Fabio Roncato from the series *Beneath the Lines*, which was created after his residence at the Museum of Geography of the University of Padua, with which the editors and two of the contributors (Giada Peterle, director of the museum, and Andrea Pase) are affiliated. The concept underlying this and other future artist residences of the *MAP-PArti* project (curated by Giada Peterle and Giovanni Donadelli since 2023) is that of creatively mobilising a storage of topographical maps edited by the Italian Istituto Geografico Militare destined to pulping. By transforming the map sheet into a sculpture realised by merging hands in clay and then filling the groove with loam and chalk, Roncato evokes not only the extractivist impulse but also the wrapping visceral relationship with the Earth that cartography performs. In the photograph, behind this unorthodox, somehow irreverent but extremely fascinating treatment of the topographic map, we see a nineteenth-century globe, the reproduction of a modern planisphere and some shelves with rare books in the background of the Sala della Musica, the museum room that preserves some of the most prestigious historical cartographic objects of our university. This image represents for us a plea for an open, plural perspective that fosters exchange between the more established and the more effervescent variations of humanistic research on maps.

The need for multidisciplinary pluralism in cartographic research was also early recognised by Monmonier (2007: 371), who wrote about the emergence of a 'humanistic turn' in cartographic scholarship with reference to a variety of 'publications reflect[ing] a dimension of cartography concerned more with the appreciation and enjoyment of maps than with the more traditional technological and methodological agendas of academia'. Yet we could also mention here the impatience expressed 20 years ago by Pickles (2004: 19), who made an inspiring claim for a multivocal, more-than-critical study of maps:

> The still deeply rooted desire for totalising monochromatic accounts that explain the map in terms of it being a socially produced symbolic object, a tool of power, a form derived from a particular epistemology of the gaze, or a masculinist representation,

> seem to me to miss the point of the post structuralist turn: that is, that not only are maps multivocal . . . but so also must be our accounts of them.

In tune with, and inspired by, such early calls, this handbook is aimed at showing how map thinking and cartographic research have benefited from the mutual exchange with the humanities in responding to such impatience and enjoyment and how the development of the cartographic humanities can provide a promising platform to further advance map theories, methodologies and empirical explorations through open-ended paths.

Handbook overview

This handbook opens with Part 1, which moves from the depths of the humans-maps relation—considering both the temporal depth of such complex relation and the implication of inner human worlds—to the posthuman dimension. Moving also from the intimate to the collective, such relation involves not only the most profound subjective experience but also society and the Earth. Taking together reflections that run from antiquity to the future of humanity, Part 1 suggests that a multidisciplinary perspective within the humanities, in dialogue also with the social sciences, helps in bringing together chronological distances, introspective instances and social engagements. This is also made possible by a post-representational appreciation of maps of the past, the present and the future, an approach that stems from the more recent map thinking and that underlies the handbook as a whole. **Veronica della Dora** offers a foundational opening to both the field of study and the approach of the cartographic humanities, taking allegorical maps of life as a site to reflect on how cartography and the humanities are in no way antithetical. **Javier Arce-Nazario, Janet Downie, Tim Shea, John Pickles** and **Toni Veneri** draw from multidisciplinary conversations to reframe the 'chorography and cartography' opposition in ways that reveal the hermeneutical potential of ancient and contemporary conceptions of chorography and revive its language in a time of spatial humanities. **Matthew H. Edney**'s chapter reflects some crucial aspects of current cartographic theory questioning the universality of mapness and cartographic language to endorse a processual approach that grasps the endless historical and cultural variations produced by mapping practices and the assemblages through which they materialise in the world. Highlighting the anthropological and phenomenological underpinnings of a conception of maps as mapping practices, **Les Roberts** elaborates on the fortunate yet elusive concept of deep mapping and the paradoxical, poetic and productive sense of *unmappability* of the human world it conveys. **Paul Schweizer** and **Severin Halder** from **kollektiv orangotango** open their intervention to social worlds and reflect on activist mapping, which is a vital trend today within the humanities that aims at contributing to the building of more just societies. From the humanistic attitude of self-reflexivity comes the need for stepping back to a truly engaged collective cartographic practice that is able to cultivate deeper caring relationships. Expressing a diverse but equally ethical preoccupation, in no less human ways, **Joe Gerlach** takes the reader to the rarefied territories of posthuman cartographies. Despite its being quintessentially anthropocentric, cartography is here considered in its capacity to generate attentiveness to the posthuman condition and transfigured into a reparative, therapeutic and caring gesture towards the Earth.

Part 2 of this handbook draws from the centrality of textuality within the fabric of the humanities, unfolding the potentialities of the Latin word *textus* as a piece of weaving. In addressing ancient, archaeological, historical or literary sources, these chapters also consider the materialities and textures embedded in such texts and in the processes of writing, reading and interpreting. Leaving aside the theoretical conceptualisation of 'the map' to reveal more nuanced engagements with 'maps' and 'mapping practices', the chapters carefully handle fictional and non-fictional cartographies that work to expand our visions to include layers of time, mobile historical phenomena, unmappable terrain features and even animal point of views. This expansion is aided by an attitude appreciating slowness, contemplation, comparison and ethical reflection that emerges in the close reading of texts, material fragments, clues and traces found in the library, the archive or the archaeological site. While lingering on the specificity of cartographic or cartographically inflected documents, such contributions pay attention to the internal and external connections they activate with multiple contexts, thus seeing maps as always mapping entanglements. **Salvatore Liccardo** focuses on the interplay of texts, diagrams, memory and travel in late antique cartography, discussing the *Tabula Peutingeriana* as a vivid example of the eclectic entanglement and coexistence of signs and performative devices within a diagrammatic map. Re-reading critically the 'mappification of archaeology' and its impact on the discipline from its very beginning until recent times, **Piraye Hacıgüzeller** discusses several notions of the map as a 'craft' with an urgent political ethos. Meditating on how archaeological mapping shapes humans just as humans shape archaeological things and maps, she provides the reader with creative and provocative examples that trouble the perduring cartographic traditionalism of the field, pointing to the material effects that archaeological maps and their mapmakers produce. Questioning the idea that historians are usually sceptical regarding cartographic language, since it seems to frame time in a static way, **Tiago Luís Gil** elaborates on the long tradition of movement maps in historical studies, showing and contextualising a wide repertoire of techniques and visual vocabularies regarding historical processes featured in printed and static maps. By exploring the fictional worlds of literary texts and paratexts through an ecocritical perspective, **Sally Bushell** highlights how cartographic contradictions have the power to disclose non-human, biocentric modes of spatial being, with the effect of generating empathetic engagement with animals. **Julien Nègre** adopts a processual approach in focusing actual cartographic documents used, read and annotated by writers, with examples from U.S. literary history. Being literary constraints or catalysts for literary creation, such maps crucially contribute to expanding the connections between the world evoked by the text and the reality inhabited by the readers. Taking inspiration from the philosophy of slowness, **Jörn Seemann** attends to the unfolding of mapping processes, which can be revealed by a slow attunement to reading maps. Stepping aside to approach maps with a slower pace may not only be an act of resistance but also the rediscovery of a pleasure, that of lingering on maps to sense their materialities, trace their histories and listen to their stories.

Part 3 of this handbook is centred on one of the most extensive areas of exchange between map studies and the humanities, one which considers maps and mappings from a medial perspective. If, on the basis of their communicative functions, maps have been thought of as media for a long time, it is only recently that the complexity of media theory and media studies has come to permeate map studies. This process has led to a growing interest not only in the medialities of maps but also in the interrelations between them and other kinds of media. The chapters included in this part oscillate between these two poles, providing both theoretical insight into cartographic mediality as well as studies of its intermedial relations with other media. Hosting a constellations of media and intermedial possibilities that reveal how a taste for maps populates imaginaries as well as cultures of consumption, this part includes not only interventions on the more established binomial of cartography and cinema but also less common areas of research that play on the borders of cartography and unexpected forms of media. **Tommaso Morawski** opens this part by offering a philosophical reflection on the concept of cartography as a matrix of human imagination. A media-anthropological approach is applied to recognise cartography as a technology, and a necessary mediation, which plays a fundamental role in the constitution of the human and its cultural forms. **Giorgio Avezzù** reverses the common idea that cinema is a fundamentally cartographic medium because of its ability to represent the visible world in order to point out how cinema can be represented 'as' the world and therefore mapped in itself. He focuses on spatial interpretive categories such as world cinema and national cinematographies, as well as on the mapping of data that show territorial differences in audiovisual consumption and regional taste cultures. **Chris Lukinbeal** mobilises the paradoxical features of cartography arguing for an antithetical cartography that merges the scopic regimes and technologies of perspectivalism and projectionism. This rapidly developing antithetical cartography, or geospatial cinema, embraces cinema and cartography as mutually inclusive to produce geovisualisations that deploy cinematic form, language and scalar multivalence to analyse and render film location, production and consumption and much more. **Tania Rossetto** proposes three acts of map thinking that proceed from an engagement with three music videos. The music video morphes into a meta-map, a theoretical object that emanates different approaches to cartographic mediality as representational machine, lived practice and visceral thingness. **Davide Papotti** catches cartography in its leisure time, in contexts where it is freed from its usual and predictable functions. By analysing examples of print advertisements that feature maps while promoting objects or services other than maps, he appreciates the ways in which a taste for maps informs the imaginaries of western societies. **Roger Paez, Manuela Valtchanova, Ferran Larroya** and **Josep Perelló** consider the operative potential of maps in design practices. As a framing device and projective media, mapping can mediate existing site potentialities, inform urban visions and test potential realities, thus prefiguring how human habitats are perceived, construed and imagined.

Part 4 of this handbook is not aimed at delineating a separate category for the digital cartohumanities, since digitality inevitably pervades all the handbook's content. Rather, this part is aimed at disclosing how a cultural cartographic perspective could be productive in researching the digital as a human experience. Digital mapping practices of the everyday, as well as those involved in research practices, are reframed under a humanistic lens that emphasises not only the force of current digital cultures but also idiosyncratic itineraries and positionalities in the digital environment. 'Cultural digitalities' in the cartographic humanities do not refer to a tech-based use of locational data in the humanities but to the development of a cultural attentiveness to a wide range of map-related phenomena, habits and events that interweave human subjectivities and non-human entities in a digital ecology. **Claire Reddleman** interprets the 'blue dot' of smartphone-based mapping apps, and the practice of making GPS selfies that record our location, in terms of a narcissistic projection of the self into the map. The 'blue dot' thus becomes a self-image, a way to depict ourselves that reveals the individualisation and solipsism of our spatial existence via maps in the digital era. **Sam Hind** considers the many forms of automation and the agencies of machinic entities that have come to characterise our navigational experience through map apps over time. He suggests how, far from de-humanising navigation, automation entails human labour, skill and activity and deserves a more-than-technical, cultural approach to be understood, also in its psychic and ecological costs. **Valentina Carraro** revisits critical cartography through a techno-feminist approach. She draws from Donna Haraway's foundational works to highlight the iteration of the god-trick performed by contemporary geospatial software. She uses the concept of map fetishism not only to create critical awareness about it but also to raise hope in the insightful, useful, profound and transformative power of (digital) maps. **Mike Duggan** offers a historical overview of the use of 'ethnographic mapping' in research practices and concentrates on the more recent role played by digital and mobile mapping technologies in engaging with human lives and stories, as well as the recent trend of ethnographising mapping practices. **Bo Zhao** argues for a GIScience that incorporates a humanistic viewpoint to better navigate today's data-intensive society. Attending to the creative forces that spread through GIS and different types of human or non-human entities, he expresses ethical interrogations and caring stances through the four categories of embodiment, hermeneutic, autonomous and background GIS. **Tadas Bugnevicius** closes this part with a chapter in memory of Roland-François Lack (1960–2021), a humanities scholar who assembled a unique online collection of thousands of screen grabs from films featuring maps known as *The Cine-Tourist* website, now archived by the National Library of France. The practice of regularly posting cine-maps is viewed here as a counterpart of a teaching mastery as well as the manifestation of a deep, extremely refined, endless fascination with the lives of cartographic objects within filmic and digital environments.

Part 5 of this handbook addresses a number of relevant issues and urgencies in the critical consideration of maps and mapping. This part is not aimed at enclosing the critical, disruptive stance of map studies within a single container, since a critical attitude permeates a great number of chapters in the handbook. Rather, this part is aimed at highlighting a series of troubling political issues, deep-rooted inequalities and ongoing discriminations that have been, and still are, at the centre of critical cartographic thinking. Applying a humanistic sensibility to such critical thinking means mobilising conceptual creativity, historical sources, subjective and collective positionalities, evocative images and different kinds of cartographic as well as non-cartographic texts from different parts of the world in order to read the past while endorsing more just futures. **Andrea Pase** moves from a famous passage from Conrad's *Heart of Darkness*, where the young Marlow places his (white) finger on the blank spaces that still existed in the atlases of the second half of the nineteenth century and analyses two cartographic documents portraying inland Africa in the 1820s and 1920s to reflect on the racial cartographic filling of those blank spaces in the process of colonial appropriation. Discussing the ways in which cartographic sciences have shaped spatial representations and landscapes in settler-colonial societies, with reference to a case in Australia, **Clancy Willmott** argues that the inherent contradictions and limits of cartography facing the force and otherness of colonised material landscapes inadvertently transformed such cartography into a counter-colonial practice that still resonates today. **Davi Pereira Junior** and **Bjørn Sletto**'s intervention (with examples from Brazil) highlights that, by exercising the power to map under their specific epistemologies, ontologies and pedagogies, Indigenous peoples and Afrodiasporic communities demonstrate that cartography can be used to protect their rights to existence, territory and collective identities, providing an intergenerational and counter-hegemonic tool for struggling against various forms of silencing and oppression. **Stephen P. Hanna** concentrates on the counter-powers emanating from a nineteenth-century hand-drawn map included in a memoir written by a once-enslaved man in North America. The memory work done by this map entails not just a personal recovery of the experience of enslavement and emancipation by a Black cartographer but also a public practice of repairing over a century of erasure of race-based slavery from local official history and landscape heritage. **Christina E. Dando** provides an overview of the gendering of cartography as a masculine knowledge, illuminates the protagonism of women's mapping in the United States since the nineteenth century and then offers a close reading of the female/feminist worldview (including the humorous nudity of a man) emerging from a pictorial map created by the American Association of University Women in a campaign to support women's education in the late 1920s. Finally, **David Chandler** discusses the process of emptying out space of its inherent inter-relationalities induced by modernist conceptions of cartography, arguing that mapping has become instead central in the Anthropocene precisely in recognition of such relationality. Seen from an ontopolitical perspective, mapping turns into a mode of governance, a tentative non-linear adaptation to emergent social, economic and environmental conditions and disturbances.

Part 6 of this handbook revolves around the growing amount of literature and creative experimentation that involve mapping in practices of eliciting individual life histories, collective identities and self-accounts. Rather than working in the archive, in the library or in media environments, humanities and social sciences scholars here work in the field, in direct contact with individuals and groups to collect—and often co-produce—spatial narrations, memories and feelings through several creative, non-Euclidean, alternative forms of mapping. Maps therefore become points of contact and of empathetic exchange of looks and interactions of bodies that stimulate listening to one another. While some of these research gestures are aimed at revealing historico-political dynamics and related collective suffering, producing social impact and creating public awareness, a humanistic attitude ensures that there is also room for purely theoretical ruminations, fictional narrations, solipsistic fantasies and inward interrogations. **Élise Olmedo, Emmanuelle Kayiganwa** and **Sébastien Caquard** trace the history of the collaboration between a Montreal-based survivor of the 1994 genocide of the Tutsi in Rwanda and two researchers in order to reflect on the notion of the cartographic co-construction of memories, thus unfolding the back-and-forth processes employed over time to involve the storyteller in a sensitive mapping practice and share her life story and difficult experiences. In reading the history of contemporary Jewish migrations from North Africa and the Middle East across the Mediterranean through an anthropological lens, **Piera Rossetto** sees cartographic research creation within ethnographic work as a way to train the art of listening to diasporic recollections, as well as an opportunity to share with wider audiences such histories and particular stories through creative collaborations and outcomes. Reporting examples of body cartography projects created by bachelor students enrolled in a humanities degree programme, **Laura Lo Presti** exalts humanistic mapping as a creative platform through which difficult stories related to psychological and bodily traumas can be more easily shared. Here, the map develops as an infrastructure of feelings, betraying the unspoken aspiration of emotional mapping in order to become a cathartic form of public intimacy, an auto-cartography where people can lay bare their problems and innermost thoughts to others through a mediating image, without exposing themselves directly. **Giada Peterle** considers instead auto-cartography as a prolific encounter between autoethnography, fictional writing and narrative approaches to maps as research methods. Two pieces of auto-cartographic and carto-fictional writing show how the *self* that is narrated and examined could be either the researcher(s) producing, analysing, using and engaging with the map in the field or the map itself, through forms of creative non-human narration. **Nancy Duxbury** and **W. F. Garrett-Petts** offer an overview of participatory cultural mapping as a conversational platform and meeting place for discussion, empowerment and co-creation between diverse stakeholders. Whereas the vernacular cartographies stemming from community–academe collaborations hold the potential to represent the cultural dimensions of a place expressed by local voices, they require critical awareness, ethical cautions and epistemic contextualisation (within the wider cultural mapping field) to be carried out in meaningful and transformative ways. Drawing from an example of guided tours in plantation museums across the southeastern United States, **Stephen P. Hanna, Amy E. Potter** and **Derek H. Alderman** propose narrative mapping as a reiterative mobile methodology that captures the bodily performances of visitors touring historical sites and the unpredictable stories that emerge as these visitors journey through curated spaces, exhibited memories, material objects and public narratives.

Part 7 of this handbook engages with a variety of ways in which we can think of maps in the public realm. Since 'going public' has become an imperative within the humanities, maps have shown an enormous potential in engaging people and producing public outreach for cultural institutions such as universities, libraries or academic associations. Mapping, then, is very much practised in community services to support public actions. This part works to widen further the notion of mapping for the public by including a humanistic reading of the role maps played and still play in the public sphere and the public space, in the commercial realm and the information system. Being alternatively—and often simultaneously—commodities, tools of emancipation, informational devices, material artefacts and carriers of political views, maps pervade our lives as well as public arenas shaping a variety of cartographic practices and cultures. **Martin Brückner** reflects on the public life of maps from a historical and material perspective, showing how from the beginning of early modern cartography printed maps emerged as social artefacts transcending the goals of spatial orientation or representation. The cartographic and non-cartographic uses of such 'cartifacts' thus intersected, as cartographic commodities—and their biographies—variously resonated with audiences through public spaces, ritual actions, social interactions, pedagogical contexts or sentimental intimate experiences. Acknowledging that maps are often a strong exhibiting medium, **Tom Harper** examines the history and present of public map exhibitions. With particular reference to the activities of the British Library, by critically considering the host institution, the exhibition curator, the exhibited collection and the audience, he interrogates the measures of their success, and their future potentialities, reflecting on how such map exhibitions create knowledge following particular strategies, can not only reproduce the biases embedded in their exhibited collections but also operate as forces for change in societies. **Barbara Brayshay** and **Aldo de Moor**'s chapter presents a participatory mapping project undertaken in the Black Caribbean community in the London Borough of Lambeth commissioned to create a map of community support available to unemployed people. While, through storytelling, the marginalised community took stock of its issues and the available support services, network mapping transformed into a practical means for public action, thus inspiring a model of bridging the divide between community members, community support services and wider networks of external institutions and agencies. **Taien Ng-Chan** elaborates on the new meaning of 'art' in relation to the discipline of cartography and shows how scientific associations find in the arts and humanities a particularly rich terrain to cultivate public outreach. Her chapter traces the development of the International Cartographic Association (ICA)'s Commission on Art & Cartography, which works to activate hybrid artistic-cartographic practices, promotes spatial awareness and facilitates public events, collaborations and the international exchange of ideas among diverse practitioners and theorists. **Laura Lo Presti** and **Tania Rossetto** adopt an ethnographic approach to dig into cartographic authorship and its public dimension. Collecting interviews with Laura Canali, a designer of maps for an Italian geopolitical magazine, they investigate how the public impact of maps in a 'hot' topic such as that of geopolitics is deeply felt by the mapmaker, as well as the decisions, ethical interrogations and precariousness always implied in the crafting of geopolitical maps. In an interview conducted by the handbook's editors, Bloomberg journalists **Laura Bliss** and **Marie Patino** talk about MapLab, a newsletter launched in 2017 and published by Bloomberg CityLab which covers the world of mapping and how it intersects with the news not only for an audience of professionals interested in geo-datavisual graphics but also for map lovers in general. They regularly highlight mapping projects that provide important perspectives on key news events and topics such as climate change, elections, geopolitics and urban planning, but they also write about how cartography illuminates everyday human experience, as in the case of the *How 2020 Remapped Your Worlds* initiative during the coronavirus pandemic.

With its desire to situate cartographic thinking and praxis historically, philosophically, empirically and in terms of changing disciplinary practices, this volume vibrantly shows how encounters between map studies and the humanities have produced 'something' that no longer belongs to them but instead exceeds them: the cartographic humanities.

Note

1 The interview is included in a recent video released for the memorial session dedicated to Yi-Fu Tuan at the Association of American Geographers' 2023 annual meeting in Denver. The video is available at www.youtube.com/watch?reload=9&v=TiAdu_DZVKc.

References

Brunn S and Dodge M (eds) (2017) *Mapping Across Academia*. Berlin: Springer.

Cerreti C (1998) In margine a un libro di Franco Moretti. Lo spazio geografico e la Letteratura. *Bollettino della Società Geografica Italiana* 1: 141–148.

Cosgrove D (2008) Cultural cartography: Maps and mapping in cultural geography. *Annales de Géographie* 660–661(2–3): 159–178.

Crampton JW (2010) *Mapping: A Critical Introduction to Cartography and GIS*. Malden, MA: Wiley-Blackwell.

Dear M, Ketchum J, Luria S and Richardson D (eds) (2011) *GeoHumanities. Art, History, Text at the Edge of Place*. London and New York: Routledge.

Del Casino Jr VJ and Hanna SP (2005) Beyond the 'binaries': A methodological intervention for interrogating maps as representational practices. *ACME: An International Journal for Critical Geographies* 4(1): 24–56.

Dodge M, Kitchin R and Perkins C (eds) (2009) *Rethinking Maps: New Frontiers in Cartographic Theory*. Abingdon: Routledge.

Dodge M, Kitchin R and Perkins C (eds) (2011) *The Map Reader: Theories of Mapping Practice and Cartographic Representation*. London: John Wiley & Sons.

Edney M (2019) *Cartography: The Ideal and Its History*. Chicago: University of Chicago Press.

Harley JB (1989) *'The Myth of the Great Divide': Art, Science, and Text in the History of Cartography*. Paper presented at the 13th International Conference on the History of Cartography.

Hawkins H (2020) GeoHumanities. In: Kobayashi A (ed.) *International Encyclopedia of Human Geography*, 2nd edition, vol. 6. Amsterdam: Elsevier, pp. 95–99.

Kent A and Vujakovic P (eds) (2018) *The Routledge Handbook of Mapping and Cartography*. London and New York: Routledge.

Kitchin R, Perkins C and Dodge M (2009) Thinking about maps. In: Dodge M, Kitchin R and Perkins C (eds) *Rethinking Maps: New Frontiers in Cartographic Theory*. London and New York: Routledge, pp. 1–25.

Lo Presti L (2018) Extroverting cartography. 'Seensing' maps and data through art. *J-Reading: Journal of Research and Didactics in Geography* 2(7): 119–134.
Mitchell P (2008) *Cartographic Strategies of Postmodernity. The Figure of the Map in Contemporary Theory and Fiction*. New York and London: Routledge.
Monmonier M (2007) Cartography: The multidisciplinary pluralism of cartographic art, geospatial technology, and empirical scholarship. *Progress in Human Geography* 31(3): 371–379.
Moretti F (1998) *Atlas of the European Novel, 1800–1900*. London: Verso.
Pickles J (2004) *A History of Spaces: Cartographic Reason, Mapping, and the Geo-Coded World*. London and New York: Routledge.
Roberts L (2012) Mapping cultures: A spatial anthropology. In: Roberts L (ed.) *Mapping Cultures: Place, Practice, Performance*. Basingstoke: Palgrave, pp. 1–25.
Rossetto T (2016) Geovisuality: Literary implications. In: Cooper D, Donaldson C and Murrieta-Flores P (eds) *Literary Mapping in the Digital Age*. Abingdon and New York: Routledge, pp. 258–275.
Travis C (2020) Digital GeoHumanities. In: Kobayashi A (ed.) *International Encyclopedia of Human Geography*, 2nd edition, vol. 3. Amsterdam: Elsevier, pp. 341–346.

PART 1

Preludes and trends

1

MAPPING INNER WORLDS

Cartography as a humanity

Veronica della Dora

He who has succeeded in seeing himself is better than he who has been graced with seeing the angels.

—*St Isaac the Syrian*

Introduction

The phrase 'cartographic humanities' evokes a tension between the order imposed by the map and the complexity of what makes us human. On the one hand is the synoptic view from above, unveiling spatial patterns and making the world legible to our eyes. On the other hand is the unfathomable mystery of the human Self. The detached cartographic view implies absolute synchronism and mastery. It presents us with a frozen image of the world, with the illusion that time has stopped, that space has shrunk before our eyes and that we are in control of both. The invisible flows of human life, emotions and desires, the endless transformations of the self, the inscrutable depths of the soul, by contrast, can be hardly pinned down or even grasped. Like time and infinitude, they elude the gaze and resist representation. Self-knowledge seems to belong to the realm of abstraction rather than to the realm of images. Yet cartography and the humanities are in no way antithetical: in helping us find our place in the world, maps tell us much about ourselves. If approached as cultural artefacts rather than simply as scientific documents or practical wayfinding devices, they can offer privileged windows on the values and beliefs of their makers and their societies. Shaping worlds and worldviews, they are moral projects at heart.

Maps enable us to visualise what our eyes otherwise fail to grasp. They create visual links between interior and exterior worlds. Nowhere are these links more explicit than in allegorical maps of life. While metaphors such as *navigatio* or *peregrinatio vitae* stretch all the way back to classical and late antiquity, they started to assume cartographic contours only in early modern times, flourished in the Enlightenment, declined after World War II and resurrected in the twenty-first century. This chapter considers maps of human life and of its regions as a starting point for reflecting on the meaning of cartography as a humanity.[1] It shows how allegorical maps not only speak of shifting social and cultural values in the West, but they also reflect changing perceptions of space, of life and of the human Self.

 DOI: 10.4324/9781003327578-3

Such aspects lie at the core of the humanities both as a set of disciplines concerned with the study of human nature and culture and as a specific approach to knowledge. Yet, what does this approach entail exactly? How are the humanities different from the sciences and the creative arts? And how have they historically related to cartography?

Cartography as an art and as a humanity

Arts and humanities designate different fields of human endeavour and achievement. The humanities emerged in the Renaissance from the study of the seven ancient liberal arts (the *quadrivium* of the mathematical disciplines of arithmetic, geometry, astronomy and music, and the *trivium* of the verbal disciplines of rhetoric, logic and grammar). They originally designated a class of secular studies concerned with human culture, in opposition to divinity. Today, the humanities are more often defined in opposition to the sciences and encompass a specific set of academic disciplines including, for example, languages and literature, history, philosophy and theology (Oxford English Dictionary—*OED*). More broadly, they designate an approach to knowledge characterised by interpretative rather than quantitative or nomothetic methods. The purpose of their study is still best encapsulated in the aphorism '*γνῶθι σεαυτόν*' (know thyself), and the belief that 'we best come to that knowledge through the reflective study of exemplary human achievement' (Cosgrove, 2011: xxii)—hence the strong historical dimension of humanistic studies.

The arts, by contrast, are a set of creative practices including, for example, creative writing, visual arts such as painting, sculpture and photography, and performative arts (drama, music, dance and architecture). As opposed to the self-reflexive and interpretative nature of the humanities, the arts place emphasis on practice rather than on commentary. In other words, they are concerned with creativity as such rather than with the interpretation of the products of human creativity. Creative arts were originally rooted in the *artes mechanicae* (or manual arts). In antiquity and in the Middle Ages, these were understood in opposition to the liberal arts, though by the mid-sixteenth century, artists were beginning to promote specific skills such as painting and sculpture as liberal arts, which were deemed intellectually superior and thus a more dignified home for artistic genius (Park, 2011). As opposed to the sciences, both the humanities and the creative arts indeed foreground the role of human authorship, either in the construction of knowledge and understanding or in the production of artistic crafts and performances.

An ambiguous term encompassing both the physical practice of mapmaking and its study, today cartography includes aspects that belong to both the arts and the humanities. Its relationship to both fields, and to the sciences, however, has changed over time. Whereas humans have produced maps since pre-historical times (Delano-Smith, 1987), 'cartography' is a relatively recent (western) invention. The word, which combines the Late medieval noun '*carta*' (the material map object) and the ancient Greek verb *γραφεῖν* (to write, but also to describe), only appeared at the end of the eighteenth century. It gained traction in the following century and was eventually raised to the status of academic discipline between the 1920s and 1950s, though clothed in the mantle of science, rather than as a creative art or a humanistic discipline (Edney, 2019).

Denis Cosgrove attributes the original appeal of the term 'cartography' over the more mundane 'mapmaking' to the professionalisation of map production at a time when 'European states were developing topographic map series for the purposes of defending territory, and using statistical mapping as a bureaucratic, regulatory and planning device'

(2008: 169). It was, however, not until the 1950s that the growth of academic cartography was accompanied by a broadened definition of the field, including writing *about* maps (Edney, 2019: 119). A decade thereafter, map historians were considering how early cartographic representations might be studied as part of the humanities, not just in providing data for other disciplines, but as objects of study in their own right (Edney, 2016).

While since the 1990s increased access to digital data, user-friendly mapping software and web resources have led to a 'democratization of mapping' (Crampton, 2010) and to a decline in the art of traditional mapmaking, map history has seen a new rapprochement to the humanities. It has expanded its focus from technical considerations to a critical reevaluation of maps as instruments of power (Harley, 1989; Wood, 1992) and cultural artefacts able to shed light on past societies and cultures in a way not dissimilar from a painting or a novel, for example (Cosgrove, 2008). Both 'critical cartography' and 'cultural cartography' rely on approaches and methods routinely employed in the humanities, such as commentary and criticism, iconographic analysis and, more recently, phenomenology (Rossetto, 2019; Jacob, 2006). The link between maps, the arts and the humanities, however, has deeper roots.

In the premodern and early modern world, mapmaking fell under the remit of Geometry, the liberal art concerned with the measurement (*μετρία*) of the earth (*γῆ*). The fifth-century polymath Martianus Capella described Geometry as a distinguished-looking lady wearing shoes reduced to shreds from her continuous wanderings around the globe, thus stressing its deep connection with the physical reality of the land (*De nuptiis* 6.583–87). By the fifteenth century, under the influence of Neoplatonism, the study of Geometry was deemed key to access the order underpinning the cosmos, which was understood as a reflection of divine love (Cosgrove, 1993; Mangani, 2018). For this reason, in the hierarchy of disciplines, Geometry was set closest to the pinnacle of Theology. Allegorical representations portrayed Geometria floating on a cloud suspended between two levels of being (physical and conceptual) (Figure 1.1).

Later representations, by contrast, featured her as a sturdy figure implanted on the ground and equipped with T-squares, plumb line, a globe, rules and castellated headdress, as to signal geometry's allegiance with architecture, survey, navigation and other 'middle sciences' or practical mathematics, which were seeing a dramatic rise at that time (Park, 2011: 363).

Maps were nevertheless much more than practical devices at the service of architects, engineers and navigators. They were (and had long been) also didactic tools and moral projects. Bound in bibles, books of psalms or even set up as stand-alone altarpieces, medieval *mappae mundi*, for example, had for centuries provided visual commentaries chronicling human salvation from Creation to Christ's Second Coming (Scafi, 2006; Woodward, 1987). *Mappae mundi* featured place events from biblical and classical pasts (from the garden of Eden to the encampments of Alexander the Great), the present (e.g. extant cities like Paris and London) and the future (e.g. the apocalyptic tribes of Gog and Magog). In this way, *mappae mundi* enabled viewers to find their place in the world and in universal history—and reminded them of the necessity of redemption.

In the Renaissance, Abraham Ortelius and other mapmakers explicitly connected their work to the humanistic tradition of self-knowledge. For example, *Theatrum orbis terrarum* (1570), the first printed atlas, presented the world as a stage for the lives, works and salvation of its human inhabitants, as witnessed from an elevated point above its surface in flux. By setting the spectator at a distance from the stage, Ortelius provided his audience with the necessary rational detachment to attain wisdom. At the same time, the opening world

Figure 1.1 *Allegory of Geometry*, Mantegna tarocchi, fifteenth century.
Source: Courtesy of British Museum

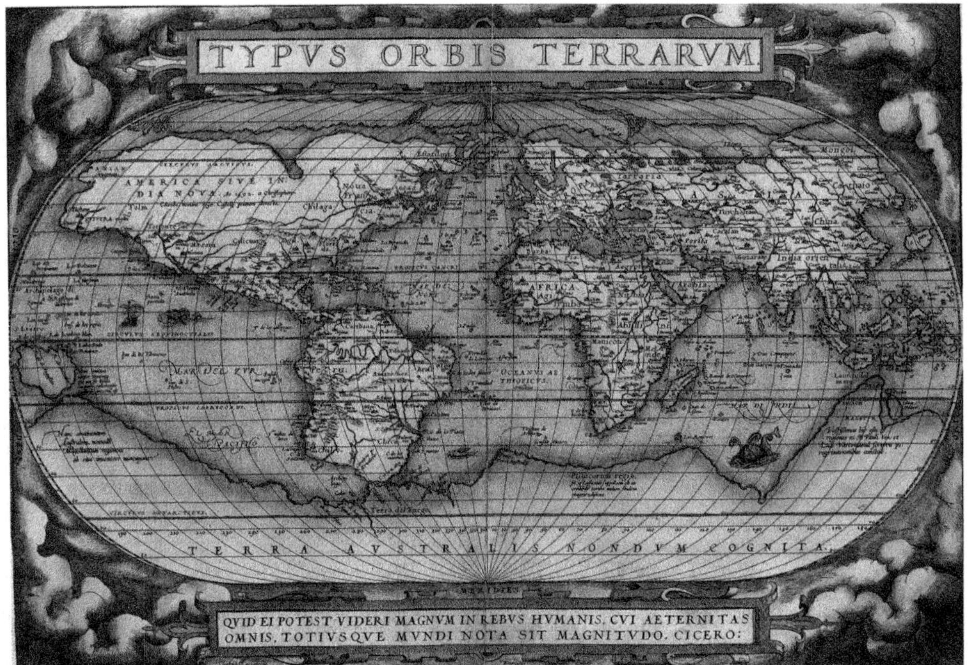

Figure 1.2 Abraham Ortelius, *Typus orbis terrarum*, 1570.
Source: Wikimedia

map in the atlas (*Typus orbis terrarum*) confronted the viewer with a vast antipodean continent yet to be discovered (*terra australis nondum cognita*) and therefore with the limits of human knowledge and the insignificance of terrestrial affairs (Figure 1.2). Under the empty white landmass, the inscription (from Cicero) reads:

> For what can seem of moment in human affairs for whom who keeps all eternity before his eyes and knows the scale of the universal world?

For Ortelius, the earthly theatre and humans' place within it were to trigger moral reflection and self-knowledge at a time of devastating conflict in Europe (Cosgrove, 2003). Less than two centuries later, maps were not simply enabling learning and contemplation, but they were routinely providing practical guidance for navigating the manifold pathways of life and its regions. Where did these maps stem from?

The journey of life

Human life has long been narrated as a journey, or a voyage. Embedded in contemporary everyday speech, the image of the journey conjures up a sense of linear progression: 'moving through life', 'passing the hill', reaching 'a crossroads' and similar idioms all imply a forward movement. Like life itself, the journey entails a beginning and an end; a starting point and a destination. Life journey metaphors thus make it possible to map time on space, and therefore visualise an invisible movement by way of familiar imagery. Here lies their timeless power to draw us near the great mystery of human temporality.

Ancient Greeks and Romans used sea voyage metaphors (*navigatio vitae*) to express the instability and fragility of the human condition or to illustrate the perils and difficulties of public and private life. Plato, the first Greek philosopher to use the metaphor, likened human doctrines to rafts upon which one precariously undertakes the voyage of life across the stormy seas of the world (*Phd.* 85c—d). Seneca likened human life to a voyage in which the soul is tossed by passions but reaches a quiet harbour towards its end, away from the turmoil and preoccupations of public life (*Brev. Vit.* 18). For Cicero, such harbour was death itself, the end of pain (*Tusc.* 5.117). For the Church Fathers, it was the kingdom of heaven and eternal life, the ultimate goal of every Christian (della Dora, forthcoming).

While they continued to employ the ancient sea imagery to describe the external contingencies of life, when it came to describe spiritual progress, the Greek Fathers turned to terrestrial topographies. They especially drew on the archetypal landscapes of the Old Testament: the austere wilderness, the fog-clad mountain, the impenetrable cloud and the cleft in the rock (Ex. 33: 21–23). Gregory of Nyssa identified these three topographic features with the purification from passions, spiritual enlightenment and the mystical union with God (*Vita Moys.*). Before him, Origen had singled out no less than 42 stops in the desert, each corresponding to a temptation to overcome or a virtue to gain on the way to God (Or., *Hom.* 27).

In the Latin West, Augustine envisaged human life as a *peregrinatio*, a journey of exile and return to the heavenly Father. His own path to conversion crossed barren 'fields of sorrow', 'valleys of tears', dangerous 'mists of passions' and the 'muddy deep of heresy', among other gloomy features (*Conf.* 1.16.26, 2.2.2). Inspired by Augustine, Dante likened life to a pathway he lost, ending up in a dreadful *selva oscura*, a dark forest where spiritual vision and the sunlight were blocked (*Inf.* 1.1–3). The Florentine poet mapped out his vertical journey of redemption through murky topographies of afterlife carved out of Italy's mountain *loci horridi* (*Inf.* 12.1–13). In his *Pilgrim's Progress* (1678), John Bunyan yet again unfolded the universal quest for salvation through an elaborated topography of hills, mountains and valleys, but this time also through the local town settings of daily life—from the village of Morality to Vanity Fair.

As all these examples show, inner journeys never take place through an empty space, but through complex, and usually vividly painted, topographies akin to the exterior worlds inhabited by the 'wayfarers'. Unlike abstract space, places and landscapes make invisibilities palpable before our eyes, be they the ineffable passing of time, temporary states of mind, or otherwise ungraspable existential conditions. Inner topographies are laid out in a sequential manner, as one encounters them on the ground during a journey. The journey of life is thus 'mapped' in the metaphorical sense of the term, that is, as a cognitive, ordering process. Order is brought by the linear narrative running through and connecting qualitatively different sceneries and places. The movement of life and its inner topographies, however, started to be fixed on actual maps only in early modern times, after Geometria had ascended on her cloud. Interestingly, these maps spread as the art of mapmaking was turning into the science of cartography.

Mapping life journeys

Initially, allegorical maps did not aim at surveying life in its totality. They rather focused on one of its central regions: love. In the literary salons of mid-seventeenth-century France, at some point, the abstract concept of love curiously started to be discussed in geographical terms. Cartographic representations were variously employed to expound the pathways leading to its idealised fulfilment and to pinpoint the complex dynamics of courtship. While debated, the origins of this genre are usually ascribed to the famous *Carte de Tendre* (Map of Tenderness) engraved by François Chauveau to illustrate Mlle de Scudéry's romantic novel *Clèlie* (1654). The strength and innovation of this map lie precisely in its cartographic status. The lack of a localised point of view asserts objectivity, the antithesis of passions and emotions—or rather, a prerequisite to tame and channel them in rational ways. By turning love into a map at the time of the rapid rise of Northern mapmaking, 'Scudery may very well have sought to appropriate the strong claim to knowledge which maps then possessed' (Brinks, 1993: 41).

The same cartographic 'claim to knowledge' underpins the countless 'maps of matrimony' circulating in eighteenth- and nineteenth-century France and England, perhaps in response to the societal changes brought by the Marriage Act of 1753 and the Divorce Act of 1857 (Reitinger, 1999). At this time, however, other regions of life were also being charted for the moral benefit of different audiences. In Auguste-Jacques Lemierre d'Argy's *Voyage of Youth to the Land of Happiness* (1802) (Figure 1.3), for example, young map

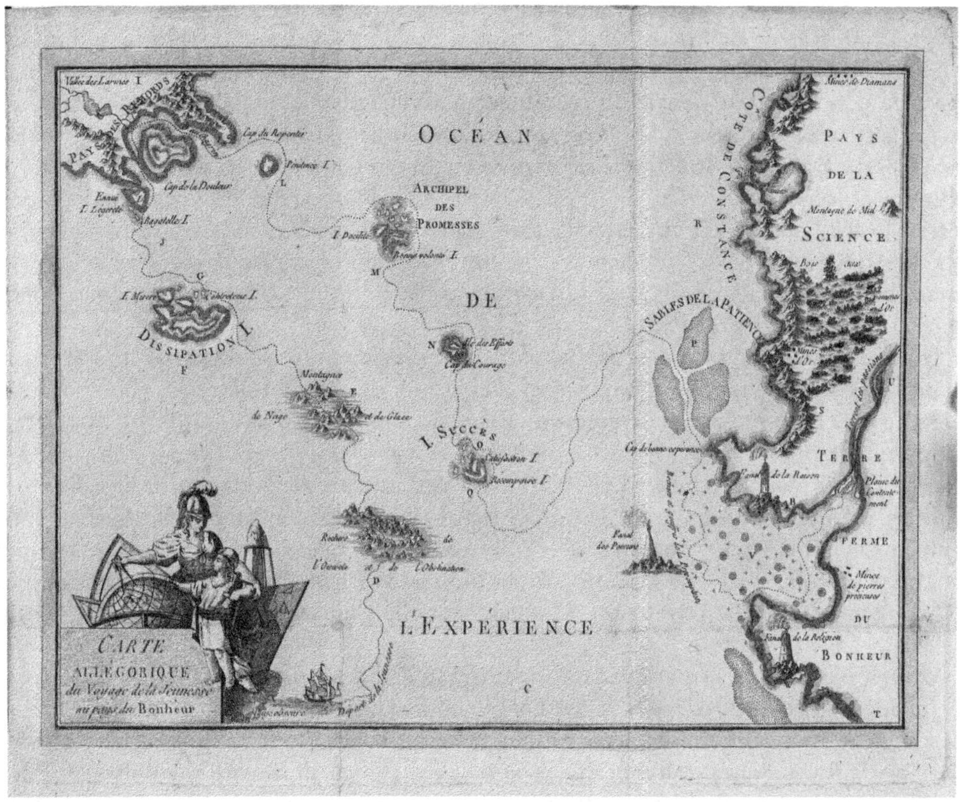

Figure 1.3 Auguste-Jacques Lemierre d'Argy, *Voyage of Youth to the Land of Happiness*, 1802.
Source: Courtesy of Cornell University—PJ Mode Collection of Persuasive Cartography

users are called to traverse the Ocean of Experience. Guided by Geometry (who is now conflated with Reason), they pass Dissipation Island, double Cape Pain and Cape Repentance in the Country of Remorse, circumnavigate the Archipelago of Promises, coast the Isles of Success and eventually reach the Mainland of Happiness and the neighbouring Country of Science. Whereas religion features as a lighthouse, science and reason are both the destinations and means for achieving happiness.

Allegorical maps of this period, however, also include a distinctively religious category. Rather than focusing on a specific region of human life, such as marriage or youth, these maps ambitiously aim at a broader overview of the world of human interiority. They ask the viewer to contemplate nothing less than the pathways to eternal salvation or damnation—and to find their place on the map. Exploiting the same combination of God's-eye view and wayfinding as their secular counterparts, these maps exhort to a kind of love different from marital love or the love for knowledge: that of a Christian life (Scafi, 2022: 49–50). In this, they mimic the journey of Bunyan's hero, Christian, who leaves the City of Destruction for the Celestial City. Like Christian, the viewer is faced with crossroads and byways. Certain paths lead to salvation, others to destruction. Unlike Bunyan's everyday hero, who finds his way through a painful process of trial and error, we, the viewers, are empowered with the map's omniscient God's-eye view and can immediately tell where each road leads.

In *The Journey of Life* (1775), for example, George Wright promises to his audience 'an accurate map of the roads, counties, towns, &c. in the ways to happiness and misery' (Figure 1.4). The centre of the map is occupied by the City of Reason or Natural Man. From it, a straight road departs in two opposite directions, upwards and downwards, setting a vertical axis to the map. The upward stretch of road is narrow (Matt. 7:14) and traverses the towns of Moderation, Hatred of Sin, Humility and other virtues. It crosses the counties of Repentance, Faith and Perseverance, runs by the mountain of Contemplation, and, after the Valley of the Shadow of Death, it eventually reaches the River Jordan, beyond which is the eternal Glory of Heaven. The lower section of the road, by contrast, is wide (Matt. 7:13) and leads to the Gulf of Death and the Bottomless Pit of Destruction, where souls are devoured by the fire of hell.

Different in content and scope as they are, sentimental, educational and spiritual charts all share the mapping impulse of their time; the desire to classify the world and make it legible to the eye. The boom of allegorical maps coincides with the democratisation of cartography. By the late eighteenth century, maps were printed, cheaper and widespread. They were no longer the preserve of an elite of princes and merchants or of sophisticated salon-goers. Allegorical maps thus drew on established cartographic conventions and were immediately legible to their viewers. They offered a common language that directly reached out to people—in the same way as vivid topographical imagery spoke directly to the audiences of the early Church Fathers, Dante and Bunyan.

In their plain 'post-cartographic reformation' style, life journey maps like Wright's spoke the unadorned rhetoric of truth of the Enlightenment. At the same time, they invited their viewers to turn their gaze inwards, to their innermost Self. In this, they aligned with the humanistic Renaissance tradition of self-knowledge, as much as with the universal salvation narrative unfolded by medieval *mappae mundi*. Their moral mission is best encapsulated in the inscription at the bottom of Wright's map:

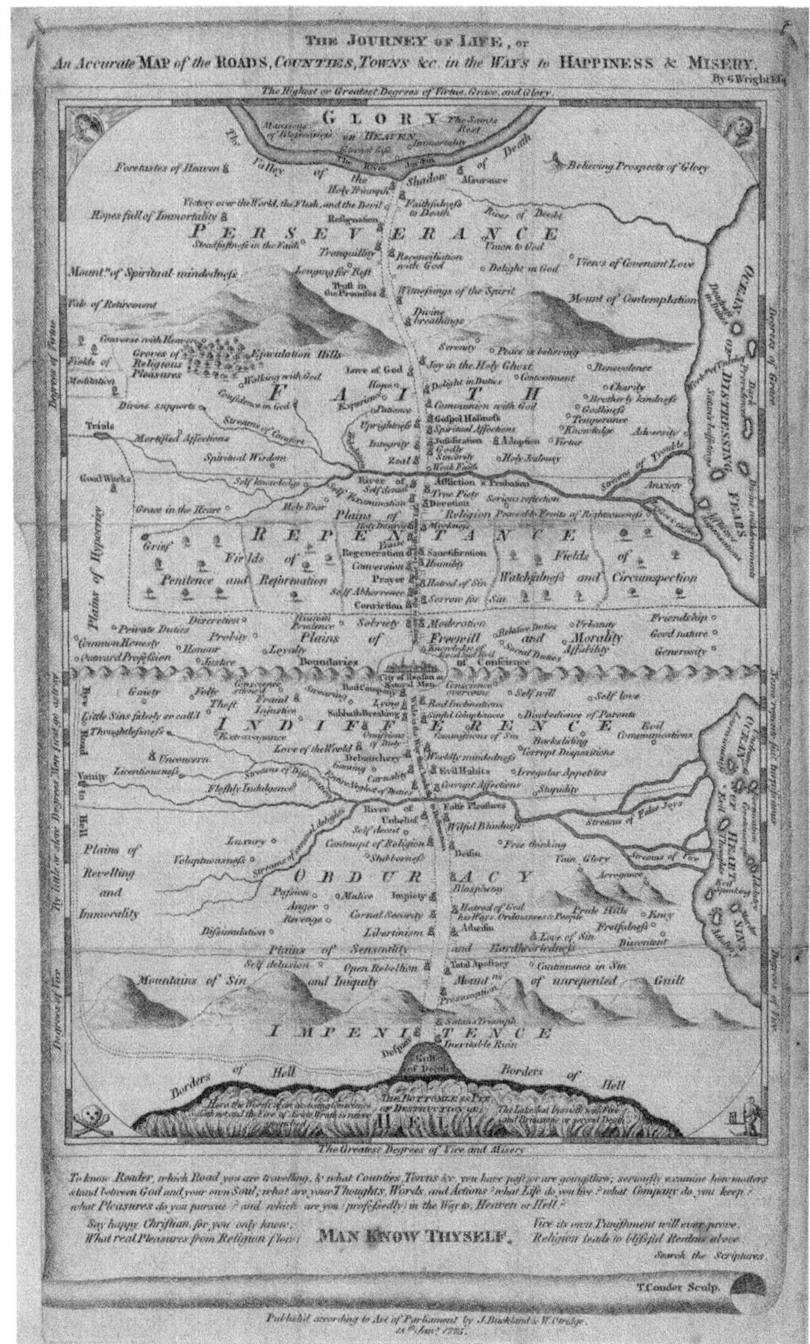

Figure 1.4 George Wright, *The Journey of Life*, 1775.

Source: Courtesy of Cornell University—PJ Mode Collection of Persuasive Cartography

> To know, Reader, which Road you are travelling, & what Counties, Towns &c. you have past or are going through, seriously examine how matters stand between God and your Soul; what are your Thoughts, Words, and Actions? What Life do you live? What Company do you keep? What Pleasures do you pursue? And which are you (professedly) in the Way to, Heaven or Hell?
>
> Say happy Christian, for you only know,
> What real Pleasures from Religion flow;
> Vice its own Punishment will ever prove,
> Religion leads to blissful Realms above.
> Search the Scriptures.
> MAN KNOW THYSELF.

Conclusions

Whereas the sciences are meant to prove and persuade, the fundamental task of the arts and of the humanities is to inspire. As crafts of human creativity, maps inspire us to look beyond our surroundings; to extend our vision where our eye cannot otherwise reach—not just outwards into the world but also inwards into the depths of the Self. Clothed in the mantle of science, maps are nonetheless inherently persuasive. Here lies the power and charm of allegorical maps, a genre born in the golden age of Northern mapmaking and spreading in the eighteenth and nineteenth centuries with the increased professionalisation of map production and popularisation of maps. Fixing emotions and charting pathways through life in their synoptic space, these maps set us the viewers at a distance and at the same time draw us in. They provide the rational detachment necessary to put life into perspective, while at the same time inviting us to zoom in and explore multiple paths through the familiar and unfamiliar topographies of the Self. Like old friends, allegorical maps of life and its regions offer counsel and warning of what lies ahead of us, beyond the ever-shifting horizon of life.

To their original audiences, allegorical maps offered tools for moral self-edification, as well as visual commentaries on their transforming societies. Taken as a whole, today they speak of changing worldviews. As fascination with sea travel and exploration gave way to the techno-optimism of modernity, maps of life started to be crossed by railways and flown over by airplanes. For example, Gospel Temperance, a social movement promoting abstinence from alcohol, produced its own *Railroad Map* (1908), which looked very much like an American railway map of its days, with lines departing from Decisionville and crossing different 'states': the State of Righteousness, of Sacrifice and Service, for example, but also the State of Vanity and, further down on the map, the State of Depravity (Ackerman, 2007). By the 1930s, the pathway to eternal life featured mapping conventions typical of the new air age. On Walter John Dittmar's map of *The Road From Earth to Heaven* (1932), the flattening cartographic view morphed into a complex three-dimensional topography of modern infrastructure, including elevated rails, buses, trucks, fast trains and aircrafts. As with their eighteenth-century predecessors, these allegorical maps presented their audiences with choices to be made and pathways to be followed (or to be avoided), while dramatising the manner of achieving salvation by way of their persuasive nature.

Contemporary maps of life present yet a different look. In the synoptic map accompanying Louise van Swaaij and Jean Klar's *Atlas of Experience* (2000), human life features as an island populated with the conventional symbols of topographic maps. This time, however, there are no straight paths nor bright beacon lights guiding the viewer. There is no paradise nor hell either, only an uncharted 'Elsewhere' across the ocean. Human life simply ends at 'Point of No Return'.

As with any other map, allegorical maps of life and its regions are mirrors reflecting and refracting cultural and social values. Ultimately, however, they are also mirrors in which to look at ourselves and reflect on the uttermost goal of the humanities: *Nosce te ipsum*.

Note

1 The research for this essay was supported by the Leverhulme Trust (Major Research Fellowship 2022-026).

List of abbreviations

Brev. Vit. = Cicero, *On the Shortness of Life*
Conf. = Augustine, *Confessions*
De nuptiis = Martianus Capella, *On the wedding of Mercury and Philology*
Inf. = Dante, *Inferno*
OED = *Oxford English Dictionary*
Or. = Origen, *Homily* 27, Numbers
Phd. = Plato, *Phaedo*
Tusc. = Cicero, *Tusculan Letters*
Vita Moys. = Gregory of Nyssa, *The Life of Moses*

References

Akerman J (2007) Finding our way. In: Akerman J and Karrow R (eds) *Maps: Finding Our Place in the World*. Chicago: University of Chicago Press, pp. 19–63.

Brinks E (1993) Meeting over the map: Madeleine de Scudéry's 'Carte du Pays de Tendre' and Aphra Behn's 'Voyage to the Isle of Love'. *Studies in English Literary Culture, 1660–1700* 17(1): 39–52.

Bunyan J (2008 [1678]) *The Pilgrim's Progress*, Pooley R (ed.). New York: Penguin Books.

Cosgrove D (1993) *The Palladian Landscape*. Leicester and London: Leicester University Press.

Cosgrove D (2003) Globalism and tolerance in early modern geography. *Annals of the Association of American Geographers* 93(4): 852–870.

Cosgrove D (2008) Cultural cartography: Maps and mapping in cultural geography. *Annales de géographie* 660(2): 159–178.

Cosgrove D (2011) Prologue: Geography within the humanities. In: Daniels S et al. (eds) *Envisioning Landscapes, Making Worlds*. London: Routledge, pp. xxii–xxv.

Crampton J (2010) *Mapping: A Critical Introduction to Cartography and GIS*. Oxford and New York: Wiley-Blackwell.

Delano-Smith C (1987) Cartography in the prehistoric period in the Old World: Europe, the Middle East, and North Africa. In: Harley B and Woodward D (eds) *History of Cartography*, vol. 1. Chicago: University of Chicago Press, pp. 54–101.

della Dora V (forthcoming) Most grievously tossed in the stormy sea of life: Mapping spiritual voyages in Byzantium. In: Mitrea M and Mullet M (eds) *Mapping the Sacred in Byzantium: Construction, Experience and Representation*. Cambridge: Cambridge University Press.

Edney M (2016) Of lectures, libraries, and maps: The Nebenzahl lectures and the study of map history. Keynote lecture, *19th Nebenzahl Lectures: Maps, Their Collecting and Study: A Fifty-Year Retrospective*, organised by Akerman J, Newberry Library, Chicago. 27 October. Text available at: www.mappingasprocess.net/blog/2018/1/27/of-maps-libraries-and-lectures-the-nebenzahl-lectures-and-the-study-of-map-history?rq=collecting (accessed 13 September 2022).

Edney M (2019) *Cartography: The Ideal and Its History*. Chicago: University of Chicago Press.
Harley JB (1989) Deconstructing the map. *Cartographica* 26(2): 1–20.
Jacob C (2006) *The Sovereign Map: Theoretical Approaches in Cartography throughout History*. Chicago: University of Chicago Press.
Mangani G (2018) *La bellezza del numero: Angelo Colocci e le origini dello stato nazione*. Ancona: Il lavoro editoriale.
Park K (2011) Allegories of knowledge. In: Dackerman S (ed.) *Prints and the Pursuit of Knowledge in Modern Europe*. Cambridge, MA: Yale University Press, pp. 360–366.
Reitinger F (1999) Mapping relationships: Allegory, gender and the cartographical image in Eighteenth-century France and England. *Imago Mundi* 51(1): 106–130.
Rossetto T (2019) *Object-Oriented Cartography: Maps as Things*. London and New York: Routledge.
Scafi A (2006) *Mapping Paradise: A History of Heaven on Earth*. London: British Library.
Scafi A (2022) *L'uomo con le radici in cielo*. Milan: SEM.
Wood D (1992) *The Power of Maps*. London and New York: The Guildford Press.
Woodward D (1987) Medieval *mappae mundi*. In: Harley B and Woodward D (eds) *History of Cartography*, vol. 1. Chicago: University of Chicago Press, pp. 286–370.

2

CHOROGRAPHY, CARTOGRAPHY AND THE GEOSPATIAL HUMANITIES

Javier Arce-Nazario, Janet Downie, Tim Shea, John Pickles and Toni Veneri

Introduction

For the past two years, we have been investigating how theories and practices of cartography and chorography have shaped visual and verbal landscape description in several historical and disciplinary contexts, and how these might inform the spatial humanities.[1] How do different historical contexts and disciplinary imaginations of place and space draw differentially on concepts of cartography and chorography? What are the conditions and contexts for these hermeneutics? And what kind of practices might they inform? This has involved ancient and contemporary conceptions of chorography and the utility of terms like mapping, cartography, chorography and geography for describing what we as scholars interested in space, place and landscape do in our respective fields of inquiry.

In this chapter, we draw on examples from the literature and archaeology of the ancient and late-ancient Mediterranean, Renaissance and Venetian Studies, contemporary cultural studies and critical cartography, and environmental studies and critical approaches to geospatial representation to think about mapping and cartography as comparative tools and strategies for understanding particular regional settings and formations. In each context, we are concerned both with mathematical approaches to mapping that use coordinate points to situate the mapped area precisely in the context of a wider world-system—what is generally called cartography—and with other approaches to mapping that focus on the local and regional, the descriptive, the qualitative, the verbal—often described as chorography. Cartographic and chorographic approaches to mapping have coexisted since ancient times and occasionally have been pitted against each other. But for us, every map, wherever it is situated on the mathematical/discursive spectrum, represents a strategic selection of data reflecting the interests and agenda of its authors and commissioners (Pickles, 2004).

Here, we work with three main lines of enquiry. First, because one of the major ways we access historical geographies is through texts—archival, documentary and literary—we are interested in the relationships between space and language, and between space and text; how do maps function to construct (as well as represent) imaginative worlds and real-world practices, and how do texts construct and encode imaginative geographies? How have the twin abstractions of text and image changed over time in different historical and epistemic contexts?

 DOI: 10.4324/9781003327578-4

Second, we are interested in the practice of spatial humanities and the ways in which the convergence of digital tools allows us to visualise, reconstruct and analyse ancient, medieval and modern modalities of rendering human landscapes; how can we explore techniques for the creation, curation and visualisation of digital spatial data, and how—as scholars from different traditions of interpretation and practice—can we collectively engage with them?

Third, we explore selected applications for non-Cartesian maps, verbal descriptions of landscapes and archaeological remains as they relate to modern urban environments; how do we understand and represent both the past and the efforts of those in the past to understand and represent their own worlds?

Revisiting Ptolemy: reframing geography and chorography

We began our investigation with the Roman mathematician Claudius Ptolemy's *Geographical Guide*, or *Guide to Depicting the World* (*geôgraphikê hyphêgêsis*). Described (rather optimistically) by modern scholars as a 'mapmaking kit' (Berggren and Jones, 2000: 5), Ptolemy's *Guide* aimed to provide the tools for creating a set of visual maps of the inhabited world as it was known to his second century CE contemporaries, when the Roman Empire was at its height. In two books, he explained the theoretical basis for creating projections, and in six books, he provided the data: latitudinal and longitudinal coordinates for more than 8,000 cities, landscape features and landmarks—'crowdsourced', corrected and refined through a process of diachronic collaboration over centuries of geographical writing in the Greek tradition (Weitzke, 2017: 356, 363).

Ptolemy's project was unique in the ancient world for its scope and its aims. The *Geography* of his predecessor Strabo, produced several generations earlier, was an equally ambitious undertaking: a 17-book presentation of world regions full of topographic, demographic and ethnographic detail, assessing spatial knowledge across the whole of the Greek tradition, reaching all the way back to the epic poems of Homer, whom Strabo considered the 'first geographer'. But Strabo's was a work of cultural geography—a multifaceted textual project directed towards a literary audience (Dueck et al., 2005). Ptolemy, by contrast, approached his world-geography as a mathematician, with a project of visual representation. His aim was to systematise historical data and establish a theoretical framework so that readers could consistently depict relative position with an ever-expanding set of coordinates.

Nevertheless, Ptolemy was keenly aware of human limitations when it came to representing the world and the cosmos:

> These things belong to the highest and most beautiful theoretical contemplation (*theôria*): to display to human understanding by means of mathematics both the heavens themselves, and also the earth. In the case of heaven, how it is by nature (*physis*)—because it can be seen revolving around us. In the case of earth, through a likeness (*eikôn*)—because the true earth, since it is enormous and does not surround us, cannot be inspected by any one person, either as a whole or part by part.
>
> *(Ptolemy in Grasshoff and Stückelberger, 2006: 1.1.9, trans. J Downie)*

This aspiration to a mathematical understanding of all space, both cosmic and earthly, poses special challenges. While distance and movement, in Ptolemy's view, allow mathematicians to diagram the heavens, earth-dwellers can never get far enough away to comprehend the earth. Thus, the best representations of space always present an approximation, an image, a likeness (*eikôn*).

In his *Guide to Depicting the World*, Ptolemy attempted to find a practical solution to these earthbound limitations. Appropriating for specific purposes a term that had a much wider purview—including for his predecessor Strabo—and a long history in the Greek world, Ptolemy defined *geôgraphia* as 'an imitation through drawing (*mimêsis dia grafês*) of the entire known part of the world', the aim of which is 'to show the known world as a single and continuous entity'. In the process of laying claim to the term *geôgraphia*, he also defined its other, *chôrographia*: 'the impression of a part' that takes 'a divided-up approach and sets each different place separate from the other by itself'. Characterising chorography's approach to spatial representation as 'qualitative' (*to poion*), 'attend[ing] everywhere to likeness, and not so much to proportional placements' (Ptolemy, 1.1.1–2 in Berggren and Jones, 2000: 57–58), he declared his understanding of geography to be thoroughly 'quantitative' (*to poson*), concerned with proportionality of distances, or relative position—a polemical dismissal of chorography that has had a distorting effect on our understanding of the history of western European geographical thought (Simon, 2014: 24–25).

However, the underlying tensions between mathematical theory and earthbound data—the problem of perspective Ptolemy identified—remain unresolved. In fact, although he rejected the qualitative, descriptive approach he associated with chorography, Ptolemy's *Guide* in fact humanises space in its own way: the regional maps for which he provides instructions themselves entail a contingent understanding of space, aligning as they do with the provincial divisions of the Roman empire and reflecting a contemporary view from Roman imperial Alexandria (Tolsa, 2013). Ptolemy's restrictive, mathematical redefinition of *geôgraphia* cast a long shadow: Matthew H. Edney (2019: 189–190) notes that the translation of Ptolemy's *Geography* into Latin in the early fifteenth century was a key moment for 'modern cosmographical geometry'. Ultimately, however, it may be the unresolved tension Ptolemy identified between *geôgraphia* and *chôrographia* that is more enduring and productive. Surely with every representational project, we ask ourselves Ptolemy's question: What is the conceptual whole of which this is a part? Or, put another way, what is the story our map is trying to tell?

Renaissance mapping and the arts of place description

As we compared our initial ideas about chorography and cartography, we found this return to a Ptolemaic framework helpful. The resulting variety of historically situated debates and practices in our own respective fields opened questions of interpretation and translation.

A first philological assessment of Ptolemy's authority in the history of western geographical thought could easily have confirmed the traditional linking of Ptolemy's *Geography* and modernity—its arc of scientific culmination in Antiquity, disappearance in the Middle Ages, revival in the Renaissance, incorporation and ultimate obsolescence in the early modern period. Instead, we found a history of competing theories and agendas disputing to this day Ptolemy's legacy in shaping our ideas of how we can represent the world and its parts.

Driving this history was the asymmetry of Ptolemy's opposition, not only because the geographer had deemed chorography a project of lesser value but also because his definition was a decidedly loose one. This asymmetry and indeterminacy have recently allowed historians to reconsider a variety of Classical and Medieval texts, literary and pictorial, that fell outside of the mathematical scope of Ptolemaic geography. Countering claims

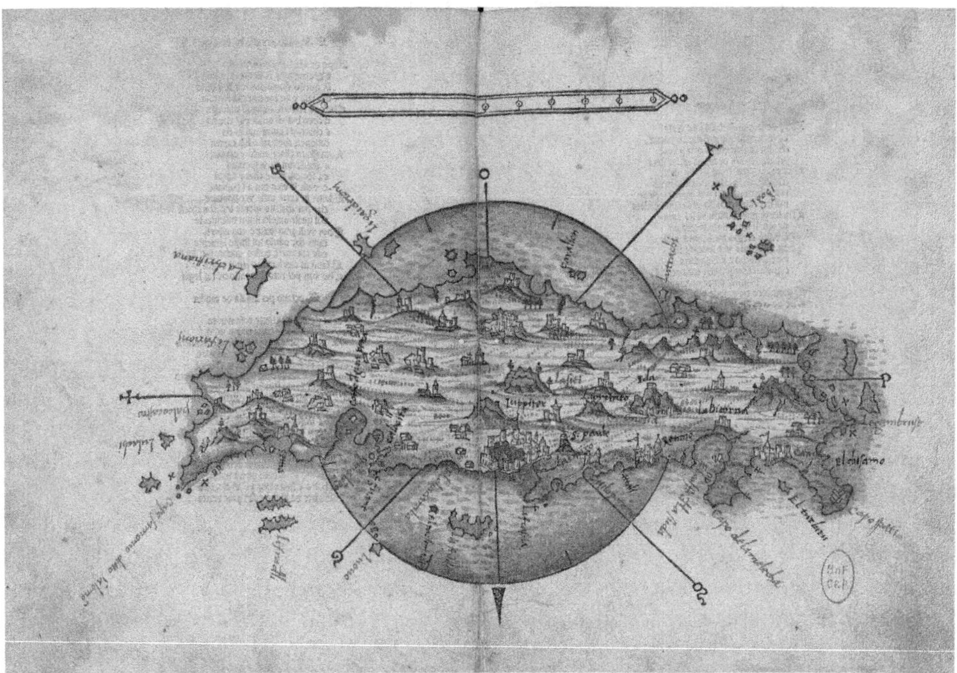

Figure 2.1 Map of the island of Crete (Candia) in the *Isolario* compiled by the Venetian shipmaster Bartolomeo dalli Sonetti (1485). The first printed book of islands, as well as the first printed collection of nautical charts, this publication 'contains the first printed maps supposedly based on actual observation; and it is the first printed collection of maps to owe no debt to Ptolemy' (Campbell, 1987: 90).

Source: gallica.bnf.fr/Bibliothèque Nationale de France

of intellectual stagnation in the Middle Ages, a whole chorographic tradition predicated on perceived reality rather than absolute accuracy has resurfaced. Conceptually rooted in the works of Strabo and Pomponius Mela rather than in Ptolemy's quantitative models, disparate strands of premodern mapmaking have been reassessed in light of their emphasis on the empirical origin of their sources, in contrast with Ptolemy's disembodying extraction of geographical data (Simon, 2014). In parallel with a continuing tradition of written descriptions grounded in travel, itinerary maps and nautical charts, these chorographies gave visual texture to a hodological (pathway or route) sense of space associated with practices of pilgrimage, trade, navigation and war. At opposite scales, illustrated *mappaemundi* dramatised the believed emplacements of scriptural and secular narratives while city plans were arguably embedded in the experience of visitors, with edges and location of objects defined in relation to the compass of the viewer, what Matthew Boyd Goldie (2019: 1–17) has termed '*estral* space'. . . . At the dawn of the Renaissance, the much-debated genre of the *isolario*, electing the island as 'an archetypal unit of chorography' (della Dora, 2016: 189), provided an alternative model to the Ptolemaic atlas in the mapping not only of the Mediterranean but also of a far-flung world expanding across the oceans (Figure 2.1).

The unresolved tension created by Ptolemy between geography and chorography became freshly productive when Renaissance humanists found in it a blank slate to accommodate their theories, or instead develop and legitimise emerging practices of place description. Their responses were far-ranging and contradictory. In Venice, Benedetto Bordone sought to bridge the authorities of Ptolemy and Strabo to bring the book of islands into the realm of mathematical sciences (1528) while Giacomo Gastaldi (1561: 2–3) associated chorography with the production of nautical charts. This last association has been recently revived by Edward S. Casey (2002: 154–193), in the effort to free early portolan charts from modern expectations of resemblance and spatial projection and to highlight instead their value as embodied chorographical representations of coastal places.

Yet other humanists and mapmakers lingered more thoroughly on Ptolemy's thorny passage. Girolamo Ruscelli (Ptolemy, 1561: 6), the editor of a new edition of the *Geography*, speculated upon the loss of drawings and paintings from ancient chorographers, while Egnazio Danti (1577: 44), the pontifical mathematician who designed the 'Gallery of Maps' in the Vatican, emphasised chorography as a distinctively pictorial endeavour. In the Netherlands, chorography as a detailed inventory of proximate realities prompted a formidable convergence of cartography and landscape painting (Besse, 2000: 35–68). Italian mapmaker and surveyor Cristoforo Sorte insisted on the visibility of landmarks, crafting chorographical maps in 'a way that people who are familiar with their countries can recognise the places without reading the letters of their names' (Nuti, 1999: 93). Pietro Coppo (1540) and Johannes Honter (1532) resorted to both text and map to draw attention to the natural and historical richness of their respective regions of origin, Istria and Transylvania. In Renaissance England, chorography emerged as an alternative historical mode of narration where particulars could be organised spatially and not chronologically, often in relation to a real or imagined journey (Helgerson, 1986; Vine, 2017). In France and in Germany, the genre of the town book, launched by Guillaume Guéroult (1553) and brought to success by Georg Braun and Frans Hogenberg (1572), advertised itself as a product of chorography, complementary rather than subordinate to geography.

Favouring the qualitative content and true likeness of chorography, these texts exceeded the realm of cartography and mobilised one or multiple genres—including maps, urban views, surveys, itineraries, illustrations and literary narratives—with the aim of providing the portrait of a region, a province, a kingdom or a city.

Cartography, chorography and chora in Athens and Attika

In the early modern period, we see a variety of cartographic and chorographic approaches responding to the interests of nascent states—what about in the ancient Mediterranean? How might we reconstruct the cartographic norms and hegemonic uses of survey and mapmaking in ancient Athens and Attica to better understand the development of cartography alongside other concepts like geography and chorography? Did nascent political entities and eventually formal states (*poleis* and empires) have visual and/or descriptive paradigms for delineating territory and asserting autonomy and hegemony? Would we consider any visual representations of territories to be the equivalent of what we consider to be maps? And how might these be recovered? This is a matter of debate. Take the example of the relationship between the city-state (*polis*) of Athens and its surrounding territory (*chora*) in the fifth century BCE. Were the smaller political units of the countryside (*demes*) surveyed and marked, or did these territorial boundaries simply emerge organically, based on hereditary political groupings?

The idea that surveyed territories emerged alongside Greek poleis (city-states) has been and continues to be contested in scholarship. Athens was the largest polis in the ancient

Greek world and comprised 139 smaller political units called *demes*. These *demes* likely took the form of villages or clusters of villages that were later incorporated into the Athenian state. In general, there are two camps: those that argue that Athenian *demes* were separated from one another by surveyed territorial boundaries, and those that argue that demes were simply administrative units with hereditary membership (by birth or adoption) (Lalonde, 2006: 194). Boundary inscriptions (*horoi*) carved into bedrock have been identified in Athens and the Attic countryside and have been proposed to correspond to boundaries between *demes*. These are often found along the ridges of mountains—logical and functional boundaries between territorial units.

Sylvian Fachard approached the question of *horoi* and their relationship to demes using cost-path analysis in the software platform ArcGIS which accounts for the effort to travel across different elevations to model possible deme territorial boundaries (Fachard, 2016: 199–201, fig 9.2). In Fachard's study of deme borders in Attika, he found that, in many cases, the *horos* inscriptions on bedrock correspond to the cost-path model's hypothetical territorial boundary, thereby supporting the idea that demes and their territories were formally and somewhat equitably surveyed units (Figure 2.2).

This use of mapping by a centralised state, in this case a polis, to delineate and assert its boundaries would be described in later historical contexts as a form of cartography, but our discussion has prompted us to ask what might territory drawing and any attendant illustrations—that unfortunately do not survive—have been called? Chorography is a likely candidate. A diagram of territorial units showing their relationship to one another within the region of Attica would certainly fit Ptolemy's definition of chorography, the most thoroughly explicit and extant definition from antiquity. It is possible that Ptolemy considered chorography a visual medium as well, since he claims that only a *graphikos* would do chorography (Ptolemy, Geography 1.1.5). While Berggren and Jones translate *graphikos* as 'a man skilled in drawing', the term is ambiguous and could instead refer to 'a man skilled in writing' (2000: 58). This implies that by the second century CE, visual representations of cities, territories and regions—what we would call maps—were considered an area of study under the umbrella of chorography.

There is some evidence that visual illustrations of regions controlled by states existed in antiquity and were used to promote and assert authority and control over a territory. In this way, a cartography in Edney's sense did seem to exist in the ancient world, and in one case, the term used to describe the document produced when drawing territories was chorography. In the second century BCE, an inscription relates a dispute between two poleis on Crete, Itanos and Hierapytna (*Inscriptiones Creticae* III 4.9, ll in Guarducci, 1942: 65–66). Rome sent arbiters from Magnesia on the Maeander to settle the dispute. Among the documents presented were chorographies (*chôrographiai*) and a *periorismos*, which is a written description of boundary lines and their respective *horoi* (Matijašić, 2022: 407). What form would these documents called *chôrographiai* have taken? It is difficult to know since papyri, the likely medium on which these documents were written and/or drawn, does not survive archaeologically in Greece, except in very rare cases. Ivan Matijašić (2022: 408) admits that even if these *chôrographiai* contained graphic diagrams or something akin to maps, they were still meant to accompany what was stated in writing in the *periorismos*—image and text were fundamentally linked, and the former may have been used to supplement the latter, which had greater legal weight.

Fachard is at the forefront of a recent trend in regional and urban archaeology to use Geographic Information Systems (GIS) to consider disparate forms of historical and

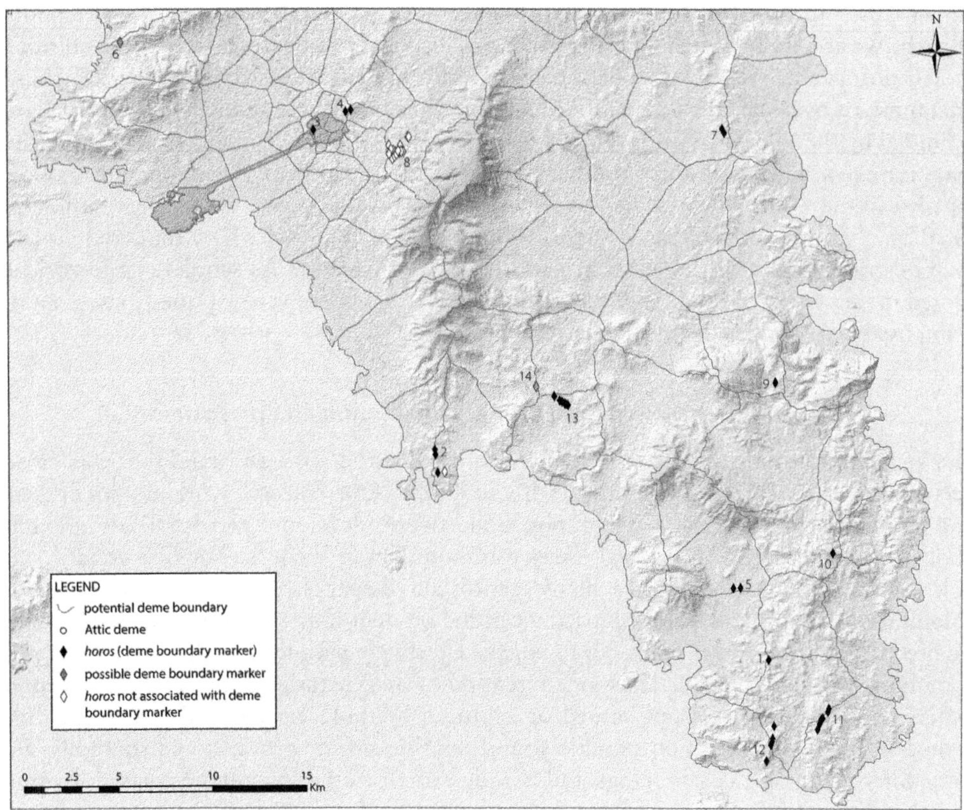

Figure 2.2 Map showing central and southern Attica with deme centres and modelled deme boundaries using cost-path analysis in GIS alongside *horos* inscriptions possibly related to deme boundaries.

Source: Image by Sylvian Fachard, originally published in Fachard, 2016: 202, fig. 9.3, with permission from Sylvian Fachard

archaeological evidence spatially in a digital working environment. By using software—in this case, ArcGIS Pro—he is able to combine the locations of inscribed *horoi* he collected in the field with *deme* centres proposed by John Traill (1975) and with elevation data for Greece to consider the relationship between settlements, their boundaries and the natural environment around them. Another excellent example of this movement in classical archaeology is the Dipylon Society's WebGIS visualisation of 19th-century maps of Attica in the Karten von Attika (https://dipylon-kartenvonattika.org/webgis) and of the rescue excavations in the urban centre of Athens, complete with references to original publication, chronology of remains and categorisation of sites (https://map.mappingancientathens.org/). In both cases, Leda Costaki, Annita Theocharaki and the rest of the Dipylon team have attempted to systematically engage with past work on Athens and Attica, from surveys of the countryside by Curtius and Kaupert between 1885 and 1903 to excavations reports and plans drawn by hundreds of archaeologists from over a century of archaeological fieldwork in city of Athens. As historians and archaeologists interested in historical landscapes, we are perpetually confronted with the duality of image and text and the utility and shortcomings of each. We are often limited by the tools at our disposal and the historically and culturally rooted norms and modes of rendering spatial information in word and image. Fachard's attempt to recover the norms and modes by which ancient Athens manifested its boundaries is limited by the software's and algorithm's capacity to reconstruct the results of an ancient mapping event. Even with these results, it is difficult to reconstruct the mapping behaviours that led to this conclusion—were border inscriptions made after a verbal negotiation and listed on a papyrus document or one in another medium? Or were they rendered visually, even abstractly, on a document that had official legal weight? As scholars of historically rooted modes of spatial thinking, we find it useful to ask these sorts of questions, even if it is impossible to answer them conclusively.

Travelling Vieques: chorography, visualisation and presentation

In a collaborative mapping project about Vieques, Puerto Rico—an island that was mostly occupied by the U.S. Navy from the 1940s to 2003—Arce-Nazario worked with students and alumni in the Carolina Cartography Collective to develop a geovisualisation exhibit about land tenure and other social issues for a museum in Vieques. The historically available cartography of Vieques, like many territorialised spaces, is mostly a cartography of colonisation, *latifundista* and occupation centred on quantitative assessments and geometric precision. These power maps fail to engage effectively with the expressed needs of local communities (Wood, 2010). Thus, one intention of the Vieques exhibit was to foreground experiences and processes not valued or captured by those cartographies, with methods rooted in a growing tradition of activist and decolonising responses to cartography such as counter-mappings, co-mappings and a wide array of other community-based mapping practices.

The histories of the role of oblique and human-centred perspectives in the chorographic tradition and how they differed from the 'god's eye' views typical of cartography directly influenced the Collective's decision to incorporate historical, oblique air photographs of the Vieques coastline into the geovisualisations (Figure 2.3). These choices were strategically important because of the ways in which oblique images facilitated apparently more 'natural' engagement by visitors with the sugar cane plantation lands and the history of appropriation of them by the U.S. Navy.

Chorography's embrace of text and voice also provided a useful tool in the Vieques exhibit, encouraging visitors to become 'authors' in conversation with different representational forms. The Vieques exhibit sought to re-work this notion of authority through a polyvocal layering of local narratives. A chorography of oblique mappings (Figure 2.3) sought to encourage Viequense exhibit visitors to enter into the process of producing maps and images for the island's future (Figure 2.4). The different ways that voices emerged in these encounters provided an inspiration and a frame for further Collective mappings and visualisations. Together the Collective and visitors created Vieques chorographies that offered different relationships between words, places and active interventions in reading landscapes. One of the most successful visualisations in the exhibit was a historical map of urban Vieques embossed on cork which allowed visitors to incorporate their narratives about the island and its residents directly by adding text, drawings and collages to the map. The result is a collage of impressions and conversations, loosely held to the same frame, and illustrating our understanding of a polyvocal imagining of place.

Figure 2.3 Chorographic visualisation of the sugarcane plantations using oblique pictures taken in 1941. The visualisation is layered with waypoints, lines and transparent hues, and a legend is given to facilitate the reading of the landscape. These pieces were placed near a window looking towards the areas pictured, to provide more reference to the oblique viewpoint.

Source: The original photographs are from a U.S. Navy survey of available land and are now housed in the Bonnie Donohue Collection in the Vieques Historical Archive. With permission from Javier Arce-Nazario.

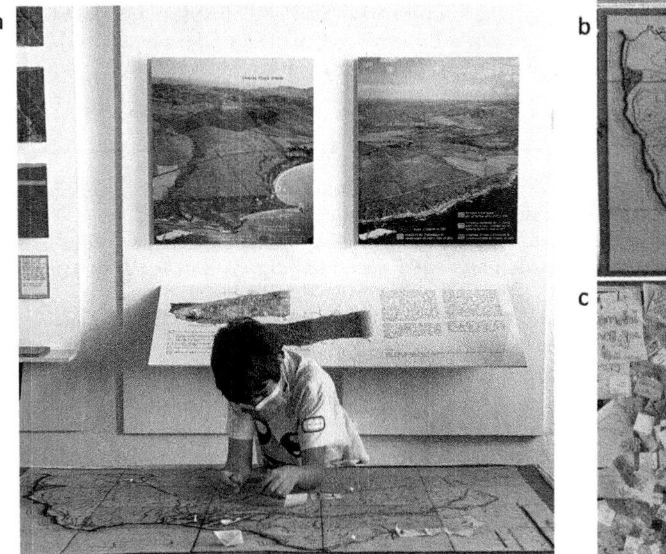

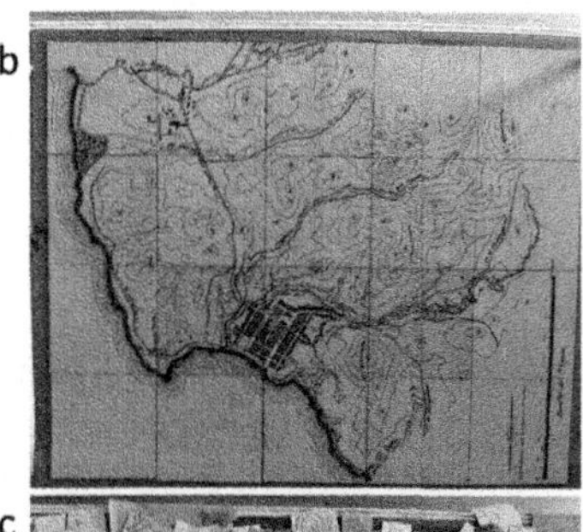

Figure 2.4 (a) One of the first visitors interacting with a visualisation based on an 1883 map by the Spanish Military Engineers. (b) The map several weeks after the opening of the exhibit. The multiple layers are narratives that converse with narratives left by other visitors as well as with the spaces in the base map and elsewhere in Vieques.

Source and permission: Javier Arce-Nazario; also, with written permission of the parents of the child in the image

Conclusion

Widening the horizon of deployments of 'chorography' as well as 'cartography' beyond Ptolemy's reductive opposition of these two concepts and their nineteenth-century accretions, our discussion has explored a variety of approaches to mapping over time. In particular, reviving the language of chorography—a multidimensional and undisciplined terrain that has resisted normalisation—seems to offer distinctive hermeneutic tools for an interdisciplinary conversation about humanistic perspectives on cartography. Chorography draws strength from the uniqueness of the place with which it engages, in comparison with more established models of representation and mapping, yet it makes room for a wide variety of representational projects, with varying aims, in various historical contexts. Chorography is a mode rather than a discipline. It is trans-historical, transmedial and transdisciplinary. It thrives outside major descriptive paradigms, while also providing material for thinking about how parameters of visual representation and political messaging are developed at different historical moments. Chorography forces us to question binary distinctions between text and map, art and science, space and place, and it can still provide tools to apprehend, teach and explore new ways to represent place and to contest outdated ones. In this sense, chorography may provide even more capacious spaces for interdisciplinary discussion to complement those of counter-cartography, counter-mapping or affective geographies.

Note

1 The authors express their gratitude to the UNC Carolina Seminar Series for support over the past two years. We would also like to thank Denis Wood, Matthew H. Edney and Ed Triplett for their engagement in the seminar. They bear no responsibility for the claims we make in this chapter. Arce-Nazario would like to acknowledge the support from the UNC Institute for the Humanities and the members of the Carolina Cartography Collective, in particular Klaus Mayr, Isabelle Smith and Tomas Roy who worked tirelessly on the design and installation of the exhibit.

References

Berggren JL and A Jones (2000) *Ptolemy's Geography. An Annotated Translation of the Theoretical Chapters*. Princeton: Princeton University Press.

Besse J-M (2000) *Voir la terre, six essais sur le paysage et la géographie*. Arles: Actes Sud.

Bordone B (1528) *Libro di Benedetto Bordone nel qual si ragiona di tutte l'isole del mondo*. Venice: Nicolò d'Aristotile detto Zoppino.

Braun G and Hogenberg F (1572) *Civitates orbis terrarum*. Cologne: Theodor Graminaeus.

Campbell T (1987) *The Earliest Printed Maps 1472–1500*. Berkeley: University of California Press.

Casey ES (2002) *Representing Place: Landscape Painting and Maps*. Minneapolis and London: University of Minnesota Press.

Coppo P (1540) *Del sito de listria*. Venice: Francesco Bindoni.

Curtius E and Kaupert JA (1885–1903) *Karten von Attika*. Berlin: Reimer.

Danti E (1577) *Le scienze matematiche ridotte in tavole*. Bologna: Compagnia della Stampa.

della Dora V (2016) Between the garden and the island: Mirror images and imaginative geographies of Greece in Thomaso Porcacchi's *L'isole piùfamose del mondo*, 1572. In: Gerstel SEJ (ed.) *Viewing Greece: Cultural and Political Agency in the Medieval and Early Modern Mediterranean*. Turnhout: Brepols, pp. 185–206.

Dipylon Society (2020) *Mapping Ancient Athens*. Available at: https://map.mappingancientathens.org/ (accessed 18 May 2023).

Dipylon Society (2023) *Karten von Attika in the Era of Digital Humanities*. Available at: https://dipylon-kartenvonattika.org/webgis (accessed 18 May 2023).

Dueck D, H Lindsay, and S Pothecary (2005) *Strabo's Cultural Geography: The Making of a Kolossourgia*. Cambridge, UK: Cambridge University Press.
Edney MH (2019) *Cartography: The Ideal and Its History*. Chicago: University of Chicago Press.
Fachard S (2016) Modelling the territories of Attic demes: A computational approach. In: Bintliff J and Rutter NK (eds) *The Archaeology of Greece and Rome. Studies in Honour of Anthony Snodgrass*. Edinburgh: Edinburgh University Press, pp. 192–222.
Gastaldi G (1561) *La universale descrittione del mondo, descritta da Giacomo de' Castaldi Piamontese*. Venice: Matteo Pagano.
Goldie MB (2019) *Scribes of Space. Place in Middle English Literature and Late Medieval Science*. Ithaca: Cornell University Press.
Grasshoff G and Stückelberger A (eds) (2006) *Klaudios Ptolemaios Handbuch der Geographie*, vols. 1–2. Basel: Schwabe Verlag.
Guarducci M (ed.) (1942) *Inscriptiones Creticae*, vol. 3, Tituli Cretae orientalis. Rome: Libreria dello Stato.
Guéroult G (1553) *Épitomé de la chorographie d'Europe illustré des pourtraitz des villes plus renommées d'icelle*. Lyon: Balthazar Arnoullet.
Helgerson R (1986) The land speaks: Cartography, chorography, and subversion in renaissance England. *Representations* 16: 50–86.
Honter J (1532) *Chorographia Transylvaniae Sybembürgen*. Basel.
Lalonde G (2006) *IG* I^3 1055 B and the boundary of Melite and Kollytos. *Hesperia* 75: 83–119.
Matijašić I (2022) *Common sense geography* nelle iscrizioni greche di età ellenistica. Rappresentazione e definizione dello spazio geografico. In: Sørensen S (ed.) *Sine Fine: Studies in honor of Klaus Geus on the occasion of his sixtieth birthday*. Stuttgart: Franz Steiner Verlag, pp. 393–412.
Nuti L (1999) Mapping places: Chorography and vision in the Renaissance. In: Cosgrove D (ed.) *Mappings*. London: Reaktion, pp. 90–108.
Pickles J (2004) *A History of Spaces: Cartographic Reason, Mapping and the Geo-Coded World*. London and New York: Routledge.
Ptolemy C (1561) *La Geografia di Claudio Tolomeo alessandrino*. Venice: Vincenzo Valgrisi.
Simon J (2014) Chorography reconsidered: An alternative approach to the Ptolemaic definition. In: Lilley K (ed.) *Mapping Medieval Geographies: Geographical Encounters in the Latin West and Beyond, 300–1600*. Cambridge: Cambridge University Press, pp. 23–44.
Tolsa C (2013) *Claudius Ptolemy and Self-Promotion: A Study on Ptolemy's Intellectual Milieu in Roman Alexandria*. PhD Dissertation, Universidad de Barcelona.
Traill J (1975) *The Political Organization of Attica: A Study of the Demes, Trittyes, and Phylai, and Their Representation in the Athenian Council*. Hesperia Supplements 14. Princeton: American School of Classical Studies at Athens.
Vine A (2017) Travel and chorography. In: Lee J (ed.) *A Handbook of English Renaissance Literary Studies*. Chichester: Wiley-Blackwell, pp. 411–425.
Weitzke J (2017) The public face of expertise: Utility, zeal and collaboration in Ptolemy's *Syntaxis*. In: König J and Woolf G (eds) *Authority and Expertise in Ancient Scientific Culture*. Cambridge: Cambridge University Press, pp. 348–373.
Wood D (2010) *Rethinking the Power of Maps*. New York: Guilford Press.

3
PROCESSUAL MAP HISTORY

Matthew H. Edney

Introduction

Originating in an attempt to escape the progressive teleology of the 'history of cartography', processual map history (PMH) is an empirically grounded implementation of post-representational theories of mapping (Rossetto, 2015; Simpson, 2020). It offers a conceptual framework for understanding and explaining mapping systems, their synchronic interactions and their diachronic development over time. While it draws inspiration from a variety of philosophical approaches, it seeks an approach to map history that is mapping-centred.

PMH is less concerned with ideas about the nature of maps than with how people have acted with maps; it seeks to understand the processes of mapping by means of a bottom-up analysis of the archive. Reflexively applied to the study of maps and map history, PMH avoids deeply rooted and misguided idealisations about the natures of 'cartography' and 'the map'. It further resolves several pervasive antagonisms drawn between the intent of a map's producer(s) and the meanings comprehended by its consumer(s); between maps' 'factual' and 'sociocultural' aspects; between western/scientific/culturally sterile maps and non-western/artistic/culturally authentic maps; and between scholars who adhere to a normative understanding of maps and those who reject it. PMH thus constitutes not a hermeneutic strategy but rather a means to establish the proper limits of interpretation.

Origins

The ethos of early twentieth-century modernism took older concepts of maps as measured and observed representations of the world and cemented them into the normative conception that *all* maps are metrical reductions of the earth's surface and as such are properly 'scientific' works. This new idealisation was sustained by new narrative structures that revamped the existing histories of cartography that traced the rise of western civilisation and of specific nations through maps. The new metanarratives demonstrated either that cartography had once been an 'art' but had long since become a 'science' or that cartography had always been a science that had steadily grown more technologically sophisticated over the centuries. The metanarratives entailed the teleological necessity of cultural and technological progress as an innate quality of western civilisation.

DOI: 10.4324/9781003327578-5

The modernist expectation that maps are works of science did not entail a precise definition of 'science' but rather worked through an oppositional relationship of 'science' with 'art'. Scientific maps were those that possessed a coherent commitment to making and remaking civilisation and that were uninfected by any self-evidently non-scientific or 'artistic' element. The last might comprise ancillary decoration on maps, propagandistic corruptions of spatial truth, unconstrained expressions of individuality or even simply crude techniques of measurement and observation. After World War II, however, scholars interested in ideas of directed communication for commercial or political purposes (propaganda writ large) and in behavioural psychology promoted studies of 'mental maps' as both personal and cultural constructs (Figure 3.1). By 1980, interest in non-scientific maps that seem to depict authentically the spatial conceptions held by individuals reached something of a tipping point, and scholars began to argue that *all* maps were consumed as social instruments and cultural documents.

A key argument justifying this 'sociocultural' turn in map studies was the sheer inappropriateness of a progressive teleology that *a priori* excluded non-western mapping from consideration, that failed to account for the wide variety of map form and function found in

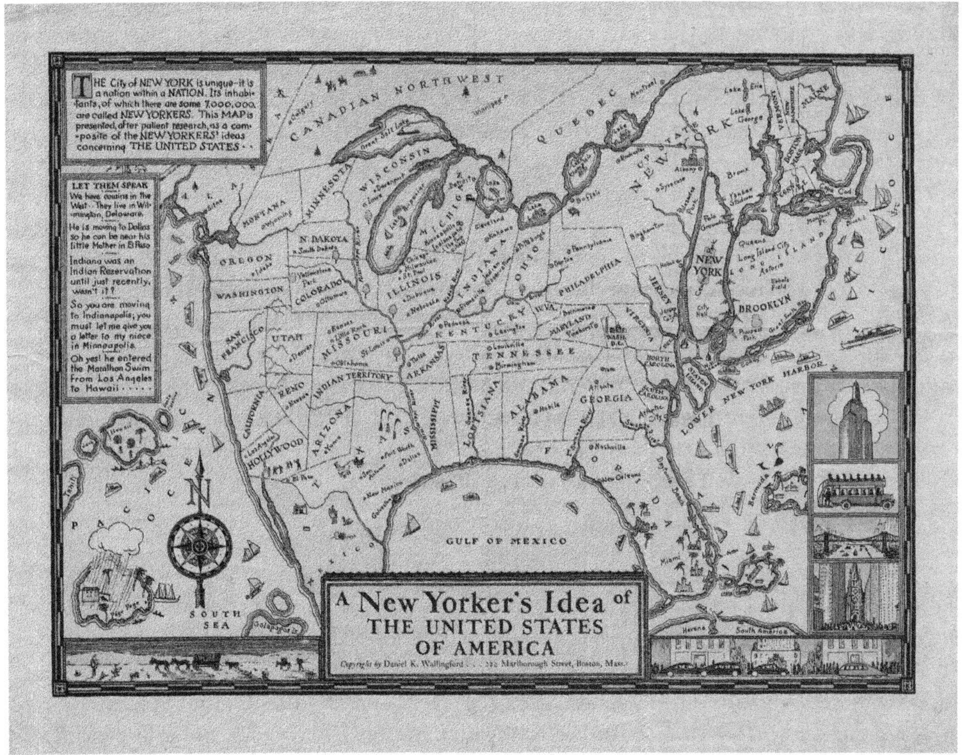

Figure 3.1 Daniel K. Wallingford, *A New Yorker's Idea of the United States of America* (New York: Columbia University Press, 1936), read by many post-war geographers as exemplifying the irrationality of cognitive maps. Colour lithograph, 19 cm × 25 cm.

Source: Courtesy of Cornell University and the PJ Mode Collection of Persuasive Cartography (item 1254.01); online at digital.library.cornell.edu/catalog/ss:3293992

the historical record and that was blatantly racist to its core. Since 1980, sociocultural map historians have generated a large and engaging literature that has spread the interpretation and analysis of maps across the humanities and social sciences. They have not, however, proposed models by which to account for how the systems and practices of mapping have changed over time. As a result, they have defaulted to older historiographies of cartography as a universal and singular endeavour and of all maps as somehow still properly algorithmic and metrical works.

This state of intellectual affairs is evident, if in nothing else, in the continued rehearsal of Alfred Korzybski's 1933 dictum, 'the map is not the territory'. Even though scholars generally quote this dictum to proper effect—to remind us that a particular map or suite of mapping practices is not mimetic or complete or even barely partial in its imaging of the world—the constant reminder of Korzybski's thoroughly modernist understanding of representation keeps us tethered to the spectre of the normative map. It remains all too easy to slip into presumptions of 'the map' as a universal category and of 'cartography' as a universal endeavour. It is too easy to pronounce that 'maps are' and 'cartography is' without empirical regard for historical and cultural variation. In deploying 'cartography' and 'map' as universals, scholars continue to hold maps up as something special and as something necessarily distinct from other texts; there remain, for example, persistent references to the distinctiveness of 'cartographic language'. Thus, even as the normative understanding of the map was challenged, the universality of mapness has persisted as an unexamined preconception (Edney, 2019, 2022).

Edney (1993) offered a preliminary attempt at a non-normative, diachronic model for map history by recognising how spatial conceptions, functionalities and institutions have tended to cluster in largely discrete ways. Each cluster constitutes a 'mode' of cartography, which is to say a way of acting cartographically. Rather than collapsing all cartography into a single endeavour with a single and inevitably progressive history, the history of cartography can be told through histories of each mode and their interactions in concert with shifting social and cultural conditions. This approach has successfully structured broad histories—notably the last three, encyclopaedic volumes of *The history of cartography*, in which entries about each mode are grouped together to draw out their interconnections while downplaying map historians' nationalistic tendencies (Harley et al., 1987–2027)—but modes prove too imprecise a heuristic for precise historical analysis. Further conceptual work and empirical investigation have produced a more precise conceptual structure for exploring and explaining maps and mapping: PMH.

Mapping, not map

The basic principle of PMH is that accounts of things in and of themselves can only be superficially descriptive. The way to explain their function, form and change is by studying the processes that generate them. That is, PMH turns away from maps as things and to mapping as sets of processes by which they represent spatial complexity. 'Spatial complexity' is a generic shorthand: if spatial relationships and differences are not complex in some way, there is no reason to comprehend and communicate them. The processes of mapping are those of the production, circulation and consumption of maps. By keeping processes in mind, map historians can trace their changes over time.

The 'process' in PMH is not the same as 'practice'. Yes, it is crucially important to explicate the practices by which people produce and consume maps, but doing so is only one step towards processual history. Nor is PMH as simple as 'putting maps into context'. Rather, PMH precisely defines the proper contexts for studying particular maps or sets of maps.

All too often, scholars treat maps as a technology that exists outside of culture, or as an overarching determinant of culture, and they argue that maps act autonomously to *do* something. But maps do not act by themselves. All maps, and not just verbal and performative mappings, such as an exchange of directions on a street corner, are 'given life' by discursive exchanges between producers and consumers. It might be argued that no one really thinks that maps are actual agents and that it is only a convenient shorthand to suggest that 'maps do *x*'. For example, since Jacob's (2006) nuanced study, it is commonly stated that maps make the invisible, visible. Setting aside questions of verbal maps and visual cognition, even if *all* maps do indeed do so, then they do not all do so in the same way and to the same ends. The shorthand generalises mapping processes into obscurity; it discourages attention to the particular formations in which maps have an effect as they are produced, circulated and consumed.

The key mapping process is circulation, the manner in which spatial texts move between producers and consumers and bind them all within specific spatial discourses (*not* the circulation of spatial data or geographical knowledge, but of maps). Each spatial discourse is a regulated network of representation concerned with comprehending spatial complexity for a particular purpose shared by the participants. Each constitutes a circuit of social and cultural exchange. Mapping is thus constitutive of social relations. Simply stating that maps 'reflect' or 'express' social conditions or that mapping is a 'social practice' does little to reveal how mapping creates inclusive groups and institutions within which participants affirm or claim a social status vis-à-vis other groups. There are innumerable examples of this principle in the historical record. In the fifteenth-century Venetian Republic, for example, newer offices of state employed graphic maps in their territorial administration as much as an effective device as a means to promote their difference from the older and more powerful offices of state that continued to rely on written lists (Marino, 1992). Various forms of map work by educated women ca. 1900 contributed to their assertion and attainment of social and political rights (Dando, 2017; see also Dando, this volume). The self-cultivation in early modern Britain of the status of 'mathematical practitioner' was a productive strategy for social advancement and so on. PMH promotes a social history of culture (Edney, 2019: 177, 47).

Maps are diffused texts

As expressions of mapping processes, maps are almost epiphenomenal. Maps are the means of mapping, not the goal. The things commonly identified as 'maps' are just the visible or sensed element of a much larger assemblage. 'Maps' are not static, fixed things; *contra* Latour, they are not immutable mobiles. No map is a stand-alone thing. There is no one primary or characteristic form of map, not even the requirement that a map must be graphic in form. There can be no 'cartographic language' that gives maps and cartography a unique character (Edney, 2019: 41, 37–39). Each map is an assemblage of signs produced from multiple strategies—verbal (oral or written), gestural and performative, tectonic, numeric and graphic—that are performatively embedded in and permeate spatial discourses. Each map is diffused across the spatial discourse that generated it.

The principle is well established for the overtly performative acts of mapping by indigenous peoples, in which the significance of any inscribed elements is embedded in their communal creation and consumption (esp. Gartner, 2011). The principle is not obviated by more heavily inscriptive mapping practices. In colonial New England, for example, property maps comprised the legal metes-and-bounds record (written and numeric signs) as well as the plan that interrelated the written description with certain dimensions (numeric), the particular landmarks at corners of the property (graphic and written) and the landmarks or monuments themselves (tectonic). In Figure 3.2, of a map from a legal archive, each side of the property is given with its length, and each corner is carefully defined; for example, the top-right corner, marked '7', is glossed in the list on the right-hand side as being marked by a 'Rocky Hill a whitt [white] oak lying on the Ground'. Traditionally understood as a simple graphic, this kind of survey map diffuses into legal descriptions in one direction, and into the terrain itself in the other. All maps extend beyond their apparently neat and well-defined physical edges into a wider discursive and performative field; their interpretations depend on how and where their consumers physically place and use them (Dym and Lois, 2021).

Map scholarship has always featured a deeply entrenched but false divide between 'the mapmaker' and 'the map user'. 'The map' appears as a strictly physical thing that instantiates, that divides and embodies the antagonism between the producer's intent in making the map (the mapmaker determines the map's meaning) and the consumer's interpretation of it (meaning is created by the reader). The divide has been replicated even by sociocultural scholars (Edney, 2022: 55). It is best to remember, however, that there are no necessary differences among the 'users' of a linguistic system. All talkers, listeners, writers and readers use and contribute to that system. More generally, there is no divide between producers and consumers: all participate in the discourse. Mapmakers use the system in creating texts just as consumers use the system in working with the texts; producers are also consumers, and consumption can also entail the modification of texts. In some discourses at certain times, professional and institutionalised practices are indeed dominant, but they cannot be presumed. To call a mapmaker a 'cartographer' before the 1820s is very much to impose a standard of behaviour and a distinction that must be demonstrated rather than just presumed.

Methodology and levels of analysis

Particular circuits can be traced by paying close attention to the (im)materiality of maps, which oftentimes provide the only source of evidence for past mapping. The task boils down to paying attention to the precise circumstances and material characteristics of the maps-as-things so as to discern how the maps actually diffuse through their generative mapping processes. Such study might be defined as answering the four Ps: participants (kinds of people; who?); purpose (why?); place (the physical sites where maps circulate; where?); and practice (strategies and techniques; how?). Participants regulate their practices (acts of representation), establishing the semiotic system and the textual strategies to be deployed, in pursuing a specific purpose. Place can vary widely, from tiny circuits of individuals making maps for themselves, to modern media circuits that widely and indiscriminately broadcast maps. Practice also embraces the ways in which techniques and strategies are regulated, whether through formal institutions (e.g. Withers, 2021) or informally through communal behaviour.

The task is to study the maps and associated texts, especially their material nature, to understand how they circulated among producers and consumers, which empirically

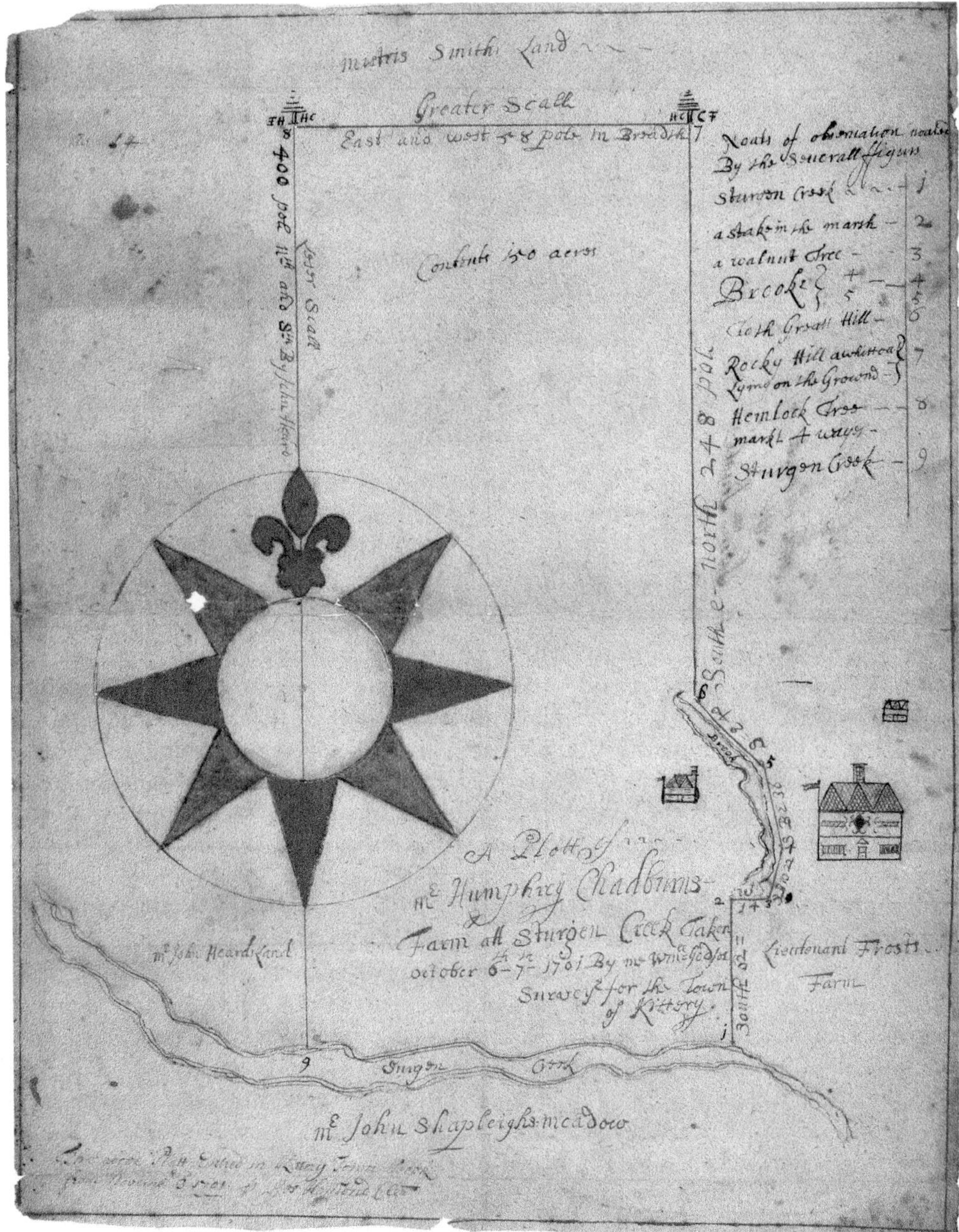

Figure 3.2 William Godsoe, 'A Plott of mr Humphrey Chadburns farm att Sturge[o]n Creek' (6–7 October 1701). Pen and ink on paper, 35 cm × 45 cm.

Source: Courtesy of the Maine State Archives (York CCP, October 1701, 6: 140, Nicholas Morrell v. Samuel Small); online at digitalmaine.com/arc_misc_maps/1/

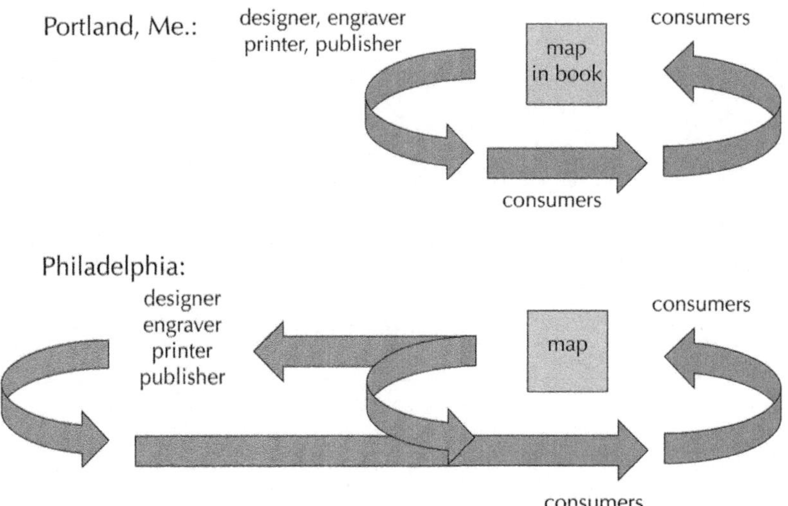

Figure 3.3 Schematic diagrams of (upper) the strictly local spatial discourse of urban maps of Portland, Maine, in the early nineteenth century and (lower) the modified discourse once Philadelphia publishers sought to meet local demand for urban maps.

Source: Elaboration of the author

defines precise spatial discourses within which we can explain changes over time. Early-nineteenth-century printed maps of Portland, Maine, for example, can be seen to fall into two circuits, two spatial discourses: initially, urban maps were produced and consumed in Portland in conjunction with the increasing challenge to the community (white, primarily male, protestant, educated, civil and mercantile) by significant population growth (Figure 3.3, upper); by 1850, Philadelphia publishers were forcing their wares into the local market (Figure 3.3, lower), in the process changing the look of the maps in order to meet local demand while still adhering to emergent national standards for engineering and urban mapping (Edney, 2022).

It is often easier, however, to work at less precise levels of analysis. If one is not able to identify a precise spatial discourse, then one might work with a thread of related discourses. Within geographical mapping for primary education in the nineteenth century, one might distinguish between maps drawn by students by hand with constantly changing pedagogic significance (see Schulten, 2017) and the printed maps enclosed within textbooks; both 'educational', but plainly working in different ways. In a sense, threads are an intermediate level of analysis between the spatial discourses and mapping modes; it is meaningful to think of the products of a thread as constituting a 'genre' of maps within a mode. The modes themselves remain a heuristic for organising and generalising large swathes of mapping practice without succumbing to the overgeneralisation of 'cartography'.

Translation

Re-situating a map artefact within a spatial discourse other than the one that gave it rise inevitably changes its function and significance. Translation is physical, intellectual and inherently diachronic (Nappi, 2021: viii). Consider the exemplar offered by Ramaswamy (2004). First, nineteenth-century biologists in Europe suggested that the discrete and otherwise inexplicable islands of certain flora and fauna found in Africa, Asia and Australasia could be explained by an ancient continent that had once filled the Indian Ocean; when this continent, which they called 'Lemuria', sank beneath the waves, it left behind the relic, isolated populations. Five decades later, new-age theosophists translated maps of the supposed Lemuria into maps of the ancient home of humanity. Many theosophists worked in southern India, studying 'eastern mysteries', and they introduced their maps in particular to Tamil elites. After Indian Independence, Tamil politicians translated Lemuria into the ancient Tamil homeland, which they populated with places from Tamil sagas, to create a nostalgic myth of lost greatness with which to counter the political dominance of north Indian Hindi-speakers. At each remove, maps with apparently 'the same' content are reconfigured within a new discourse to new ends and interpretations.

The translation of maps between spatial discourses and the participation of individuals in multiple spatial discourses are both avenues for the appropriation, adaptation and alteration of mapping concepts and ideas. Sixteenth-century Netherlandic geographers appropriated the compass rose from the marine charts that they also collected and used, but not in the same way as mariners, deploying compass roses on regional maps to signify 'ocean' rather than denote directions. English and French geographers soon followed suit. In such ways, repeated over and over, individuals have translated particular mapping practices from one spatial discourse to another, incrementally adjusting and changing them as they do so.

An end to universalisms

All told, PMH is about recognising and studying the dynamism and instability of mapping, without presuming that 'maps' are somehow coherent, stable and universal. It is an approach that topples 'the map' from the high, isolated pedestal on which western culture has placed it and it dissolves the rigid conceptual boundaries erected around 'the map' and 'cartography'. In their place, PMH offers the study of suites of mapping processes—spatial discourses, threads and mapping modes—by which people have understood and have communicated spatial complexity; maps are generated as part of those processes and must be understood accordingly. PMH rejects the unwarranted universalisms of 'map' and 'cartography'.

Instead, PMH calls for map historians to analyse maps, and their subtle differentiations, to reveal the processes by which they were produced, circulated and consumed. From this vantage point, one starts to see how and why mapping practices changed over time. Processual studies do not have to explicitly trace diachronic change, of course, but they must be sensitive to how the maps under study differ from other, seemingly similar works.

Ultimately, processual studies are about empirical analyses that shun overly grand statements. PMH is a pragmatic framework by which to assess mapping in any society or culture in any period without reliance on any universalisms. It requires that scholars establish the character of the mapping in which they are interested without relying on any preconceptions. It offers a framework that cuts through entrenched nonsense about cultural 'styles' (see Cams and Papelitzky, 2024) and about maps as being determined by the territory they represent.

References

Cams M and Papelitzky E (eds) (2024) *Remapping the world in East Asia: Toward a Global History of the 'Ricci Maps*. Honolulu: University of Hawai'i Press.

Dando CE (2017) *Women and Cartography in the Progressive Era*. London: Routledge.

Dym J and Lois C (2021) Bound images: Maps, books, and reading in material and digital contexts. *Word & Image* 37(2): 119–141.

Edney MH (1993) Cartography without 'progress': Reinterpreting the nature and historical development of mapmaking. *Cartographica* 30(2–3): 54–68.

Edney MH (2019) *Cartography: The Ideal and Its History*. Chicago: University of Chicago Press.

Edney MH (2022) Making explicit the implicit, idealized understanding of 'map' and 'cartography': An anti-universalist response to Mark Denil. *Cartographic Perspectives* 98: 51–60.

Gartner WG (2011) An image to carry the world within it: Performance cartography and the Skidi star chart. In: Brückner M (ed.) *Early American cartographies*. Chapel Hill: University of North Carolina Press, pp. 169–247.

Harley JB, Woodward D, Lewis GM et al. (eds) (1987–2027) *The History of Cartography*. 6 vols. in 12 books and parts. Chicago: University of Chicago Press. Available at: www.press.uchicago.edu/books/HOC/.

Jacob C (2006) *The Sovereign Map: Theoretical Approaches in Cartography throughout History*, trans. by Tom Conley. Chicago: University of Chicago Press.

Marino J (1992) Administrative mapping in the Italian states. In: Buisseret D (ed.) *Monarchs, Ministers, and Maps: The Emergence of Cartography as a Tool of Government in Early Modern Europe*. Chicago: University of Chicago Press, pp. 5–25.

Nappi C (2021) *Translating Early Modern China: Illegible Cities*. Oxford: Oxford University Press.

Ramaswamy S (2004) *The Lost Land of Lemuria: Fabulous Geographies, Catastrophic Histories*. Berkeley: University of California Press.

Rossetto T (2015) Semantic ruminations on 'post-representational cartography'. *International Journal of Cartography* 2(1): 151–167.

Schulten S (2017) Map drawing, graphic literacy, and pedagogy in the early Republic. *History of Education Quarterly* 57(2): 185–220.

Simpson P (2020) *Non-representational Theory*. London: Routledge.

Withers CWJ (2021) Map making, defamation and credibility: The case of the *Athenaeum*, Charles Tilstone Beke, and W. & A. K. Johnston's *Edinburgh educational atlas* (1874). *Imago Mundi* 73(1): 46–63.

4

SPATIAL ANTHROPOLOGY AND DEEP MAPPING

Les Roberts

Introduction: killing space/giving life to space

In the book, *Spatial Anthropology: Excursions in Liminal Space* (Roberts, 2018), I suggest that, in order to give life to space, it is at the same time necessary to be in the business of killing it off. In this prescription, *killing space* does not refer to a negation of space (cf. Ingold, 2009) or a dismissal of the diverse assemblage of practices that are variously engaged with what 'it' is and does. What it is pointing to is the divesting of an idea of space that is held above, apart, or at arm's length from the warmth and fleshiness of space as it is lived. Killing space, by this reckoning, is the adoption of a necessarily reflexive disposition towards the object of study; a stepping precariously into uncertain, or *uncharted* terrain and being alert to what continually falls away and reconstitutes itself therein: 'in the breath of a moment that is hatched in the space that *becomes*' (Roberts, 2018: 263, emphasis in original). If we accept that what might be understood as a 'map' can be as vaporous, malleable and contingent as the spaces being mapped, then this throws attention towards the anthropological and phenomenological entanglements that are constitutively bound up with the open-ended practices of mapping. A map can only ever be as good as the mappings it sets in train: a device to be squinted at when absolutely necessary but otherwise best left scrumpled in the back pocket, the better to assist in the task of giving life and corporeality to everyday spaces that are otherwise destined to be consigned to abstraction.

In pursuing further these lines of enquiry, critical reflection on recent scholarship on deep maps and deep mapping—not least the tensions and contradictions between the two—offers valuable insights into the ways in which qualitative and humanistic forays into the representation, imagining and practice of space are invariably multifaceted, undisciplined and irreducible to formal and programmatic design. As we will see, to flesh out the *doing* of deep mapping necessitates engaging with the same performative dynamics by which its various iterations are made manifest in practice. As itself a form of spatial practice, deep mapping is at best a convenient label to reach for when necessity demands but which quickly needs ditching the minute it conspires to mould itself into anything that starts to resemble a model or paradigm.

DOI: 10.4324/9781003327578-6

In this regard, it is instructive to look upon deep mapping in similar terms to those that define endeavours to formalise a discourse of psychogeography. Which is not to suggest that the concepts are necessarily related (although there are obvious correlations and enticing points of overlap—see Modeen and Biggs, 2020: 61–62), but that they each represent labels that work best when they succeed in eluding containment or pat definition. Those comfortable with the tag 'psychogeographer' would doubtless wince at attempts to shoehorn what it is they do into a kind of 'how to . . .' guide: 'Psychogeography for Beginners'; 'An A-Z of Psychogeography' (A is for Aragon, D is for dérive, S is for Situationist), and so on. This instrumental approach is as problematic when it comes to deep mapping as it is for psychogeography. The elemental and very reasonably put question 'what is deep mapping?' is best tackled not by outlining a set of defining characteristics and features (an exercise that is unavoidably weighted by the ballast of disciplinary persuasion), but by surveying the various precincts by which, as a coagulation of approaches and (inter)disciplinary interventions, it is performatively put to work. If, along the way, the sustainability or epistemological coherence of deep mapping—or of the deep map, its artefactual product—is called into question then that itself may be a worthwhile and productive outcome of these deliberations.

Diving within

But it equally well may be the case that the scrutiny and attention afforded to deep mapping have the effect of providing a fresh set of insights by which otherwise different research practices may be tentatively brought into critical alignment. In this sense, deep mapping may be regarded as a statement of intent insofar as what it is *not* can at least be evinced and a certain familial resemblance correspondingly transacted. What it is not—or at least what it *should* not be—is irreducibly representational if by this we mean a process that is predicated on stemming the flow of spatial and temporal vitality that bleeds into and through the 'map' as a cartographic abstraction. It is on account of this necessarily processual underpinning to deep mapping practices that the very notion of a 'deep map' becomes problematic. *PrairyErth*, its writer William Least Heat-Moon tells us, is a 'deep map' (1991). But while I have no issue with the suggestion that the book may be the creative outcome of a process of deep mapping, I am less sold on the idea that the text itself constitutes a map. Although, as coiner of the term, Heat-Moon's name is routinely rehearsed in discussions of deep mapping as part of a preliminary conceptual backstory, Heat-Moon himself was not consciously laying the foundation stones for something that others would go on to pick up as 'deep mapping'. As a dense, *deeply* layered and richly textured literary survey of Chase County in the U.S. state of Kansas, *PrairyErth* is a deep map of sorts; an entirely fitting metaphorical description of a textual cartography that aspires to yield what a conventional map or guide cannot even come close to conveying. What it is not is a representational device to which we can ascribe a set of formal and reproducible cartographic features that *project* Chase County or which provide a serviceable locative function (beyond that of a rudimentary stitching of narratives—however deep—to place). But, as ever with these things, it does depend on what is meant and understood by the term 'map'.

When we start to think about the ways *PrairyErth* may be considered a deep map, there are a number of key touchpoints from which we can extrapolate a broader outline of analysis. As a self-styled 'secretary of under-life' (1991: 367), Heat-Moon is desirous to dig deeper in his research studies: to burrow down from the surface in order to *excavate* that which is hidden or buried beneath thinly layered deposits of topsoil or asphalt. Deep mapping in this sense is as much a process of archaeology as it is cartography. With this comes an emphasis on *verticality* (Schiavini, 2004–5): the 'plumbing of a place's depth' (Gregory-Guider, 2005: 5). Horizontality is for the thin mappers (Harris, 2015, 2022); those who hold back from peeling off the surface layer and who, in the process, thus allow limited space for *time*. The temporal configurations that anchor places in turf that have been synchronically as well as diachronically ploughed are the stuff from which the deep mapper fashions their craft. Our role—as readers, viewers, consumers, users—is to take up the invitation to 'dive within', as artist, filmmaker and transcendental meditator David Lynch might put it (2006). Wydeven writes that Heat-Moon 'encourages us to fit ourselves in the creases [of maps]' (1993: 134), a nice turn of phrase which neatly captures the materiality and performativity that goes with the act of wayfinding: of exploring and placing oneself within the multi-scalar locative dimensions that are opened up through the act of deep mapping.

Another important touchpoint, one that casts a quizzical spotlight on the abstracted notion of a 'deep map', is that deep map*ping* necessarily entails what Schiavini refers to as 'deep travel' (2004–5). I do not wish to over labour the 'deep' terminology here, but what Schiavini rather usefully points to is the performative work that goes into both the production of deeply configured spatial knowledge (what it is that the deep mapper *does*) and what is precipitated by way of action performed *in response to* the production of such spatial knowledge. Were someone sufficiently inspired by *PrairyErth*—as the prototypical literary deep map—to *dive within* the folds and creases of Chase County then they very well might find themselves tramping across the same geographical terrain that Heat-Moon's literary excavations have turned over. Deep mapping, in other words, cannot be reduced to the otherwise a-spatial and a-temporal domain of the (deep) *map*. It denotes an anthropology of practice. People are doing things when they engage in deep mapping; what it is they are doing becomes the focus of a spatial anthropology: a culture of mapping practice (Roberts, 2012, 2018).

Undisciplining the deep map

The important emphasis placed on performance is most notably explored by the archaeologists Mike Pearson and Michael Shanks, whose book *Theatre/Archaeology* (2001) distils (by its title alone) a re-oriented and quintessentially interdisciplinary view of landscape, one that pays heed to 'the grain and patina of place . . . [the] interpenetrations of the historical and the contemporary, the political and the poetic, the factual and the fictional, the discursive and the sensual' (2001: 64). For Pearson and Shanks, deep mapping extends to 'everything you might ever want to say about a place'(2001: 65). Of course, everything you might want to say may be voluminous, polyvocal or open ended (as any deep mapping worth its salt should be aiming for anyway). Unlike the surface dimensions that delineate and give shape to the locational properties of place, verticality and depth denote a comparative absence of limitations. The deeper you go the more layers you accrue. The problem becomes how to hold it all together: How to *frame* it as a map. The performativity and theatricality of place that might accompany a walker in Heat-Moon's Chase County, or which might give flavour to his or her practice, is not predicated on there being a material cartographic resource as a necessary reference point when out 'in the field'. The 'map' is lodged in the more im/material spaces of the body and imagination. Its performativity is made flesh in the way the walker inhabits and dwells within the space that both map/book and walker conjure into being.

There is, then, a fundamental creativity at work in the practice of deep mapping, both on the part of the mapper and that of the map reader/user. Given this, it is not all that surprising to discover that, alongside the proponents of a literary deep mapping—chiefly, but by no means exclusively originating from the United States (Maher, 2014; Modeen and Biggs, 2020: 59–61)—the most notable traffic of activity conducted under the banner of deep mapping has been initiated by visual and performance artists. Two of its most eloquent champions are Clifford McLucas (2000) and Iain Biggs. The latter in particular is at pains to stress the interdisciplinarity or post-disciplinarity of deep mapping. For Biggs, one of the defining ingredients of an 'open' deep mapping is the extent to which it is able to avoid 'becoming complicit in its "disciplining"' (2010: 21). This echoes the point made earlier about resisting the formalising of a language or method of deep mapping that in some way reins it in as an otherwise 'knowledgeable, passionate, polyvocal engagement with the world' (2010: 8). Cultivating what Biggs refers to as a *metaxy* of practice—a 'space between' (2010: 8) in which to pitch a precarious and purposefully indeterminate sense of a deep mapping practice—is to tread a fine line between complicity and creative dissolution. The creative efficacy of open deep mapping is co-extensive with that which underpins an artistic praxis that is operative outside of the tramlines of disciplinary or institutional orthodoxy. The complicity comes in the form of challenges that are posed in having to dance around a discursive object—deep mapping—whose constitutive openness is itself open to the dangers of disciplining. In other words, the process of framing an open deep mapping runs the risk of a sort of inverse disciplining on account of the very fact that it *is* an object of discourse, even if it is trying its best not to be. The paradox is that Biggs's call for an 'open deep mapping' only makes sense insofar as its openness is sufficiently diffuse as to do away with the very idea of deep mapping in the first place. The challenge of balancing these contradictory facets means questioning the coherence and validity of deep mapping on the one hand and maintaining a loose, plural and open application of the term on the other.

The representational constraints attached to the idea of a deep map as something that aspires to be open, performative and *more-than-representational* (Lorimer, 2005) are analogous to those that are routinely confronted by ethnographers tasked with the translation of *experience* (the flux and messiness of everyday life) into *narrative* (the ordered and disciplined fieldwork monograph). The fixity and abstraction of the cartographic frame (the map) belie the unboundedly complex, contingent and temporal spatialities of 'the field'. The deep map is a utopian imaginary of space inasmuch as it strives to frame or in some way open itself up to that which is *lived*. By contrast, the *thin* map (if we can accept, for a moment, this oppositional conceit) is unapologetically representational: it is a representation of space that is ill- or, at least, under-equipped when it comes to servicing the needs of those whose inclinations are to *dive within*. The writing culture debates that surfaced in anthropology in the 1980s, and which precipitated much hand-wringing in respect of a perceived crisis in ethnographic representation, drew closer attention to the interpretative mechanics of thick description in the writing-up of fieldwork data. One of the consequences of this was to raise the question as to whether a sharply observed and experientially immersive literary description of a given sociocultural landscape could offer up as much if not more than a disciplinary-framed ethnographic account. A similar question could be posed in relation to cartography once the epistemological consequences of depth have been factored into the equation. If, as an exemplar of a geo-literary thick description, *PrairyErth* can be considered a deep map then might we not correspondingly draw the conclusion that a writer or poet (or, indeed, filmmaker, artist, musician, or performer if we extend this to other branches of the arts) could be considered a deep cartographer on literary (or cinematic, artistic, musicological, or performative) terms alone? And, if so, doesn't this risk spreading what we might think of as the art of mapping too thinly? Put another way: does deep mapping need to be discursively labelled as such for it to qualify as deep mapping? And if the answer is 'no', then might not the cartographic hoops through which one might otherwise be required to jump be dispensed with altogether without any significant detraction in terms of what or how a place is being mapped?

These are questions I raise more by way of problematising deep mapping as an open and undisciplined field of spatial thinking and practice rather than to render a partisan position as to its sustainability or viability as an object of discourse. Although there is unevenness and variability in terms of what deep mapping extends to in practice, there are certainly some common threads that can provisionally be woven together: a concern with narrative and spatial storytelling; a multi-scalar and multi-layered spatial structure; a capacity for thick description; a multimedial navigability; a spatially intertextual hermeneutics; an orientation towards the experiential and embodied; a strongly performative dimension; an embrace of the spatiotemporally contingent; a compliance with ethnographic and autoethnographic methods and frameworks; an 'undisciplined' interdisciplinary modality; a time-based cartographics; an open and processual spatial sensibility; and, perhaps most telling, a reflexive sense of the fundamental *unmappability* of the world the deep map sets out to map.

The instrumentalism of the deep map

When we relate this all to developments in geospatial computing and the increasingly migratory domain of geographic information systems (GIS), then the idea of what a deep map might look or act like takes on more concrete form, and therein lies the problem. Responding to the challenge to create a model of a deep map and to 'explore how digital tools and interfaces can support ambiguous, subjective, uncertain, imprecise, rich, experiential content alongside the highly structured data at which GIS systems excel', Ridge et al. conjure the notion of a 'greedy deep map' (2013: 176, 181). This rather intriguing and suggestive metaphor presents us with an image of a data-rich and data-hungry geospatial resource whose value lies in its capacity to outstrip the ability—and agency—of its human counterparts in terms of a spatial praxis sublimated towards more computational ends: the provision of a *potentiality* of retrievably layered data. To conceive of the deep map as 'a space in which a near limitless range and quantity of sources can be included, interrogated, manipulated, archived, analysed, and read' (Ridge et al., 2013: 184) is to imagine what the realisation of a deep map is or could be as a big data-driven, totalising model. The question this raises for those invested in the development of a digital spatial humanities is whether the acquisition of the prized goal of a digitally limitless deep map comes at the cost of jettisoning the more anthropological, embodied and performative spatialities that are bound up with the practice of deep mapping.

Although, as geographer David Harvey observes, 'maps are typically totalizing, usually two-dimensional, Cartesian, and very undialectical devices' (in Bodenhamer, 2015: 18) that does not, of course, mean that digital deep maps—or, rather, deep mapping practices that exploit the many possibilities and advantages offered by digital and geospatial technologies—are necessarily cut from the same Cartesian, undialectical cloth. As David Bodenhamer notes (2015: 23), at its best GIS-based deep mapping is an 'ideal storyboard for humanists', offering a conceptual, technological and spatial framework adapted to the need to tell spatial stories that are harvested from 'experiential as well as objective space' and which are replete with the 'rich contradictions and complexities' (Bodenhamer, 2015: 23) that ordinarily, as abstract representations of space, maps fall short of conveying.

Yet while the centrifugal pull of the digital world will continue to shape new ways of qualitatively mining the layered and experiential history of places, this should not be at the expense of a deep mapping praxis that is: (a) entirely at ease with the dispensing of programmatic labels (such as 'deep mapping praxis'); (b) informed but not slavishly driven by digital tools and geospatial technologies; and (c) capacious enough to accommodate a diverse constituency of voices, actors, stakeholders, communities and performers whose resonant clamour—the 'multitonal chorus' (Maher, 2014: 22) of everyday spatial dialogue—is not muted by the dead hand of corporate instrumentalism (as manifested by an increasingly audit- and impact-driven culture of academic research). Critical vigilance is no less necessary in response to a creeping instrumentalism whereby, discursively, the deep map—as distinct from practices of deep mapping—is privileged in ways that '[fail] to acknowledge the poetics of paradox and ambiguity essential to open deep mapping' (Modeen and Biggs, 2020: 53). An example of this is Bodenhamer's framing of the deep map as one or all of the following: a platform, a process and a product. As a *platform*, Bodenhamer suggests that a deep map can be thought of as 'an environment embedded with tools' that allows for the interrogation of data; as a *process*, a deep map 'engages evidence . . . [to elicit] a spatial narrative and ultimately a spatial argument'; and as a *product*, a deep map is a means of visualising 'the results of our enquiry and [of sharing] the spatially contingent argument enabled by the deep map' (2022: 7). While there is no suggestion that this observation and other contributions to the book *Making Deep Maps: Foundations, Approaches, and Method* (Bodenhamer et al., 2022) are being held up as the sum of all that can be corralled under the deep maps/deep mapping label, there is nevertheless the implication that it is necessary to assuage those less taken with the idea of an open deep mapping praxis by affirming the utility and instrumental value of the deep map as something that can be directly pointed to and positivistically *applied* as a fixed methodological framework.

From deep mapping to spatial anthropology

It is precisely the contention that deep mapping can be made to fit into a uniform methodological structure that the idea of open deep mapping, or what I would prefer to term a *spatial anthropology* of mapping cultures (Roberts, 2012: 11), is weighted against. In this respect, the attempt to tramp out some form of common ground might be a legitimate, if necessarily broad-brushed, way of approaching the breadth and diversity of deep mapping practice. Another approach, one that is arguably more productive and *undisciplined*, is to take heed of the loosely anthropological underpinnings that root deep mapping in the performative and processual flux of everyday life.

In their ethnographically informed case study based in rural North Cornwall, Bailey and Biggs describe a deep mapping process that consists of 'observing, listening, walking, conversing, writing and exchanging . . . of selecting, reflecting, naming, and generating . . . [and] of digitizing, interweaving, offering and inviting' (2012: 326). While this full set of verbs will not apply to all variations and permutations of deep mapping practice, what they do usefully signpost is the way that very little of what deep mappers are *doing* is in fact oriented towards the production of maps so much as immersing themselves in the warp and weft of a lived and fundamentally intersubjective spatiality. It is from that performative platform—that *space*—that the creative coalescence of structures, forms, affects, energies, narratives, connections, memories, imaginaries, mythologies, voices, identities, temporalities, images and textualities starts to provisionally take shape. Whether or not we wish to call what emerges from this process a 'map'—or the process itself 'mapping'—seems to me less important than the fact that it is taking place at all. In its most quotidian sense, then, deep mapping can be looked upon as an embodied and reflexive immersion in a life that is lived and performed spatially. A cartography of depth. A *diving within.*

References

Bailey J, Biggs I (2012) 'Either Side of Delphy Bridge': A deep mapping project evoking and engaging the lives of older adults in rural North Cornwall. *Journal of Rural Studies* 28(4): 318–328.

Biggs I (2010) The spaces of 'Deep Mapping': A partial account. *Journal of Arts and Communities* 2(1): 5–25.

Bodenhamer DJ (2015) Narrating space and place. In: Bodenhamer DJ, Corrigan J and Harris TM (eds) *Deep Maps and Spatial Narratives.* Bloomington: Indiana University Press, pp. 7–27.

Bodenhamer DJ (2022) The varieties of deep maps. In: Bodenhamer DJ, Corrigan J and Harris TM (eds) *Making Deep Maps: Foundations, Approaches, and Methods.* London: Routledge, pp. 1–16.

Bodenhamer DJ, Corrigan J and Harris TM (eds) (2022) *Making Deep Maps: Foundations, Approaches, and Methods.* London: Routledge,

Gregory-Guider CC (2005) 'Deep Maps': William Least Heat-Moon's psychogeographic cartographies. *eSharp*, 4: 1–17. Available at: www.gla.ac.uk/media/Media_41154_smxx.pdf (accessed 15 February 2023).

Harris TM (2015) Deep geography—Deep mapping: Spatial storytelling and a sense of place. In: Bodenhamer DJ, Corrigan J, Harris TM (eds) *Deep Maps and Spatial Narratives.* Bloomington: Indiana University Press, pp. 28–53.

Harris TM (2022) Deep mapping the lived world: Immersive geographies, agency, and the virtual umwelt. In: Bodenhamer DJ, Corrigan J and Harris TM (eds) *Making Deep Maps: Foundations, Approaches, and Methods.* London: Routledge, pp. 112–131.

Heat-Moon WL (1991) *PrairyErth (a Deep Map).* Boston: Houghton Mifflin.

Ingold T (2009) Against space: Place, movement, knowledge. In: Kirby P (ed.) *Boundless Worlds: An Anthropological Approach to Movement.* Oxford: Berghahn Books, pp. 29–43.

Lorimer H (2005) Cultural geography: The busyness of being more-than-representational. *Progress in Human Geography* 29(1): 83–94.

Lynch D (2006) *Catching the Big Fish: Meditation, Consciousness and Creativity.* London: Penguin.

Maher SN (2014) *Deep Map Country: Literary Cartography of the Great Plains.* Lincoln: University of Nebraska Press.

McLucas C (2000) *Deep Mapping.* Available at: http://cliffordmclucas.info/deep-mapping.html (accessed 23 December 2015).

Modeen M and Biggs I (2020) *Creative Engagements with Ecologies of Place: Geopoetics, Deep Mapping and Slow Residencies.* London: Routledge.

Pearson M and Shanks M (2001) *Theatre/Archaeology.* London: Routledge.

Ridge M, Lafreniere D and Nesbit, S (2013) Creating deep maps and spatial narratives through design. *International Journal of Humanities and Arts Computing* 7(1–2): 176–189.

Roberts L (2012) Mapping cultures: A spatial anthropology. In: Roberts L (ed.) *Mapping Cultures: Place, Practice, Performance.* Basingstoke: Palgrave, pp. 1–25.

Roberts L (2018) *Spatial Anthropology: Excursions in Liminal Space.* London: Rowman and Littlefield.

Schiavini C (2004–5) Writing the land: Horizontality, verticality and deep travel in William Least Heat-Moon's *PrairyErth. Revista di Studi Americani* 15–16: 93–113.

Wydeven JJ (1993) Review of *PariryErth (a Deep Map). Great Plains Quarterly* 763: 133–134.

5

DON'T BELIEVE THE MAPPING HYPE! THREE STEPS BACK FOR AN ENGAGED CARTOGRAPHY

Paul Schweizer, Severin Halder and kollektiv orangotango

Introduction

Critical cartography with participatory, activist and counter-mapping—to state just the most-used labels—are in vogue in social sciences and humanities, as well as in cultural and educational projects and even public policies. As kollektiv orangotango, we look back on a decade of, how we like to put it, 'collective critical mapping'. In short, our ambition is to apply collective mapping as a tool for awareness-raising and community activation, as a joint educational process in which one's relationship to space is reflected upon and in which different intersubjective perspectives as well as different types of knowledge can flow together and spaces for action emerge. Yet, as Kelly and Bosse (2022: 4) affirm in their recent call for reflexivity in mapping processes, 'good intentions are not immune from failure and can even cause inadvertent harm'. Thus, we take this occasion to scrutinise the criteria that make up the classification of a mapping practice as critical, counter-hegemonic, participatory, or activist. In analogy to Sasha Costanza-Chock's (2020: xviii) work on design justice practices, we argue that it is imperative to explore the pitfalls and possibilities of mapping 'as a tool for social transformation'. But let us take two steps back and trace critical cartography's path from the critique of modern western cartography to critical and participatory mapping, in order to point out where mapping might, indeed, have the potential of being a tool for progressive activism and popular appropriation.

From critique of cartography to critical cartography

Maps have long been an important tool of research and a central medium through which knowledge transfer between scientific geography and society takes place. The question of the process of mapping—the actors involved, methods and means of design—has long been ignored, as the monopoly of map production has been (almost) unchallenged by state, university and military institutions. Since the 1990s, debates on critical cartography have challenged conventional cartography. In addition to a perspective that focuses strongly on maps as social texts and enacted discourses (Harley, 1989), process-oriented approaches to research have emerged (Dodge et al., 2009). A canon of common map criticism, map

DOI: 10.4324/9781003327578-7

deconstruction and critical analyses of cartographic and mapping practice has developed, as have diverse forms of critical mapping practices in science, art, educational work and political intervention.

Since the 2000s, there has been a remarkable expansion of participation in cartographic production. The production of maps has been the domain of scientific and institutional cartographers since the nineteenth century, with artistic and amateur cartographic practices marginalised. However, an expansion of the production and use of maps is now observed with the participation of a wide range of actors. While participatory mapping had been used in the last decades of the twentieth century as a common tool in development cooperation projects and as a participatory research method, an expansion of contexts has taken place. Today, participatory mapping is an integral part of participatory processes in both government and civil society initiatives worldwide. In this context, the concept of 'participation' usually remains blurred insofar as it is declared as 'all participatory', but a detailed derivation and reflection of the specific mapping methods and means of design through which this participation is achieved is omitted. To what extent are these mapping processes designed in such a way that a low-hierarchy participation of diverse actors is possible? Are the chosen means of design suitable for grasping and conveying multiple spatially related knowledges on an equal footing?

Since the 1990s, technological developments have made collaborative and participatory forms of mapping increasingly widespread, confirming Wood's statement that 'anybody can make a map' (1992: 184). Authors such as Parker (2006), Perkins (2007) and Crampton (2009) made participatory mapping practices a field of investigation in cartographic research in the 2000s. Since then, a range of scientific works has investigated questions of participatory cartography. A number of studies have dealt with participatory mapping with indigenous communities, for example in Brazil (Almeida, 2013), Chile (Mansilla Quiñones and Imilan, 2020), Ecuador (Scazza and Nenquimo, 2021), different African contexts (Dieckmann, 2021), Australia (Robinson et al., 2016) and Canada (Neilson et al., 2020). Other works explore the potential of mapping as a tool for action research with housing activists and urban political struggles (Graziani and Shi, 2020) or investigate mapping practices with migrants (Gangarova and Unger, 2020), children (Derr et al., 2019; Gryl et al., 2022; Pettig, 2019; Sobel, 1998), drug users (Germes et al., 2021) or sex workers (Gangarova and Unger, 2020).

Yet again critical inquiry poses questions of how the claim of inclusion, transparency and empowerment (Parker, 2006: 470) made in these processes can be guaranteed. According to Wood (2010: 162), when participatory mapping aims at building consensus, for example, for a particular political or urban planning project, it always tends to treat the 'involved' group as uniform and with common interests. Accordingly, divergent interests are made invisible, just as different forms of experience and knowledge are not captured. For example, Lucchesi (2020: 165) complains that participatory mapping projects with indigenous communities often employ traditional mapping styles and epistemologies that are unable to visualise other forms of knowledge, let alone actually benefit the community in question. Building on broader social science critiques of the participatory paradigm (see Cooke and Kothari, 2001; Hickey and Mohan, 2004), recent contributions have questioned the quality of participation in participatory mapping processes, examined power relations within their own action research (Graziani and Shi, 2020) and called for the development of a post-colonial participatory cartography (Barella, 2020) which, they argue, must be able to develop theoretical and practical tools capable of incorporating multiple perspectives.

Bittner and Michel (2018: 309) differentiate between degrees of participation in mapping processes, from approaches that work rather top-down and allow participation only within a narrowly defined framework to collective mapping, in which the focus is less on the map as a product, or as an intermediate product for later scientific analysis, than on the mapping process itself and the collective knowledge production to which it gives space (see Halder, 2018: 271). Groups such as the Argentinian collective Iconoclasistas (2013), the U.S.-based Counter Cartographies Collective (Counter Cartographies Collective et al., 2012) or kollektiv orangotango (Halder and Schweizer, 2020) conceive mapping processes as forms of collective knowledge production. When the Iconoclasistas (2013) call for the incorporation of everyday knowledge 'from below' into mapping, the question arises as to which method is capable of retrieving this knowledge as equally valued and which means of design are capable of representing it. Feminist critiques of traditional cartographic modes of representation and mapping practices also question how other forms of knowledge can be visualised (D'Ignazio and Klein, 2020). Considering the potential that artistic methods offer for the transmission and exchange of unconventional forms of knowledge (Michelkevičius, 2018), a stronger integration of artistic elements into mapping practices seems obvious.

Indeed, while various studies have found that artistic engagement with maps has multiplied in recent decades (D'Ignazio, 2009: 205) and that artists combine maps with different aesthetic forms and content (Cartwright et al., 2010; Lo Presti, 2018), interdisciplinary projects show how a map practice understood as artistic also enriches collective mapping with new ways of knowing and levels of meaning (Sletto, 2009). These works experiment with and reflect on new materialities beyond paper and digital mapping (Mekdjian and Olmedo, 2016), emotional mapping (Caquard and Griffin, 2018), and the performative dimension of mapping (Olmedo and Christmann, 2018; Sotelo Castro, 2009) or make the body itself the object and tool of collective mapping (Colectivo Miradas Críticas del Territorio desde el Feminismo, 2017; De Jager et al., 2016).

With such a well-documented and scientifically published cartographic cacophony, one might think that there is surely a right mapping method for any setting for a critical, participatory, counter-hegemonic, activist or you-name-it labelled mapping process. Indeed, as kollektiv orangotango, our access to this cartographic knowledge is especially privileged. Yet, we feel, it is not always an abundant choice of tools that is decisive for the quality of mapping. But again, let us take one and a half steps back to see how we got here.

Planetary relations: engaging in cartographic knowledges

Almost two decades ago, personal encounters brought us into contact with activist mapping practices. Enthusiastic about meeting critical geographers, politically engaged artists and community leaders, some of whom would become close friends, we soaked up the social movement and militant research techniques that these encounters opened up to us, among which included the first experiences of collective mapping. At this point—the beginning of our cartographic journey—we as geography students and activists were determined to create maps to support emancipatory struggles and desired to learn more from and with fellow activist mappers. Looking back, we affirm that not only did we acquire mapping skills but also we experienced our own becoming as individuals and as kollektiv orangotango in relation to a wider network of cartographic soulmates and to struggles in which we co-created maps and which appear in these maps.

The most visible result of these years of becoming, in dialogue with befriended activists, artists and cartographers is *This Is Not an Atlas* (kollektiv orangotango+, 2018). Yet the heavy atlas-style book cannot adequately represent the relationships from which it sprang. Since 2018, we have also co-created the *Not-an-Atlas* platform in order to share cartographic knowledges and methods and to collaborate with activist mappers in events and common projects—an attempt to represent and reciprocally enrich the many worlds of critical and activist mapping. While the book is a collection showing the diverse ways in which maps are created as part of political struggle, for critical research, or in art and education, the *Not-an-Atlas* platform is no longer a 'global collection' but a planetary meshwork of counter-cartographies. By sharing experiences and materials, we continue our engagement in this collective learning process. Yet we realise that the knowledge transmitted in this dialogue cannot be detached from the relations through which it is transmitted. Thus, our aim as curators of the platform could be described in Vazquez's (2017: 247) words as the creation of cartographic 'knowledge as relationality'. In short, we feel that cartographic activism takes shape, above all, in the relations in which we learn and exchange cartographic knowledges. Or, to frame it through the words of Bini Adamczak (2017), in the 'Beziehungsweise'—the kind of relation, or the way of relating. Indeed, this becomes especially apparent when we look at the actual cartographic practice that we conduct as kollektiv orangotango on the ground, as we will understand by taking yet another half step back.

Local relations: engaged mapping

In the first years of our work in facilitating mapping processes, we were invited mostly by befriended activists with whom we already shared a basis of common practice, topics, relations and trust, but this changed considerably after the *Not-an-Atlas* hype. As the book sold out in eight weeks and the second edition in only two months, we—as editors—received, and continue to receive, considerable requests from scientific, education and cultural institutions to conduct mapping processes as part of their respective short- to middle-term projects. Depending on how convincing the critical approach a project presents as well as on our limited resources, we occasionally accept this kind of request. In these cases, frequently, we exert ourselves to meet the necessities of a context unknown to us by choosing the right mapping method, that is (to put it simply), the one that is most capable of including the diverse participants on an equal footing, of mediating relevant knowledges and avoiding eventual biases. We have often learned a great deal and enjoyed the challenge of further developing methods in order to adapt them to diverse contexts and involved subjects as well as to the critical ambitions that we feel like meeting in each mapping site. Yet we have realised that it is not as much the chosen tools, but rather the relations—ways of relating, *Beziehungsweisen*—on which the mapping is built, that are decisive for the process' success.

Our experience shows that critical, participatory mapping with a transformative aspiration is hard, if not impossible, to realise within the framework of institutional projects driven by a necessity to generate a presentable output within a rather short duration. Even if the vocabulary of many projects uses terms coined by social movement practices and organisers continuously reaffirm their commitment to community and transformation, their ignorance of the actual interests and urgencies of the involved community are, repeatedly, astonishing and, in some cases, even include irresponsible, harming project planning. Regardless of the immense variety of already existing or yet to be developed methods of mapping, it is the relations between organisers, facilitators and co-mappers that are, above all, decisive for the legitimate qualification of a mapping process as critical, participatory, collective, activist and so on. To specify the criteria we consider relevant when assessing these relations, consistent with our conception of mapping as a collective dialogical learning process, we rely on bell hooks' notion of engaged pedagogy.

For hooks (1994, 2010), engaged pedagogy demands that educators should be authentic, committed and present in the learning environment as whole persons—not just as professionals reduced to the role of teachers—as a basis for opening oneself to the specific learning setting and experience of the given learning community (Asher, 2003: 239). Analogously, we argue that engaged cartographers must commit to not reposing on their critical methods but rather exposing themselves as whole persons within the mapping context. One must make oneself vulnerable as a critical cartographer, reveal one's own embeddedness and positionality, and set aside one's 'expertise' in order to listen deeply to and learn from co-mappers and become other. This becoming takes time. Thus, the commitment that genuine critical mapping requires is also a commitment to take the time for slow processes and to openness to not producing expected outputs in pre-established schedules. Concretely, in our experience, this may imply the time to get to know a neighbourhood by spending many days in it, playing soccer with the kids, chatting at the corner store, getting invited to have coffee at participants' places and, possibly, even accepting that there will not be a mapping at all.

hooks (2010: 22) emphasises the caring and responsible relations as the basis for an holistic progressive pedagogy (Madge et al., 2009: 37). Similarly, we believe that the cartographies in which we design the worlds that we are keen to fight for must not involve (self-) exploitative, hurtful relations or conditions. Nor can they be constructed in dry, mechanical processes without genuine affect. Critical mapping necessarily involves practices of physical and mental care, as these are the bases of trust and of fearless discussion for all participants. As in hooks' (2010: 19) engaged pedagogy, exercises of getting to know one another and building trust within the group are crucial to any genuine learning, and mental and corporal exercises become equally important elements of the mapping process. Only by opening up to bodily and affective practices and cultivating shared values such as friendship, humility and care (see Freire, 2000: 89) can hierarchies be broken down. Moreover, such exercises can help flatten the divisions of roles implied in the logic of most institutional projects. As called for by Colectivo Situaciones (2007), these roles should lose their importance in common practice and be replaced by friendly and caring relationships.

Finally, for hooks, engaged pedagogy is genuinely committed to social transformation and to anti-oppressive struggle for a diverse and just society (hooks, 1994: 33). In this sense, we call for critical cartography to commit to never becoming a broken promise or a fancy label. Engaged cartography is committed to actually serving the interests of the communities with whom mapping is done, even if that means disregarding project guidelines. Engaged cartography commits itself in the sense of a militant cartography. It commits itself to long-term, (self-)reflexive engagement with social movements and resistant practices.

References

Adamczak B (2017) *Beziehungsweise Revolution: 1917, 1968 und kommende*. Suhrkamp 2721. Berlin: Suhrkamp.

Almeida AWB (2013) Nova cartografia social. In: Almeida AWB de and Farias Júnior E de A (eds) *Povos e Comunidades Tradicionais. Nova Cartografia Social*. Manaus: UEA Edições, pp. 157–173.

Asher N (2003) Engaging difference: Towards a pedagogy of interbeing. *Teaching Education* 14(3): 235–247.

Barella J (2020) Ramener la justice sociale au centre de la carte. *Geographica Helvetica* 75(3): 271–284.
Bittner C and Michel B (2018) Partizipatives Kartieren als Praxis einer kritischen Kartographie. In: Wintzer J (ed.) *Sozialraum erforschen*. Berlin, Heidelberg: Springer, pp. 297–312.
Caquard S and Griffin A (2018) Mapping emotional cartography. *Cartographic Perspectives* 91: 4–16.
Cartwright W, Gartner G and Lehn A (eds) (2010) *Cartography and Art*. Berlin, Heidelberg: Springer.
Colectivo Miradas Críticas del Territorio desde el Feminismo (2017) *Mapeando el cuerpo-territorio—Guía metodológica para mujeres que defienden sus territorios*. Quito: Territorio y Feminismos.
Colectivo Situaciones (2007) Something more on research militancy. In: Shukaitis S and Graeber D (eds) *Constituent Imagination: Militant Investigations*. Oakland: AK Press, pp. 73–93.
Cooke B and Kothari U (eds) (2001) *Participation: The New Tyranny?* London and New York: Zed Books.
Costanza-Chock S (2020) *Design Justice: Community-Led Practices to Build the Worlds We Need*. Cambridge: The MIT Press.
Counter Cartographies Collective, Dalton C and Mason-Deese L (2012) Counter (mapping) actions: Mapping as militant research. *ACME: An International Journal for Critical Geographies* 11(3): 439–466.
Crampton JW (2009) Cartography: Performative, participatory, political. *Progress in Human Geography* 33(6): 840–848.
De Jager A, Tewson A, Ludlow B and Boydell K (2016) Embodied ways of storying the self: A systematic review of body mapping. *Forum Qualitative Sozialforschung* 17(2): 1–31.
Derr V, Corona Y and Gülgönen T (2019) Children's perceptions of and engagement in urban resilience in the United States and Mexico. *Journal of Planning Education and Research* 39(1): 7–17.
Dieckmann U (ed.) (2021) *Mapping the Unmappable? Cartographic Explorations with Indigenous Peoples in Africa*. Bielefeld: transcript.
D'Ignazio C (2009) Art and cartography. In: Kitchin R and Thrift NJ (eds) *International Encyclopedia of Human Geography*. Amsterdam: Elsevier, pp. 190–206.
D'Ignazio C and Klein LF (2020) *Data Feminism*. Cambridge, MA: The MIT Press.
Dodge M, Kitchin R and Perkins CR (2009) Mapping modes, methods and moments. In: Dodge M, Kitchin R and Perkins CR (eds) *Rethinking Maps: New Frontiers in Cartographic Theory*. New York: Routledge, pp. 220–243.
Freire P (2000) *Pedagogy of the Oppressed*. New York: Continuum.
Gangarova T and Unger H von (2020) Community Mapping als Methode Erfahrungen aus der partizipativen Zusammenarbeit mit Migrant*innen. In: Hartung S, Wihofszky P and Wright MT (eds) *Partizipative Forschung*. Wiesbaden: Springer, pp. 143–177.
Germes M, Klaus L and Steckhan S (2021) Mapping 'drug places' from below. The lived cities of marginalized drug users. *Drugs and Alcohol Today* 21(3): 201–212.
Graziani T and Shi M (2020) Data for justice: Tensions and lessons from the Anti-Eviction Mapping Project's work between academia and activism. *ACME: An International Journal for Critical Geographies* 19(1): 397–412.
Gryl I, Lehner M and Pokraka J (2022) Kritisches Kartieren in Bildungskontexten—zwischen Erkenntnismittel und politischer Kommunikation. In: Dammann F and Michel B (eds) *Handbuch Kritisches Kartieren*. Bielefeld: transcript, pp. 223–237.
Halder S (2018) *Gemeinsam Die Hände Dreckig Machen: Aktionsforschungen Im Aktivistischen Kontext Urbaner Gärten Und Kollektiver Kartierungen*. Bielefeld: transcript.
Halder S and Schweizer P (2020) Von Aktivismus, Geographien und dem Dazwischen—Überlegungen anhand der Praxis von Kollektiv Orangotango. *Standort* 44(4): 255–261.
Harley JB (1989) Deconstructing the map. *Cartographica: The International Journal for Geographic Information and Geovisualization* 26(2): 1–20.
Hickey S and Mohan G (2004) *Participation, from Tyranny to Transformation? Exploring New Approaches to Participation in Development*. London and New York: Zed Books.
hooks bell (1994) *Teaching to Transgress: Education as the Practice of Freedom*. New York: Routledge.
hooks bell (2010) *Teaching Critical Thinking: Practical Wisdom*. New York: Routledge.
Iconoclasistas (2013) *Manual of Collective Mapping—Critical Cartographic Resources for Territorial Processes of Collaborative Creation*. Buenos Aires: Tinta Limón.
Kelly M and Bosse A (2022) Pressing pause, "doing" feminist mapping. *ACME: An International Journal for Critical Geographies* 21(4): 399–415.

kollektiv orangotango+ (ed.) (2018) *This Is Not an Atlas: A Global Collection of Counter-Cartographies*. Bielefeld: transcript.
Lo Presti L (2018) Extroverting cartography. 'Seensing' maps and data through art (J-Reading). *Journal of Research and Didactics in Geography* 2(7): 119–134.
Lucchesi AH (2020) Spatial data and (de)colonization. *Cartographica: The International Journal for Geographic Information and Geovisualization* 55(3): 163–169.
Madge C, Raghuram P and Noxolo P (2009) Engaged pedagogy and responsibility: A postcolonial analysis of international students. *Geoforum* 40(1): 34–45.
Mansilla Quiñones P and Imilan W (2020) Colonialidad del poder, desarrollo urbano y desposesión mapuche: urbanización de tierras mapuche en la Araucanía chilena. *Revista electrónica de geografía y ciencias sociales* 24: 1–23.
Mekdjian S and Olmedo É (2016) Médier les récits de vie. Expérimentations de cartographies narratives et sensibles. *Mappemonde* 118: 1–16.
Michelkevičius V (2018) *Mapping Artistic Research: Towards Diagrammatic Knowing*, Dobriakov J (trans.). Vilnius: Vilnius Academy of Arts Press.
Neilson M, Charles J, Daniels K et al. (2020) Embodying emergence: Reclaiming ŁÁU, WEL̲NEW̱. *Cartographica: The International Journal for Geographic Information and Geovisualization* 55(3): 177–182.
Olmedo É and Christmann M (2018) Perform the map: Using map-score experiences to write and reenact places. *Cartographic Perspectives* 91: 63–80.
Parker B (2006) Constructing community through maps? Power and praxis in community mapping. *The Professional Geographer* 58(4): 470–484.
Perkins C (2007) Community mapping. *The Cartographic Journal* 44(2): 127–137.
Pettig F (2019) *Kartographische Streifzüge: Ein Baustein Zur Phänomenologischen Grundlegung Der Geographiedidaktik*. Bielefeld: transcript.
Robinson CJ, Maclean K, Hill R and Rist P (2016) Participatory mapping to negotiate indigenous knowledge used to assess environmental risk. *Sustainability Science* 11(1): 115–126.
Scazza M and Nenquimo O (2021) *From Spears to Maps: The Case of Waorani Resistance in Ecuador for the Defence of Their Right to Prior Consultation*. London: IIED.
Sletto BI (2009) 'We drew what we imagined': Participatory mapping, performance, and the arts of landscape making. *Current Anthropology* 50(4): 443–476.
Sobel D (1998) *Mapmaking with Children: Sense of Place Education for the Elementary Years* (Varner W, ed.). Portsmouth, New Hampshire: Heinemann.
Sotelo Castro LC (2009) *Participation Cartography: Performance, Space, and Subjectivity*. PhD Thesis, University of Northampton.
Vazquez R (2017) Precedence, earth and the Anthropocene: Decolonizing design. *Design Philosophy Papers* 15(1): 77–91.
Wood D (1992) *The Power of Maps. Mappings*. New York: Guilford Press.
Wood D (2010) *Rethinking the Power of Maps*. New York: Guilford Press.

6
POSTHUMAN CARTOGRAPHIES

Joe Gerlach

Introduction

To think of cartography, the art of space, not merely from a posthuman perspective, but in itself as posthuman, will doubtlessly seem strange. There is perhaps nothing more quintessentially human, after all, than the charting of bodies in the middle of things, especially when that middle is itself the upshot of, among other things, an egotistical geodesy, a colonial imagination or a geopolitical imposition by imperial fiat. Even if a body is placed outside of the middle, perhaps at the frayed edges and hinterlands of a map, the desire *to be placed*—somewhere, anywhere—is an obdurate one. For the avoidance of doubt, this desire emanates not from a sense of placelessness nor from a reactionary unease at lacking a feeling of rootedness, but instead emerges from a machinic sense of geography; the active becoming of encounter. A repetitive cartographic encounter, as it were, with what Aït-Touati et al. (2022: 188) describe as the Earth as skin: 'We have drawn its points, traces, holes, scares, excavations, cracks, frostbite, and sunburn'. The inclination to map and the urge to orientate bodies in a certain order and ratio, too, seem peculiarly human dispositions. Indeed, the very notion of 'disposition' is a cartographic one, a simultaneous arrangement and orientation of mood, temperament, (dis)inclination, and somatic capacity. On this prospectus, cartography and mapping stand at the apex of anthropocentrism. The point of this chapter is not to deny or apologise for cartography's humanist conceits. To do so would be to re-instantiate the conceptual error of characterising posthumanism and its attendant scholarship as a pointed and deliberate erasure of the human from accounts of existence, and indeed an exiling of the human from manifestos tilted to an earthly becoming. Of course, that is not to say there are not genres of posthumanism that lean explicitly into extinctionist narratives and eschatological nihilisms (see MacCormack, 2020), but these are categories of the posthuman from which this chapter demurs, if only because of their ironic valorisation of the human whose own species demise would somehow, in the eyes of extinctionists, constitute an unprecedented threshold event, one to be feared, perhaps, perversely, even to be mourned.

What, then, is meant here by 'posthuman cartographies'? The term might be understood, in this instance, along four lines. First, as an analytical posture, or as a posthuman perspective on cartography, one that acknowledges, along actor-network, speculative-realist, and

 DOI: 10.4324/9781003327578-8

non-representational tracks, the non-human, affective, virtual and immaterial coordinates of mapping (Gerlach, 2014; Rossetto, 2019). Second, and by way of inversion, the term might also allude to an intellectual cartography of the posthuman, a gazetting and taxonomy of the various scholarly genres of posthumanism (see e.g. Sharon, 2013). These two versions of a posthuman cartography are, to be sure, important endeavours, but they merely foreground its third and fourth iterations; those that this chapter argues are the least developed and yet most compelling vectors. Specifically, this is to figure cartography and mapping as generative of the posthuman, or propagative of posthumanity. This, in turn, precipitates an ethical gesture, namely the promulgation of a cartography not of, but for, the posthumanities. In short, cartography here performs a speculative one-two manoeuvre; one, it constitutes a diagnostic function, a surveying of what Rosi Braidotti (2019a) describes as the posthuman convergence. Two, cartography exceeds its representative vocation and becomes a machine for cultivating subjectivities and dispositions that are geared to navigating the complexities of this posthuman moment. Having traced the Earth's skin, its lines, its sunburn and its frostbite, and having done so, over and over again, the task now is to learn from these inscriptions and to shift cartographic attentiveness to the becoming of the posthuman.

Diagnosis

The Earth persists, but the world is imperilled. If the Earth is comprised by the geo and all its more-than-human forces, capacious bodies and their manifold concatenations and incipiencies, then the world, in contrast, comprises the narrow envelope of human, techno-political activity. Accounts of its rampant denudation and degradation are recitable by collective memory. Cartography has, of course, been central in surveying how bad things are planetary-wise, and indeed just how much worse things are set to become, mapping, even, the end of the world as we know it (Lo Presti, 2022). Who, for example, has not seen maps illustrating how their own homes, towns and cities are set to be inundated by sea level rises in the next century? Or who has not been witness, now, and in recent months, to real-time mapping of armed incursions into sovereign territory; offensive and counter-offensive oscillating cartographically in different shades of red. Portents of extinction and bellicose geographies, these maps are frightening diagnoses of an ecological and societal malaise, albeit ones that have yet to succeed in shaking certain human bodies out of affective hesitation or willfully ingrained inertia. Running alongside this ongoing and continued ecocide is another crisis that has gone largely ignored, and yet it is one that sets apart the posthuman convergence from what has gone before, namely, a crisis in subjectivity. Put differently, this is a posthuman cartography characterised by an admixture of a crisis in human psychology and a simultaneous interdiction of desire. On this score, philosopher and activist Franco Berardi (2021: ix) describes the 'new psychopathological regime' as 'the age of panic, depression and, ultimately, psychosis'. Maps play a part here, too, in inscribing psychological dispositions—panic or otherwise—much as they have, as mentioned, propagated pessimistic forecasts and attendant geographical imaginations of a future planet in dire biophysical and climatic terms. This is precisely because of cartography's renowned and well-rehearsed performative capacity to fuse the power-knowledge dyad to such a degree that maps and mappings can circumscribe movement, mobility and sense itself. Even when cartographies are disabused of their presumed ontological security, they retain a remarkable, and at times hegemonic, epistemic capacity to crowd out subjectivities and subjective formations that cut against the grain of a prevailing mood, affect, or structural injunction. It is not all repressive, however, given that the corollary of cartography's onto-epistemic ambit is its generative capacity to engender new lines of thought and affinity, and new modes of existence. In this respect, in its ability to conjure novel values and likewise novel systems of valuation, cartography is always, already, an ethics.

Yet the problem, or concern, at the heart of the posthuman convergence remains intact: the problem of disorientation. As the late philosopher-cartographer Bruno Latour (quoted in Aït-Touati et al., 2022: 4) remarks, '[t]his is a strange moment when people are beginning to wonder where, when, and who they are'. The questions of 'when' and 'who' are to be blunt, old hat and perhaps not as pressing or as strange as Latour makes out; history, anthropology and sociology have made sure of that. The ultimate geographic enquiry of 'where', on the other hand, is arguably the concern that unlocks posthuman subjectivity, if only capital-G Geography had not lost sight of that which mattered most: Earth. According to Latour (quoted in Ait-Touati et al., 2022: 6),

> the hard, complicated, boots-on-the-ground, contradictory, specific, tailor made attention to the "geo-" that the suffix "-graphy" underlined was jettisoned in the name of a more '"scientific", data-driven management to help outsiders drive through a land which they had no real interest—except for locating resources to be exploited. Geographers have lost the Earth in the process.

Posthuman disorientation is borne out of alienation. A mode of alienation rendered not merely in the classic sense of an occlusion of, or detachment from social and metabolic relations, but an alienation that untethered bodies and their dispositions from the immanence of the Earth; a denaturalisation of humans. To that end, disorientation denotes an alienation from the Earth's excessiveness and exuberance, and by the same token, from its diminution and mundanity. This alienation can be explained, in part, by a long-held obsession in understanding the Earth by way of cartographic neutrality, whether expressed in scientific or artistic terms. And yet, as psychoanalyst Félix Guattari (2009: 54) states in plainly obvious terms, 'we are ourselves always mixed up in the situation. And we will do better to realize this so that our interventions will be a little alienating as possible'. Instead, then, of thinking of cartography as a humanist mirror or as a Euclidean semiotics of representation, to consider cartography in posthumanist terms is to think of it as a conceptual machine; an inventive, productive and generative set of techniques for re-cuing subjectivities and dispositions. Conceptual insofar as concepts can be conceived as active, open-ended and indeterminate assemblages of thought, and machinic to the extent that cartography hinges neither on structural nor individual frameworks of reference, nor reproduces such strictures.

The coronavirus pandemic (declared to be no longer a public health emergency of international concern by the World Health Organization in May 2023) to some extent represents an enforced return to the immanence of Earth, and likewise to an all too visceral and viral evocation of posthuman entanglements. COVID-19, on the one hand, reterritorialises the self, delimiting the ambit and potential of encounter through the imposition of social distancing. Strangely enough, albeit perhaps none too surprisingly, such distancing engendered its own intimacies, in love, in grief, in proximity and, for many, in separation. On the other hand, and at the same time, the virus deterritorialises the social, picking away at its ontological foundations, unsettling, in turn, what it means to be immunised, to be human, to be protected, to be cared for, and now for what it means to live in its uncertain aftermath.

Deploying the language of territorialisation is to court, deliberately, a cartographic vocabulary, one that featured heavily in a concomitant proliferation of cartographic expressions during the pandemic. Here cartography again played its one-two role in simultaneously surveying everything from epidemiological exigencies to cultivating lines of collective solidarity among newly locked-down bodies (see e.g. Pase et al., 2021). But this explosion in cartographic output and proclivity, for all its energetic attempts at bringing order to the unruly and unseemly, far from diminishing extant disorientation, served only to augment it. Latour's posing of the question 'where' we are in this posthuman convergence has not been resolved by maps or by cartography. But this is not the same thing as arguing that this specific instance or moment of disorientation is qualitatively unique to others, or that disorientation is a peculiar characteristic or hallmark of the posthuman convergence. Indeed, to purloin and interpolate another Latourian refrain, we have always been disorientated. It is only our strenuous efforts to orientate the Earth, in part through dampening modes of base-grid cartography, that can be argued to be a curio of modernity. Deepening cartographic attachments to certain places on the map has only served to demonstrate how lost humans truly are. Having said this, it is at the juncture of the pandemic that has wrought a different kind of posthuman disorientation, one that registers at the level of the unconscious. For Berardi, the pandemic marks the advent of what he labels the Third Unconscious, a psycho-spatio-temporal cut in the subliminal. If the First Unconscious was characterised by Sigmund Freud's mapping of dark intensities repressed by clinical regime and psycho-sexual morality, and if the Second Unconscious was the laboratory of creative and cartographic schizo and psychic invention for the likes of Gilles Deleuze and Félix Guattari (2004), then the Third Unconscious denotes the present and futurity of the unconscious, or, as Berardi (2021: xi) puts it when speaking of this third age: 'I refer to an open future that will be shaped by our consciousness, by our political action, by our poetic imagination and by the therapeutic activity that we'll be able to develop during this transition'. The Third Unconscious, then, is properly a posthuman unconscious, one that both diagnoses a cut in the world and that at the same time seeks to cultivate and nurture a therapeutic cartography, seemingly by way and motive of reparative gesture in response to the immensity of the pandemic's shock to the threads of social, environmental and mental ecology.

Posthumanity

What, then, would comprise and play into a therapeutic cartography, a cartography for the posthuman? Answering this question necessitates the posing of another, namely, what counts? Mapping always involves a series of choices and decisions concerning what matters, what gets included, occluded and excluded. The first decision in this instance is to establish what counts as immanence, and by extension, to fathom what it means to cultivate a cartography in an immanent register, to consider alongside Lucretius' (1951: 187, emphasis in the original), '*how the earth remains fixed in the middle of the world*'. Critically, a cartography of immanence is emphatically not a return to the primordial or a nostalgic throwback to the land, to the soil, or indeed back to nature. Nor is such a cartography a return to geometric rationality. Indeed, mapping is rarely concerned with regression, even if the prevailing temptation is to regard the act of cartography as a charting of an external reality that has already passed. Instead, a therapeutic, posthuman cartography is necessarily anticipatory in posture, such that it elides replacing one cartographic dogma with another.

And this presents, alongside disorientation, another quandary, specifically, the task of how to undo the ossified ethics of a humanist cartography, one in which, as Berardi (2021: 124) claims, 'our relation with the other has been reduced to competition with an Other that has become disembodied'. Epistemically, it is easy to see how cartography induces competition. It is the very motor of both a logic and an affect of territorialisation. Moreover, humanist cartography has promulgated a certain fixity in worldly imaginations in which the Earth is rendered mute, static and elegiacally passive. This, of course, forms the basis for cartography's transcendentalist inclinations. What would it take, against Lucretius' physics, to unfix a posthuman Earth from the middle of an obstinately human world? One tactic, in response, entails the development of a geophilosophical cartography. Taking a lead from Deleuze and Guattari's (1994) assaying of a thinking, productive and excessive Earth, a geophilosophical, that is to say a posthuman, cartography would be one that ditches a preoccupation with charting the world 'as is', in the present tense, in favour of surveying an Earth capable of generating thought. Importantly, this involves an abandonment of the cartographic logic of the subject-object relation, while at the same time accentuating the significance of non-human forces, and not least the matters of desire and the unconscious, in the modulation of existence. In tarrying in matters and registers non-representational, a geophilosophical cartography is, almost by default, a cartography of and for immanence, whereby immanence denotes not a proximity to the Earth (Colebrook, 2022), but instead speaks of cartography's capacity to generate pure potential—to cultivate the virtual. The upshot is nothing less than a new sense of the Earth (Keating and Williams, 2022; Roberts et al., 2022).

Geophilosophical and posthuman cartography only function effectively when plugged into specific matters of concern, into controversies and into things happening. It is simply not enough to use cartography to expose and critique human exceptionalism. To do so, aside from risking rank repetition, only tells half the story and limits cartography to its 'major' mode, one of gazetting and spotlighting stories that are already eminently visible to collective sense. A posthuman cartography, by contrast, is one that labours at a minor register. To map in the minor, or to minoritise, is to unpick habits of thought, to generate novel geographical sensibilities and, importantly, to work at the register of affect (Gerlach, 2015). A minor cartography is plugged into desire and, therefore, at the same time, power. For Braidotti (2019b: 33), such a cartography is unremittingly practical, one that 'aims at tracking the production of knowledge and subjectivity . . . and to expose power as both entrapment (*potestas*) and as empowerment (*potentia*)'. This is a posthuman cartography that, far from abandoning critical accounts of repressive power, works to make visible and sensible the intricacies of power in its authoritarian guise, but one that recognises and charts along Foucauldian lines, the distributive capacity of power as potential. In this respect, a posthuman cartography is always provisional, oscillating on the cusp of at least two genres of power, and at the same time anticipating an Earth in flux, tentatively coming into being. It is a cartography shorn of any preoccupation with universality or with a ground-truthing impetus, not least owing to the ground's own groundlessness.

Conclusion

The second part of the chapter opened by recalling the importance of the question of 'what counts' in the ambit of a posthuman, geophilosophical mode of mapping. It then swiftly outlined what does *not* count in a posthuman cartography. So much for a posthuman ethic and ethos of affirmation. Not to lose sight of the question, therefore, and on the present matter of what comprises a posthuman cartography, one might follow Félix Guattari (2013: 18) in understanding that such a cartography can imbibe all manner of concerns, from a 'clinical tableau, an unconscious phantasm, a diurnal fantasy, an aesthetic production, a micro-political fact'. He continues, 'what counts here is the idea of an existential circumscription that implies the deployment of intrinsic references—one might also say, a process of self-organization or singularization'. Guattari's jargon machine here is working at full throttle, but the intimation is clear enough. First, a posthuman cartography can appeal to and comprise anything at all, but the more marginal the focus, the better. It is not a matter of making cartographic calls on the back of aesthetic judgement or political snobbery. Second, that a posthuman cartography can be conceptualised as an 'existential circumscription', that is to say it is an aesthetic act of geometry (a mode or craft of measuring of the Earth). It is existential insofar as the act of circumscription, not to be confused with limiting or circling around a problem, is a beckoning of sense, an invocation of existence, and not of its representation. Cartography not as simulacra, but as sensibility. Precisely because of cartography's aesthetic charge does it bear, at the same time, the demand for an ethical accountability and a reckoning of responsibility for that which is brought into being or indeed extinguished by the map. On these grounds alone, there is a need to be clear eyed about the consequences of developing existential circumscriptions; as much as these posthuman cartographies might foreground discursive exchange and coalitions of affinitive concerns, they can also promulgate spaces of antagonism. Tension and contestation are inconstant features of a posthuman cartography, and the point is not to avoid them, but to invite their inconstancy in disrupting humanist cartography's territorialising habits. Cartography, as such, 'loses its primary vocation of having to represent the Territory' (Guattari, 2013: 35). Instead, cartography, in a posthumanist register, becomes embroiled in processes of re-singularisation, in other words, in the revolution and transformation of subjectivities. Given that questions of desire, subjectivity and the unconscious have for so long been marginalised or regarded as trivial aspects of existence, it might seem that a posthuman cartographic concern for these matters is traversing on the non-political or worse still, on the apolitical. Yet a posthuman cartography is precisely the intensification of a mapping's political valence, potential and vocation. What is afoot here is a shift in political register, and to one that is no less pressing in its emancipatory ambit. To return to the schizoanalytic cartographies of Guattari (2009: 54) for a final time:

> Instead of conducting a politics of subjection, of identification, normalization, social control and setting the people we are dealing with along a semiotic track, it is possible to opt for a micropolitics that at least takes into account our own humble participation in the story; it is possible to work in the direction of dis-alienation, of a liberation of expression, of opening 'exit doors', if not 'lines of escape', from oppressive social stratifications.

Put differently, in contrast to re-instantiating a humanist cartography through which bodies are pre-assigned meaning and signification, and thus subjectified and subjected, a posthuman cartography is one that draws and diagrams lines of flight and lines of escape that encourage a deterritorialisation of bodies, of all kinds and types, consonant with an immanence of becoming. The only outstanding challenge of a posthuman cartography is to not lose the Earth along the way.

References

Aït-Touati F, Arènes A and Grégoire A (2022) *Terra Forma: A Book of Speculative Maps,* DeMarco A (trans.). Cambridge, MA: The MIT Press.

Berardi F (2021) *The Third Unconscious: The Psychosphere in the Viral Age.* London: Verso.

Braidotti R (2019a) *Posthuman Knowledge.* London: Polity.

Braidotti R (2019b) A theoretical framework for the critical posthumanities. *Theory, Culture & Society* 36(6): 31–61.

Colebrook C (2022) Geophilosophy as the end of philosophy. *Subjectivity* 15(3): 169–186.

Deleuze G and Guattari F (1994) *What Is Philosophy.* London: Verso.

Deleuze G and Guattari F (2004) *Anti-Oedipus.* London: Continuum.

Gerlach, J (2014) Lines, contours and legends: Coordinates for vernacular mapping. *Progress in Human Geography* 38(1): 22–39.

Gerlach J (2015) Editing worlds: Participatory mapping and a minor geopolitics. *Transactions of the Institute of British Geographers* 40(2): 273–286.

Guattari F (2009) *Soft Subversions: Texts and Interviews 1977–1985,* Wiener C and Wittman E (trans.). Los Angeles: Semiotext(e).

Guattari F (2013) *Schizoanalytic Cartographies,* Goffey A (trans.). London: Continuum.

Keating T and Williams N (2022) Geophilosophies: Towards another sense of the earth. *Subjectivity* 15(3): 93–108.

Lo Presti L (2022) One map closer to the end of the world (as we know it): Thinking digital cartographic humanities with the Anthropocene. In: Travis C, Dixon D, Bergmann L, Legg R and Crampsie A (eds) *Routledge Handbook of the Digital Environmental Humanities.* London: Routledge, pp. 388–403.

Lucretius (1951) *The Nature of the Universe,* Latham RE (trans.). London: Penguin.

MacCormack P (2020) *The Ahuman Manifesto: Activism for the End of the Anthropocene.* London: Bloomsbury.

Pase A, Lo Presti L, Rossetto T, Peterle G (2021) Pandemic cartographies: A conversation of mappings, imaginings and emotions. *Mobilities* 16(1): 134–153.

Roberts T, Lapworth A and Dewsbury JD (2022) From 'world' to 'earth': Non-phenomenological subjectivity in Deleuze and Guattari's geophilosophy. *Subjectivity* 15(3): 135–151.

Rossetto T (2019) *Object-Oriented Cartography: Maps as Things.* London and New York: Routledge.

Sharon T (2013) A cartography of the posthuman. In: Sharon T, *Human Nature in an Age of Biotechnology: The Case for Mediated Posthumanism.* Dordrecht: Springer, pp. 17–56.

PART 2

Textural connections

7

IN BREVI TABELLA. THINKING WITH DIAGRAMS IN LATE ANTIQUITY

Salvatore Liccardo

Measuring, travelling and recollecting the world in late antiquity

A passage from the Book 6 of the *De nuptiis Philologiae et Mercurii* (c.420–490) condenses into a few sentences the content and main forms of late antique geography:

> Immediately there came into view a distinguished-looking lady, holding a geometer's rod in her right hand and a solid globe in her left. . . . This tireless traveller was wearing walking shoes, to journey through the world, and she had worn the same shoes to shreds in traversing the entire globe. . . . Her hair was beautifully groomed, but her feet were covered with dust. . . . She began drawing diagrams on the powdery surface of her abacus and spoke: '. . . I am called Geometry because I have often traversed and measured the earth, and I could offer calculations and proofs for its shape, size, position, regions, and dimensions. There is no portion of the earth's surface that I could not describe from memory'.
>
> *(Mart. Cap. 6.580–588, trans. Stahl, 1977: 218–220)*

The text describes the allegory of geometry and the first words spoken by this personification. The apparent confusion between geography and geometry might seem odd to a modern reader, who considers them distinct branches of knowledge. However, the taxonomies of the late antique literati do not always coincide with ours. In the eyes of the contemporaries of Martianus Capella, *agrimensores* (land surveyors), geometers and geographers were all experts who measured the Earth according to a similar methodology and with the help of analogous technical instruments. As proposed in the 550s by Cassiodorus in his *Institutiones* (*Inst.* 1.25), geography could be distinguished from geometry, when interpreted as a tool for locating the toponyms mentioned by the sacred Scriptures. However, the scheme proposed by *De nuptiis Philologiae et Mercurii* enjoyed wide popularity (Lozovsky, 2000: 113–138; Teeuwen and O'Sullivan, 2011), and the two disciplines remained closely interconnected throughout the Middle Ages.

Set in an allegorical framework, that is, the wedding of Mercury and Philology, Martianus Capella's work offers an overview of the seven liberal arts, which are personified as

 DOI: 10.4324/9781003327578-10

female bridesmaids given to Philology. Building on different older traditions and adapting Varro's systematisation of knowledge (Gerth, 2013: 119–155), the author puts *Geometria* first among the maidens representing the sciences with a mathematical basis. In a scheme highly influenced by Neoplatonic ideals, *Geometria* marks the first step in a general transition from earth to heaven, from bodily to incorporeal. She represents the inferior beginning of the natural sciences, that is, their most practical and material side, while *Harmonia*, or music, who comes last, is the ultimate abstract dimension (Ramelli, 2001: VIII—XIII and XXIX—XXXII).

The allegory serves to define the research methods and objectives of this liberal art. With a geometer's rod in one hand and a solid globe in the other, the allegory returns an image of an *ars* caught between the two extremes of land surveying and astronomical calculation. In Martianus Capella's mirror game between earthly and heavenly dimensions, the globe of *Geometria* recalls the mystical sphere that permits Jupiter and Juno to see and control the universe (1.68). In a sense, it is its miniature version. *Geometria* combines sensorial perceptions and mathematical methods to create an image of the world, which the Gods can integrate in the general picture of the cosmos.

The groundwork of this enterprise is the measurement of the world. This is the essential trait of the discipline, to the point of being enshrined in her name. *Geometria* comes from the Greek words γῆ (earth) and μετρία (measure). Accordingly, she takes care of the practical and legal matter of determining the boundaries of properties and administrative entities. Furthermore, in order to measure it, *Geometria* travels around the world. Thus, her feet are covered with dust, she is referred to as a *viatrix infatigata* (tireless traveller) and she calls herself a *squalentior peregratrix* (a rather run-down itinerant). Beyond the mathematical basis, *Geometria* is mostly an empirical form of knowledge, which consists in gathering the information made available by travelling the world. The metaphor of the wanderer recalls the Roman tradition of the *itineraria* and connects knowledge of space with its experience through motion. The *Geometria* of Martianus Capella explains in a few words what shines through most common geographical texts of the time. Often known as *periploi* or *itineraria*, these texts are lists of cities, ports, landmarks and staging posts (Salway, 2007). Regardless of their specific aim and context—some of these works might have a clear pedagogical intent while others a more literary purpose—they are all linear versions of potentially infinite journeys. They might refer to a scaled representation of space, but the main function of these texts is to mirror a specific 'lived space' that is made significant by mobility.

On the other hand, the allegory also personifies another form of geography, one that has attracted less scholarly attention. *Geometria* is a traveller, but she is also an archivist. She keeps track of all the information that she has gathered. She has collected material from the entire world on her trips, and now she can describe any portion of earth's surface from memory. Her ability to recollect data is the foundation of her knowledge. In this allegory, the world becomes a sort of a reference work, a container of data that the learned person can later retrieve, while the geography taught at school turns into an archive of culturally meaningful names and images.

Following the model of Pliny the Elder, Martianus Capella provides an exposition of geographical knowledge with the intent of being exhaustive, even encyclopaedic. The notions are set in a broader educational system and narrative frame. Nevertheless, toponyms and ethnonyms, which are the outline of this understanding of the world, could also be transmitted as dry lists of names with almost no authorial comment or literary embellishment.

Other 5th-century sources, such as the *Cosmographia* of Julius Honorius, the *Dimensuratio provinciarum* and the *Divisio orbis terrarum* gather data in the form of lists, providing archives of geographical names organised by region (Monda, 2008; Fuhrmann, 2020). As sequences of names, these texts were an essential instrument for training the memory of anyone who wanted to learn about geography. A list engages memory and, being suitable for non-sequential readings, it invites the reader to a more proactive posture. The lack of narrative content requires more dynamic input from the reader, who is called to impose on the list a certain framework and to establish connections both between the items and between the list and the surrounding context (Von Contzen, 2017, 2018). Allocating a specific place to each piece of information, late antique catalogues are mental templates that disassemble subjects into smaller, easy-to-remember conceits; all the more so if the catalogues, as the three mentioned above, are based on drawn templates, that is, on maps of the world. In this sense, they mediate between the map and the reader, facilitating the association of names and figures as well as the recollection of data in the same or a different order.

Small-sized diagrams and general assumptions about the world

The reciprocal influence between images of space, whether mental or sketched on a support, and the memorisation of toponyms and ethnonyms as a method for learning about geography shines through the use of the analogy between a brief summary of facts and a small-sized representation of the oikumene—the latter usually called *In brevi tabella*. Found in some late antique authors and most frequently used by Jerome (Diederich, 2018: 126–127, 2019: 109–110, 121–124), this comparison indicates that maps were not uncommon among the ranks of the literary elite, for whom maps had become synonymous for conciseness. Moreover, it suggests the interplay of images and words in the learning techniques of the time. In a letter from 396 CE, Jerome tries to console his old friend Heliodorus for the loss of his nephew Nepotian by offering him a eulogy of the departed. Before listing Nepotian's virtues, Jerome writes: 'Like those who paint maps on small tablets, in this little book you will see indications of his virtues drawn in outline, not depicted in detail' (Jer. *Ep.* 60.7.3, trans. Scourfield, 1993: 53).

After an analogous allusion to mapmaking, in another letter (*Ep.* 123.15), Jerome focuses on geography in providing a list of the cruel peoples (*ferocissimae nationes*) that had invaded the empire and an overview of cities and provinces which had been ravaged by these incursions. Since all the listed toponyms appear in the *Itinerarium Antonini*, the *Tabula Peutingeriana* or both, it is possible that in this case with the expression *In brevi tabella* Jerome is referring to the tradition of the *itineraria* specifically. However, regardless of the nature and level of detail of these *tabellae*, the evidence indicates that maps and texts could be regarded as complementary devices for transmitting and especially summarising data. This reveals both the textualisation of territories, meaning the understanding of space as the sum of their toponyms and ethnonyms, and the transformation of linear text into diagrammatic maps to aid visual memory.

Similar conclusions can be drawn from sources that do not refer to this specific analogy, such as a letter written by Emperor Julian in 358/359 (Julian *Ep.* 10.403c–d). In this text, which is a thank-you letter to a certain Alypius, Julian expresses his joy at receiving a drawing table (*πίναξ*) containing diagrams (*διαγράμματα*). Julian describes these illustrations as more precise in comparison to previous editions and mentions a poetical composition that was appended to the maps. From the epistle, one can draw three conclusions.

First, the emperor's allusion to previous maps indicates the existence of a series of similar illustrations and Julian's familiarity with such maps. Second, the office that Alypius held at the time—*vicarius Britanniarum*—might suggest that the maps in question represented Roman Britain. Third, the inclusion of a poetical text hints at the coexistence and interplay of texts and images in the work presented to the emperor. Written in iambic verse (*ἴαμβοι*), one could imagine that the poem was dedicated to Julian, who, as the emperor, embodied Roman claims of universal dominion and was consequently the ultimate patron of geographical inquiry. In 435, another emperor, Theodosius II, was praised in declamatory verse for commissioning a world map (Salway, 2005: 128; Lozovsky, 2006: 355–362). Alternatively, the text might have served to facilitate comprehension of the map, functioning as a sort of didactic poem. In an educational system that relied heavily on repetition and memorisation, poeticised geographical descriptions of the world, such as the work of Dionysius Periegetes and its Latin translations, became popular handbooks that ensured the preservation of the bulk of ancient geographical data across the Middle Ages.

The familiarity that late antique intellectuals had with maps suggests that representations of the world, which were both textual and figurative, were relatively widespread. These maps could be transportable and even pocket-sized and made a great impression as special gifts. This hypothesis finds confirmation in the later manuscript tradition. In the Early Middle Ages, small diagrams began to be included in texts of geographical, historical, exegetical and poetical nature (Gautier Dalché, 1994, 2003). Developed in Late Antiquity (for the debate on the origins of these diagrams, see Mauntel, 2021), these illustrations, almost as if they were drafted following the analogy of the *In brevi tabella*, reflect in a schematic form the information found in the body of the text. Drawn as T-O maps (Edson, 2008), these heavily stylised geographical drawings translated into images general assumptions regarding the Earth and its parts (its roundness and division into three continents), thus facilitating the understanding of such knowledge. They were intelligible representations of the world that engaged readers by means of similarity, that is, through the likeness of the sign and its subject (Ljungberg, 2016). For Early Medieval copyists and readers, this type of map was the first design that came to mind when outlining geographical knowledge. As evidenced by a sketching of a T-O map found in a 9th-century manuscript (Figure 7.1), the mere engagement with geographical material could trigger the recollection of such a design even if the text copied or read did not support a tripartite world partition.

The *Tabula Peutingeriana*

The *Tabula Peutingeriana* (henceforth the *Tabula*) is an extraordinary example of late antique cartography and serves as a perfect case study for this brief survey of the interaction of texts, diagrams, memory and travel in Late Antiquity. Preserved today at the Austrian National library, this unique artefact survives as an incomplete medieval copy of a Roman world map. The *Tabula* is composed of 11 leaves, which were originally bound together in a long and narrow parchment roll (6.75 m long and between 32.8 and 33.7 cm wide). While it was intended to represent the entirety of the oikumene—from the Atlantic Ocean to India—the extant copy lacks its western end, which portrayed Northwestern Africa, the Iberian Peninsula, most of Britain, Ireland and perhaps other islands, such as Thule and the Isles of the Blessed.

As a unique wealth of geographical and ethnographic data, the *Tabula* attracts scholars with varied interests—from Hellenistic to Renaissance geography, from Roman archaeology

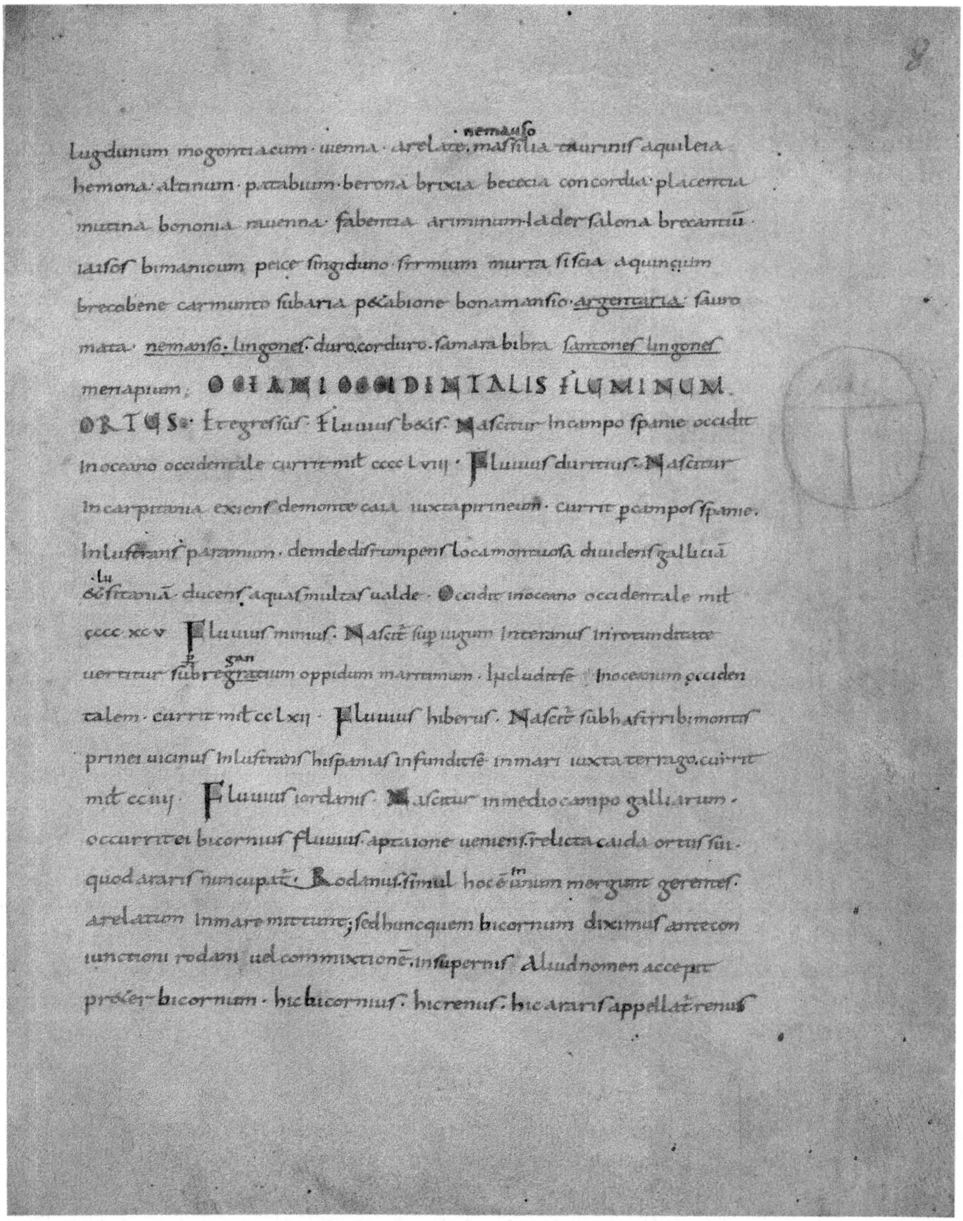

Figure 7.1 Pal. Lat. 973, f. 8r. (Reims, 2nd half of the ninth century). Sketched T-O map drawn next to the text of the *Cosmographia* of Julius Honorius.

Source: By permission of Biblioteca Apostolica Vaticana

to medieval art. Today scholarly debate concentrates on the dating of its archetype, the specific kind of geography reflected by the map and the primary function of the *Tabula*. Concerning the crucial debate over its dating, one can list four main theories. The *Tabula* has been analysed as a product of Hellenistic geography (Rathmann, 2018), a revision of the Map of Agrippa made by order of Theodosius II (Weber, 2016), a spatial representation of the Tetrarchs' dominion (Talbert, 2010) or an artefact produced at the Hohenstaufen court based on a Carolingian mapping tradition (Albu, 2014). Aside from the supporters of this last, idiosyncratic theory (Johnson, 2016), most scholars today date its final editorial work to Late Antiquity.

Regarding its main purpose, the *Tabula* would have been of little use for travellers. Its scale, its extreme distortion, the focus on ornamental elements, the unsystematic use of measurement units, as well as the lack of grid lines or axes suggest that the map had no practical application. Although structured around the Roman road network, the *Tabula* offers little help for actual travellers, while it focuses on historical and literary journeys, such as the campaigns of Alexander in Asia. Similarly, concerning the nature of the map, in spite of the centrality of the public road network, the *Tabula* is not just an illustrated itinerary but an actual map, one which provides not always reliable but mostly consistent cartographic information on the shape and size of geographical areas and their relative position.

Its nature and objectives become clearer if one considers the *Tabula* as an expression of Late Roman antiquarianism and cataloguing tradition (Drijvers et al., 2018). The coexistence of place names, ethnonyms and figurative elements drawn from a variety of sources and chronological periods makes the map a both textual and graphic compendium of Greco-Roman geography. The *Tabula* embodies the late antique concern for preserving older knowledge. Whether it was a direct expression of imperial authorities or the result of a less official enterprise, the *Tabula* is first and foremost a very eclectic archive of geographical and topographical data. It is the result of patching together public and private documents and, as such, has both a political meaning and, broadly speaking, an educational purpose.

Like the allegory of Martianus Capella, the map makes use of the memory of its viewers to make sense of the world. Most often, the *Tabula* includes ethnonyms for illustrating the peripheral areas of the inhabited world, which are mostly devoid of roads and urban centres—elements that are predominant at the centre of the map. These ethnic names function as placeholders with cultural and historical significance (Liccardo, 2020). By highlighting lifestyles deemed backward and barbarian (e.g. *AMAXOBIISARMATE* and *SARMATEVAGI*, meaning 'Sarmatians living in wagons' and 'wandering Sarmatians'), ethnonyms mirror the otherness of distant lands. Nevertheless, an audience equipped with the necessary knowledge could associate these names with known tales and myths and thus understand these exotic lands and peoples in familiar terms. Similar to the bystanders of triumphal ceremonies, who knew or made up stories about the captives paraded through the city of Rome (Östenberg, 2009; Merrills, 2017: 86–91), viewers of the map could translate the labels into ethnographic narratives. In this sense, compared to a triumph the *Tabula* offers a more challenging and yet more rewarding interaction by including a variety of captions and illustrations that provide several layers of interpretation.

Although agency and motivation are not yet determined, the degree of detail shown by the *Tabula* makes it probably inappropriate for public display. The *Tabula* might well include material drawn from monumental maps, but the characteristics of the depiction reflect a passion for minutiae typical of intellectuals and bureaucrats, which would have hardly attracted illiterates. As an expression of antiquarianism, the map does not solve the contradictions arising from collecting names, symbols and designs taken from a variety of sources. Nevertheless, if we judge the *Tabula* against the backdrop of contemporary geographical knowledge and educational programmes, the map reveals itself as a most remarkable instance of the Late Antiquity encyclopaedic tradition. With Rome at its centre, its tentacular road network and the barbarians mostly relegated to its edges, the map preserves a specific worldview and idea of community that both mapmakers and viewers could understand and support.

Conclusions

In Late Antiquity, geography was integrated into the school curriculum. It was taught increasingly with the aid of diagrammatic maps, and its political value remained essential, in spite of the decline of a unifying political power. Similar to the uninterrupted use of Latin for high literature or the preservation of the historical memory of the Empire, the conservation of a thoroughly Roman worldview and geographical idiom continued to feed all-encompassing notions of identity throughout the post-Roman kingdoms. If the geography taught at school did not provide late antique pupils with many theoretical updates, it preserved the bulk of Greco-Roman geographical knowledge, and, by relying more systematically on graphic representations of space, it helped generations of literati memorise a wealth of information.

By collecting as much material as possible, as in the case of the *Tabula*, or by following primarily the principle of selection, to the point of abstraction, as in the case of the T-O and zonal maps, late antique maps created an intelligible structure for understanding the world. They made visible, and thus more graspable, the relations between geographical names and, consequently, geographical places. Since maps and geographical texts were essentially conceived as archives of names and images worth remembering, late antique geography became an organising principle for classifying knowledge. Maps and catalogues were collections of place names that evoked a sense of continuity with the past and represented not only tangible symbols of collective memories but also archives of *loci* (places), intended as visualisations that triggered mnemonic-meditative reactions from readers. Although these traits are not exclusive to the period, the learning technique and the epistemological approach of the time emphasised the role of memory and diagrams in understanding and reflecting on geographical knowledge (Carruthers, 2020; Kupfer, 2020).

References

Albu E (2014) *The Medieval Peutinger Map*. New York: Cambridge University Press.

Carruthers M (2020) Geometries for thinking creatively. In: Chajes JH, Cohen A and Kupfer M (eds) *The Visualization of Knowledge in Medieval and Early Modern Europe—Studies in the Visual Cultures of the Middle Ages*. Turnhout: Brepols, pp. 33–44.

Diederich S (2018) Kartenkompetenz und Kartenbenutzung bei den römischen Eliten—Teil 1. *Orbis Terrarum* 16: 55–136.

Diederich S (2019) Kartenkompetenz und Kartenbenutzung bei den römischen Eliten—Teil 2. *Orbis Terrarum* 17: 101–184.

Drijvers W, Focanti L, Prat R and Van Nuffelen P (eds) (2018) *Mapping Antiquarianism in Late Antiquity*. Bruxelles: Société pour le progrès des études philologiques et historiques.

Edson E (2008) Maps in context: Isidore, Orosius, and the medieval image of the world. In: Unger R and Talbert RJA (eds) *Cartography in Antiquity and the Middle Ages*. Boston: Brill, pp. 219–236.

Fuhrmann M (2020) Die Demensuratio provinciarum und die Divisio orbis terrarum. In: Berger JD et al. (eds) *Handbuch der lateinischen Literatur der Antike Bd. 6: Die Literatur im Zeitalter des Theodosius (374–430 n.Chr.)*. München: C.H. Beck, pp. 47–48.

Gautier Dalché P (1994) De la glose à la contemplation. Place et fonction de la carte dans les manuscrits du haut Moyen Âge. In: *Testo e immagine nell'alto medioevo: 15–21 aprile 1993*, vol. 2. Spoleto: CISAM, pp. 693–771.

Gautier Dalché P (2003) Les diagrammes topographiques dans les manuscrits des classiques latins (Lucain, Solin, Salluste). In: Lardet P (ed.) *La tradition vive: Mélanges d'historire des textes en l'honneur de Louis Holtz*. Turnhout: Brepols, pp. 291–306.
Gerth M (2013) *Bildungsvorstellungen im 5. Jahrhundert n. Chr. : Macrobius, Martianus Capella und Sidonius Apollinaris*. Berlin: De Gruyter.
Johnson SF (2016) The medieval Peutinger map: Imperial Roman revival in a German empire. by Emily Albu. *Imago Mundi* 68(2): 242–243.
Kupfer M (2020) The rhetoric of world maps in late antique and the Middle Ages. In: Chajes JH, Cohen A and Kupfer M (eds) *The Visualization of Knowledge in Medieval and Early Modern Europe—Studies in the Visual Cultures of the Middle Ages*. Turnhout: Brepols, pp. 259–290.
Liccardo S (2020) Geography of otherness. Ethnonyms and non-Roman spaces in the Tabula Peutingeriana. *Orbis Terrarum* 18: 147–165.
Ljungberg C (2016) The diagrammatic nature of maps. In: Krämer S and Ljungberg C (eds) *Thinking with Diagrams*. Berlin and Boston: De Gruyter, pp. 139–160.
Lozovsky N (2000) *The Earth Is Our Book: Geographical Knowledge in the Latin West ca. 400–1000*. Ann Arbor: University of Michigan Press.
Lozovsky N (2006) Roman geography and ethnography in the Carolingian Empire. *Speculum* 81(2): 325–364.
Mauntel C (2021) The T-O diagram and its religious connotations. In: Mauntel C (ed.) *Geography and Religious Knowledge in the Medieval World*. Berlin and Boston: De Gruyter, pp. 57–82.
Merrills AH (2017) *Roman Geographies of the Nile: From the Late Republic to the Early Empire*. Cambridge: Cambridge University Press.
Monda S (2008) *La "Cosmographia" di Giulio Onorio. Un exceptum scolastico tardo-antico*. Roma: Aracne.
Östenberg I (2009) *Staging the World*. Oxford: Oxford University Press.
Ramelli I (2001) *Marziano Capella. Le nozze di Filologia e Mercurio*. Milano: Bompiani.
Rathmann M (2018) *Tabula Peutingeriana : die einzige Weltkarte aus der Antike*. Darmstadt: Philipp von Zabern.
Salway B (2005) The nature and genesis of the Peutinger Map. *Imago Mundi* 57(2): 119–135.
Salway B (2007) The perception and description of space in Roman itineraries. In: Rathmann M (ed.) *Wahrnehmung und Erfassung geographischer Räume in der Antike*. Mainz: Philipp von Zabern, pp. 181–209.
Scourfield JHD (1993) *Consoling Heliodorus: A Commentary on Jerome, Letter 60*. Oxford: Clarendon Press.
Stahl WH (1977) *Martianus Capella and the Seven Liberal Arts. 2, The Marriage of Philology and Mercury/Transl. by William Harris Stahl and Richard Johnson with EL Burge*. New York [u.a.]: Columbia University Press.
Talbert RJA (2010) *Rome's World: The Peutinger Map Reconsidered*. Cambridge [u.a.]: Cambridge University Press.
Teeuwen M and O'Sullivan S (2011) *Carolingian Scholarship and Martianus Capella : Ninth-century Commentary Traditions on De Nuptiis in Context*. Turnhout: Brepols.
Von Contzen E (2017) Die Affordanzen der Liste. *LiLi, Zeitschrift für Literaturwissenschaft und Linguistik* 47(3): 317–326.
Von Contzen E (2018) Experience, affect, and literary lists. *Partial Answers* 16(2): 315–327.
Weber E (2016) Die Datierung des antiken Originals der Tabula Peutingeriana. *Orbis Terrarum* 14: 229–258.

8
ARCHAEOLOGY, CRAFTING MAPS AND POLITICAL CHANGE

Piraye Hacıgüzeller

Archaeology and crafting maps

Archaeology is often referred to as a spatial discipline.[1] What this means is that much of the information, knowledge and insights in archaeology has a spatial dimension, a consideration for where things are located. It is, therefore, not surprising that the production of maps and map-like images in archaeology has a long history, tracing back to European antiquarianism in the seventeenth century (Morgan and Wright, 2018: 137–139; see Flexner, 2009). As Lucas (2023: 58) put it: 'Maps are one of the most basic records archaeologists make—from individual feature plans to landscapes, the representation of space in two dimensions is a core product of archaeological fieldwork'.

A detailed study of what archaeologists map, along with how and why they map them, still awaits thorough research. From general observations, it can be deduced that a certain traditionalism exists in the domain, underscored by a loyalty to Cartesian geometry and its applications. This is a mindset that has largely dominated archaeological cartography up to the present day (see also Flexner, 2009; Hacıgüzeller, 2017), with the exception of pioneering applications since the 2000s, which I discuss below. Take, for instance, landscape maps, which are likely the most common spatial scale in archaeological mapping, where archaeological sites are the predominant spatial entity. The conceptual and cartographic definitions of an archaeological site have been surprisingly difficult to pin down, even though they have played a foundational role in the discipline. Nonetheless, the manner in which these entities are presented on maps has remained resilient to change: since the very early days of archaeological cartography, sites have been presented as either a dot or an area on a map (see McCoy, 2020). This is, I contend, a typical example of archaeological cartographic traditionalism, widespread across the discipline.

Archaeological mapping is a unique craft, especially when it refers to mapmaking with pen and paper, and using paper maps or their skeuomorphs (like a scanned paper map on a tablet screen). Crafting involves things shaping the human body and mind (I separate the two here for clarity in the argument) just as much as the human body and mind shape things (Ingold, 2013; Sennett, 2008). This suggests that archaeological mapping, at least in its traditional form with pens, papers and their skeuomorphs, shapes humans just as humans shape archaeological things and maps during the processes of mapping. Thus, mapping

DOI: 10.4324/9781003327578-11

in archaeology is fundamentally a process of (re-)gaining familiarity with archaeological materials and the map itself through both body and mind while establishing new material and cartographic relations each time mapping takes place. Put differently, archaeological mapping involves knowing archaeological materials anew, following what is going on at an archaeological setting while going along with the map (Aldred and Lucas, 2019: 23). It is also a rearrangement of archaeological things both empirically (on the map) and conceptually in this process. As Morgan and Wright (2018: 142) note for the case of stratigraphic excavations: 'drawing by hand invokes an intimate interaction with the materiality of the archaeological record, forcing the archaeologist to observe stratigraphic relationships'.

All these can be compared to a potter throwing a ceramic vessel on a wheel. The act of wheel throwing forms the potter (e.g. their muscles, their hand-eye coordination, their ways of thinking) as much as it forms the pot (Malafouris, 2013). In the process of throwing on their wheel, in that generative dynamics of 'embodied sensibility', the potter is 'bound to respond to the moment-by-moment variations in the environmental conditions' that arise (Ingold, 2011: 2). This improvisation also characterises the craft of archaeological mapping when it involves tracing lines by hand, where observations and descriptions of archaeological materials become one with the inventive improvisatory movements of drawing (Ingold, 2011). A potter cannot make the same pot twice (even if the wheel, clay, amount of water and any other material condition one can think of are the same) since making the pot is an intuitive process responding to those moment-by-moment variations. Similarly, it would not be possible to make the same archaeological map twice despite all efforts. This is because each archaeological mapping process, no matter how familiar the material relations are from the previous go, requires building new familiarities with the archaeological materials. It demands new realisations, new learnings, new things to come to terms with, new intuitions enacted at the moment and new improvisations for tracing lines; 'every [map] is realised in practice' (Ingold, 2011: 7).

The majority of archaeological mappings were paper based until the 1990s. However, since then, technologies involved in archaeological cartography, along with those in other forms of archaeological record-making (including databases, audio-visuals and even daily excavation diaries), have witnessed rapid changes, driven by the digital transition. Digital maps have since become the preferred end products of both hybrid and fully digital cartographic processes in archaeology. Given the skilled and embodied nature of mapping by hand discussed above, the transition to digital media in archaeological cartography has not been straightforward. Mapping by hand and mapping with fully digital workflows, such as those based on orthophotographs created with 3D techniques, are two related but distinct practices.

As Sapirstein (2020: 136) notes, drawing by hand in archaeology is 'a complex process coordinating eye, mind, and hand which cannot be offloaded unproblematically to mechanical recording systems'. James (2015: 1200) adds:

> The basic experience of drawing things by hand, accurately, to scale . . . offers something unique and invaluable. Drawing makes you examine the subject much more closely than simply pointing a camera at it, or even just handling it and inspecting it by eye.

Elsewhere (Hacıgüzeller, 2019), I have written 'small stories of mapping people', narrating how different digital and analogue mapping workflows are situated in archaeological practice in terms of bodily movements and interactions with materials during fieldwork. According to Caraher (2016: 436; see also Morgan and Wright, 2018),

> The removal of the time-consuming illustration process [by hand] from excavation work . . . transforms a crucial step in the [archaeological] documentation process from one requiring detailed and careful knowledge of the features in a trench and of the conventions of illustration to one requiring the understanding of a digital camera and relevant software. The former is vital to the archaeological process whereas the latter is not.

The proliferation of archaeological digital maps and mappings since the 1990s is rightly attributed to widespread advancements in the domain of digital geospatial data and technologies, including their user-friendliness and 'mappification' of daily lives through mobile web applications (see McCoy, 2021). The media in question here include geographical information systems (GIS), global navigation satellite systems (GNSS), online (mobile) mapping, airborne and space-borne remote sensing, geophysical survey and publicly available (though not necessarily free) data platforms sharing geospatial and related information (e.g. Google Earth Engine, EOSDIS, ArcGIS Online). Beyond the investigations into the changes in skilled and embodied dimensions of analogue and digital archaeological mappings I detailed earlier, it is important to further reflect on the characteristics of digital geospatial advances and their agency within archaeology. So far, such reflections have been predominantly limited, focusing mainly on the new workflows facilitated by these advances and the novel knowledge that the new workflows generate (e.g. McCoy, 2017: table 1). While these assessments are informative, answers to many key questions relating, directly or more broadly, to digital transition in archaeology and mappings remain less clear. To enumerate a few: what are the intersectional identities of mapmakers in archaeology considering gender, sexuality, ethnicity, race, ability, class, language, religion and academic age, among others? More broadly, throughout the history of archaeology, who has taken on the role of the mapmaker, and at the expense of whose exclusion and dispossession (see Morgan and Wright, 2018; see also e.g. Cook, 2019; Gupta and Nicholas, 2022)? As I previously noted, "sites" have been dominant features on archaeological maps. However, which other spatial features have recurrently appeared on maps in archaeology, and for which implicit or explicit reasons? Who has owned archaeological spatial data sets, or possessed the privilege to retrieve, access and reuse them throughout the discipline's history (see Gupta et al., 2020)? Those who have consistently had easy access to archaeological maps and related media will readily recognise the increasing ubiquity of these maps since the discipline's digital transition—a phenomenon I would like to term the 'mappification of archaeology'. Yet, how ubiquitous have archaeological maps and mappings truly become and with what implications for archaeological knowledge production? For example, can it be suggested that our knowledge of archaeological places has been enriched (and if so, in which ways?), given the unprecedented surge in archaeological map production and sharing? That is, when it comes to archaeological mapping, is more genuinely more (see Hacıgüzeller, 2019)? Or, should archaeological mappings—especially in the privileged contexts of Europe and North America characterised by power, affluence and prestige, and the intersectional identities associated with these—occur less frequently, adhering to a principle of 'minimal mapping' (imbued with political goals akin to 'minimal computing', see Wythoff, 2022)?

Can the 'deliberate silences' (after Jungkunz, 2013) of the privileged—when it comes to mapping archaeology—serve to amplify the voices of others and make room for 'other' cartographic representations?

Mapping archaeology differently, mapping archaeology for change

Since the mid-1980s, a range of political discourses have proliferated in archaeology, aiming to locate the accumulation and distribution of power in/through archaeological practices and their resulting effects. The discourses have centred on inclusiveness, inequality, justice, diversity, degrowth and conflict, among others. This ethical-political turn in archaeology marks a crucial and welcome development in archaeological thought and practice during which politics and related ethical issues have 'abandoned their niche status to become a shared concern' across the discipline (González-Ruibal, 2018: 346; see also Hamilakis, 2007). Importantly, the politicisation of archaeological practices should be contextualised within the broader domain of humanities and social sciences, encompassing intersectional post-representational, feminist, Marxist, posthumanist, participatory, anti-capitalist, decolonial, indigenous, activist, slow science, punk and anarchist movements (see González-Ruibal, 2018).

The ethical-political discursive shift has had a profound impact on archaeological cartography, primarily from the 2000s onwards. It has ushered in a rethinking of archaeological mappings, influenced also by what has been happening in other domains, primarily human geography (Gillings et al., 2019; Hacıgüzeller, 2017; see e.g. Kitchin and Dodge, 2007). While earlier attempts existed calling for non-traditional engagements with archaeological maps (e.g. Tilley, 1994), new discourses emerging in the 2000s have been particularly inspiring in regards to alternative mappings (cf. Gillings et al., 2019; Hacıgüzeller, 2017 for detailed overviews and references; see Álvarez Larrain and McCall, 2019 for an extensive overview of archaeological and historical participatory mapping). They provided a welcome diversion from the usual preoccupations of archaeological cartography. These preoccupations primarily involve the challenge of "truthfully" depicting archaeological findings and the past on maps, a task frequently accompanied by epistemological anxieties concerned with how to access past 'realities'.

The discursive change has created fertile ground for archaeologists to map in ways that are provocative, proactive, prefigurative, imaginative, playful, enchanting and more. The primary objectives in the new archaeological mappings have been to challenge established power structures, amplify nondominant voices and present places in ways that diverge from the dominant narratives. There has been heightened awareness of the agency of maps and their affordances for positive political change and archaeological knowledge production. More attention thus has been given to the qualities of the change a map enacts in the world, and its potential to better the world for related actors and produce different types of archaeological knowledge, rather than how well maps represent existing archaeological knowledge and the past. In Álvarez Larrain and McCall's (2019: 645) words, the pivotal questions about archaeological mappings has become: 'mapping for which purposes and for whose benefit?'

Given the limited space available, it is not possible to meaningfully summarise the numerous studies that self-consciously map archaeology for change (for overviews, see bibliographical references above). However, it is crucial to provide a sense of what *mapping archaeology differently* might involve. Therefore, I will discuss two selected examples in some detail before concluding this chapter.

The first is an exceptionally creative and powerful mapping performance by Helen Wickstead and visual artist Janet Hodgson from 2009. Titled 'The Uber Archaeologist: Art, GIS and the male gaze revisited', it has since become a trailblazing example of archaeological feminist mapping. Through their mapping performance at Stonehenge, the duo aimed to critically reflect on the 'gaze', a significant point of critique in both feminist thinking and modernist mapping. The gaze, often associated with masculinity, is critiqued for observing from a distance and treating 'the viewed' as merely an object of vision (see also Dando and Carraro in this volume). This distant gaze assumes surveillant power, visual pleasure, privilege and ownership, as Wickstead (2009: 252–253) also explains. In the mapping performance at Stonehenge, Hodgson engaged in both map- and 3D image-making to highlight the gender imbalance on site. They directed a film titled 'The Uber Archaeologist', featuring the archaeological team at Stonehenge as its cast. In the film, Dr. Kate Welham, the only woman on the directorial team and head of the geoinformation team on site, plays Dr. Frankenstein. Hodgson and Welham used laser scanning on the bodies of the other site directors to generate a composite body—a Frankenstein's monster, named the Uber Archaeologist, composed of the body parts of men in authority. They also designed a costume for the Uber Archaeologist in the 3D digital environment. Regarding the laser scanning of the four male site directors for their film, Hodgson remarked, 'I wanted to make them a bit uncomfortable. . . . I wanted to objectify them' (Wickstead, 2009: 264). During their mapping performance on site, Hodgson purposefully utilised GIS and 3D scanning—tools that were some of the most criticised in archaeology for their association with the gaze, especially in the 1990s and 2000s, which was the time of the performance. Their aim was to show that these tools could be repurposed to challenge established power structures, becoming part of cartographic performances that benefit the nondominant. As Wickstead (2009: 264) writes, 'Hodgson was interested in how Welham, as a visualizer, held a kind of behind-the-scenes power essential to the operation of the entire [archaeological] project, but still somehow understated in the everyday social dynamics of the site'. Also in the movie, this gaze-related power of Welham is accentuated: the audience sees the Uber Archaeologist through Welham's eyes. Through these new power dynamics and other strategies detailed by Wickstead (2009: 265), 'Hodgson's projects play with visualities, undermining any essential male or female gaze. Hodgson insists on the importance of this freedom to occupy multiple gazes, for herself and other image makers'.

The second mapping is a playful performance I personally carried out. I have chosen to showcase this map here to narrate the performance as the performer. Moreover, my objective is to emphasise the archaeological opportunities that arise at the intersection of playfulness and archaeological mapping and align them with other emerging playful discourses in archaeology that aim 'to make play into one of the pillars of archaeological practice' (Politopoulos et al., 2023: 1). The performance involves telling spatial narratives of everyday practices by manually drawing lines on GIS maps, presenting potential everyday movements in a Bronze Age building, Zeta Beta, in Malia, Crete. Within the narrative (as explained in the caption of Figure 8.1), a person moves throughout the building, carrying out a sequence of cooking-related tasks using materials reported from the excavations (see Figure 8.1). Maps featuring two, three, or more individuals can also be crafted through similar cartographic performances (see Hacıgüzeller, 2017: fig. 2). The lines and associated narratives are not the outcomes of strictly 'scientific' archaeological processes in the traditional sense. They are only partially based on empirical evidence and, as such, can be considered hyper-interpretative within the discipline of archaeology. In other words, these

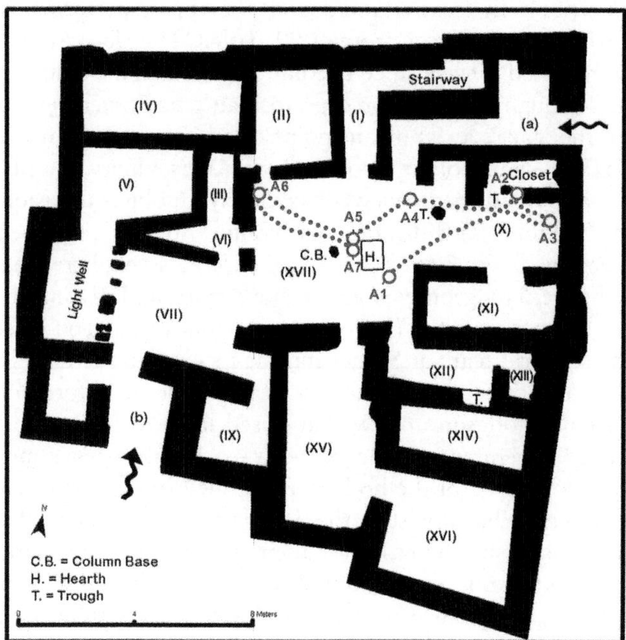

Figure 8.1 Single-person playful narrative map for cooking at the Bronze Age building of Zeta Beta in Malia (Crete), by the author. In the narrative, an individual picks up the lamp in Space XVII (A1) to venture into the possibly dark closet located northeast of Space X. Upon reaching the entrance of the closet (A2), they set the lamp on the partitioning wall to their right and take a bowl and a cup. They then approach the two pithoi situated along the east wall of Space X (A3), use the cup to scoop out foodstuff to fill the bowl, and subsequently place the cup back into the pithos. The narrative extends to other areas of the building and can also be broadened to feature two, three or more individuals (Hacıgüzeller, 2017: figure 2), presenting the building as lived-in, humming with people and daily buzz.

playful mappings do not have strong claims to 'truthfulness' as understood in terms of 'empirical truth' in the western scientific discourse. As such, while carrying out the mapping performance, I was not distressed about whether the events the lines present and the narratives describe actually took place in the past. Instead, I crafted the lines and narratives with intentional playfulness, akin to puzzles connecting dots with numbers. This is an explicitly *enchanting* way to engage with archaeological materials through mappings (see also Perry, 2019). In the process, I made use of imagery familiar to many, including myself (i.e. the Cartesian map of the building in the background), which aided in spatially orienting the mapping performance. Thus, the narratives and mapping performance provided ways for me to playfully engage with, relate to, and reflect on archaeological materials and places, along with their affordances. I also followed 'what is going on at an archaeological setting while going along with the map' (see above) while drawing the lines during the performance, connecting objects and spaces with one another. Furthermore, the performance violated and destabilised the Cartesian map's authoritative surface, challenging its claim to objectivity and truth by hacking it for fun and imagination.

Despite the exciting developments in archaeological cartography since the 2000s that the two performances presented here hopefully exemplify, many archaeological contexts, particularly in fieldwork settings, maintain the expectation that an archaeological map 'truthfully' represents a material reality, detached from the map itself (see also Barnard, 2023). In my opinion, such Cartesian/modernist expectations haunting archaeological cartography are acceptable, provided they also accommodate other types of mappings. The principles upon which traditional Cartesian maps are built are useful—they are well-known, familiar and widespread conventions for documenting archaeological material settings—but they are not the exclusive 'truthful' or 'best' way to map archaeology (Hacıgüzeller, 2017). It is imperative that this insight gain more prominence in archaeological discourse in the upcoming years. At any rate, I believe the future looks bright when it comes to mapping archaeology differently and effecting positive political change with archaeological maps. Relying exclusively on Cartesian cartography in the discipline is becoming increasingly challenging. After all, the use of mapping in archaeology to destabilise existing power relations and accumulations has demonstrated immense potential, thanks to the quality and quantity of pioneering work that has emerged within the last two decades. Much promising terrain remains for future explorations.

Not delaying change, acting now

In this brief overview, my aim has been to position archaeological cartography within the discipline and to summarise how, over the past 20 years, various methods for engaging differently with archaeological mappings have emerged. Some of these methods are particularly effective in advocating for, demanding, and effecting political change. I contend that, whenever possible, those in privileged positions with the power to do so should facilitate or even push for change through archaeological mappings, as they can act with less personal risk (see Cook, 2019). A critical component of this effort can be involving nondominant individuals and communities, and their allies in collaborative engagement with archaeological mappings. There is no need to wait to become more tech-savvy or better funded, or for the next user-friendly method, technology, or tool to emerge before teaming up to map archaeology differently for a fairer power distribution.

Do-It-Yourself (DIY; Caraher, 2019; Morgan, 2015) or minimal computing (Wythoff, 2022) mappings can be powerful, even if (and sometimes particularly because) the related applications are humble in technical complexity. They may involve hacking accessible components of more complex workflows, such as free and relatively user-friendly digital databases, GIS, and tools for earth surveying and 3D visualisation. However, I concede that there can be initial barriers. For instance, initiating and conceptualising these DIY and minimal computing processes may still require expert knowledge or, at the very least, some familiarity with archaeological cartography and the affordances of digital components, regardless of their user-friendliness. Challenges also may arise when tackling steps further in the cartographic workflow where geospatial expertise seems indispensable but is not accessible. Further challenges may arise when it comes to being and feeling heard in the mapping process, securing widespread visibility for the resultant map, and addressing issues related to governance, streaming speed and the size of the involved data sets. Moreover, in many parts of the world, state-induced political pressures can stifle communities' free expression and meaningful engagements with their identity and heritage. Neoliberal demands and exploitation may also render finding the time, resources and enchantment required to join up in archaeological mapping processes difficult or impossible.

Still, it remains essential for nondominant individuals and communities to receive the message that archaeological maps can be powerful allies in making claims about the past, present, and future, and associated places. Targeted conversations may be necessary here, aimed at understanding what actors actually want to achieve with mapping and for whose benefit. With targeted conversations, I am not alluding to a top-down critique or control mechanism over the intended uses of archaeological mappings. Rather, in alignment with González-Ruibal (2018: 355), I believe that for a political-ethical archaeology, what we, as archaeologists, need to assess is 'the effects of behaviors that may be virtuous in theory and perfectly compliant with the highest ethical standards of the moment'. Such conversations and assessments can occur through dialogue, where a diverse stakeholder group is involved without a hierarchical structure. Some of the questions that need to be asked and discussed in such a context are Who produces the map and with what intended purposes? Who benefits from the map? What are potential uses and misuses of the map, beyond its intended purposes? Who is included and excluded from the map? What can the map tell us that other forms of maps and media cannot? What are the political risks and advantages of using the map (see Mah, 2017: 125)?

Note

1 Being an archaeologist trained in Europe, my research expertise in archaeological cartography primarily focuses on Anglo-American and European traditions, as far as language barriers allow. Therefore, the observations I present in this chapter are unavoidably intended mainly for this archaeological sphere. However, I hope that the points I raise will have relevance in the broader global archaeological landscape.

References

Aldred O and Lucas G (2019) The map as assemblage: Landscape archaeology and mapwork. In: Gillings M, Hacıgüzeller P and Lock G (eds) *Re-Mapping Archaeology Perspectives, Alternative Mappings*. New York: Routledge, pp. 19–36.

Álvarez Larrain A and McCall MK (2019) Participatory mapping and participatory GIS for historical and archaeological landscape studies: A critical review. *Journal of Archaeological Method and Theory* 26(2): 643–678.

Barnard H (2023) *Archaeological Mapping and Planning*. Cambridge: Cambridge University Press.

Caraher W (2016) Mobilizing past for a digital future: The potential of digital archaeology. In: Walcek Averett E, Gordon JM, and Counts DB (eds) *Slow Archaeology: Technology, Efficiency, and Archaeological Work*. North Dakota: The Digital Press @ The University of North Dakota, pp. 421–441.

Caraher W (2019) Slow archaeology, punk archaeology, and the 'Archaeology of Care'. *European Journal of Archaeology* 22(3): 372–385.

Cook K (2019) EmboDIYing disruption: Queer, feminist and inclusive digital archaeologies. *European Journal of Archaeology* 22(3): 398–414.

Flexner J (2009) Where is reflexive map-making in archaeological research? Towards a place-based approach. *Archaeological Review from Cambridge* 24(1): 7–21.

Gillings M, Hacıgüzeller P and Lock G (2019) On maps and mapping. In: Gillings M, Hacıgüzeller P and Lock G (eds) *Re-Mapping Archaeology Perspectives, Alternative Mappings*. New York: Routledge, pp. 1–16.

González-Ruibal A (2018) Ethics of archaeology. *Annual Review of Anthropology* 47(1): 345–360.

Gupta N, Blair S and Nicholas R (2020) What we see, what we don't see: Data governance, archaeological spatial databases and the rights of Indigenous peoples in an age of big data. *Journal of Field Archaeology* 45(sup1): S39–S50.

Gupta N and Nicholas R (2022) Being seen, being heard: Ownership in archaeology and digital heritage. *Archaeologies* 18(3): 495–509.

Hacıgüzeller P (2017) Archaeological (digital) maps as performances: Towards alternative mappings. *Norwegian Archaeological Review* 50(2): 149–171.
Hacıgüzeller P (2019) Archaeology, digital cartography and the question of progress : The case of Çatalhöyük (Turkey). In: Gillings M, Hacıgüzeller P and Lock G (eds) *Re-Mapping Archaeology Perspectives, Alternative Mappings*. New York: Routledge, pp. 267–280.
Hamilakis Y (2007) From ethics to politics. In: Hamilakis Y and Duke PG (eds) *Archaeology and Capitalism: From Ethics to Politics*. Walnut Creek, CA: Left Coast Press.
Ingold T (2011) *Redrawing Anthropology: Materials, Movements, Lines*. Farnham: Ashgate.
Ingold T (2013) *Making: Anthropology, Archaeology, Art and Architecture*. London and New York: Routledge.
James S (2015) 'Visual Competence' in archaeology: A problem hiding in plain sight. *Antiquity* 89(347): 1189–1202.
Jungkunz V (2013) Deliberate silences. *Journal of Deliberative Democracy* 9(1): 1–32.
Kitchin R and Dodge M (2007) Rethinking maps. *Progress in Human Geography* 31(3): 331–344.
Lucas G (2023) *Archaeological Situations: Archaeological Theory from the inside Out*. Abingdon and New York: Routledge.
Mah A (2017) Environmental justice in the age of big data: Challenging toxic spots of voice, speed, and expertise. *Environmental Sociology* 3(2): 122–133.
Malafouris L (2013) *How Things Shape the Mind: A Theory of Material Engagement*. Cambridge, MA: The MIT Press.
McCoy MD (2017) Geospatial big data and archaeology: Prospects and problems too great to ignore. *Journal of Archaeological Science* 84(1): 74–94.
McCoy MD (2020) The site problem: A critical review of the site concept in archaeology in the digital age. *Journal of Field Archaeology* 45(sup1): S18–S26.
McCoy MD (2021) Defining the geospatial revolution in archaeology. *Journal of Archaeological Science* 37: 102988.
Morgan C (2015) Punk, DIY, and anarchy in archaeological thought and practice. *AP: Online Journal in Public Archaeology* 5: 123–146.
Morgan C and Wright H (2018) Pencils and pixels: Drawing and digital media in archaeological field recording. *Journal of Field Archaeology* 43(2): 136–151.
Perry S (2019) The enchantment of the archaeological record. *European Journal of Archaeology* 22(3): 354–371.
Politopoulos A, Mol AAA and Lammes S (2023) Finding the fun: Towards a playful archaeology. *Archaeological Dialogues* 30(1): 1–15.
Sapirstein P (2020) Hand drawing versus computer vision in archaeological recording. *Studies in Digital Heritage* 4(2): 134–159.
Sennett R (2008) *The Craftsman*. New Haven: Yale University Press.
Tilley CY (1994) *A Phenomenology of Landscape: Places, Paths, and Monuments*. Oxford and Providence: Berg.
Wickstead H (2009) The Uber archaeologist: Art, GIS and the male gaze revisited. *Journal of Social Archaeology* 9(2): 249–271.
Wythoff G (2022) Ensuring minimal computing serves maximal connection. *Digital Humanities Quarterly* 16(2): 1–6.

9

CHARTING MOVEMENT THROUGH HISTORICAL SOURCES

Tiago Luís Gil

Introduction

The relationship between history and cartography has an irregular trajectory, and the two areas are pretty distant at present. Being published from the 16th century onwards, historical atlases had a particular projection in the 19th and 20th centuries, when several projects of this nature were conceived (Hofmann, 2000). History and cartography were once more connected, especially in the 1970s. Nevertheless, since then, this closeness has been in decline, and even with the stimulus provided by the diffusion of digital cartography, there is still a long way to go to regain a stable foothold.

The extent to which historians have used cartographic language in recent decades could be taken as a reference to evaluate this trajectory. Considering just one journal, the prestigious French journal *Annales*, which is close to completing a century of existence, the following scenario is visible: 59 maps in the 1950s; 125 in the 1960s; 327 in the 1970s, falling to 150 in the 1980s and 46 in the 1990s; finally, 77 and 28 in the 2000s and the 2010s, respectively.[1] Of course, these data have biases, but they can serve as a guide to observing trends.

These data, however, hide a crucial fact: historians have not always used maps to represent historical processes. In particular, they did not use cartography to express ideas of motion. In truth, one of the best-known historians of the 20th century, Fernand Braudel, made extensive use of cartographic language in famous works such as *The Mediterranean* (1949) and *Material Civilization, Economy and Capitalism* (1979). Many of the maps created for these works were prepared by Frank Spooner, who also used them profusely in his classic *Risks at Sea* (Braudel, 1949, 1980; Spooner, 2002). For his part, Janus Szego made significant studies on various ways of representing movement and displacement by analysing dozens of historical atlases, emphasising Hagerstrand's tube (as will be seen below), of which he was a leading defender (Szegö, 1987).

The purpose of this chapter is to evaluate the most widespread forms of cartographic representation of movement among historians in order to present a set of ideas on how to portray social dynamics through maps, taking several examples of historical sources as references. The intention is not to exhaust the subject but to present ideas that may be relevant for experimenting with the use of cartographic language to represent time.

DOI: 10.4324/9781003327578-12

Motion mapping

The example of the *Annales* was not accidental. Besides being a prominent journal which had an undeniable influence throughout the 20th century, it was based at the École des Hautes Études en Sciences Sociales (EHESS), where Jacques Bertin, a cartographer who contributed significantly to the discipline, worked. In his *Sémiologie Graphique* (1963), Bertin relied on the idea of 'maps to see' as opposed to 'maps to read', the latter requiring a great deal of effort to understand. On the other hand, 'maps to see' would be easily understandable and very expressive. Bertin worked to radicalise the language of cartography, opening a way to thinking about different communicative forms.

Across the pages of the numerous articles published in the *Annales* that contained maps, a significant number of them were signed by the Laboratoire de Cartographie de L'École Pratique des Hautes Études or simply with the acronym EPHE. It was a reference to Bertin's laboratory, and the maps were made by him or under his guidance. Beyond the ideas contained in his book, this set of maps available in the *Annales* articles presents a proper dictionary of cartographic phraseology to refer to movement, transition and displacement. Considering these maps, typologies of printed or static representation can be pointed out as follows: the 'comic', 'stage', 'arrow', 'contour line' maps and 'place and date' (see Figure 9.1).

The maps that can be called 'comic' (Figure 9.1A) are those inspired by the so-called sequential arts, a formula popularly consecrated by comics and marked by the sequence of images (maps, in this case) where some change occurs between frames, denoting motion. The form of this transition is meaningful since the subsequent frame can mean a gradual or abrupt change or even convey an idea of simultaneity or context. For instance, a map can show a scenario of a year of population growth in many cities to, the subsequent year, a deceleration represented by a new frame, which could continue in the third frame/year. In contrast, a war scenario can be represented in the first frame, followed by another with a concluded battle ended and a third with a retreat, without the time between frames necessarily being the same. The play of transition is fundamental in comic maps, as it is in comics themselves.

Within the idea of frames, the 'stage' type maps (Figure 9.1B) are pretty helpful when the proposal is further directed to compare two or three moments (more than three would be visually tricky). These are maps in which most of the 'set', to use Szego's term, remains unchanged while only one variable is changed, highlighting the variations between two or three snapshots. It is a very convenient alternative to highlight small (and relevant) changes which would hardly be perceived at a glance between two images/maps, as in the case of comic maps.

The use of arrows (Figure 9.1C) is a widespread solution to represent movement, especially displacement. This cartographic 'theme' also has its own nomenclature: arrows can play with thickness and colour, and they can, together and in great quantity, create a sensation of imprecision, contrary to the generally idealised image of arrows, used to point out specific things (Tobler, 1987). In the great theme of lines, contour lines (Figure 9.1D) are less frequent but very powerful. They can be employed very well to express the concept of an oil slick, with relatively homogeneous dispersion, being used in the opposite direction of the arrows. Finally, 'place and date' maps (Figure 9.1E) are perhaps the most primitive. Nonetheless, they can be helpful if used to represent a minimal number of relevant places, as a vast accumulation of textual information would make it difficult to interpret the dynamics.

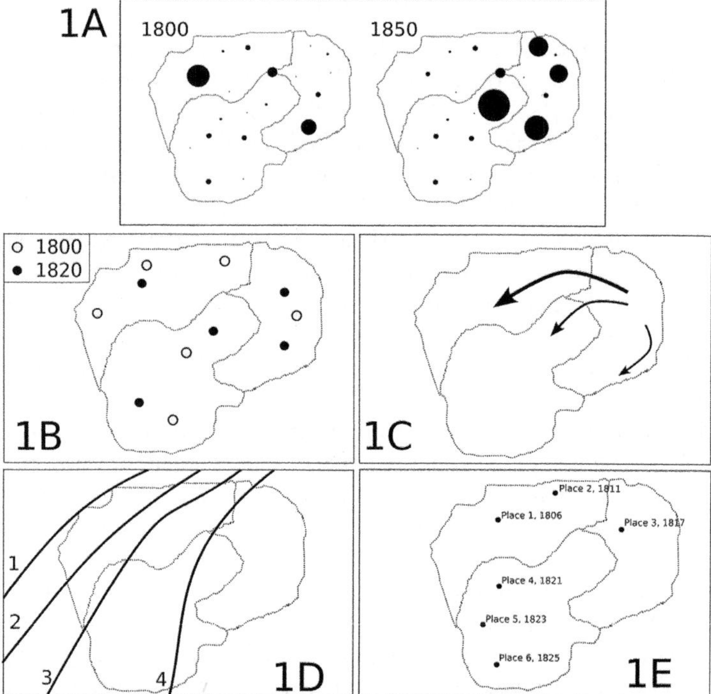

Figure 9.1 Varieties of motion maps. 1A—Comic map inspired by Guerreau Alain. L'atelier monétaire royal de Mâcon (1239–1421). In: Annales. 29e année, 1974 (2): 377; 1B—Stage map inspired by Bautier Robert-Henri. Feux, population et structure sociale au milieu du XVe siècle: l'exemple de Carpentras. In: Annales. 14e année, 1959 (2): 262; 1C—Arrow map inspired by Lombard Maurice. La chasse et les produits de la chasse dans le monde musulman (VIIIe-XIe siècle). In: Annales. 24e année, 1969 (3): 586; 1D—Contour line map inspired by Ruffié Jacques, Bernard Jean. Origine du polymorphisme hématologique chez l'homme et dynamique des populations. In: Annales. 34e année, 1979 (6): 1329; 1E—Place and date map inspired by Jansen P. Ph. Aux Pays-Bas, une prodigieuse conquête: le plan Delta. In: Annales. 16e année, 1961(4): 660.

Source: Elaboration by the author on the basis of Annales' articles

The possibilities pointed out so far were largely collected from works that had Bertin's intervention, but their use can be found in many other authors. A rather original cartographic vocabulary was created by the Swedish geographer Torsten Hägerstrand with the concept of a space-time cube (or aquarium). This three-dimensional figure kept the Cartesian plane for geographical data and used the third dimension to represent displacement and speed, as well as to compare different simultaneous motion without visually overloading the map (Hagerstrand, 1967). Cartographers such as Szegö (1987), Kwan and Ding (2008) and Kwan et al. (2015) later continued the spatio-temporal cube idea.

More recently, another relevant innovation is the space-time mosaic created by Pearce (2008), in which a map is overlaid by fragments of other maps which vary in size according to the route taken by a traveller along a day's journey. The great innovation of this map rests in presenting the description of a route without using lines but instead a set of map fragments with colouring and appearance consistent with the weather described in the travelogue (e.g. darker for rainy days). It is a fascinating technique, as it creates a solid sense of transition between days and even speed since the size of a fragment is proportional to the distance travelled in a day. Thus, at least three variables are in dialogue: time, space and travel conditions, the third being perfectly replaceable by another criterion.

We still need to talk about a resource that has been historically disregarded but is being well used: the anamorphic map. The usage of anamorphosis is a technique that considers the distortion of Euclidean space to accentuate a variable, such as time series. One of the first uses for this purpose was Émile Cheysson's map, *Accélération des voyages en France depuis 200 ans*, from 1888 (Palsky, 1999), which sought to create an image of changing speeds possible on French roads over time. More recently, Reuschel and other researchers (Reuschel and Hurni, 2011; Reuschel et al., 2014) have been using these techniques for motion cartography, although more for their interest in discussing simultaneity (equally relevant when movement cartography is considered).

Finally, it would not be possible to think in terms of movement cartography and motion without mentioning the idea of animation, including both the sequence of images, with the idea of frames per second from cinema, and the more modern vectorial animation, with the use of computing and attributing behaviour to a digital vector. These also have their language, which must reflect the various action rhythms and the visual arrangement inside the frame. Despite being very beautiful and exciting, animations are often classified as not very effective in the visual communication of data (Tversky et al., 2002).

What was presented above sought to create a repertoire of communicative techniques of cartography and to create conditions for thinking about a visual vocabulary of maps regarding historical processes. There are several ways of representing movement, and each involves many choices and produces various results. The proposal now is to examine these resources in consideration of different historical sources.

Historical sources and geographical imagination

Thus far, many possibilities for visual communication have been explored; their potential application can now be assessed on historical sources. This apparent inductive perspective does not mean that the assessment is detached or can be disconnected from theoretical issues. The way space can be conceived and, finally, mapped has a direct connection with the notion of geography (even implicit) that organises the whole study, from the definition of the research problem, through the selection of data and finally in the analysis and writing. Mapping parish registers, for example, involves a worldview that believes there is a connection between religious liturgies and the broader social space. At the same time, it involves the idea that accumulating these individual or family experiences (e.g. a baptism) can disclose relevant mid-term or structural elements. None of this, however, is neutral or commonplace.

Understanding that the cartography of historical documents is not neutral and is always guided by theoretical questions does not mean abandoning any initiative of the inductive approach. Different historical sources are the product of the record of different social practices and, consequently, have different geographies. In this sense, their relevance can be utilised to think about inherited structural limits maintained by a society and thereby present spatial interactions that would be unexpected by a scholar.

The data, thus, should not be taken as a homogeneous outcome since different people give different meanings to (apparently) similar practices. In the same example of parish registers, one map may show a case of people who were baptised with great devotion alongside people who did so without great engagement, in an almost secular way and out of a diffuse notion of tradition. Both cases coexist in time and physical space but can also be interpreted as belonging to entirely different worlds.

Geography of dialogic sources: testimonies and judiciary records

So-called dialogical sources—those produced by dialogue—are persistently used in 20th-century historical research, especially in social history (Thompson, 1977; Ginzburg and Prosperi, 1975). Examples of this type of source are records of police or inquisitorial interrogations. Therein lies their importance: privileged sources detailing the accounts of ordinary people are usually rare. In a certain sense, mapping these sources means noticing the geography of the people giving testimony and, as far as possible, observing how they organise and create the space in which they live. This type of information can have many biases and always will, as there is always a 'case' to be unravelled, and witnesses often try to avoid trouble or lie deliberately. In any case, these testimonies tend to bring the experiences of everyday life and its geography into the text.

There are many ways of mapping this source, but only two will be explored here. One is to treat the testimonies as 'travelogues', that is, treating the journey reports in the testimonies (be they unfaithful but feasible) as if they were a travel diary. In this case, Pearce's proposal (2008) would be a highly relevant reference since it focuses precisely on this type of source. In the same way, the use of arrows could be relevant for certain journeys, depending on the situation. Another possibility would be to map the personal relationships referred to in the reports as if they were items of a graph of social network analysis but one based on the Euclidean plane. Carvalho and Moraes (2016) did such an exercise with gossip accounts about heresies at the time of an inquisitorial visitation in 16th-century Bahia, Brazil (Carvalho and Moraes, 2016; Carvalho, 2018) (Figure 9.2A). Lines connecting points were used to represent connections, in a clear inspiration of the use of lines proposed by Bertin (1967).

Geographical imagination of parish records

Parish registers are traditional sources and have been used particularly for reconstructing the demography of pre-industrial societies (Henry and Fleury, 1985; Macfarlane, 1977; Wrigley et al., 1997). The possibilities for mapping these records are extensive. However, they require a great deal of work, since these sources rarely provide data on the location of the participants in the recorded acts, subsequently requiring the consultation of many other sources. Sometimes the location of participants is available, and it is possible to map the residences of the most coveted godfathers and those of their godchildren. This type of source brings forth a scenario of trivial everyday life and permits the following of the slowest rhythm of social life: that of long-term social relations, far from the rumours and accusations of dialogical sources, for example, and with a much broader social universe, since it includes almost the entirety of society and not only people involved in denunciations or witnesses thereof. In a study on the theme, Sirtori and Gil (2012) analysed the geographical limits of the godparenthood of enslaved people in Brazil in the modern era, ascertaining the extent to which enslaved people were capable of inviting godfathers from distant regions, thereby measuring an area of possible daily action of those individuals.

2A

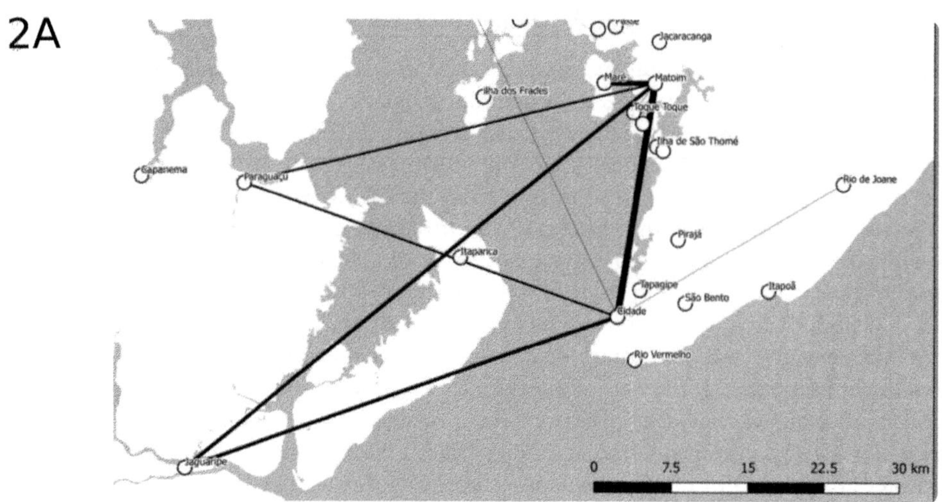

2B

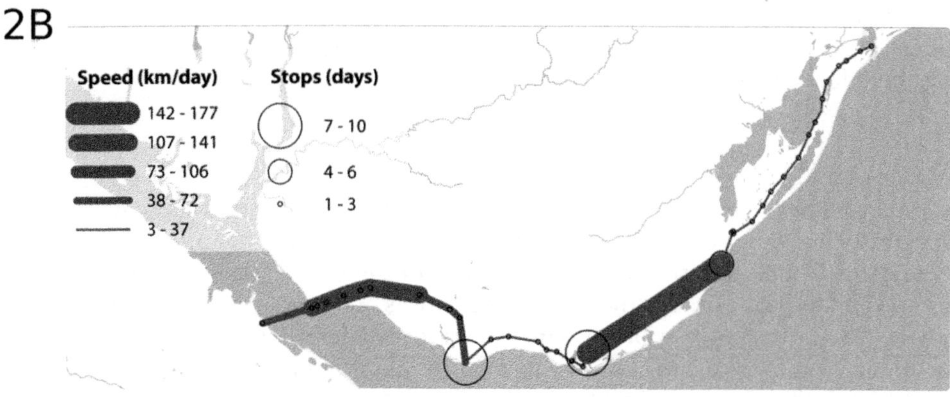

2C

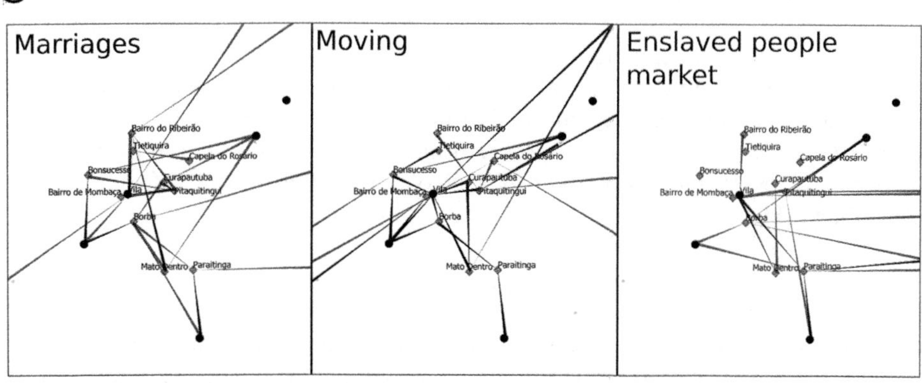

Figure 9.2 Charting movement with historical sources.

Source: 2A—Carvalho, Carlos Antonio Pereira de (2018) 'Espacializando as conversas na Cidade da Bahia do século XVI'; 2B—Prepared by the author based on Cabrer JM 'Diários . . .'; 2C—Prepared by the author based on Arquivo do Estado de São Paulo (Brazil). Lista Nominativa de Pindamonhangaba, 1802. All permissions granted.

Geography of narrative: travel diaries

Textual narratives are among the sources researchers have most mapped, and in many ways, whether they be narratives with fictional intent or with pretensions of truth, such as travel diaries (Caquard and Cartwright, 2014; Ryan, 2003; Pearce, 2008; Rossetto, 2014; Reuschel and Hurni, 2011). There are many ways to map these stories, and the works cited here present varied and creative solutions, from the use of arrows to comic maps, mosaics, the space-time cube and anamorphoses. Notwithstanding this, some routines are pertinent to remember when facing historical sources. It would be very convenient to evaluate some variables: the source's writing rhythm (over the space), where the author is when writing and if at high or low speed (always relative to the displacement itself). This information, even if it seems to break the beauty of the story, helps in understanding something hidden, the silences in the text. If the author speaks a lot in some regions and is discreet in others, it may be necessary to consider the reasons for this silence. It is not enough to regard what is written; it is also recommended to map the unsaid. Likewise, placing a narrative on the map reinforces the notion of displacement since precisely this change of coordinates is the central object of narrative cartography. This attribute presents another problem: stops along the way, more difficult to map and less attention-grabbing, are often neglected. Nevertheless, this can be very relevant for research. Evaluating speed means not only calculating the mileage in movement but also estimating the extent of stops, which can modify the author's textual production in turn (see Figure 9.2B).

The multiple geographies of a census

Some possibilities of working with censuses, a source type more commonly related to demography but which may have multiple readings, can be seen. They allow an assessment of a snapshot of a place (city or country). Furthermore, in some exceptional cases, the route journeyed by the census taker, the order of the streets and other precious information are available. Moreover, it is possible to map the narrative of daily life: people's working spaces and the places newlyweds, workers, single mothers and the descendants of enslaved people inhabited, thereby considering wildly different contexts.

In Brazil, some censuses from the beginning of the 19th century mention the people who had—when this was the case—recently left a house, as well as those who had just moved in, in addition to explaining the reason, origin or destination of each one. With this information, it is possible to evaluate migrations or simply understand that the eldest son had left the parental home to get married or to study in another city. With the configuring of these data, a series of maps was prepared that takes this information as if it contained micro-travel reports and considers it as a whole, according to the different purposes of displacement (Figure 9.2C).

Conclusion

The connection between history and cartography is promising, despite the distance with which the former has been treating the latter. Cartography has great communicative potential and allows for the visualisation of unexpected or even silent elements of historical sources. It also permits, and perhaps this is its most outstanding contribution, the consideration of questions that otherwise would have been unimaginable, thus allowing the development of historical studies. Historians are usually sceptical regarding cartographic language, since it seems to frame time in a static way. However, as this chapter tried to present, there are many sophisticated ways to convey historical processes in maps. There is an old tradition of movement maps in historical studies, and this approach might be considered an inspiring path to better use this powerful language.

Note

1 A map-per-article ratio would not bring a different scenario: around 0.2 in the 1950s and 1960s, going to approximately 0.6 in 1970, 0.32 in 1980 and below 0.2 since 1990, the 2010s being the worst decade. See Annales, in *Persée* [persee.fr].

References

Bertin J (1967) *Sémiologie graphique: Les diagrammes, les réseaux, les cartes.* Paris: Mouton.

Braudel F (1949) *La Méditerranée et le monde méditerranéen à l'époque de Philippe II.* Paris: Colin.

Braudel F (1980) *Civilization matérielle, économie et capitalisme, XV-XVIII siecle.* Paris, France: Armand Colin.

Caquard S and Cartwright W (2014) Narrative cartography: From mapping stories to the narrative of maps and mapping. *The Cartographic Journal* 51(2): 101–106.

Carvalho CAP de (2018) *Espacializando as conversas na Cidade da Bahia do século XVI : redes de informação e mobilidade geográfica de escravizados.* Master Thesis, Universidade de Brasília, Brazil. Available at: https://repositorio.unb.br/handle/10482/34029.

Carvalho CAP de and Moraes LS de (2016) Geoprocessando as relações sociais na cidade da Bahia-século XVI. In: Valencia Villa C and Gil TL (eds) *O retorno dos mapas: sistemas de informação geográfica em História.* Porto Alegre: Ladeira Livros.

Ginzburg C and Prosperi A (1975) *Giochi di pazienza: un seminario sul Beneficio di Cristo.* Torino: Einaudi.

Hagerstrand T (1967) *Innovation Diffusion as a Spatial Process,* Pred A, ed. Chicago: University of Chicago Press.

Henry L and Fleury M (1985) *Nouveau Manuel de Dépouillement et d'exploitation de l'état Civil Ancien.* Paris: L'Institut National d'Études Démographiques.

Hofmann C (2000) La genèse de l'atlas historique en France (1630–1800): Pouvoirs et limites de la carte comme "œil de l'histoire". *Bibliothèque de l'École des chartes* 158(1): 97–128.

Kwan M-P and Ding G (2008) Geo-narrative: Extending geographic information systems for narrative analysis in qualitative and mixed-method research. *The Professional Geographer* 60(4): 443–465.

Kwan M-P et al. (eds) (2015) *Space-Time Integration in Geography and GIScience.* Dordrecht: Springer Netherlands.

MacFarlane A (1977) *Reconstructing Historical Communities.* Cambridge: Cambridge University Press.

Palsky G (1999) The debate on the standardization of statistical maps and diagrams (1857–1901). Elements of the history of graphical semiotics. *Cybergeo: European Journal of Geography.* Available at: https://journals.openedition.org/cybergeo/148?lang=en (accessed 29 January 2023).

Pearce MW (2008) Framing the days: Place and narrative in cartography. *Cartography and Geographic Information Science* 35(1): 17–32.

Reuschel A-K and Hurni L (2011) Mapping literature: Visualisation of spatial uncertainty in fiction. *The Cartographic Journal* [Online] 48(4): 293–308.

Reuschel A-K et al. (2014) Data-driven expansion of dense regions: A cartographic approach in literary geography. *The Cartographic Journal* [Online] 51(2): 123–140.

Rossetto T (2014) Theorizing maps with literature. *Progress in Human Geography* 38(4): 513–530.

Ryan M-L (2003) Cognitive maps and the construction of narrative space. In: *Narrative Theory and the Cognitive Sciences.* Stanford: Publications of the Center for the Study of Language and Information.

Sirtori B and Gil T (2012) A geografia do compadrio cativo: Viamão, Continente do Rio Grande de São Pedro, 1770–1795. In: Xavier R (ed.) *Escravidão e Liberdade: temas, problemas e perspectivas de análise*. São Paulo: Alameda, pp. 123–142.
Spooner FC (2002) *Risks at Sea: Amsterdam Insurance and Maritime Europe, 1766–1780*. Cambridge: Cambridge University Press.
Szegö J (1987) *Human Cartography: Mapping the World of Man*. Stockholm: Swedish Council for Building Research.
Thompson EP (1977) *Whigs and Hunters*. London: Penguin Books.
Tobler WR (1987) Experiments in migration mapping by computer. *The American Cartographer* 14(2): 155–163.
Tversky B et al. (2002) Animation: Can it facilitate? *International Journal of Human-Computer Studies* 57(4): 247–262.
Wrigley EA et al. (1997) *English Population History from Family Reconstitution 1580–1837*. Cambridge: Cambridge University Press.

10

ZOOCENTRIC TEXTS AND CARTOGRAPHIC CONTRADICTIONS

Sally Bushell

Introduction

The use of a map hierarchises, asserting man's superiority over other beings who can never experience the world in this way and whose sense of geography, place and relative distance is far more limited. At the same time, cartographic knowledge alters one's relationship with the environment. Does it create a kind of false knowing then—a knowledge that distances and reduces not just visually, but ontologically? This chapter is concerned with the kind of existence represented within literature which privileges a more direct, experiential mode of spatial being, one which concerns a more purist dwelling in place, and that potentially extends beyond the human.

In the late twentieth and early twenty-first centuries, increased environmental concern has led to a shift from an anthropocentric world view to a more biocentric one that values all forms of life equally. The phenomenological concept of 'dwelling' articulated by philosophers such as Martin Heidegger (Heidegger, 1971: 343–364) is enlarged to encompass all life-forms, human and non-human. For literary studies, the emerging subfields of eco-criticism and posthumanism have allowed critics to address environmental issues powerfully through representation of animal life. In works of this sort, the presence of a map (obviously intended for the only life-form able to read it), alongside a narrative with a cast of non-human characters, creates a cartographic contradiction between text and paratext that I seek to explore here.

Cartographic contradiction often reflects a tension between two dominant spatial modes: itinerary route mapping and abstract use of a plan view. *Not* being able to read a map for humans presupposes geographical knowledge of a restricted extent (the distance which can be travelled in a day or so) and is strongly associated with a premodern mindset. Jeremy Harwood reminds us that 'the majority of medieval people were unlikely to have travelled more than 15 miles from their birthplaces during the whole of their lifetimes' (Harwood, 2006: 51). He continues: 'If they did venture outside their own localities, medieval travellers seem to have used descriptions and gazetteers rather than maps' (Harwood, 2006: 51). In medieval times, the itinerary, involving grounded experience of landmarks and waypoints along a route, was the preferred means of negotiating place and space. In a literary work,

 DOI: 10.4324/9781003327578-13

this tour/map distinction corresponds to readerly immersion in the sequential unfolding of a text and the (relative) simultaneity of an endpaper map.

Rabbits can't read maps

The potential for contradiction between endpaper and text is felt most strongly in zoo-centric texts such as Richard Adams' *Watership Down* and Henry Williamson's *Tarka the Otter*. In *Watership Down*, a group of rabbits are forced from their home (as a result of human development) and, following a vision, set off to find a new Eden in a heroic rabbit-epic. However, the moment they behave in an *untypical* way, by leaving their warren and travelling a relatively long distance in order to establish a new one, they immediately run into trouble: 'this was a strange, forbidding land. Trees, herbage, even the soil—all were unfamiliar. . . . The very plants were unknown to them' (Adams,1973 [1972]: 60). For them, 'home' represents a constant point of return that itself determines how far they will venture. Without the protection of the burrow, and with nowhere to escape to, they frequently panic, either scattering wildly or freezing in the face of danger. To a large extent, then, they are, *where* they are, (a literal 'being-there') and once they lose that reliance upon location, or become able to shift location, they also begin to lose their identity as 'rabbit'.

At the front of *Watership Down*, a small-scale map provides a total picture of the rabbits' movements and habitats. It is worth comparing the map that appears in the first edition (1972) with the map that replaces it in the second and all subsequent editions (1973).

Both contain largely the same information, but the first map is extremely realist in style, whereas the second is far more literary. The original map by Marilyn Hemmett (Figure 10.1) is presented in colour on gridded paper with numbered squares. For the hardback first edition (the only place it appears), it is given at the back of the book on a large, glossy, fold-out A4 sheet emphasising it as materially distinct from the text. Features such as railway lines, roads and boundaries appear as on a standard Ordnance Survey map with key sites and places named upon it—and of course many of the fictional locations do correspond to actual sites in the world ('*Great Litchfield Down*'; '*Ashley Warren Farm*'), as does '*Watership Down*' itself. Other elements, corresponding to details of the narrative that might appear on a literary map (e.g. 'Where they crossed the railway line'), are given alongside the map as 'Map References' with a corresponding letter and number: 'D 16'. This requires the reader to read the map in a standard way by cross-referencing.

The map for the second edition (Figure 10.2) was designed by famous illustrator, Pauline Baynes. It is clearly derived from the first map and represents the same area, but is given in black and white within the body of the text (between Contents page and narrative) at the front of the book. It displays greater selection of information and a more illustrative style, felt most strongly in the depiction of the woods. References to events are now much more explicitly linked to different parts of the story and circled on the map, making it easy to trace the rabbits' journey. All of this makes the visualisation feel far more like a natural *extension* of the narrative, rather than a map of another real-world place that corresponds to the fictional one, as in the first example.

When we compare these two maps directly, we can see that they draw attention to the 'world-within a world' nature of the narrative and the way in which it embeds an anthropomorphised rabbit fantasy within human geography. From this perspective, the first map is emphatically realist, but the second points towards a different experience of the same place, for example in the use of a rabbit head for the compass points (non-standardised with East

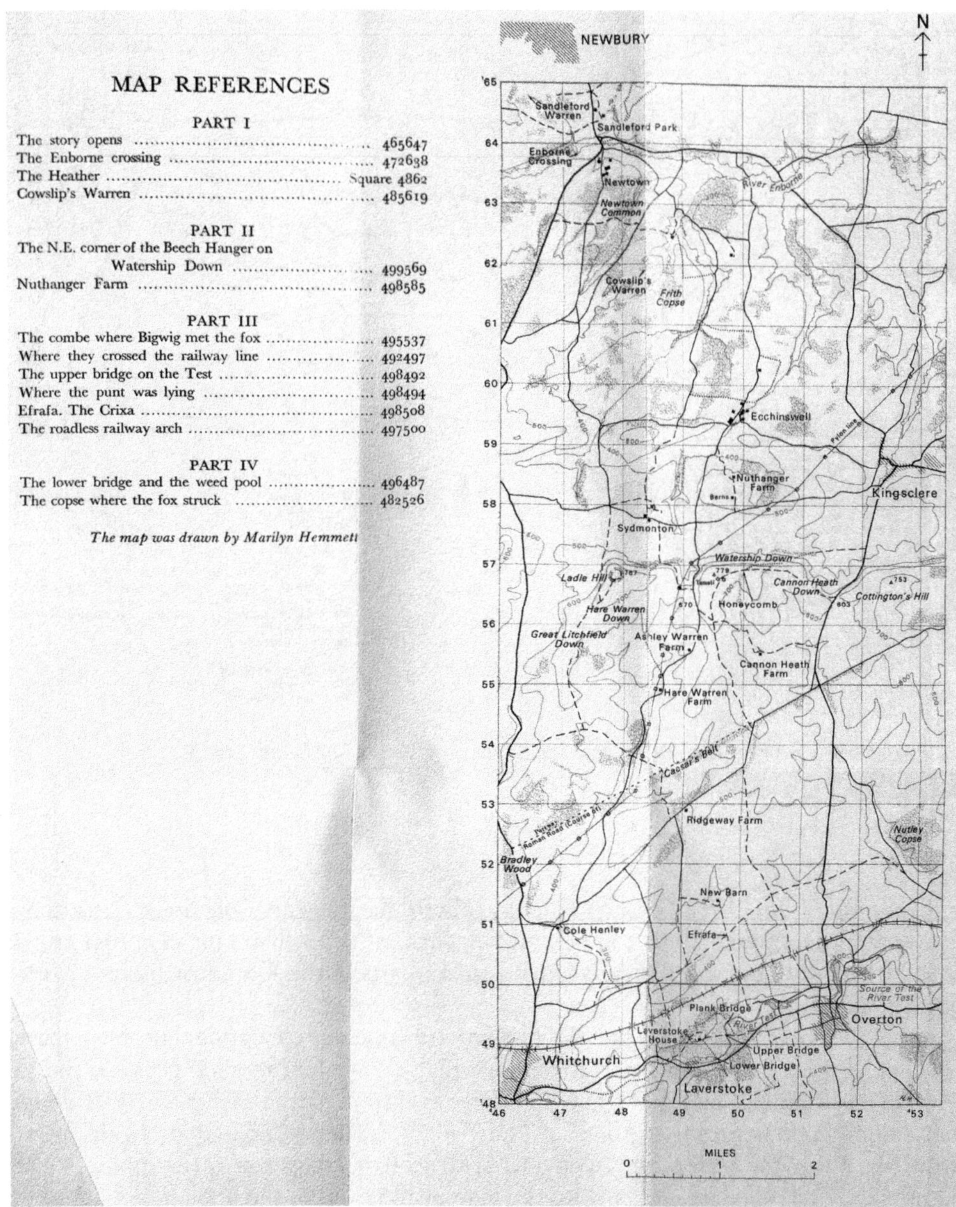

MAP REFERENCES

PART I

The story opens 465647
The Enborne crossing 472638
The Heather .... Square 4862
Cowslip's Warren 485619

PART II

The N.E. corner of the Beech Hanger on Watership Down 499569
Nuthanger Farm 498585

PART III

The combe where Bigwig met the fox 495537
Where they crossed the railway line 492497
The upper bridge on the Test 498492
Where the punt was lying 498494
Efrafa. The Crixa 498508
The roadless railway arch 497500

PART IV

The lower bridge and the weed pool 496487
The copse where the fox struck 482526

The map was drawn by Marilyn Hemmett

Figure 10.1 First Edition Map for *Watership Down* (1972).

Source: Reproduced under Fair Use

at the top) or the warren/carrot design of the scale bar. Nevertheless, both maps remain a form of complex higher level knowledge only of use to mankind and the situating of them at the front of a story that is narrated from a 'rabbit's eye view' is inherently problematic. It draws attention to the deeper contradiction involved in anthropomorphic representation of the animal world by the author, in and through human communicative structures. In

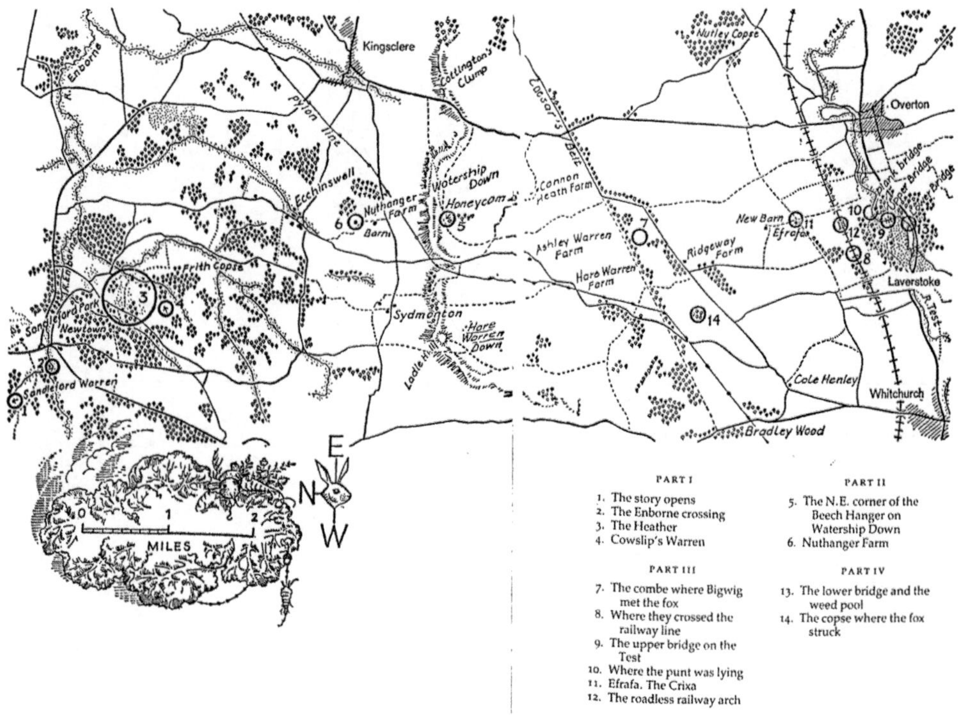

Figure 10.2 Second Edition Map by Pauline Baynes for *Watership Down* (1973).
Source: Reproduced under Fair Use

his discussion of *Tarka the Otter* as biopolitical critique, Stewart Cole draws attention to the 'essential unrepresentability, in human language, of other-than-human consciousness' (Cole, 2019: 549), and this difficulty extends to all parts of the zoocentric literary work—both text and map.

Within the narrative, Adams handles this unresolvable representational problem mainly through the use of a doubled narrative voice that allows him to slide between the two perspectives of animal and man. The first voice is distanced and factual (in part derived from a non-fiction source that Adams acknowledges).[1] The second is that of third person narration. When the two perspectives merge this often involves a cartographic register and implicitly refers the reader back to the map at the front of the book. For example, as the rabbits undertake a second journey away from Watership Down, in search of females for their new warren, description of the landscape begins in one register, (that of the rabbit's viewpoint), and then opens up into larger geographical knowledge and human place naming:

> Two or three hundred yards away and directly across their line, a belt of trees ran straight across the down, stretching in each direction as far as they could see. They had come to the line of the Portway—only intermittently a road—which runs from

north of Andover, through St Mary Bourne with its bells and streams and watercress beds, through Bradley Wood, on across the downs and so to Tadley and at last to Silchester.

(*Adams, 1973 [1972]: 249*)

Here, the more contained world view of zoocentric dwelling slides into the second as it hits its physical limits 'as far as they could see'. The narrator then imposes larger human cartographic knowledge onto the scene, pulling back to enlarge and relocate them.

Another example occurs at the start of Part 2 when the rabbits first arrive at their mythological place of safety and climb the hill, unable to see what lies ahead of them. The passage begins in the voice of factual commentary and directly addresses the reader:

> A man walks upright. For him it is strenuous to climb a steep hill, because he has to keep pushing his own vertical mass upwards and cannot gain any momentum. The rabbit is better off. His forelegs support his horizontal body and the great back legs do the work. . . . On the other hand the man is five or six feet above the hill-side and can see all round. To him the ground may be steep and rough but on the whole it is even; and he can pick his direction easily from the top of his moving, six-foot tower. The rabbits' anxieties and strain in climbing the down were different, therefore, from those which you, reader, will experience if you go there . . . they were feeling the strain of prolonged insecurity and fear.
>
> (*Adams, 1973 [1972]: 128*)

Here, the narrative makes a physical and perceptual comparison between the moving body of the two animals in a way that cleverly distances the reader from both, but it ultimately enables us to visualise an alternative bodily perspective from our own. The description of the walking man as a 'six-foot tower' functions transitionally to move into the rabbit's-eye view, which allows the reader to imagine the radical spatial redefinition created by walking upright. In turn this reminds us that the *reason* Watership Down is such a wondrous place to the rabbits—'Oh Frith . . . are you sending us to live among the clouds?' (Adams, 1973 [1972]: 129)—is because it provides them with a perspective that is unachievable to them ordinarily, although taken for granted by man: 'They were on top of the down . . . where the turf ended, the sky began' (Adams, 1973 [1972]: 129, 130).

The map at the front of the book also partakes of this doubled perspective that is deeply embedded in the text. This has a practical purpose in tracking the rabbit's journey across the narrative. Key points within the rabbit micro-epic correspond to natural (rivers) and man-made barriers shown on the map and referenced in the key below: 'The Enborne Crossing'; 'Where they crossed the railway line'. At first sight, this could be understood to *diminish* the rabbits' achievement—since it powerfully reminds us that what is an epic trek to a rabbit is less than a day's walk to a man (the map scale suggests they move 8 miles at most). But such distancing seems to be at odds with the microcosmic nature of the rabbit world in which the text immerses the reader so powerfully. In fact, the reader is rendered complicit by the presence of the map. The re-situating of Watership Down within the context of man reminds us of the fragility of this rabbit paradise which is always subject to human acts upon the landscape. Since the map is purely readerly, it functions to implicate the reader by creating a split self in terms of empathetic engagement with the fictional non-human characters and the harsher truth that, 'The refugee rabbits, who are in search of a different home, reveal the devastating impact of modernization upon the nonhuman world' (Battista, 2012: 158).

What should not be on the map

The second example of cartographic contradiction as a result of a zoomorphic perspective that I want to consider here occurs in *Tarka the Otter* by Henry Williamson (1927). Representation of the animal world is a lot less anthropomorphic than *Watership Down*, but this makes the presence of a map in later editions all the more surprising.

Roglan and Scholtmeijer, in an article comparing environmental and animal advocacy approaches to literature, define *Tarka* as an example of the latter and argue that it moves beyond an ecocritical approach since 'humankind is cast as the enemy . . . appearing incomprehensibly vicious from the animal's perspective' (Raglan and Scholtmeijer, 2007: 132). Animals are not given speech, only make natural sounds, and the book is predominantly descriptive in mode and unemotive in its depictions of otter life. Stewart Cole claims that 'in its apparent decentering of the human subject in favour of the agential otter, Tarka stands as an early posthumanist corrective to the seeming inherency of anthropocentricism to literary endeavor' (Cole, 2019: 543).

Within the text, the relationship between man and otter is treated as a cyclical occurrence, but one with human rather than natural agency. It is something to be endured and hopefully survived, but it is as much a life of the river as the coming of winter. The communal event of the hunt replays itself down the river, recurring eight times throughout the book, starting at the moment of Tarka's birth and ending with the 'long hunt' of ten hours at his death. At such points, the natural beauty of the surroundings is juxtaposed with the organised, sinister determination of the human game and the spatial nature of the otters' habitat (embedded within an anthropocentric world but coexisting with it) is turned against them. So, bridges become places for motorcars to park and look-outs to be set, while the shallower parts of the river are deliberately blocked by a line of men to stop the otter escaping. The slow swim along the river with regular stopping points at various known holts, which characterises the otter's natural existence, is transformed into a complex negotiation of the known space, looping back to confuse the trail or attempting to break through:

> Up and down the pool he went, swimming in midstream or near the banks, crossing from side to side and varying his depth of swimming as he tried to get away from his pursuers. Passing under the legs of hounds, he saw them joined to their broken surface-images. From under water he saw men and women, pointing with hand and pole, as palsied and distorted shapes on the bank.
>
> *(Williamson, 1978 [1927]: 245)*

The unequal nature of the pursuit, and the casual on-looking of the humans, who enable the kill but do not undertake it themselves, is profoundly discomforting for the reader—as is the fact that one animal is killed by another bred for this purpose by man. This is enhanced still further by the otter's fatalistic, unemotive attitude towards the events that he is unwittingly caught up in: 'Tarka felt neither fear nor rage against the hound. He wanted to be left alone' (Williamson, 1978 [1927]: 244).

Where a more cartographic sense of the world occurs within the narrative, it is presented in highly naturalised terms, as in *Watership Down*. So, for example, we are given a brief, bird's-eye view of the otters:

> Old Nog the Heron, beating loose grey wings over Leaning Willow Island as the sun was making yellow the top of the tall tree, saw five brown heads in the salmon pool.
>
> *(Williamson, 1978 [1927]: 66)*

At one point, as a young Tarka plays in the pool a local account of orientation based on direct, bodily experience is given:

> He turned and turned many times in his happiness; east towards Willow Island and the water-song, west towards the kingfishers' nest, and Peal Rock below Canal Bridge, and the otter-path crossing the big bend. North again and then south-west, where the gales came from.
>
> *(Williamson, 1978 [1927]: 63)*

Again, as for *Watership Down*, the narrator's cartographic knowledge (east, west) is layered onto the experiential and sensory geography of the animal. The young otters sometimes remember places they have been before, but frequently they do not. Age brings a degree of habitual remembered experiential geography to the creatures, so that Greymuzzle is introduced in the following terms:

> She knew every river and stream that flowed north into the Severn Sea. She had roamed the high cold moors of three counties, and had been hunted by four packs of otter hounds. Her name was Greymuzzle.
>
> *(Williamson, 1978 [1927]: 103)*

As this concise account of an otter's life in the region suggests, the dominant spatial mode in the narrative is that of a naturalised itinerary mapping of the river, with its uppermost zone in the moors, the dangerous, human-occupied, middle section and the lower tidal and coastal region. This region of 'The Two Rivers' is represented in detail on an endpaper map (Figure 10.3).

The 1964 Nonesuch Press edition was illustrated at the author's request by a well-known wildlife artist (Barry Driscoll) who also designed the map.[2] This map pushes cartographic contradiction to its limits. On the one hand, it is entirely accurate to the North Devon coastline that corresponds to Tarka's habitat, and to the geographic region—The Land of the Two Rivers'—near Torrington (where Henry Williamson studied the otters on which he based the book). On the other hand, the map is extremely stylised, in two colours with wild animals roaming across it at a disproportionately large scale—deer, foxes, birds, otters and hounds. The compass bearing is represented by a raven with outspread wings.

These creatures dominate the visual image as if asserting their ownership of place over and against that of the human and threaten to turn the map into an illustration, to be viewed rather than read. Visually then, the map respects the zoocentric nature of the text. However, it also contains language.

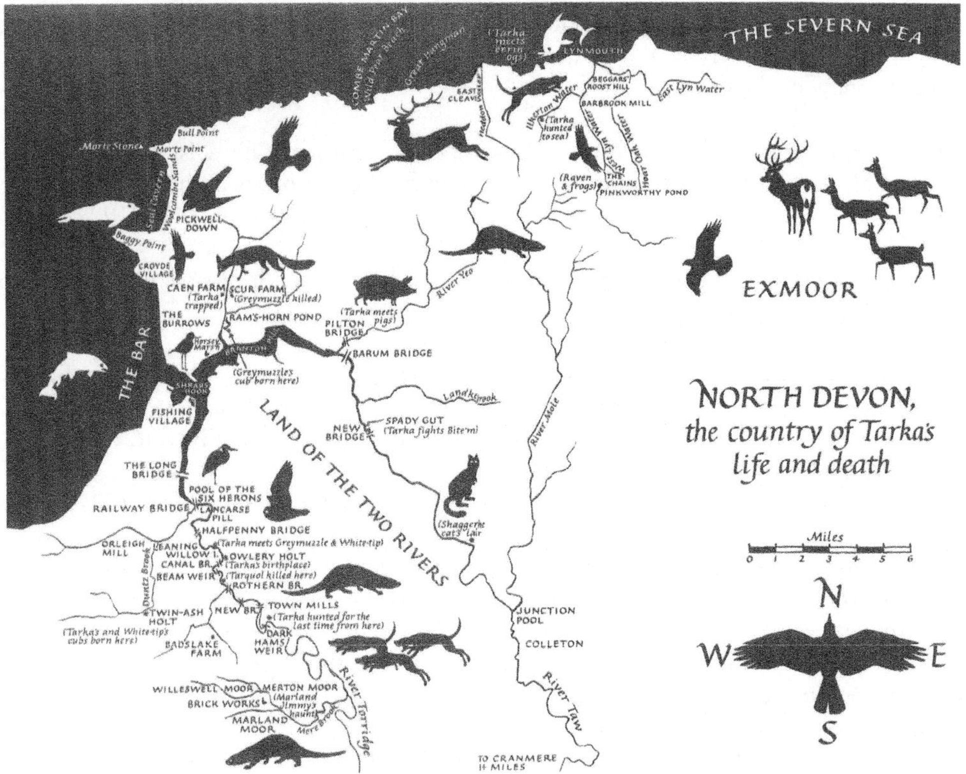

Figure 10.3 Endpaper Map for *Tarka the Otter* Nonesuch Edition (1964).
Source: Reproduced with kind permission of the Henry Williamson Literary Estate

As is characteristic of a literary map, the information given on the map is determined by reference to key events across the narrative. This is strongly centred on the otters' lives lived up and down the rivers, with comments such as: 'Tarka hunted to sea' or 'Tarka's and White-tips cubs born here'. A potential problem immediately arises here in terms of the simultaneity of map representation as opposed to the sequentiality of narrative revelation. This is also felt in the title of the map ('North Devon, the Country of Tarka's Life and Death') which itself echoes the full title of the book: *Tarka the Otter: His Joyful Water-Life and Death in the Country of the Two Rivers*. This map does not hold back at all, and also contains stark entries such as: 'Greymuzzle killed' and 'Tarquol [Tarka's cub] killed here' at the specific locations of Scur Farm and Owlery Holt. At such moments it is as if the language on the map (emphatically human) partakes of the non-emotional acceptance of death by the animal. For the reader, it is surprisingly shocking to read such entries. This suggests that, even for a *literary* map, we expect the information it provides to be controlled and objective and when it is *not* the charge of emotion is powerful.

Tarka's own death eludes both visual and verbal representation. The final entry for him on the map reads only 'Tarka hunted for the last time from here'. This respects the fact that in the narrative his end is depicted as an act of reclamation by the river, witnessed by men:

> And while they stood their silently, a great bubble rose out of the depths, and broke, and as they watched, another bubble shook to the surface, and broke; and there was a third bubble in the sea-going waters, and nothing more.
>
> *(Williamson, 1978 [1927]: 251)*

Conclusion

The resolution of cartographic contradiction ultimately lies with the reader who can unite the juxtaposed meanings held in text and paratext. The reader is capable of moving repeatedly between imaginative immersion in the fictional world and abstract visualisation of that world provided by the endpaper image. Without a doubt, the presence of the map heightens and draws attention to the fact that zoocentric narratives are an impossible representational paradox—'to portray an otter's agency in so inescapably anthropocentric a medium as human language involves taking translational leaps at every juncture' (Cole, 2019: 543)—but this is not necessarily a bad thing. It reminds us that, however much we identify with them imaginatively in the experience of reading fiction, we remain part of the problem for the creatures of the text and it is our anthropocentric worldview that threatens and impinges on theirs. The map faces outward in a desire to act beyond the limits of merely literary place and bring about change in the world beyond. Thus, the very contradiction that such a map presents, when set alongside the text, becomes a powerful part of the overall meaning of such a work and of the 'animal advocacy' (Raglon and Scholtmeijer, 2007: 121) that lies at its heart.

Notes

1 Lockley, 1954.
2 In fact, two maps were initially created for the first edition of *Tarka the Otter* by T. A. Falcon with the author's approval and liking, but these were not published. See Williamson, 1987.

References

Adams R (1973 [1972]) *Watership Down*. London: Penguin Books.
Battista C (2012) Ecofantasy and animal dystopia in Richard Adams' *Watership Down*. In: Baratta C (ed.) *Environmentalism in the Realm of Science Fiction and Fantasy Literature*. Newcastle upon Tyne: Cambridge Scholars, pp. 157–167.
Cole S (2019) The animal novel as biopolitical critique: Henry Williamson's *Tarka the Otter*. *Interdisciplinary Studies in Literature and the Environment* 26(3): 540–569.
Harwood J (2006) *To the Ends of The Earth: 100 Maps That Changed the World*. London: Marshall Editions.
Heidegger M (1971) Building, dwelling, thinking. In: Heidegger M, *Poetry, Language, Thought*, Hofstadter A (trans.). New York: Harper Colophon Books, pp. 343–364.
Lockley RM (1954) *The Private Life of the Rabbit*. Meopham, Kent: André Deutsch.
Raglon R and Scholtmeijer M (2007) Animals are not believers in ecology: Mapping critical differences between environmental and animal advocacy literatures. *Interdisciplinary Studies in Literature and the Environment* 14(2): 121–140.
Williamson A (1987) A note on the Falcon maps. *Henry Williamson Society Journal* 16: 19–21.
Williamson H (1978 [1927]) *Tarka the Otter*. London: The Bodley Head.

11
WRITING WITH MAPS

Julien Nègre

Introduction

Within the cartographic humanities, the field of literary cartography covers an increasingly wide range of topics, genres and types of interactions between maps and literary texts (Engberg-Pedersen, 2017; Tally, 2017). Since the 1970s (Muehrcke and Muerhcke, 1974), several studies have offered useful typologies of these interactions (Rossetto, 2014), depending on *when* the map interacts with the text (before, during or after the writing process) and *where* the interaction takes place (on the frontispiece, inside the book or on online blogs, for instance; Mantzaris, 2022: 4). In this chapter, I am specifically interested in what happens when writers use real maps in the course of their work. By 'real' maps, I mean pre-existing cartographic documents of real-world places—as opposed, for example, to the fictional maps of imaginary places studied by Bushell in *Reading and Mapping Fiction* (2020) (see also Bushell, this volume).

Until the late 2000s, reflections on the cultural and literary significance of maps were largely dominated by what Rossetto calls a 'Foucaultian-Harleian' deconstructive approach (2014: 523), mainly characterised by suspicion and critical distance. In the meantime, the spatial turn in the humanities led many scholars to adopt an increasingly 'spatialised' idiom to discuss phenomena that, in essence, had little to do with space. The term 'mapping', specifically, has at times been used as a broad metaphor to describe a diverse range of textual practices and critical pursuits. As noted by Brückner, the use of a cartographic lexicon in literary studies became increasingly common, even as scholars discussed 'the map' in often abstract terms, to the point that 'only a small number of studies engage[d] with actual maps' (2016: 184 endnote 41). The interactions between cartographic documents and literary texts have been viewed primarily as confrontational. Maps were approached with scepticism—as untrustworthy tools of domination and manipulation—and, very often, in an abstract manner that subsumed all maps under the vague appellation 'the map' (Edney, 2019: 20).

Critical cartography played a key role in teaching literary scholars to be wary of maps. It has now become well internalised that a map's 'supposed stability and self-containedness' (Edney, 2019: 40) and its 'seductive transparency and alleged scientific objectivity'

DOI: 10.4324/9781003327578-14

(Engberg-Pedersen, 2017: 5) should not be taken at face value. Such a critical stance, however, might prove too restrictive when it comes to discussing literary map use and unpacking the different effects touched off by the contact between map and text. As early as 1995, Lawrence Buell hinted at this problem in a passage of *The Environmental Imagination* devoted to 'Map Knowledge and Place-Sense':

> the challenge of present-day interpretation is to deconstruct the official map. Set the white man's maps against each other. Oppose the official version with the map of Indian claims. . . . My own chief concern, here, however, is not with mapping or official geography as the site of clashing political or cultural systems. I am more concerned with its role as a provoker of environmental consciousness . . . official maps look more complexly productive than when seen merely as agents of cartographical imperialism.
>
> *(Buell, 1995: 270)*

Buell recognised that the utilisation of cartographic documents by writers generates a plethora of significant effects and representations that cannot be solely reduced to a clash between the map's 'imperialistic' discourse and the text's assumed anti-hegemonic stance. Recent paradigm shifts in the field of map studies have enabled literary scholars to adopt a more positive, creative and 'complexly productive' approach towards map-text interactions. While this list is not comprehensive, there are at least three recent orientations that provide new opportunities to think about what it means to write with maps. First, a renewed focus has been placed on a 'processual' reading of mapping practices (Edney, 2019: 44), which takes into account a map's creation and how its agency is defined not only by the maker but also by the user (Kitchin et al., 2009). Second, attention has been directed towards the materiality of maps, which are viewed not only as repositories of spatial knowledge but also as objects or 'things' with specific physical properties (Jacob, 2006: 98; Rossetto, 2019). Finally, scholars have gained more insight into the trajectories followed by maps after they are created and how they are produced within a specific 'spatial discourse' (Edney, 2019: 45), yet are often used and read in different contexts, and how the way they circulate amounts to a 'social life' (Brückner, 2017b, see also Brückner, this volume).

Such insights allow us to move away from a hermeneutics of suspicion and confrontation and to unpack the effects generated by the interactions between maps and texts in a more fruitful manner. Instead of discussing the relations between literature, on the one hand, and cartography or the map, on the other hand, we are encouraged to examine the specificity of cartographic documents and pay attention to their trajectories. What is their history? How and why were they used originally, and how do we perceive them today? What kind of physical and sensorial engagement did they require from their users, and how does that shape the way they produce(d) meaning? While it is true that 'maps are semiotically stable only within specific spatial discourses' (Edney, 2019: 44), this observation should not be interpreted as a wholly unfavourable evaluation of their inherent limitations. Rather, the same assessment can be viewed in a positive light as an affirmation of a map's ability to elicit multiple meaningful representations.

In the following pages, I examine some of the most salient effects produced by the interaction between a literary text and a pre-existing map, drawing most examples from U.S. literary history. Maps have played a central, yet ambivalent, role in American history, serving as both instruments of exploration and appropriation. At times, they have also been used as tools of resistance or to promote environmental awareness. Due to this intricate role, maps have profoundly influenced the manner in which literary texts envision space and geographical knowledge within the North American context. The nature and the intensity of this interaction may vary depending on the situation, from texts that contain no reference to specific maps but are 'cartographically inflected' (Brückner, 2017a: 325), like Kerouac's road novels, to texts that are known to have been written with the help of maps, like James Fenimore Cooper's *The Prairie*, published in 1827, to those in which precise maps are explicitly mentioned and/or discussed (*Moby-Dick*, published in 1851; *Trace*, published in 2015). These effects can be broadly categorised as: anchoring; (playing with) authority; and redefining navigation and regimes of visibility. I conclude by arguing that cartographic interactions establish a continuum between the text and the readers.

Anchoring

The most obvious effect produced by a text based on a map is a sense of anchoring. Whether explicitly mentioned in the text or never acknowledged, the use of maps enables the writer to conform to the parameters of the real world, providing the narrative with credibility, verisimilitude and accuracy. One typical example is Andy Weir's *The Martian* (2011), which belongs to the genre of hard science fiction. Readers of this genre expect a high level of scientific and factual accuracy, and the storyline must be perfectly plausible. As expected, the novel includes an actual map of the Martian surface, and multiple details in the narrative show that Weir used maps of Mars to situate his story. The area in which the action takes place is presented as credible (it would make sense for NASA to send a mission to this geologically significant region—see Weir, 2011: 121), topographically convenient (the Mawrth Valley allows the protagonist to leave the plain and reach the crater where he will be rescued—2011: 287), and even already historicised (the character retrieves the nearby Pathfinder lander sent to Mars in 1997 and uses it to communicate with Earth). Such accuracy is part of the reading contract that implicitly binds authors of hard science fiction to their readers, and the cartographic foundation of the narrative plays a key role in serving that purpose. However, the maps mentioned in the novel also play a more symbolic role. For example, when the protagonist, Mark Watney, finds himself navigating on Mars using crude maps and a homemade sextant, he replicates the colonialist experience of European explorers and settlers placed in a seemingly 'virgin' land but having access to the technological resources of their human/western civilisation. *The Martian* thus portrays Watney as a 'colonist' (Weir, 2011: 171)—a potentially problematic characterisation that, despite the humorous tone of the passage, suggests that beyond the heroic and entertaining Robinsonade focusing on survival, isolation and rescue, his story also raises important questions regarding the environmental and appropriative impact of human (space) exploration. *The Martian* shows that map use provides narratives with a distinct sense of anchoring even if the real places in question remain out of reach of and unfamiliar to the immense majority of readers.

Consider two other radically different examples: James Fenimore Cooper's *The Prairie*, published in 1827, and Judith Schalansky's widely popular *Atlas of Remote Islands: Fifty Islands I Have Never Set Foot On and Never Will* (2010). Cooper resorted to Major Stephen Long's maps of the Platte River region, published in 1822, to situate his novel in an area he (like most of his contemporaries) had never visited (Nègre, 2016). Schalansky, on the other hand, painstakingly designed the maps of her atlas and used these unreachable (as the subtitle of the volume makes clear) islands as a creative surface onto which she projects the micro-narratives printed on each opposite page.[1] The two projects are radically different, but they make use of maps in a similar way: in each case, maps enable the writers to create a fictional universe grounded in the physical parameters of the real world. The purpose and effect of this cartographic interaction, however, differ significantly in each case. For Schalansky, the fact that these distant islands are unreachable is what liberates her creativity and enables her to craft her stories. Cartographic anchoring elicits literary inventiveness and poetic exploration.

For Cooper, the stakes are more closely historicised and rooted in a larger reflection on the identity and the future of the young American nation, a few years before the beginning of its rapid expansion westward in the 1840s. Contrary to what he does in *The Pioneers* (located in the more familiar and already storied setting of the New England forest), in *The Prairie*, Cooper uses what is described on Long's maps as the inhospitable 'Great American Desert' as the backdrop of his story in order to explore the limits of settler colonialism. Just as on Long's maps, the setting of the novel is a barren and nearly abstract landscape that looks like an empty theatrical stage on which the plot of the novel is acted out (Cooper, 2002: 107). At the end of the book, all the white characters are expelled from the prairie and travel back east, which suggests that this geographical space is used by Cooper as a symbolical space to think about the conditions in which American westward expansion will take place. Only Natty Bumppo, a model of virtue, dignity and restraint, is allowed to remain. In such instances, the use of maps serves as a literary constraint, compelling writers to shape their narratives to the parameters of the world represented on the maps. However, as with many constraints, this limitation becomes a catalyst for literary creation.

(Playing with) authority

These texts are concerned with the world as it is. They rely on cartographic documents to establish their connection to real-world geographies, thus indicating their intention to actively engage with the world they share with their readers. In doing so, they exploit the perceived authority of maps, which possess the potent ability to appear transparent and present themselves as precise and unbiased representations of reality. However, other texts deliberately challenge this authority, incorporating a playful reevaluation of maps into their overall project and interconnected themes and depictions. Chapter 44 of *Moby-Dick*, for example, contains a footnote, seemingly added by Melville in the later stages of writing the novel, in which the narrator (or the author himself) rejoices in the similarities between Ahab's (fictional) method of hunting one single whale around the globe (compiling onto a chart a vast amount of data on whale sightings and sea currents) and the techniques used by Matthew Fontaine Maury (now considered the father of modern oceanography) to construct his soon-to-be-published 'whale chart' (Melville, 2001: 199). Melville uses Maury's chart as an argument of authority: he provides the exact references to Maury's announcement, thereby encouraging his readers to retrieve the original document, in order to establish the verisimilitude of the entire hunt. By 'pointing' at a cartographic object outside itself, the text performs a literally *deictic* gesture that relies on the chart's 'scientific' authority. But the reference to the chart also creates other echoes, especially for modern readers who interact with the novel from the perspective of an entirely different spatial discourse. For instance, the connection between Ahab and Maury created by the footnote has recently been interpreted in light of the expansion of capitalism and the extractivist overconsumption that led to the near extinction of whales (Sugden, 2014). This, in turn, reverberates with certain motifs or themes that can be found in other parts of the book, and that might appear unrelated to the issue of map use at first sight—for example, when Ishmael reflects upon the plausible extinction of the great Sperm Whale in chapter 105 of the novel.

Furthermore, the deictic gesture performed by the footnote is not as straightforward and limpid as it might seem. Before quoting from Maury's description of the projected chart, Melville calls his published announcement a 'circular', a term that is repeated three times and that implicitly connects his cartographic endeavours to the circular images used in the novel to describe Ahab's madness. The most significant detail of the footnote, however, might be the fact that the chart is described as being 'in course of completion'. Even though the footnote appears to rely on a cartographic document as on a stable and trustworthy object, in reality it is playfully pointing at an absence, a map that does not exist and that, interestingly, never coalesced into a stable and satisfying form (Maury experimented with different ways of representing his data but never settled on a definitive version). This, in turn, replicates Captain Ahab's yearning for an object that constantly eludes him (Nègre, 2022). Far from being a passing allusion to a cold and stable exterior object, the reference to Maury's chart is thus carefully woven into the symbolic and stylistic texture of Melville's writing.

Redefining navigation and regimes of visibility

Maps can also serve as crucial facilitators for the emergence of novel spatial practices, particularly by eliciting modes of navigation that would be otherwise inconceivable. Thoreau's *Cape Cod* is a good example. In a passage of his *Journal* that is related to the Cape but that Thoreau did not incorporate in his final manuscript, he notes:

> With my chart and compass I can generally find a shorter way than the inhabitants can tell me . . . when I show them a shorter way on the map, which leaves their village on one side, they shrug their shoulders, and say they would rather go round than get over the fences. I have found the compass and chart safer guides than the inhabitants, though the latter universally abuse the maps.
>
> *(Thoreau, 1906: 428)*

Cape Cod itself can be read as the textual space where this new spatial thinking is put into practice. Just as the map allowed Thoreau to identify new lines of sight and new lines of flight along which he crisscrossed the Cape on his own terms, the text becomes a meditation on the experience of limits (of human life, of the land), defamiliarisation and the extent to which Thoreau can 'put all America behind him' (the famous final words of the book; Thoreau, 1988: 215). In *Cape Cod*, as in most of Thoreau's other texts, maps become useful tools of clarification that allow him to understand his position in the world and determine his own trajectories—spatially and philosophically (Nègre, 2017).

Maps, when they join forces with literary texts that seek to question conventional representations and consensus, reveal their potential for subversive spatial thinking and resistance. Another instance of this phenomenon is Kerouac's road novels. The enumeration of place names provides the narratives with a distinct mappiness and a characteristic sense of being anchored in 'real' space. All the toponyms in *On the Road*, for instance, can be placed on a map, thus making visible the four distinct road trips narrated in the novel (Nègre, 2011). Maps, however, are rarely mentioned in the road novels; when they are, they are more commonly found in the hands of bourgeois characters like 'motorists' looking for touristic 'sights' (Kerouac, 1976: 19)—the literary ancestors of the drivers who, in DeLillo's *White Noise*, search for the 'most photographed barn in America'. Kerouac's protagonists, by contrast, know exactly where they are and which city they want to reach, but they navigate across the country by hitch-hiking, train-hopping and walking (*The Dharma Bums*), thereby delineating a non-consumerist, free and detached mode of travelling that is not submitted to the constraints and costs of modern-day motoring.

These texts are firmly grounded in the geographical knowledge provided by cartographic documents and use it to creatively explore alternative modes of navigation—even when the maps they interact with were originally part of a conventional historical narrative. In *Trace* (2015), for instance, Lauret Savoy, a scholar of Environmental Humanities and Geology and an African-American woman, blends personal memories of family road trips and the description of ancient American maps to scrutinise the points of convergence and divergence between individual identity and a corpus of shared representations of the American landscape that were historically shaped by a Eurocentric interpretation of American history (Savoy, 2015: 6, 22, 70, 94). Savoy is able to bring these conventional maps into a conversation on the intersection between racial issues and territorial history because she operates within a completely different spatial discourse than those in which the maps were created.

In the above quote, Buell notes that mapping could be 'a provoker of environmental consciousness'. Taken in a broad sense, 'environmental' can refer to everything that *environs* us. Indeed, the interaction between map and text can make visible phenomena that previously went unnoticed, and usher in a new regime of visibility. Thoreau's late manuscripts on seed dispersal and on wild fruits, for example, have brought to light some of Concord's neglected spots, like its swamps, by underlining their ecological importance and their complexity (Nègre, 2017). William Least Heat-Moon similarly reveals the stratified intricacy of a portion of the U.S. territory with his 'deep map' of Chase County in *PrairyErth*, published in 1991 (see also Roberts, this volume). Heat-Moon uses USGS maps as a structuring device to make perceptible the web of geological data, geographical knowledge, Native American legends, historical anecdotes, subjective perceptions, local gossips, etc., that define this area's identity (Heat-Moon, 1999: 15).

Cartographically inflected texts are all place-based in one way or another, but the contrary is not always true: texts with a strong and identifiable sense of place are not always shaped by cartographic interactions—they don't feel 'mappy'. *Pond* by Claire-Louise Bennett, published in 2015, for instance, is profoundly shaped by the narrator's subjective perception of the place where she lives, but neither the location nor the configuration of this specific spatiality is ever clarified in a 'cartographic' manner. Even though the protagonist does come across an ancient map of the village she inhabits, what she sees on the map is not described and this spatial knowledge does not become part of the text's parameters (Bennett, 2016: 39). Cartographically inflected texts, on the contrary, are characterised by a distinct sense of anchoring in real-world spatiality. Be they conventional narratives of 'heroic' survival (*The Martian*) or retrospective and critical assessments of how maps shape representations (*Trace*), these texts constitute a body of works that are preoccupied with the world as it is. Beyond the differences in literary genre, tone, style and orientation, the interaction with maps establishes a continuum between the texts and the readers: writing with maps signals that there exists no fault line between the world evoked by the text and the reality inhabited by the readers. They point at the existence of a shared and common place.

Note

1 I am grateful to my colleague Mandana Covindassamy for bringing Schalansky's book to my attention.

References

Bennett CL (2016) *Pond*. London: Fitzcarraldo Editions.

Brückner M (2016) The cartographic turn and American literary studies: Of maps, mappings, and the limits of metaphor. In: Blum H (ed.) *Turns of Event. Nineteenth-Century American Literary Studies in Motion*. Philadelphia: University of Pennsylvania Press, pp. 40–60.

Brückner M (2017a) Popular map genres in American literature. In: Engberg-Pedersen A (ed.) *Literature and Cartography. Theories, Histories, Genres*. Cambridge, MA: The MIT Press, pp. 326–360.

Brückner M (2017b) *The Social Life of Maps in America, 1750–1860*. Williamsburg, VA: Published by the Omohundro Institute of Early American History and Culture; Chapel Hill: University of North Carolina Press.

Buell L (1995) *The Environmental Imagination: Thoreau, Nature Writing and the Formation of American Culture*. Cambridge, MA: The Belknap Press of Harvard University Press.

Bushell S (2020) *Reading and Mapping Fiction: Spatialising the Literary Text*. Cambridge: Cambridge University Press.

Cooper JF (1992) *The Prairie*, Ringe DA (ed.). Oxford: Oxford University Press.

Edney MH (2019) *Cartography. The Ideal and Its History*. Chicago: The University Press of Chicago.

Engberg-Pedersen A (2017) Introduction. Estranging the map: On literature and cartography. In: Engberg-Pedersen A (ed.) *Literature and Cartography. Theories, Histories, Genres*. Cambridge, MA: The MIT Press, pp. 1–18.

Heat-Moon WL (1999) *PrairyErth (a Deep Map)*. Boston: Mariner Books/Houghton Mifflin Company.

Jacob C (2006) *The Sovereign Map. Theoretical Approaches in Cartography throughout History*, Conley T (trans.). Chicago: The University of Chicago Press.

Kerouac J (1976) *On the Road*. New York: Penguin Books.

Kitchin R, Perkins C and Dodge M (2009) Thinking about maps. In: Dodge M, Kitchin R and Perkins C (eds) *Rethinking Maps: New Frontiers in Cartographic Theory*. London and New York: Routledge, pp. 1–25.

Mantzaris T (2022) Understanding maps after multimodal literature: A new taxonomy. *Cartographic Perspectives* 101: 1–15.

Melville H (2001) *Moby-Dick, or The Whale*, Hayford H, Parker H and Tanselle GT (eds). Evanston: Northwestern University Press.

Muehrcke PC and Muehrcke JO (1974) Maps in literature. *The Geographical Review* 63(3): 317–338.

Nègre J (2011) Jack Kerouac—*On the Road*. Online Google Map. Available at: http://tinyurl.com/KerouacMap or www.google.com/maps/d/u/0/edit?mid=18baLVLlQRPRPPgyjldtz_0QJoLk&usp=sharing

Nègre J (2016) Écriture romanesque et inscription spatiale: *The Prairie* et les cartes du major Long. In: Derail A and Roudeau C (eds) *James Fenimore Cooper ou la frontière mélancolique: The Last of the Mohicans et The Leather Stocking Tales*. Paris: Éditions Rue d'Ulm, pp. 63–80.

Nègre J (2017) From tracing to writing: The maps that Thoreau copied. *Nineteenth-Century Prose* 44(2): 213–234.

Nègre J (2022) The world in a footnote: Examining Ahab's chart in chapter 44 of *Moby-Dick*. *Miranda* 26: 1–21.

Rossetto T (2014) Theorizing maps with literature. *Progress in Human Geography* 38(4): 513–530.

Rossetto T (2019) *Object-Oriented Cartography. Maps as Things*. London and New York: Routledge.

Savoy LE (2015) *Trace: Memory, History, Race, and the American Landscape*. Berkeley: Counterpoint Press.

Schalansky J (2010) *Atlas of Remote Islands: Fifty Islands I Have Never Set Foot on and Never Will*, Lo C (trans.). New York: Penguin Books.

Sugden E (2014) An oceanic modernity: Matthew Fontaine Maury, Ahab, and the white whale. *Leviathan* 16(2): 23–37.

Tally RT Jr (2017) (ed.) *The Routledge Handbook of Literature and Space*. London and New York: Routledge.

Thoreau HD (1988) *Cape Cod*, Moldenhauer JJ (ed.). Princeton: Princeton University Press.

Thoreau HD (1906) *The Journal of Henry David Thoreau, Vol. IX*. Boston: Houghton, Mifflin and Co.

Weir A (2011) *The Martian*. New York: Crown Publishers.

12
A PLEA FOR SLOW MAPPING

Jörn Seemann

Introduction

The image in Figure 12.1 is a large-size drawing board explaining hachuring techniques and grey-tone graduation for the representation of relief and landforms in maps. The author was the little-known Austro-Hungarian cartographer João Soukup (?—1967) who emigrated to Brazil shortly after World War I and produced 25 thematic drawing boards for his lessons in general cartography between 1949 and 1955 (Vasques, 2023). The drawings tell stories about themes such as the history of cartography, map types, cartographic conventions and projections. For each artwork, Soukup calculated the time that it took him to finish it, which was between 60 and 206 hours. In the case of the hachured relief poster from 3 December 1951, he needed 155 hours for its composition, compilation and confection. Soukup's artistic hand-drawn maps may seem unfeasible in a world of data-driven maps on the computer screen, but they are a good starting point to reflect on the speed of maps.

Can maps be measured in terms of time? Can they be fast or slow? What is a 'good' speed to make or read a map? Different from the debates on how maps can visualise time in space and space in time (e.g. Wigen and Winterer, 2020) and what characterises cartographic temporalities (Lammes et al., 2018), these questions require a deeper reflection on aspects of general map communication, mapping processes and the purposes and meanings of maps.

This chapter is not a plea to uncritically and automatically deaccelerate cartographic communication and do everything more slowly, but to find the right pace to enjoy making and reading maps instead of conceiving maps as mere objects or tools. Inspired by more recent trends in some subfields related to geography and cartography that make a call for a slower reading, mapping and conscientious appreciation of reality such as travel and tourism (Dickinson and Lumsdon, 2010), computing and the use of digital technologies (Kitchin and Fraser, 2020) and academic scholarship (Keighren, 2017), this chapter defends the idea of slow mapping that is based on the unhurried appreciation of maps on paper and in digital format and that recognise the essentially humanistic value of visual narratives and story maps (Seemann, 2023)—an approach that runs against tenors of the neoliberal fast-paced world of mapmaking. The term 'slow mapping' entails multivocal cartographies

DOI: 10.4324/9781003327578-15

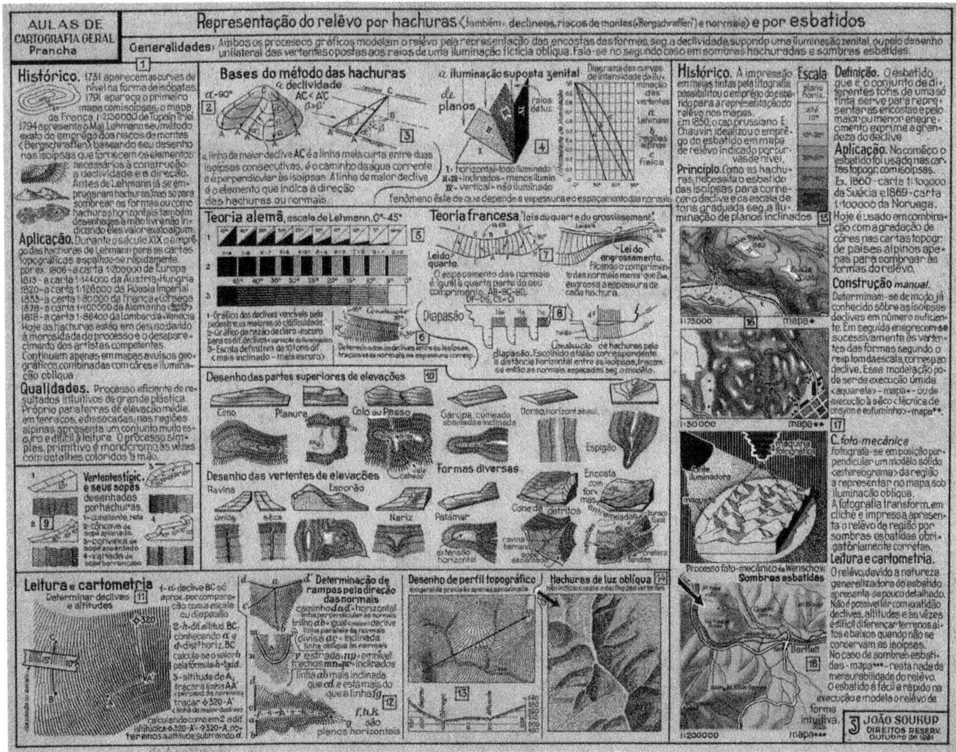

Figure 12.1 *Representação do Relevo por Hachuras e por Esbatido* (relief representation by hachuring and graduation).

Source: Soukup (1951), with permission by Antonio C. B. Vasques and Cristina Vasques

(in the plural) as unfolding and ambiguous multilayered processes considering multiple contexts and involving diverse social actors rather than static and immutable products (Cosgrove, 1999; Kitchin and Dodge, 2007).

The main argument is that a slow(er) approach to cartography can contribute to a better understanding of maps as powerful visual forms of communication through time and space that provoke emotions, give pleasure and connect us to the world—in opposition to the maps-as-objects or mere-data-repository conceptions in scientific cartography.

Maps are not what they were a few decades ago. The definitions of what a map should be and what it can do have undergone dramatic changes due to technological advances and innovations that have promoted digital and virtual forms of mapping. This has resulted in new ways of making and reading maps. Maps as forms of communication are now evaluated by their effectiveness and appear as sequences of zoomable images on the computer screen or visual versions of enormous data spreadsheets that do not give the user much time to process the information. The time required for reading a map and extracting details and messages becomes more important than space and place and their unique representation. In an age of information anxiety (Wurman, 1989), effectiveness is prioritised over affect (Wilson, 2017: 140): the faster the mapmaking and the map reading, the more time can be saved, without looking into aesthetic details. Though the benefits of the technology age cannot be denied, the acceleration of these processes has impoverished the understanding of what a map can be and what it can be used for.

Slow movements

It is generally agreed upon that the beginning of the slow movement goes back to 1986, when the Italian activist Carlo Petrini organised a protest against the opening of a McDonald's restaurant near the Spanish Steps in the centre of Rome. The initial idea was to counter fast-food culture and propose a slower (and healthier) lifestyle that allowed people to enjoy authentic food without hurry. Over the next decades, the philosophy of slowness reached other aspects of daily life, work and culture. For example, *Cittaslow* (Slow Cities) expanded the idea of slowness to high-quality urban life including the reduction of noise and traffic, the increase of green spaces and the support for local business to preserve traditions. The label 'slow' is now used in many different areas, from arts, movies, gardening, music (playing classical pieces in *tempo giusto*), religion and travelling to reading, journalism, education, science, fashion and even gaming.

Already in the nineteenth century, in a passage from an unsent letter to Georg Heinrich Ludwig Nicolovius, dated 25 November 1825, German writer and philosopher Johann Wolfgang von Goethe (Mandelkow, 1967: 159) complained about the accelerated pace of life, using the term *veloziferisch* (velociferian), a neologism that combines velocity and the devil (Lucifer) to point out the infernal consequences of an accelerated lifestyle in the age of industrialisation and capitalism:

> Just as little as the steam wagons can be dampened, this is also not possible in moral matters: the liveliness of action, the rushing through of paper money; the mounting of debts to pay debts, these are all the enormous elements which a young man currently faces. Happy he is if he is endowed by nature with a moderately calm mind.
>
> *(Mandelkow, 1967: 159–160)*

People are forced or force themselves to rush their thoughts and actions (Norton, 2021). This does not mean that slowness is the only option. It is a call for finding the right pace in one's life, or, using Thoreau's words, in a passage from Walden, 'if a man (*sic*) does not keep pace with his companions, perhaps it is because he hears a different drummer. Let him step to the music which he hears, however measured or far away' (Thoreau, 1854: 348).

Slowness does not mean deliberately slowing down and doing everything at a snail's pace. It is a reaction to the agitated life in post-industrial societies where faster is almost always considered better since speeding up means gaining time. Slowness and fastness are not necessarily opposites, where fast corresponds to 'busy, controlling, aggressive, hurried, analytical, stressed, superficial, impatient, active, quantity-over-quality' and slow means 'calm, careful, receptive, still, intuitive, unhurried, patient, reflective, quality-over-quantity' (Honoré, 2004: 14–15). Slowing down is about taking time to ponder and make '*real and meaningful connections*—with people, culture, work, food, everything' (Honoré, 2004: 14–15, my emphasis).

At first glance, it seems to be difficult to link a more relaxed lifestyle, appreciation of beauty and sense of pleasure to maps. Can maps really improve the quality of life like good food or a positive environment? Maybe not, but what is proposed in this chapter is to rediscover the essence of maps as enjoyable visual narratives (or 'real and meaningful connections'), not so different from reading a good book, cooking and eating a tasty meal or listening to relaxing music. In short, why not take pleasure in maps (Wood, 1987)?

Maps are made and read. From a philosophical point of view, cartographic slowness can be conceived as a virtue that entails processes of writing, reading and mindful interaction (Burbules, 2020). Whereas slow writing can refer to 'self-discipline and focus to force one's self to slow down, to develop as a habit a certain pace of writing and rewriting' (Burbules, 2020: 1447), slow reading is a dialogue between the reader and the author, 'a virtual, indirect and partly imagined dialogue, mediated by the present text, over which both reader and author have a legitimate claim' (Burbules, 2020: 1448) and 'a way of reflecting on what we want from reading and the possibilities and responsibilities as a reader that go with that' (Burbules, 2020: 1449). Writing and reading are inseparable parts of a slow philosophy that can be compared, metaphorically, to the use of maps or a GPS receiver:

> Following a map or using a GPS will help you get from point A to point B, but by itself it will never *teach you your way around*. To do that, Wittgenstein says, you must actually walk around the city. You must get lost a bit. You must meander, apparently directionless. You must transverse the same paths multiple times, while noticing something different about them each time you take them. You must suddenly approach a familiar spot from an entirely unexpected direction.
>
> *(Burbules, 2020: 1450, original emphasis)*

'Walking around' does not have to be a physical process. It can equally be imagined, as long as a connection is established between the map and the maker and/or user. 'Getting lost' is not a reason to panic and feel anxious but an invitation to discover and explore—the map as a terrain that can be walked.

Arts and design, as creative forms of expression, can add to a tentative definition of slow maps. Going beyond simple writing and reading, the idea of slow design addresses communication, visualisation and engagement. Strauss and Fuad-Luke (2008) have proposed six principles of slow design that can be compared to processes of mapping and mapmaking. In both fields, makers, doers and users *reveal* (pay attention to missed or forgotten experiences in everyday life), *expand* (perceive environments beyond their functionalities, physical attributes and lifespans), *reflect* (induce contemplation and provoke thoughts), *engage* (collaborate and share information), *participate* (actively engage in the design process) and *evolve* (understand products as temporal phenomena whose functions and meaning change over time).

The conception of slowness in (cartographic) design is not only limited to speed but also refers to perception, contemplation, interaction and temporal dynamics and entails '(e) motion' (feelings and movement) transforming maps into performances and embodiments rather than conceiving them as representations (Perkins, 2009).

Slow communication

Reflecting on a slow movement in cartography means revisiting the essential functions of maps in the broadest sense. Maps are a form of visual communication that involves two main actors and agents, the maker and the user. Someone creates a message and someone else receives it, reads it and interprets it. This simple idea was the dominant paradigm in cartography in the 1960s and 1970s and led to the elaboration of map communication models (Kent, 2018). Among the different proposals, Koláčný's (1969) graphic scheme of the communication of cartographic information is one of the most frequently cited examples. The graphic shows a complex system of interactions starting with the cartographer's reality and ending with the map user's reality, passing through seven filters, from the perception, selection and symbolisation of information to produce a map to the contents in the map user's mind that depend on his or her map readings skills, needs and knowledge.

Koláčný (1969: 47) criticised the separation of 'the production of the map and its utilization' showing a concern with the map user. However, his model falls short of a cultural and inclusive approach to maps and is still centred on the cartographer and the technical and formal aspects of mapmaking despite Koláčný's plea to

> study the function of cartographic information and its practical importance for society; . . . define the demands which various social groups make on maps; . . . [and] popularise the employment of maps and encourage a *fancy for maps*.
>
> *(Koláčný, 1969: 49, my emphasis)*

Recent assessment of map communication models takes into account new mapmaking technologies and the impacts of social media to transform cartography into an open-ended, accessible and democratic toolkit that establishes a 'dialogue [between mapmaker and user] through the processes of creation and engagement, which are symbiotic and interdependent' (Kent, 2018: 98). This new model of communication recognises that maps are in a constant process of becoming or unfolding (Kitchin and Dodge, 2007) that celebrates 'creativity and diversity in cartography as maps evolve and flourish' (Kent, 2018: 107). However, once again, this vision narrowly focuses on effectiveness and cognition and dismisses the wider context of distinct and alternative cultures of map use (Perkins, 2008). Understanding maps from a cultural perspective means 'allow[ing] us to answer different questions about mapping and to explore different aspects of the ways in which our society deploys the map' (Perkins, 2008: 150).

Slow mapping in practice

How does slow mapmaking work in practice? What are the qualities of slow map communication? How to read a slow map? How to read a map slowly? The answers to these questions depend much on the type of map, its format and context. For instance, in the pre-computer age, the production process was a labour-intense work that required patience and artistic skills. Maps were not created in one day. Prominent examples are the highly sophisticated relief maps by the Hungarian-born cartographer Erwin Raisz that point out the role of speed (or the lack of it) in cartographic practice (Wilson, 2021a). Every detail is thought out, every line and every pattern have a specific meaning conferring an artistic character to the map, as in the case of his 'Landforms of the Northwestern States' (Figure 12.2).

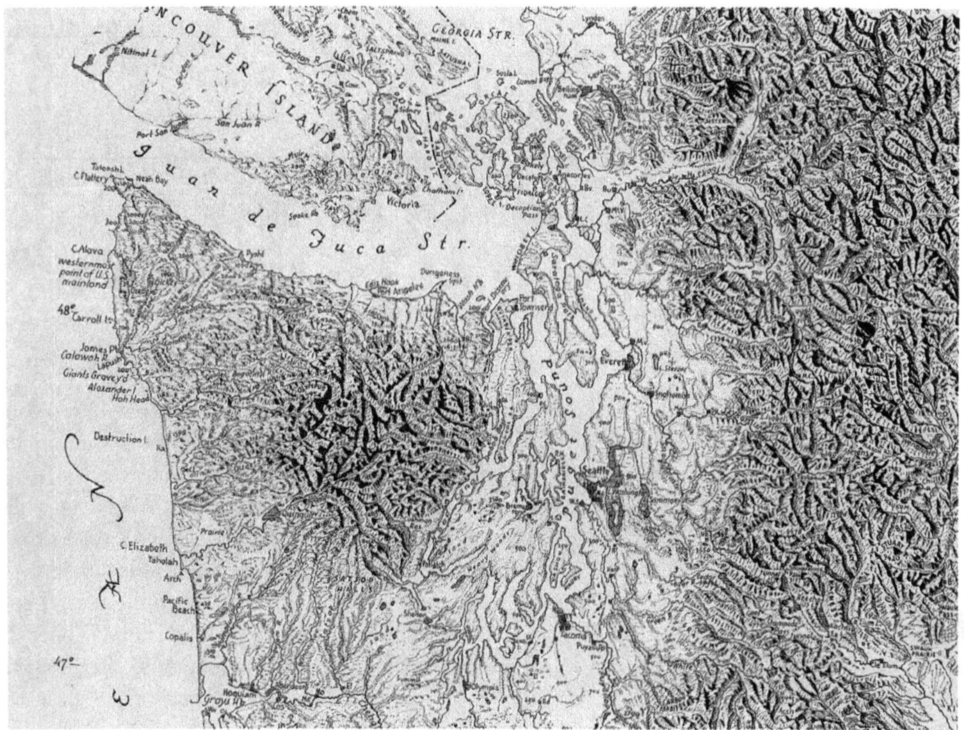

Figure 12.2 Detail from *Landforms of the Northwestern States*.
Source: Raisz (1965). Permission by Mali Raisz, www.raiszmaps.com

When looking at Raisz' maps, the reader can literally follow his tracing of lines and point out the affective dimension of drawing when the mapmaker almost lies down on the map, almost hugging it, to craft lines and other features:

> To examine a piece of hand-drawn cartography, like Raisz's Landforms, is to seemingly witness the movement of the cartographer's pen, the force of their body against the manuscript. These are the maps that compel our attention, to attend to the movements of their lines amid the static. We trace them with fingertips and eyes until they come to ends. We are moved by these lines, affectively, just as we move about through their instruction, effectively. Beyond appreciation, their grasp is elemental, a most basic correspondence between craft and utility.
>
> *(Wilson, 2017: 140)*

In poststructuralist philosophy, tracing gains a yet different meaning beyond the mere drawing of lines. For Deleuze and Guattari (1987: 12), a map is rhizomatic, 'open and connectable in all of its dimensions: it is detachable, reversible, susceptible to constant modification. It can be reworked by an individual, group or social formation'. In addition, something else that continues being ignored in scientific cartography emerges: feeling emotions, attachment and even love for maps, mapmaking and map reading—making a case for affective geovisualisation, as described in the following passage:

> Think back to the last time you lost yourself in a series of contours, shadings and cartographic symbols. Perhaps you have a treasured map at hand that will do the job for you right now (indeed, the boundlessness of imagination enables continuous return to the same map for fresh marvels). This process involves a particularly quirky engagement, pouring over the image to the extent that we may lose a conscious connection to our corporeality.
>
> *(Aitken and Craine, 2006: no page number)*

One of the most prominent manifestoes of slow mapping is a piece of fiction. In his short story *The Mappist*, writer Barry Lopez (2000) tells the story of a quest for a mapmaker named Corlis Benefideo who made maps that expressed the ephemeral, human agency and the history or histories of the environment. Benefideo settled down in North Dakota where he started working on a decades-long project to map the state and its geography in more than 1,600 maps addressing themes such as ephemeral streams, fence lines along the Missouri River, bedrocks and soils of McIntosh County, eighteenth- and nineteenth-century foot trails and the summer distribution of Swainson's hawks and their main prey, the Richardson ground squirrel (Lopez, 2000: 52–53), all maps based on fieldwork, personal observations and conversations. Benefideo draws the conclusion that maps mean illumination giving access to the world: 'The world is a miracle, unfolding in the pitch dark. We're lighting candles. Those maps—they are my candles. And I can't extinguish them for anyone' (Lopez, 2000: 55).

In 2013, the North American Cartographic Information Society (NACIS, 2023) created the Corlis Benefideo Award to recognise mapmakers and artists for their creative work on imaginative cartography that goes beyond conventional maps and methodologies to

> venture into the world of the possible . . . and explore real places from perspectives that allow us to see it fresh and full of possibility, and . . . take our established traditions of mapmaking, and indeed take fully-constructed maps themselves, and turn them on their heads to make us see ourselves anew.
>
> *(NACIS, 2023: no page number)*

The story of the mappist is closely related to the concept of deep map (see Roberts, this volume) which is almost impossible to grasp and visualise since 'deep maps are not confined to the tangible and material, but include the discursive and ideological dimensions of place, the dreams, hope, and fears of residents—they are, in short, positioned between matter and meaning' (Warf, 2015: 135).

Cultural geographer Richard Francaviglia adds to this 'deepist' perspective when he writes about how mapping a place (in his case, the Great Basin in the United States) is related to knowledge and understanding of sensory experiences such as colours, smells or the sound of wind. He affirms:

> [U]ntil we can map all the other aspects that make up this magnificent place—the erratic motion of a dust devil moving across a playa, the pungent smell of sagebrush wafting on a cool breeze after a rain shower, the sizzling feel of chocolate-colored

> basalt under a Great Basin sun—imagination will always be one step ahead of, and yet will always inspire, mapmaking.
>
> *(Francaviglia, 2005: 186)*

The passage above shows that maps do not necessarily emerge in map form or format. What matters is that they connect people and become meaningful to the user, beyond effectiveness and cognitive approaches.

Cartographic deacceleration

Slow maps are not about deaccelerating the map communication process, but about taking one's time to create maps and enjoy reading them. They imply awareness of the time employed to make and use maps without setting rigorous limits to the duration of these processes and invite both makers and readers to linger on, lean back and explore stories depicted in a paper map or a computer screen, an activity that is closely related to deep maps and mapping. Slow can mean deep. It is a matter of intensity and engagement, affect instead of effect, not of speed, maybe a return to a more leisurely conception of cartography and definitely part of the emerging subfield of Cartographic Humanities (see Introduction of the volume; Rossetto, 2021; Lo Presti, 2022).

Deep mapping, humanistic and time-consuming (i.e. slow), may be too far-fetched for ordinary mapmaking, especially in the capitalist world driven by money and determined by deadlines. However, what looms through this slow and deep approach is the possibility of affect, sense of place, attachment, empathy and admiration for beauty, giving a humanistic touch to mapmaking beyond the cold and objective lens of scientific cartography and its conventions and mask of objectivity. Evidently, some maps do not render much substance or even must be made or read speedily (thinking about a plain choropleth map of population density or a twenty-first-century interactive online map of airplane movements in real time), whereas others are like a complete book, for example, a historical map like Fra Mauro's *mappamundi* and its visual power to tell stories. Each view, each angle reveals new details, resulting in a pleasure in making and reading a map,

> in its honesty, its excitement, its subtlety, its respect for details, its consistencies, its ironies, its enthusiasms; and . . . this is as true for the reading of dictionaries and encyclopedias as it is for the reading of travel literature and biographies.
>
> *(Wood, 1987: 26)*

Mappings (i.e. everything 'around' a map), similar to the processes of writing and reading, belong to the art of 'stepping aside, taking one's time, staying quiet and becoming slow', as observed by the German philosopher Friedrich Nietzsche (1887: xi). Both philology (based on words) and cartography (based on maps) can be conceived 'as a goldsmith's art and experience with words, which have nothing but purely fine and careful work to do and which achieve nothing if they do not achieve it slowly' (Nietzsche, 1887: xi). This way, mapmaking and map use can be considered conscious acts, activities and actions that counter speed in cartography, taking into account that

> [s]low maps value the intense concentration and investment of attention necessary to understand and learn about geographical relations. . . . The slow map intensifies and

produces desire. As maps of traces, slow maps resist the temptation of a fast-map, neo-Robinsonian functionalism, in order to foster different thought and action.

(Wilson, 2017: 139)

Ultimately, the debate on slow maps must go one step further by thinking not only about cartographic slowness as a personal option or attitude immersed in subjective phenomenologies, but also about how slowness can gain relevance and be related to pressing issues in the social world (Wilson, 2021b; Lo Presti, 2022). How can a slow cartographic mode help transform society and fight injustice, prejudice and exclusion? How to counter the prejudice against slowness? Does speed atrophy our map skills? In this sense, having control of one's speed means that slowness can be 'an act of resistance, not because slowness is a good in itself but because of all that it makes room for, the things that don't get measured and can't be bought' (Solnit, 2007: no page). Making room, resisting and diving into the immeasurable, after all slow maps can be a cartographic lifestyle.

References

Aitken S and Craine J (2006) Affective geo-visualizations. *Directions: A Magazine for GIS Professionals* 4(1). Available at: www.directionsmag.com/article/3015 (accessed 8 February 2023).

Burbules NC (2020) Slowness as a virtue. *Journal of Philosophy of Education* 54(5): 1443–1452.

Cosgrove D (1999) Introduction: Mapping meaning. In: Cosgrove, D (ed.) *Mappings*. London: Reaktion Books, pp. 1–23.

Deleuze G and Guattari F (1987) *A Thousand Plateaus. Capitalism and Schizophrenia*. Minneapolis: University of Minnesota Press.

Dickinson J and Lumsdon L (2010) *Slow Travel and Tourism*. London and Washington, DC: Earthscan.

Francaviglia RV (2005) *Mapping and Imagination in the Great Basin. A Cartographic History*. Reno: University of Nevada Press.

Honoré C (2004) *In Praise of Slowness. How a Worldwide Movement Is Challenging the Cult of Speed*. New York: Harper Collins.

Keighren I (2017) History and philosophy of geography I: The slow, the turbulent and the dissenting. *Progress in Human Geography* 41(5): 638–647.

Kent AJ (2018) Form follows feedback: Rethinking cartographic communication. *Westminster Papers in Communication and Culture* 13(2): 96–112.

Kitchin R and Dodge M (2007) Rethinking maps. *Progress in Human Geography* 31(3): 331–344.

Kitchin R and Fraser D (2020) *Slow Computing: Why We Need Balanced Digital Lives*. Bristol: Bristol University Press.

Koláčný A (1969) Cartographic information—a fundamental concept and term in modern cartography. *The Cartographic Journal* 6(1): 47–49.

Lammes S, Perkins C, Gekker A, Hind S, Wilmott C and Evans D (eds) (2018) *Time for Mapping: Cartographic Temporalities*. Manchester: Manchester University Press.

Lopez B (2000) The mappist. *The Georgia Review* 54(1): 45–55.

Lo Presti L (2022) One map closer to the end of the world (as we know it). Thinking digital cartographic humanities with the Anthropocene. In: Travis C, Dixon DP, Bergmann L, Legg R and Crampsie A (eds) *Routledge Handbook of the Digital Environmental Humanities*. London: Routledge, pp. 388–403.

Mandelkow KR (1967) *Goethes Briefe. Volume 4: Briefe der Jahre 1821–1832*. Hamburg: Christian Wegener Verlag.

NACIS [North American Cartographic Information Society] (2023) *Corlis Benefideo Award for Imaginative Cartography*. Available at: https://nacis.org/awards/corlis-benefideo-award (accessed 8 February 2023).

Nietzsche F (1887) *Morgenröthe. Gedanken über die moralischen Vorurteile*. Leipzig: Verlag von E.W. Fritsch.

Norton, B (2021) Veloziferisch (velociferian). *Goethe-Lexicon of Philosophical Concepts* 1(1): 113–120.
Perkins C (2008) Cultures of map use. *The Cartographic Journal* 45(2): 150–158.
Perkins C (2009) Performative and embodied mapping. In: Kitchin R and Thrift N (eds) *International Encyclopedia of Human Geography*, vol. 8. Amsterdam and Boston: Elsevier, pp. 126–132.
Raisz E (1965) *Landforms of the Northwestern States*. Third revised edition. [Map]. Scale 1:1,300,000. Cambridge, MA: Harvard University.
Rossetto T (2021) Not just navigation: Thinking about the movements of maps in the mobility and humanities field. *The Cartographic Journal* 58(6): 1–13.
Seemann J (2023) Story maps and visual narrative. In: Richardson D, Castree N, Goodchild MF, Kobayashi A, Liu W and Marston RA (eds) *International Encyclopedia of Geography: People, the Earth, Environment, and Technology*. Chichester: John Wiley and Sons. DOI:10.1002/9781118786352.wbieg2023.
Solnit R (2007) *Finding Time*. Available at: https://orionmagazine.org/article/a-fistful-of-time (accessed 20 January 2023).
Soukup J (1951) *Representação do relevo por hachuras e por esbatidos*. Available at: www.cartografiajoaosoukup.com.br/prancha/prancha-ix (accessed 6 February 2023).
Strauss C and Fuad-Luke A (2008) *The Slow Design Principles: A New Interrogative and Reflexive Tool for Design Research and Practice*. Paper presented at the Changing the Change Conference, Turin, July. Available at: http://files.cargocollective.com/653799/CtC_SlowDesignPrinciples.pdf (accessed 30 January 2023).
Thoreau HD (1854) *Walden or Life in the Woods*. Boston: Ticknor & Fields.
Vasques ACB (2023) *João Soukup: vida e obra*. Available at: www.cartografiajoaosoukup.com.br/vida-e-obra/ (accessed 6 February 2023).
Warf B (2015) Deep mapping and neogeography. In: Bodenhamer D, Corrigan J and Harris TM (eds) *Deep Maps and Spatial Narrative*. Bloomington: Indiana University Press, pp. 134–149.
Wigen K and Winterer C (eds) (2020) *Time in Maps. From the Age of Discovery to Our Digital Era*. Chicago: University of Chicago Press.
Wilson M (2017) *New Lines. Critical GIS and the Trouble of the Map*. Minneapolis: University of Minnesota Press.
Wilson M (2021a) *Slow Maps: Cartographic Attention and the Question of Geography*. Available at: www.youtube.com/watch?v=OPfs4DzDB4Q (accessed 6 February 2023).
Wilson M (2021b) GIScience III: Questions of time. *Progress in Human Geography* 46(6): 1431–1438.
Wood D (1987) Pleasure in the idea. The atlas as narrative form. *Cartographica* 24(1): 24–45.
Wurman RS (1989) *Information Anxiety. What to Do When Information Doesn't Tell You What You Need to Know*. New York: Bantam Books.

PART 3

Mediations and intermedialities

13
A MEDIA THEORY OF (WESTERN) CARTOGRAPHIC IMAGINATION

Tommaso Morawski

Preliminary concepts for a media theory of space (without media)

In this chapter, I will explore the relationship between space, imagination and cartography with a media-anthropological (Glaubitz et al., 2014) approach, that is, according to a perspective that assigns an active and primary role to technology and media in the constitution of the human and its cultural forms. The main thread of my reflections will be the concept of cartographic imagination, and the idea of the map as both a cultural technique (Siegert, 2011) and a reference system for a media-history of space. My reflection on the concept of cartographic imagination and the map's mediality begins with a question that is as general as it is essential: What is space?

It is well known that critical discourse on space has undergone profound changes over the past decades. A series of conceptual and methodological *turns* has transformed space into one of the most dynamic and influential categories in the human and social sciences (Merriman, 2022). Yet, upon closer inspection, one could say about space the same thing that Wittgenstein, and before him Augustine, said about time: If no one asks me, I know what it is. If I wish to explain it to one who asks me, I do not know. But why is it so difficult to remind ourselves what space actually is, and to give an account of it to those who ask us? What are the anthropological reasons behind this difficulty? And what does it mean to 'see right into' phenomena in order to trace their 'possibilities' (Wittgenstein, 2009: 47)? Some good indications for possible answers to these questions are to be found in the reflections that Immanuel Kant devoted to the concept of space as a transcendental condition for the possibility of human knowledge and experience. His theory of space is, in fact, 'a media theory, though without media' (Doetsch, 2004: 24).

In the *Critique of Pure Reason*, Kant (1998: 157–159) claims that space, like time, is one of the *a priori* forms of our sensibility. That is, it is a subjective, absolutely universal and necessary condition which is the 'ground of all outer intuitions', and by virtue of which sensation can be related to 'something outside me', and objects—that is, the 'outer appearances'—can be ordered in certain relations. For Kant, then, space as a condition of possibility is constitutively paradoxical: a form of sensibility, which remains constitutively indeterminate although it is presupposed, as its most original condition and determination, by every concrete spatial experience, whatever its further determinations (geometrical and

 DOI: 10.4324/9781003327578-17

otherwise). For this constitutive indeterminacy of space, Kant provides convincing evidence when he discusses its ontological nature and argues that space (and the same is true of time) is, as a pure form of intuition, an *ens imaginarium*: 'if extended beings were not perceived, one would not be able to represent space' (1998: 383). Thus, paraphrasing Kant, space as a condition of experience in general only obtains concrete meaning by differentiating itself in the innumerable spatial contents of individual experience, that is, in particular constructs, or specific practices, which may occur in the form of both spontaneous and conscious cultural strategies. But what is the relationship between the spatial condition and what is conditioned by it, that is, the multiplicity of spatial forms of our concrete experience? Is it a strict ontological distinction, or are there implications between the two terms of the relationship that can help us better define it? In this regard, it is peculiar that in Kant's critical philosophy, so careful to preserve the purity of the transcendental condition, certain forms of spatiality nevertheless seem to claim a special status. In particular, those associated with the 'feeling of a difference' (Kant, 1996: 8) in my own body, whose activities not only decide what is above and below, left and right, in front and to the side, but also form the basis, the subjective ground, of our sense of orientation and its metaphorical meaning. Right, left, above and below are not absolute, objective and pre-existing spatial regions but the result of somatically based spatialisation activities that are 'the originals for all imaginings' (Kant, 2005: 229). In summary, Kant's philosophical investigation deserves credit for shedding light on the fundamental paradox between indeterminacy and determinacy that underlies the human standpoint on space, and for highlighting the role of the body in our spatialising activities (primarily orientation) and the connection they have with the imagination, whether they relate to our perception of space or are merely symbolic and metaphorical.[1]

In addition to showing that it is, above all, the 'intuition of *space* in which the interpenetration of sensible and spiritual expression in language most thoroughly proves itself' (Cassirer, 2021a: 148), Cassirer was the first to argue that the constitutive paradox between indeterminacy and determinacy not only concerns space as a transcendental form of sensible intuition but also invests the very historicity of space as the 'general medium' (2021a: 176) of every symbolic or cultural form:

> We cannot show how something that was previously as such aspatial achieves the quality of spatiality—we can and must, however, inquire into what way and by virtue of what mediations mere spatiality passes over into space, how pragmatic space transitions into systematic space. For it is a long way from the primary mode of *spatial lived-experience* to formed space as the condition of the intuition of *objects* and further from this intuitive-objective space to the mathematical space of measure and order.
>
> *(Cassirer, 2021b: 175)*

Cassirer is not interested in a critique of reason according to the Kantian model but in a 'critique of culture'. His goal, then, is to show how 'the content of culture, insofar as it is more than merely individual content, insofar as it is grounded in a general principle of form, presupposes an original act of spirit' (2021a: 9). For him (2013: 325–326), space as such 'does not possess an absolutely given, final, and fixed structure' but gains this structure only by virtue of the general coherence of meaning within which its construction is accomplished. That which links all these spaces is only a 'pure formal determination that is expressed most clearly and concisely in Leibniz's definition of space as the "possibility of coexistence" and as the order of possible coexistences (*l'ordre des coexistences possibles*)'. This is a purely formal characterisation that undergoes 'different kinds of realization, actualization, and concretization'. Indeed, there is not *one* space, independent of who experiences it and already determined in its structure, but *many* spaces, for as many articulations and constructive viewpoints are possible. At the same time, however, every spatial arrangement of the manifold in a coordinated whole is subject to a definite law of form, without which the single construction would not be possible. Hence, the speculative challenge of his critique is to analytically hold together the original determination of space and the history of its multiple mediations and cultural forms. But how to break such a difficulty? Given that all understanding of spatial forms is 'ultimately bound to the activity of their inner production and to the lawfulness of this production' (2021a: 19), the solution Cassirer proposes is to examine the evolution of spatial forms of order, by focusing on that 'force of creative imagination' (2013: 324) and that 'particular schematism of presentation' (2021b: 175) on which the diversity and heterogeneity of its configurations depend.

From this brief review, it emerges that spatiality, space, spatialisation, body, orientation and imagination constitute the key terms, the preliminary concepts of a media-philosophy of space without media. I will now try to take a step further and explore their intertwining from a media-anthropological perspective, that is, starting from the idea of an '*instrumental constitution* of the original *intuitive* space (through the tools of orientation)' (Stiegler, 1996: 271).

Space between *aisthesis* and *techne*

The most fruitful way to address the original instrumental constitution of intuitive space is to start from an anthropological, and more precisely palaeoanthropological framework. Within this framework, it should be easy to see that the question of technique constituted an essential adaptive resource for *Homo sapiens*, a peculiar survival strategy that defines one of the distinctive features of the human standpoint on space. Indeed, we are the only species that evolves not simply genetically but extra-genetically—or as Stiegler (2009: 4) puts it, 'epiphylogenetically': by means of 'other than life'.

For Leroi-Gourhan (1993), we perceive the surrounding world essentially in two ways: a dynamic one and a static one. Similar to other animals, early *Homo sapiens* individuals were essentially mobile. Their perception of the world was therefore linked to displacements and the ability to travel through space while becoming aware of it. Surviving and adapting in this 'itinerant space' (1993: 325) meant being able to identify with easy representational sequences a different variety of paths, objects and places, to profile, through the schematic selection of experimental data and discrete units, a complex reality that could not be directly experienced. In short, it meant developing early forms of mapping skills. Building on similar premises, Lewis (1987) in his contribution on the origins of cartography accurately illustrates the link between the emergence of *Homo sapiens*' spatial consciousness and the development of cognitive mapping systems:

> Like all animals, but far more so than most, early *Homo sapiens*, of 40,000 or more years ago, was mobile. . . . Consciousness of the world involved monitoring it for novelty for both unanticipated events in time and unexpected objects and conditions in space, which might constitute hazards or, alternatively, afford opportunities. In either case, they compelled attention. More than in other primates and far

> more than in other animals, the well-developed eyesight of *Homo sapiens* provided the necessary sensory basis for developing a spatial mental schema against which to relate these hazards or opportunities. . . . Not surprisingly, therefore, 'spatialization' was probably the 'first and most primitive aspect of consciousness', so much so that attributes of space such as distance, location, networks and area continue to pervade many other areas of human thought and language. . . . However, survival and success were not dependent only on consciousness and on response in individuals. They also depended on cooperation between individuals and within the society and on the ability to communicate between individuals and within the group, to store and transmit information and to decode it in message form. Hence the development of the several forms of language—including those for communicating spatial information—which ensured the emergence of society and the handing on of its accumulated culture to later generations.
>
> *(Lewis, 1987: 51)*

The development of early forms of cognitive mapping constitutes the most primitive aspect of our spatial consciousness. Yet, the ability to convey information about the surrounding and spatial relationships among phenomena is not unique to *Homo sapiens*. Studies of animal behaviour have shown that animals also have systems for deciphering and transmitting something like a map. What differentiates the development of human mapping skills from those of animals is above all the fact that the latter are inherited exclusively through genetics, while human ones are from the very beginning tied to technical consciousness.[2] In short, animals' mapping skills are not susceptible to rapid adaptation, remaining tied to the slow development of the biological conditions of the species to which they belong. In contrast, we humans are able to communicate and refine, depending on need and contingency, our mapping techniques, moving on a level beyond the simple biological programme. In this sense, then, humanity's specific ability to produce maps through the use of graphic methods, to translate and communicate ephemeral spatial information into a permanent and irreducible material form, marks another fundamental difference between the development of human and animal spatial consciousness. The material map performs the function of a '*sensorium communis*' (Neve, 2005: 10) that pragmatically organises, stores and transmits information in view of possible forms of collaboration. It is 'a tool for the *orientation* of knowledge' (Stiegler, 2009: 78), which makes the devolution of orientation into (technical) prosthesis possible. And, as a prosthesis, the material map 'is not a mere extension of the human body; it is the constitution of this body *qua* "human"' (Stiegler, 1998: 152).

So, to recapitulate, in line with German media theory turning Cassirer's critique of culture into a critique of media (Siegert, 2011, 2015), what I would like to suggest by adopting a media-anthropological perspective[3] on space is to rethink the problematic dualism between media and culture. Indeed, according to Siegert (2015: 9), humans as such 'do not exist independently of cultural techniques of hominization', and 'space *as such* does not exist independently of cultural techniques of spatial control'. In general terms, my thesis is that the way in which Cassirer's law of the form schematically operates, that is, by articulating a multiplicity of syntheses and mediating different spatial representations, should be examined by assuming an essential correlation between *aisthesis* and *techne*, human and non-human, organic (life) and organological (the technical extension of life in prostheses, even inorganic ones). If we assume with Kant (1998: 158) that we can 'speak of space, extended beings, and so on, only from the human standpoint', we must also recognise that humans' most common spatial experiences require 'not only a biological body, but in addition to it, or even in its place, a technical, semiotic, and artifactual body' (Siegert and Engell, 2013: 5). It is precisely the processes of spatialisation mediated by this body—a hybrid of nature and culture, biology and technology—that determine, through a feedback loop, the historicity of space as a condition of human experience. As we have seen, technology is not something that is simply added to the human and extends a ready-made and finished body but is an integral part of humanity's constitution and development. It is, yes, a prosthesis, but a prosthesis that constitutes the human as such:

> The technical inventing the human, the human inventing the technical. Technics as inventive as well as invented. . . . Or again: the human invents himself in the technical by inventing the tool—by becoming exteriorised technologically. But here the human is the interior: there is no exteriorisation that does not point to a movement from interior to exterior. Nevertheless, the interior is inverted in this movement; it can therefore not precede it. Interior and exterior are consequently constituted in a movement that invents both one and the other: a moment in which they invent each other respectively, as if there were a technological maieutic of what is called humanity. The interior and the exterior are the same thing, the inside is the outside, since man (the interior) is essentially defined by the tool (the exterior).
>
> *(Stiegler, 1998: 137–142)*

Thus, a media-anthropological approach to the question of space implies that if we want to analyse the transformative processes that invest our capacity to produce spatial syntheses—that is, to follow the historical evolution of the order of space through its different configurations—we have to start by considering our natural predisposition, rooted in the very process of hominisation, to interact and dominate the world environment through the spontaneous displacement and extension of our sensibility into something external and inorganic: an artefact. And the map, as a 'technical mediation', is an 'artefact': it is both a 'technical prosthesis that extends and redefines the field of sensorial perception' (Jacob, 2006: 11), presenting 'a schema, visual as well as intellectual, that takes the place of an impossible sensorial vision' (Jacob, 2006: 29), and a 'support of inscription that guarantees the empirical-transcendental possibility of communitisation [*communautisation*]' (Stiegler, 1996: 274).

Given these general premises of a media-anthropology of space, I can now proceed to the analysis of the mediality of (western) cartographic imagination.

The mediality of (western) cartographic imagination

The concept of 'cartographic imagination' was introduced by British geographer Denis Cosgrove (2001) in order to trace how ideas of globalism and globalisation have shifted historically in relation to changing images of the spherical earth, from antiquity to the Space Age and GIS. Cosgrove's 'cultural history of imaging, seeing, and representing the globe' (2001: 3) aims to highlight the media-historical foundation of the concept of globalisation. Underlying his genealogical approach is a simple but important observation: until the first photographs taken from space, the unity of the globe was not perceptible from its surface, and to represent it required an act of the imagination that translated, by selecting and schematising certain salient features, the invisible into the visible; an imaginative effort that had to be embodied into a technical prosthesis in order for it to give a legible graphic guise to the holistic space that remained by definition inaccessible to human eyes. It is by virtue of the anthropological necessity of such mediation that cartographic representations have emerged as the graphic-material reference system of our 'mythical space' (Tuan, 1977): a 'trace' (Derrida, 1974: 70) of our 'cartographic impulse'—that is the ability to transform graphic structures in virtual spaces of intellectual activity, and conceive cognition as a 'spatially oriented movement' (Krämer, 2016: 19–20)—that embodies and fixes the historically determined schematisations of our spatial consciousness.

This idea that mapping is an act of the imagination and that maps are the material product of this act embodied in a graphic synthesis suggests that we adapt to the graphic representations of the globe what Benjamin (2008: 23) observed in the early 20th century about the relationship between cinema and aesthetics, media and the historicity of perceptual processes: that the 'way in which human perception is organised—the medium in which it occurs—is conditioned not only by nature but by history'. Indeed, as we know from Leroi-Gourhan, graphism, understood as the capacity 'to express thought in material symbols' (1993: 207), is an exclusively human phenomenon, a direct consequence of the erect posture and of the freeing of the hands from the tasks of locomotion. It is an adaptive resource that has progressively transformed humans' creative performances and their engagement with the environment 'from technically assisted to technically dependent' (Montani, 2017: 40). The concept of technical dependency here refers to the fact that technical prostheses are not mere exterior aids but also transformations of consciousness associated with the environment and emerging in specific historical practices. Vogl (2001) points out with reference to Galileo's prospective tube that this is the transformative process that makes a simple instrument become a medium.

In line with the perspective I have adopted so far, by the term 'medium', I refer to the 'material substrate of culture' (Siegert, 2015: 3), that is, to the tools and operations that constitute the conditions of possibility—the technical-material *a priori*—of every form of knowledge and experience as well as of every cultural process. As media, maps, in fact, are 'instruments', 'spaces of representation' that 'are themselves agents of subject constitution. . . . The cartographic operations produce a subject, which correlates to them'. The analysis of maps as cultural techniques, that is, as conditions of possibility of a cultural system of meanings, thus concerns 'the way changes in cartographic procedures give rise to various orders of representation' (Siegert, 2015: 13–14). Now, it is worth pointing out that these changes in the orders of representation not only are about the processes of territorialisation and the economisation of space, or the geographical icon of the globe studied by Cosgrove, but they also relate to other media operations, such as writing processes, for which the map serves as 'an operational or imaginative matrix' (Dünne, 2011: 44).[4] This transmedial perspective on cartographic imagination, which holds together technological changes and imaginative transformation, is legitimised by the fact that we could never recognise the representational power of a medium if we grasp it in isolation, as if it were a single element. A medium, in fact, 'is that which remediates. It is that which appropriates the techniques, forms and social significance of other media and attempts to rival or refashion them in the name of the real' (Bolter and Grusin, 1999: 65). Consider as a valid example the *pinax*, the table on which the philosopher Anaximander is said to have first drawn the ecumene (Purves, 2010: 109) and compare it with other cultural techniques that were used in antiquity for the visual production and systematic organisation of knowledge, like the *tabula* on which, according to the Platonic *Theetetus*, memories were imprinted, the Babylonian *mappamundi* of the city of Sippar, or the clay tables on which scribes recorded the daily life of the Akkadian people. Well, all these artefacts, when considered together, suggest placing the space of representation of Anaximander's map in a broader medial constellation, in which different tools for the orientation of knowledge hybridise, starting from a similar material support (wax or clay). Considering its technical-material condition, then, the cultural technique of the *pinax* seems to spring from remediation processes that have deep time scales. In conclusion, as this example teaches, to take the map as the reference system for a media-history of space is to write the history of its remediations: a history that knows not only linear progress but also upheavals and the recoveries of fragments and experiences that come from the past.

Notes

1 As Montani (2021: 253) points out, the examples of symbolic hypotyposis discussed by Kant (2000: 226) show 'the *somatic rooting* of this indirect and analogical mode of schematism, as if the imagination had to attach itself firstly to a typical corporal and sensorimotor experience. . . . The nature of the intuitive reference is therefore . . . a bodily experience and *a somatic aptitude to schematization*'.

2 For Stiegler (1998: 151), technical consciousness 'means anticipation without creative consciousness. Anticipations means the realization of a possibility that is not determined by a biological program'.

3 For German media theory's media-anthropological turn, see Schüttpelz (2006).

4 One notable example is Smith's work (2008: 10), which focuses on 'how the new techniques of surveying and mapping produced a fundamental shift in the way space was imagined and the effects, in turn, of that new spatial consciousness in the ways people thought and wrote'.

References

Benjamin W (2008) *The Work of Art in the Age of Its Technological Reproducibility, and Other Writings on Media*. Cambridge and London: Cambridge University Press.

Bolter JD and Grusin R (1999) *Remediation. Understanding New Media*. Cambridge and London: The MIT Press.

Cassirer E (2021a) *The Philosophy of Symbolic Forms. Vol 1: Language*. London and New York: Routledge.

Cassirer E (2021b) *The Philosophy of Symbolic Forms. Vol 3: Phenomenology of Cognition*. London and New York: Routledge.

Cassirer E (2013) Mythic, aesthetic and theoretical space. In: Loft SG and Calcagno A (eds) *The Warburg Years (1919–1933). Essays on Language, Art, Myth and Technology*. New Haven and London: Yale University Press, pp. 317–333.

Cosgrove D (2001) *Apollo's Eye. A Cartographic Genealogy of the Earth in Western Imagination*. Baltimore and London: The Johns Hopkins University Press.

Derrida J (1974) *Of Grammatology*. Baltimore and London: The Johns Hopkins University Press.

Doetsch H (2004) Intevall. Überlegungen zu einer Theorie von Räumlichkeit und Medialität. In: Dünne J, Doetsch H and Lüdeke R (eds) *Von Pilgerwegen, Schriftspuren und Blickpunkte. Raumpraktiken in medienhistorischer Perspektive*. Würzburg: Könighausen & Neumann, pp. 23–56.

Dünne J (2011) *Kartographische Imagination. Erinnern, Erzählen und Fingieren in der Frühen Neuzeit*. München: Willhelm Fink Verlag.

Glaubitz N et al. (2014) Medienanthropologie. In: Schröter J (ed.) *Handbuch Medienwissenschaft*. Stuttgart-Weimar: JB Metzler, pp. 383–392.

Jacob C (2006) *The Sovereign Map. Theoretical Approaches in Cartography throughout History*. London and Chicago: University of Chicago Press.

Kant I (1996) What does it mean to orient oneself in thinking? In: Wood AW and Di Giovanni G (eds) *Religion and Rational Theology*. Cambridge: Cambridge University Press, pp. 1–18.

Kant I (1998) *Critique of Pure Reason*. Cambridge: Cambridge University Press.

Kant I (2000) *Critique of the Power of Judgment*. Cambridge: Cambridge University Press.

Kant I (2005) Notes on metaphysics. In: Guyer P (ed.) *Notes and Fragments*. Cambridge: Cambridge University Press, pp. 68–404.
Krämer S (2016) *Figuration, Anschauung, Erkenntnis. Grundlinien einer Diagrammatologie*. Berlin: Suhrkamp Verlag.
Leroi-Gourhan A (1993) *Gesture and Speech*. Cambridge, MA and London: MIT Press.
Lewis MG (1987) The origins of cartography. In: Harley B and Woodward D (eds) *The History of Cartography I. Cartography in Prehistoric, Ancient and Medieval Europe and the Mediterranean*. Chicago and London: University of Chicago Press, pp. 50–53.
Merriman P (2022) *Space*. London and New York: Routledge.
Montani P (2017) *Tre forme di creatività: tenica, arte, politica*. Napoli: Cronopio.
Montani P (2021) Techno-aesthetics and forms of the imagination. In: Chiodo S and Schiaffonati V (eds) *Italian Philosophy of Technology*. Cham: Springer, pp. 247–261.
Neve M (2005) "Milieu", luogo e Spazio. L'eredità geoestetica di Simondon e Merleau-Ponty. *Chiasmi international* 7: 153–169.
Purves AC (2010) *Space and Time in Ancient Greek Narrative*. Cambridge: Cambridge University Press.
Schüttpelz E (2006) Die medienanthropologische Kehre der Kultutechnicken. *Archiv für Mediengeschichte* 6: 87–110.
Siegert B (2011) The map is the territory. *Radical Philosophy* 169: 13–16.
Siegert B (2015) *Cultural Techniques. Grids, Filters, Doors, and Other Articulations of the Real*. New York: Fordham University Press.
Siegert B and Engell L (2013) Editorial. *Zeitschrift für Medien- und Kulturforschung 1. Schwerpunkt Medienanthropologie*: 5–10.
Smith DK (2008) *The Cartographic Imagination in Early Modern England: Re-writing the World in Marlowe, Spenser, Raleigh and Marvell*. Aldershot: Ashgate.
Stiegler B (1996) Être-lá-bas. Phénoménologie et orientation. *Alter*, 4: 263–277.
Stiegler B (1998) *Techniques and Time 1. The Fault of Epimetheus*. Stanford: Stanford University Press.
Stiegler B (2009) *Techniques and Time 2. Disorientation*. Stanford: Stanford University Press.
Tuan YF (1977) *Space and Place: The Perspective of Experience*. London and Minneapolis: University of Minnesota Press.
Vogl J (2001) Medien Werden. Galileis Fernrohr. *Mediale Historiographien* 1: 115–123.
Wittgenstein L (2009) *Philosophical Investigations*. Chichester: Blackwell Publishing.

14
THE MAP IN CINEMA AND CINEMA ON THE MAP

Giorgio Avezzù

The map in film: cinema as cartography

The relationship between cinema and the map is a long-standing one. It is an articulate and complex kind of connection, but I will try here to identify what seems to me to be its fundamental terms. While some simplification is inevitable, I am equally committed to preserving the sense of interdependence between the various dimensions at stake. For indeed, talking about cartography in relation to film is not quite the same as talking about it in relation to any other cultural product: several discourses that have surrounded the medium over the years have claimed the existence of a 'special' connection between the moving image and the map.

First and foremost, maps can be found in films and thus be the object of representation. They can be part of the narrated world and be 'used' by the characters in a scene—and then be 'diegetic' maps—or they can be 'extra-diegetic' and represented in graphic form, for example in animated sequences (as, sometimes, in the opening credits), outside the space of the story but for the purpose of locating it, that is, giving spectators the narrative's coordinates. There are countless famous examples throughout the history of cinema, from *Casablanca* (1942) to *Raiders of the Lost Ark* (1981), from *Fitzcarraldo* (1982) to *The Lord of the Rings* film series (2001–2003). In fact, as it has been argued, cinema's ability to make fluid zooms in some of its animated maps, to smoothly traverse different scales and to dissolve maps into establishing shots or vertical tracking shots, would even have prefigured features of virtual globes and functions of digital cartography (Caquard, 2009). And it did so many decades in advance, as early as the 1930s, for example the film *M* (1931). On the other hand, arguably, the possibility of 'varying focal length' has been a feature of geographical atlases for centuries, allowing the viewer to move progressively closer to or further away from the terrain, and it is precisely because of this peculiar way of guiding vision that the atlas has been defined as a form of cinematic 'découpage' *ante litteram* (Jacob, 1992: 106), almost suggesting an affinity with the rules of continuity editing of classical Hollywood cinema.

Therefore, there are those who argue that it is cinema that foreshadows cartography, while others appear to argue rather the opposite. Obviously, the relationship is not so linear and unambiguous, and nevertheless, it seems very close and almost intimate. Cine-maps

 DOI: 10.4324/9781003327578-18

themselves are not interesting merely because they would anticipate something that cartography will only experiment with later. The presence of maps in the body of films is never neutral nor inert: rather, it hints at something intrinsic to the medium itself. At least, that is a common claim among those who acknowledge a special connection or even a 'collusion' between geography and cinema, as French geographer Lacoste (1976) posited. Some argue that there's actually a relationship of 'co-extension' between maps and films: 'A map underlies what a film is and what it does. . . . A corollary is that films are maps insofar as each medium can be defined as a form of what cartographers call "locational imaging"' (Conley, 2007: 2). The belief that cinema, since its origins, and perhaps especially then, has been a geographic medium is held by many scholars and commentators, up to the present and for more than a century, namely since Häfker (1914) devoted a first volume to the subject. The invention of cinema should be placed in the context of the spatio-temporal compression of modernity, that is, within a series of technical achievements in the conquest of world space that contributed to the creation of an image of the globe as an explorable and representable totality. This would have turned the viewer into an all-seeing subject, in an age characterised by a frenzy of vision that was precisely geographical in nature. Therefore, from this perspective, cinema and geography share a planetary, globalist mission, traces of which remain, for example, in the large number of globes that appear in the logos of film production and distribution companies—think of the Universal logo. According to Shohat and Stam (1994: 106), cinema remedies the cartographic *horror vacui* of modernity and has sought to fill in the blank spaces of the globe, sometimes even with funding from geographical societies, managing to 'transform the obscure *mappa mundi* into a familiar, knowable world'.

Such is the thesis of cinema's 'cartographic impulse' (Castro, 2009) or its '*furor geographicus*' (Bruno, 2018), that is, essentially, the thesis of its mapping vocation. Maps and cartographic animations, in cinema, then serve to evoke and implicitly celebrate the technological—and geographical—qualities of the medium. Maps in film are the metaphorical manifestation of cinema's ability to represent the whole world, in other words, its cartographic mission. Castro (2011), who authored a seminal volume on the subject, elsewhere argues that 'the map *in film* would also account for this particular power shared by the cartographic apparatus and that of cinema' (Castro, 2008: 17). The argument can also be found elsewhere quite frequently: in including maps in films, 'cinema represents itself as the contemporary heir of a more ancient visual medium: cartography' (Shohat and Stam, 1994: 147). Shohat (1991: 46) insists:

> Geography was reflected . . . in map-based adventures such as travel narratives and fiction of exploration, which involved drawing or deciphering a map—often used as an instrument for the telos of rescue—and its authentication through physical contact with the 'new' land. Western cinema, from the earliest anthropological films through the *Indiana Jones* series, has also relied on map imagery for plotting the Empire, while simultaneously celebrating its own technological power . . . to illustrate vividly the topography with which the hero comes into touch. By associating itself with the visual medium of maps, cinema represents itself as the twentieth-century continuation of cartographic science.

Ultimately, it is not surprising that there may be some deep connections between cinema and cartography, as both have traditionally embraced a rhetoric—an ideology—of transparency, neutrality and objectivity. Nevertheless, the possibility that cinema's use of maps may also not be celebratory (i.e. self-celebratory) has not been much discussed. That is to say, there has not been much discussion of the fact that cinema may instead have made critical (and thus self-critical) use of maps, similar to the way geography as a discipline has critically reflected on itself, its own discourses and ideologies, over the past few decades. Indeed, upon closer inspection, several contemporary films, including mainstream and Hollywood cinema, have reflected on the malfunctioning of some of their diegetic maps and on their obsolete, artificial, or flawed character, for example in westerns and in historical, war and science fiction films (Avezzù, 2019). Arguably, the cultural role of the medium has changed, and its planetary mission needs to be rethought, as do its promises of description and intellectual, if not downright material, appropriation of the world. Films themselves seem to realise this at times, and question their own cartographic capabilities. Of course, a strong and more classical use of cartographic iconography continues to survive in contemporary cinema as well, from *Lord of the Rings* to apocalyptic films, to ecological documentaries, to films that despite being far removed from the problematic (and imperialist) genealogy of globalist imagery nonetheless believe that cinema can and should continue to provide an overview of the planet and visualise it as a whole, also in order to cope with current global ecological upheaval (de Luca, 2022).

Mapping film: cartography of cinema

The presence of the map in film then implies an idea of cinema as a fundamentally cartographic medium due to its privileged, realistic, objective relationship with the visible world: in other words, it implies that cinema should be considered as the *subject* of cartography, or as a medium that produces maps. On the other hand, cinema can also be considered, conversely, as an *object* of cartography. That is, just as films can contain maps within them—or 'be' maps as Conley puts it—cinema can also, in turn, be placed on the map and be charted. Of course, this is also the case for any kind of cultural product of which a 'distant reading' is attempted (Moretti, 2013), but for cinema, the matter takes on different nuances, perhaps precisely because of the special relationship that is believed to exist between films and specific places, since they represent them.

For example, a methodological perspective that has enjoyed some success in recent decades, one aiming at the systematic study of films from all over the planet, is the so-called *World Cinema* approach. It is the expression of a true mapping effort intended to explore and place on the maps of a veritable 'atlas of world cinema' the film cultures of the entire Earth, including those only partially and fragmentarily known in the West, also in search for 'a different cinema, whether in the hope that a purer vision may be available, or a purer people' (Andrew, 2004: 18). It seems evident that such a cartographic project of reordering and (intellectual) domination of the planet's audiovisual cultures can only end up taking upon itself the 'whole-earth' and 'one-world' rhetorics that, as Cosgrove (2001) has shown, lie at the root of any globalist discourse. That the world cinema project has a cartographic imagination in mind is also evidenced, for example, by the fact that the beautiful book series titled *World Cinema* edited by Nagib and Ross for I. B. Tauris and Bloomsbury contains before the title page of each volume the image of a planisphere. Specifically, and consistently with the egalitarian goals of the editors and authors, it is the famous Arno Peters equal-area projection (curiously, Peters, who was also a film director, obtained a PhD in Berlin in 1945 on cinema as a means of propaganda). These are publications that often emphasise the dimension of realism, from Greek to Argentinean cinema, from Iran to Thailand, which possibly demonstrates how discourse on cinema as the subject of cartography can end up reinforcing and justifying discourse on cinema as the object of a mapping operation.

It is likely that the aspiration to put cinema on the map also stems from a new concern about the invisibility and virtualisation of films and their circulation. That is, it may be an aspiration that arises from a preoccupation with the consistency of films as physical entities, made up of images with a referential nature and projected in designated places: 'Today, amidst digital confections tempting filmmakers and audiences to escape into the air of the virtual, world cinema brings us back to the earth, this earth on which many worlds are lived and perceived concurrently' (Andrew, 2004: 21). However, the duplicity, which we have acknowledged, between cinema as a subject and as an object of cartography can also be found in geographers' publications about film since at least the middle of the last century. For example, of the two series of articles on cinema edited for *The Geographical Magazine* by Roger Manvell beginning in 1953, the first consisted in a series of contributions devoted to various national cinematographies, for example, French, Italian, Indian or Japanese cinema (introduced by Manvell, 1953), while the other series comprised articles focused on the documentary use of cinema for depicting Australia, Canada, the United States and the British Overseas Territories (introduced by Manvell, 1956).

As is evident, the desire to place cinema on the map is certainly not a recent thing, nor is it solely motivated by a reaction to the 'digital confections' that would threaten the visibility of cinematic phenomena. In fact, a basically geographical paradigm has always characterised the traditional approach of film historiography, namely the one that proceeds by national cinematographies. That is, films are usually said to be good objects for thinking and visualising nations, expressing their distinctive characteristics. They are believed to construct, and at the same time reflect, in terms of content, theme and style, different national identities. The reference, typical of recent decades, to a different scale and to a transnational or supranational dimension—'world cinema', but also 'European' cinema, etc.—has not eradicated this well-established approach, nor the deep-seated conviction that there exists a necessary—and cartographic—link that connects film images to the places where they were shot and circulate best. Therefore, the notion that the former can impart pertinent information about the latter, that is, that film images can shed light on the territories, remains steadfast and widely held.

In recent years, there appears to have been an increased interest in mapping cinema, and it also seems to have declined in different ways, with a particular emphasis by film scholars on the usefulness of GIS for visualising various aspects of cinematographic phenomena, hence for a true cartography, literally speaking (e.g. Hallam and Roberts, 2014; see also Lukinbeal, this volume). In other words, and most notably by advocates of so-called *New Cinema History*, cartographic tools have been utilised to visualise production, distribution and consumption data and even to make cinema-going practices visible. For example, the film production centres and peripheries of various nations have been mapped, as well as the structure of cinematographic exhibition in various eras; even the routes taken by cinema-goers walking to the cinema have been traced, for example, in 1950s Italy (see also Treveri Gennari et al., 2021).

One trend that seems most interesting to me is the one that, by also using cartographic tools of this kind, turns its attention to the subnational dimension of analysis and which considers a 'regional turn' in film studies to be necessary (Marlow-Mann, 2017). It is an approach that goes beyond the traditional 'national' one to the study of film cultures and takes a different, and in a sense opposite, path from those which attempt a similar overcoming but on a supranational scale instead. For it is true, as mentioned just above, that supranational approaches have not usually censored the relevance of a proper geographical dimension in changing their scale of reference, and that they have often continued to take for granted the existence of a relationship of necessity between images and places. But it is also a fact that one of these approaches, and by far one of the most popular and powerful, argues exactly the opposite. This approach contends that the planetary success of Hollywood cinema derives from its textual 'transparency', namely, from being geographically neutral and thus allowing anyone around the world to see themselves reflected and to project something of their own culture into its films (Olson, 2009). This would enable American cinema to travel freely around the globe and be accepted everywhere in the same manner, homogenising tastes, creating cultural dependence and erasing local viewing inclinations (Fu and Govindaraju, 2010). By contrast, a study of the geographic specificities of audiovisual consumption at a subnational level can serve to disprove such an assumption and show how such homogeneity does not really exist, and that, for example, the penetration of what is purported to be 'international' or 'global' cinema actually encounters varying degrees of 'friction' in different regional contexts. For example, this happens in Italy, as certainly elsewhere: a science fiction film will be more widely seen in some regions, and a romantic comedy in others (see Avezzù, 2022a). A cartography of film consumption can be very useful, therefore, in inferring the diversity of identities and taste cultures of different territories based on differences in consumption behaviour—namely, to investigate the persistence of territorial differences.

A study of subnational geographies of consumption first and foremost requires a large amount of data that can describe the regional dynamics of distribution and theatrical viewings. Among the challenges of film and media studies for the coming years is undoubtedly to acquire more familiarity with data and its analysis, which has never been the focus of these disciplines, so to speak. However, when it can dispose of sufficient data with adequate granularity, such an approach can make it possible to achieve a new and deeper understanding of the industrial and economic aspects of cinema and the cultural and identity aspects that go with it. It can, for example, succeed in making visible the fragmentation of national identities from the fragmentation of consumption, thus making clear that the power of images, that is, their ability to attract visions, still remains locally differentiated. And thus emphasising, at the same time, the validity of that relationship of consequentiality, or necessity, between images and territories, which after all is also what underlies the 'logic of the map' (Harley, 1989), or 'cartographic reason' (Farinelli, 2003). In doing so, a cartography of audiovisual consumption can finally challenge concepts that we too automatically and uncritically tend to resort to. It is the case of 'international' cinema, which I have already mentioned, and which instead penetrates with different ease depending on local inclinations and attitudes. But not only that. Are we sure, for example, that there is an 'Italian cinema' today? Or that it ever existed? That is, is there a 'domestic' cinema seen in a homogeneous way throughout the country? If there is one thing that characterises the geography of film consumption in Italy, in fact, it is precisely its extreme variability and its great regional concentration, and the fact that depending on the films, different local tastes and audiences are being activated all the time, for domestic cinema (and to a lesser extent, depending on the genres, for imported films). The cartography of theatrical consumption shows us this unequivocally (see e.g. Figure 14.1). 'What is a national cinema if it doesn't

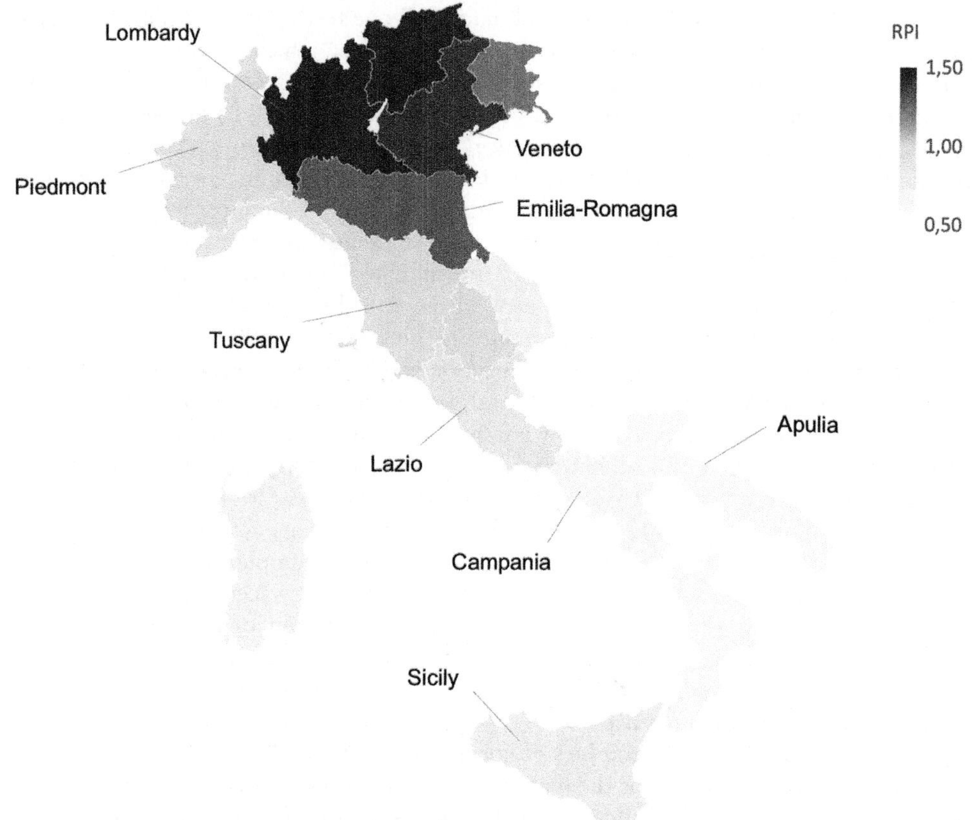

Figure 14.1 Regional popularity of *Call Me by Your Name* (Luca Guadagnino, 2017). Author's elaboration based on Cinetel data on theatrical admissions. The regional popularity index (RPI) normalises the regional share of admissions to the median figure (for each region) of the 500 most viewed domestic films for the years 2000–2020. A value of 1.5 indicates a popularity 50% higher than the median regional value, while one of 0.5 denotes a popularity 50% lower than normal.

have a national audience?' Higson (1989: 46) asked in a crucial article on the concept of national cinema. While for some territories—Italy—we have managed to outline a general picture of the situation and something of its historical evolution (Avezzù, 2022b), much still needs to be done for other national cinematographies—or rather, for other territories or for other parts of the world, since the existence of any 'national cinema' should not be taken for granted without verifying the national homogeneity of its audiences.

Conclusion

When speaking of 'cinematic cartography', we are referring to the various relationships that can occur between cinema and cartography, which, as stated from the beginning, are articulated and multifaceted. They concern both the level of representation, that is, that of so-called filmic texts, and the levels of production, distribution, consumption and circulation of images. That is, the textual dimension, the cultural dimension, as well as the one related to industry and market, etc. But these do not seem to be entirely independent planes. If cinema can be put on the map, and indeed it almost demands to be mapped, this is also because cinema has in it, intrinsically, something cartographic. If cinematic phenomena can be represented well on maps, this is because cinema happens in space and because films help us construct places and give them meaning. Recent analyses of film consumption have shown that movie viewings are never evenly distributed and that in different parts of the planet, as well as in different parts of specific countries, viewers tend to want to watch movies that show places and landscapes that are close to them. This is a theme that has prompted the interest of media economists, who have theorised concepts of 'cultural congruence', 'cultural proximity', or 'cultural discount'. But above all, this is clearly a topic that concerns the necessity of cinema from an anthropological point of view. Cinema is a fundamental orientation tool allowing people to better situate themselves in space, to feel present, to give themselves a place on the map.

References

Andrew D (2004) An atlas of world cinema. *Framework* 45(2): 9–23.

Avezzù G (2019) Cinema and the crisis of cartographic reason. In: Lukinbeal C, Sharp L, Sommerlad E and Escher A (eds) *Media's Mapping Impulse*. Stuttgart: Franz Steiner Verlag, pp. 67–85.

Avezzù G (2022a) The market for foreign cinema in contemporary Italy: A geography of film consumption. *GeoJournal* 87(Suppl 1): 73–84.

Avezzù G (2022b) *L'Italia che guarda. Geografie del consumo audiovisivo*. Roma: Carocci.

Bruno G (2018) *Atlas of Emotion: Journeys in Art, Architecture, and Film*. New York: Verso.

Caquard S (2009) Foreshadowing contemporary digital cartography: A historical review of cinematic maps in films. *The Cartographic Journal* 46(1): 46–55.

Castro T (2008) Les cartes vues à travers le cinéma. *Textimage* 2. Available at: www.revue-textimage.com/03_cartes_plans/castro1.htm (accessed 15 February 2023).

Castro T (2009) Cinema's mapping impulse: Questioning visual culture. *The Cartographic Journal* 46(1): 9–15.

Castro T (2011) *La Pensée cartographique des images. Cinéma et culture visuelle*. Lyon: Aléas.

Conley T (2007) *Cartographic Cinema*. Minneapolis and London: University of Minnesota Press.

Cosgrove D (2001) *Apollo's Eye: A Cartographic Genealogy of the Earth in the Western Imagination*. Baltimore and London: Johns Hopkins University Press.

De Luca T (2022) *Planetary Cinema: Film, Media and the Earth*. Amsterdam: Amsterdam University Press.

Farinelli F (2003) *Geografia. Un'introduzione ai modelli del mondo*. Einaudi: Torino.

Fu WW and Govindaraju A (2010) Explaining global box-office tastes in Hollywood films: Homogenization of national audiences' movie selections. *Communication Research* 37(2): 215–238.

Häfker H (1914) *Kino und Erdkunde*. M. Gladbach: Volksvereins-Verlag.

Hallam J and Roberts L (eds) (2014) *Locating the Moving Image: New Approaches to Film and Place*. Bloomington and Indianapolis: Indiana University Press.

Harley JB (1989) Deconstructing the map. *Cartographica* 26(2): 1–20.

Higson A (1989) The concept of national cinema. *Screen* 30(4): 36–47.

Jacob C (1992) *L'Empire des cartes. Approche théorique de la cartographie à travers l'histoire*. Paris: Albin Michel.

Lacoste Y (1976) Cinéma-géographie. *Hérodote* 2: 153–159.

Manvell R (1953) The geography of film-making. *The Geographical Magazine* 25: 640–650.

Manvell R (1956) Geography and the documentary film. *The Geographical Magazine* 29: 417–422.

Marlow-Mann A (2017) Regional cinema: Micro mapping and glocalisation. In: Stone R, Cooke P, Dennison S and Marlow-Mann A (eds) *The Routledge Companion to World Cinema*. New York: Routledge, pp. 323–336.

Moretti F (2013) *Distant Reading*. London: Verso.

Olson SR (2009) *Hollywood Planet: Global Media and the Competitive Advantage of Narrative Transparency*. New York: Routledge.
Shohat E (1991) Imaging terra incognita: The disciplinary gaze of empire. *Public Culture* 3(2): 41–70.
Shohat E and Stam R (1994) *Unthinking Eurocentrism: Multiculturalism and the Media*. London and New York: Routledge.
Treveri Gennari D, O'Rawe C, Hipkins D, Dibeltulo S and Culhane S (eds) (2021) *Italian Cinema Audiences: Histories and Memories of Cinemagoing in Post-War Italy*. New York: Bloomsbury.

Filmography

Call Me by Your Name (2017) Directed by Luca Guadagnino. [Feature film]. Rome, Italy: Warner Bros. Pictures.
Casablanca (1942) Directed by Michael Curtiz. [Feature film]. Burbank, CA: Warner Bros. Pictures.
Fitzcarraldo (1982) Directed by Werner Herzog. [Feature film]. Berlin, Germany: Filmverlag der Autoren.
M (1931) Directed by Fritz Lang. [Feature film]. Berlin, Germany: Vereinigte Star-Film GmbH.
Raiders of the Lost Ark (1981) Directed by Steven Spielberg. [Feature film]. Los Angeles, CA: Paramount Pictures.
The Lord of the Rings (2001–2003) Directed by Peter Jackson. [Feature film]. New York, NY: New Line Cinema.

15

THE ANTITHETICAL CARTOGRAPHIES OF GEOSPATIAL CINEMA

Chris Lukinbeal

Introduction

Cinema is the antithesis of cartography. Where cartography deploys scale to construct a memetic representation of the real, cinema takes advantage of scale's illegibility (Doane, 2022) or schizophrenia (Lukinbeal, 2012a) to distort, enhance and disrupt. Scale's multivalence produces and affects emotive and embodied representations and responses. These scopic regimes (cinema, cartography) reflect the full development of the cartographic paradox (Pickles, 2004): that the science of map projection and the development of linear perspective mutually arose together and aided each other's development. The paradox is where projectionism configures the 'god trick of seeing everything from nowhere' (Haraway, 1988: 581) while perspectivalism is a grounded topographic view from below. These two techniques are paradoxically relate to the tripartite Ptolemaic tradition of geography, chorology and topography. Whereas Ptolemy's geography is reflective of today's cartography (by dealing with the world as a whole and the view from above), cinema derives from the Ptolemaic scopic regime of topography (the grounded view from below and like chorology it deals with the world as parts), that gave rise to landscape paintings and later technologies associated with linear perspective. In contrast, GIS is more reflective of chorology as it most frequently deals with the world's regions.

Whereas cartographic scale validates scientific accuracy and contrives mimesis, cinema uses scale as a device that can appear, emphasise or project sensorial characteristics that disrupt, shock and destabilise visual meaning. Where Ptolemy's geography is reflective of cartography, or the mapping of the world, chorology maps the parts. Chorology is often associated with the scale of landscape or region and is reflective of the drone's eye or bird's-eye view from above (Kullman, 2016). Topography, or place-writing and place-making, is best captured through the cinematic event (Fletchall et al., 2012). The culmination of cartography and linear perspective are respectively geographic information systems (GIS) and cinema. This chapter further mobilises the cartographic paradox (Lukinbeal, 2010) by arguing for an antithetical cartography that merges these scopic regimes and their technologies. Geospatial cinema is already here and comes in many forms. 'Full Motion Video for ArcGIS' is a product where one can 'analyze and manage georeferenced video and motion data in your maps' (ESRI, 2022). ESRI CityEngine was also responsible for the production

 DOI: 10.4324/9781003327578-19

of Las Vegas in *Blade Runner* 2049 (ESRI, 2018). Think of all the summer blockbuster movies where towns, cities and countries are destroyed, those rely on 3D cartographic modelling.

I begin this chapter by reviewing linear perspective and relate it to the history of measure, scale and the making of accurate 2D representations. This first section lays the groundwork for a comparative discussion of scale, prospect and aspect in cinema and cartography. Following this, I examine four components of antithetical cartography including (1) producing affective moving geovisualisations that use cinematic language, form and technique; (2) conducting spatial analysis on films and film production; (3) using GIS in film permitting, production logistics and reporting; and (4) using WebGIS for film tourism and historical preservation.

Linear perspective and landscape

Linear perspective is a method of constructing a representation based on a frame and grid to produce the illusion that proportion between objects is maintained. Where Leon Battista Alberti (1435) was the first to document linear perspective, Filippo Brunelleschi (1377–1446) was a Florentine architect and engineer who observed that from a fixed point of view parallel lines appeared to converge at a single point in the distance known as the vanishing point. Brunelleschi applied this optical principle to a painting of the Florentine Baptistery. He then put a hole through the painting where the vanishing point occurred. Next, he drilled a hole in the base of a mirror. Then, holding the mirror backward in front of his face, he aligned the hole in the mirror with the hole in the picture so that he could see the painting in the mirror. After this, he could move the painting out of the way to be able to see how well his perspectival painting of the Baptistery aligned with the actual Baptistery. The remarkable demonstration allowed people to observe the powerful technique of linear perspective in constructing realistic images.

According to Edgerton (1975: 56), 'linear perspective has come to be regarded as unaesthetic since it implies the primacy of objective realism over true artistic subjectivity'. Cosgrove (1985: 45) notes that landscape, as a way of seeing, is based on linear perspective and deployed to exercise power over space: 'First applied in the city and then to a country subjugated to urban control and viewed as landscape'. Landscape as unaesthetic, as a stage, frames what the author wants us to see and frames out all else. For Mitchell (2003: 8):

> The transition, one could say, is from a city of public spaces, a city of people at home in the city, not wealthy, but not abject either, to a city as *landscape*, a place devoid of residents, a landscape in which people are only visitors and welcome mostly as consumers (especially of imagery). The city-as-landscape does not encourage the formation of community or of urbanism as a way of life rather it encourages the maintenance of surfaces, the promotion of order at the expense of lived social relations, and the ability to look past distress, destruction, and marginalization to see only the good life (for some) and to turn a blind eye towards what that life is constructed out of.

Like cartography, perspectival landscape becomes a tool to colonise and police space.

Central components of linear perspective are the frame, the horizon line, orthogonal lines and the vanishing point. The vanishing point is 'the illusion in ordinary vision that the parallel edges of objects stretching away from the eye seem to be converging at an infinite point on the horizon' (Edgerton, 1975: 25). Where the horizon line sets the balance and tilt of the representation, orthogonal lines connect the frame to the vanishing point and thereby help orient one to the optimal gaze. Scale in a perspectival system is incalculable as it is based on the distance from the viewer's eye to the ever-receding vanishing point. Cinematic language codifies scale using anthropometric measures, a system of measure based on the human body. The close-ups reference a face, medium shot being the waist up, full shot shows the full body, and the American shot cuts characters off at the knees while providing a context of the individual in the landscape. Cinematic shot scale determines how far one can view (such as in the extreme long shot) or how much of the body is shown in the frame. Anthropometric measures were widely used before the implementation of the Treaty of the Meter (Lukinbeal, 2015).

The rediscovery of Ptolemy's *Geography* in the European Renaissance is associated with the foundations of western geography and cartography. However, it is also cited among art historians for being one of the first uses of distant-point perspective. Edgerton (1975: 104) argues that Ptolemy's *Geography* was 'the first recorded instance of anybody—scientist or artist—giving instructions on how to take a picture based on a projection from a single vantage point representing the eye of an individual human beholder'. However, Alpers (1983) makes the argument that the Albertian perspective and Ptolemy's perspective are different. With Alberti, we are looking through a framed window onto the world with the vanishing point directing the gaze. In contrast, 'Ptolemy and the distance-point perspective conceived of the picture as a flat surface, unframed, on which the world is inscribed. The difference is a matter of pictorial conception' (Alpers, 1983: 138). I disagree with Alpers, because maps are 'framed' in three different ways: first, through the use of cartographic element 'map frame' which sets both the viewing extent and mapped area extent. Similarly, the medium of the map acts as a frame by setting the viewing extent. Finally, the resolution at which the map is captured and rendered for display also acts as a frame. The resolution acts as a frame because it provides the level of detail at which a mapped area can be accurately viewed. Both maps and paintings can apply distant-point perspective to create realism using representational frames, however, the results vary depending on scale, prospect and aspect.

Scale, prospect and aspect

Cartographic scale is defined as the relationship between map distance and ground distance. One inch on a map reflects one ground mile but only when the numerator and denominator are the same units of measure. Thus, a verbal scale of 'one inch on the ground equals one mile on the map', equals a fractional scale of 1:63,360. To make the numerator and denominator the same units, the conversion of miles to inches (63,360 in = 1 mi) is required. Cartographic science depends on scalar accuracy to maintain a map's authority. However, when evaluating accuracy, it is common cartographic practice to record the amount of scalar *inaccuracy* occurring on the map as a predicate for validity and precision. Furthermore, cartographic scale only allows for accuracy of angles (conformal) or equal areas because distortion is inevitable in the map projection process. In contrast, cinema will deploy scale to destabilise space. As Eisenstein (1949: 34–35) argues, true proportion in film 'is by no means the correct form of perception. It is simply the function of a certain form of social structure'. For cinema 'the laws of cinematographic perspective are such that a cockroach filmed in close-up appears on the screen one hundred times more formidable than a hundred elephants in medium-long shot' (Eisenstein, 1974 in Doane, 2003: 96).

Prospect and aspect addresses looking, locating and positioning in cartography and cinema. Prospect is the action of looking from a specified position to orient distance and direction in a scene. Cosgrove (1985: 55) notes that in Italian, the term *prospettiva* combines both prospect and perspective and involves commanding the sight/site of the view. By the mid-seventeenth century, prospect was synonymous with landscape. In cartography, landscape painting was central to topographical surveys that were prospectus of the science and wealth of a land. When a painting or film uses linear perspective, the vanishing point becomes the invisible origin for spatial organisation in the *mise en scene*. Whereas landscape paintings establish the prospects of the scene in a single shot, cinema requires the prospects of the scene be navigated in motion to allow the narrative to unfold in a temporally and spatially coherent manner. In cartography, the prospect of the view is framed by the areal extent and resolution of the map. Historically in European cartography, the map frame was decorated with filigree, landscape, people and colonial scenes, or the prospects of the map. But more broadly, the prospect of cartography is best reflected by Harley (1989: 2), who stated: 'Our task is to search for the social forces that have structured cartography and to locate the presence of power—and its effects—in all map knowledge'. With prospect, we are concerned about power, the gaze and visuality.

Aspect references a way of locating and positioning surfaces to capture the transformation of a Keplerian objective picture to an Albertian subjective one; or from the world as 'seen' to the world as 'scene' or socially constructed. A Keplerian picture 'represents the frame of the visual field and thereby encloses a representation of the world seen, or more simply, a representation of vision' (Friday, 2001: 353). Cinematic space is defined by the *aspect ratio* of the camera and structured by the *frame*. Cinematic space is in the frame (*mise en scene*), outside the frame (scene), or in-between frames (montage) (Lukinbeal, 2010).

Aspect in cartography references locating and positioning scalar accuracy on a developable surface (plane, cone or cylinder) using standard points or lines to note where there is a one-to-one relationship with a model of the world (geoid, ellipsoid or sphere). The cartographic aspect determines the reader's perspective of the global graticule, because standard lines and points shape how the graticule and subsequent landmasses will look. There are three aspects commonly used in cartography: equatorial, transverse or oblique (Figure 15.1).

We typically see a world Mercator projection with a standard line on the equator (or equatorial aspect) with Greenland portrayed as ten times the size of Mexico, Tissot circles[1] larger at the poles and smallest near the equator, and meridians and parallels meeting at right angles. The Mercator projection is most frequently used for its high levels of accuracy in a transverse aspect—following a meridian of longitude—as is the case for the global coordinate system: Universal Transverse Mercator (UTM). The oblique aspect is when standard lines are at an angle to parallels and meridians, like the Aleutian Islands in Alaska, which is the only deployment of the oblique Mercator projection in the United States Geological Survey (USGS) topographic series.

Four areas of antithetical cartographies

A pursuit of antithetical cartography is emerging research that embraces cinema and cartography as mutually inclusive and questions what could unfold from mobilising the cartographic paradox to produce spatial-visual representations that draw on the different

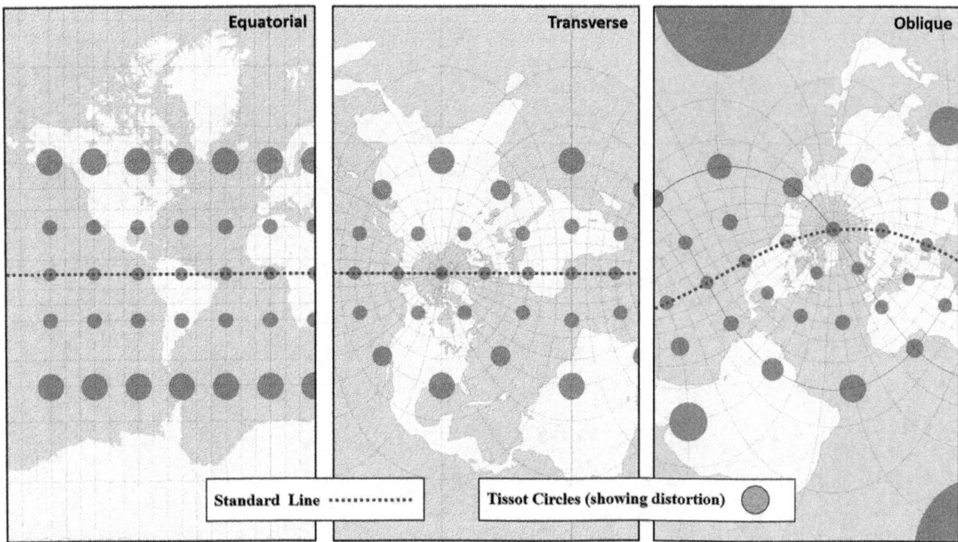

Figure 15.1 Three aspects of cartography.

Source: Adapted from: Justin Kunimune, CC BY-SA 4.0 (https://creativecommons.org/licenses/by-sa/4.0), via Wikimedia Commons

uses of scale, prospect and aspect. Antithetical cartography considers the implication and applications of commingling the different applications of prospect and aspect as well as scalar accuracy (cartography) and scalar distortion (cinema) in geovisualisations alongside montage and continuity editing. It is not too far off before ESRI-Adobe extends its cartographic collaboration to Adobe Premiere Pro, the world's most popular video editing software. Applications of antithetical cartography interweave spatial analysis, film theory and cartography. One applied outcome includes the GeoMedia lab at Concordia University Montreal's (Caquard, 2022) interactive Atlascine 4.0 that brings maps and stories together similar to ESRI's StoryMaps but using an open-source platform.

One area of antithetical cartography is the production of moving geovisualisations deploying cinematic language, form and technique. Aitken and Craine's (2006: 1) challenge for an 'affective geovisualization' that 'embrace the emotional power of cinema' can be done by linking GIS with Hollywood narrative formula models (Lukinbeal, 2018; Lukinbeal and Sharp, 2019), montage (Doel and Clarke, 2007) and accurate 3D geospatial data and display. Using Blake Synder's *Save the Cat* narrative formula for Hollywood blockbusters, Lukinbeal (2018) provides one example of how to map a movie through the production of Google Earth flythroughs of Los Angeles. Using film permit data, film stills, html coding and recorded voice-over, the *500 Days of Summer* affective geovisualisations trace the narrative formula and narrative action of this three-act film.[2] Here, the geovisualisations are both structured or 'framed' by cinematic and cartographic language, convention and form.

Antithetical cartographies deploy GIS to analyse cinematic space as well as film production and consumption. A process-based method is where film permit data was geocoded, ground-truthed, indexed to the act and scene, then analysed as a *tour* and as a *map* using ESRI ArcGIS & Business Analyst Online and feminist film theory (Lukinbeal, 2018). In

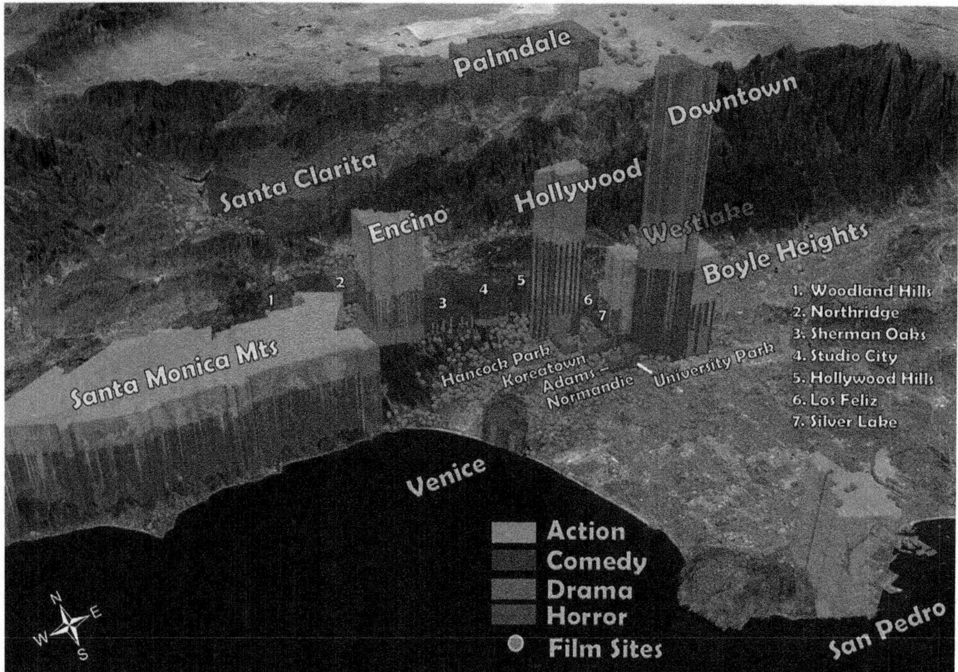

Figure 15.2 LA Feature Film Genre Density Map (color map available at: https://sites.google.com/view/lukinbeal/home/los-angeles).

Source: Cartography by Chris Lukinbeal. Data from FilmLA

contrast, Mengqian and Caquard (2019) analysed the interrelationship between social media posts from film tourists at *Shawshank Redemption* prison, while others have done spatial analysis on regional film production centres (Lukinbeal, 2012b, 2022; Roberts, 2012). Spatial film analysis of 'on-location' filming sites is possible when film-permitting information is made public. The LA Film Genre map provides an example of the types of spatial analysis that are possible when data is available (see Figure 15.2). Data for this map derives from FilmLA and 11,439 film permits from 2008 to 2012 for all feature films. Other data sources include LA GeoHub, and IMDB's (Internet Movie Database) API to gather genre and director information (Figure 15.2).

A third area of antithetical cartographies focuses on film permitting, production logistics and integration of permitting data into other smart city geospatial systems. The implementation of GIS in the film-permitting process would improve day-to-day logistics as well as producing accurate spatial data for analysis and reporting to government agencies and the public. One example of this is *NYC Filming Maps*, an 'essential tool for Producers, Location Managers and Location Scouts in the NYC Area' (NYC Film Maps, 2020). Film permit data provides the basis for a regional film GIS extending antithetical cartographies from analysis to praxis as

(1) a tool for planning and implementing logistics related to location production;
(2) integration into regional GIS databases allowing for better management,

Figure 15.3 1958 Spatially accurate 3D model of Old Tucson Studios with 3D rectified stills from the film *Rio Bravo* (color geovisualization available at https://sites.google.com/view/lukinbeal/home/old-tucson-studios).

Source: Geovisualisation by Chris Lukinbeal

> maintenance, and economic development by government agencies in a region; (3) allowing a means to produce ongoing spatially accurate reports about production like those issued by FilmLA but with more spatial detail and, (4) serve as a digital humanities archive for researchers and film buffs *if* the data is made public.
>
> *(Lukinbeal, 2022: 186)*

It is this last item of using WebGIS for film tourism, historic preservation and spatial archiving that represents the fourth area of antithetical cartography.

Film permits, once geocoded, provide a primary source of locational data for online film tourism atlases such as those found in Britain,[3] Denmark[4] and Madrid,[5] to name a few. One example that uses WebGIS as an archive for cinematic data is Jeff Klenotic's 'Mapping Movies'.[6] This site offers spatial data on historical theatre locations and where motion pictures were shown at the turn of the twentieth century in the United States. In contrast, Gleich and Lukinbeal (2022) are using geospatial technologies to model Old Tucson Studios through time. This project developed a 3D model using drones, terrestrial Lidar, historical air photos and historical film documents found at the Oscars Museum such as the main street blueprint for *Rio Bravo*. This research seeks to build online, virtual and augmented reality tours as a means of historical preservation and as an archive of film history (Figure 15.3). Narrativising these geovisualisations will include local film historians and critical post-colonial narratives to flush out the politics of the prospects of this geographical imaginary. Three different temporal models are being constructed including *Arizona* (1940), *Rio Bravo* (1959) and *Joe Kidd* (1972).

Conclusion

Cartography is the antithesis of cinema. Although cartographic science is critiqued as a mimetic device, GIS data should never be thought of as accurate but rather a matter of precision, resolution and documenting the scalar inaccuracies produced in the map projection process. Cartography strives for mimesis, but when it cannot be maintained at one-to-one, moving the distortion and inaccuracies to the terra incognita outside the map's scalar purview will suffice. In contrast, cinematography can have diegetic scalar consistency by using a smooth sequencing of shot scales in the *mise en cadre* or 'the pictorial composition of mutually independent cadres in a montage sequence' (Eisenstein, 1949: 16). With cinema, the spectator evaluates the scale of a subject in diegetic space in reference to their body (Doane, 2009). Suspension of belief can be formed or ruptured by scale, where ruptures in the pictorial composition are considered 'out of scale' to the rest of the image event within which the event takes place.

Geospatial cinema, or the deployment of antithetical cartographies, is already with us, and its further development and use will continue to expand as sensor technologies become more sophisticated. Full video GPS might not be in your local theatre just yet, but it is already in use by companies like Telemetry (https://goprotelemetryextractor.com/). While the technologies of antithetical cartographies are being developed, the deployment of cinematic form and language onto geovisualisations remains an area in need of further exploration. Antithetical cartographies include film location heritage preservation, using GIS in film production logistics in film commissions and permitting agencies (FilmLA), digital atlas production for film tourism, and spatial analysis of film(s) and the film industry. Antithetical cartography embraces cartography and cinema as mutually inclusive; it is a mobilisation of the *scalar* paradox that allows for the beginning of a geospatial cinema.

Notes

1 Tissot's circles or indicatrix is a method to represent the amount of distortion due to a 3D map being projected onto a 2D developable surface. Mexico is about 91% the size of Greenland (Mexico = 1,964,375 sq. km.; Greenland = 2,166,086 sq. km).
2 Act One: https://youtu.be/It5y3lNt8UE; Act Two: https://youtu.be/ymuuf1HoNvI; Act Three: https://youtu.be/NitU_gnPa-8
3 https://player.bfi.org.uk/britain-on-film/map
4 www.danmarkpaafilm.dk/
5 https://geocine.uc3m.es/geocine/mmm_map.html
6 http://mappingmovies.unh.edu/maps

References

Aitken S and Craine J (2006) Guest editorial: Affective geovisualizations. *Directions Magazine: A Magazine for GIS Professionals* 4(1). Available at: www.directionsmag.com/article/3015 (accessed 24 December 2022).

Alberti L (1956/1435) *On Painting*. New Haven: Yale University Press.

Alpers S (1983) *The Art of Describing: Dutch Art in the Seventeenth Century*. Chicago: The University of Chicago Press.

Caquard S (2022) *AtlasCine: An Online Cartographic Application to Map Your Story*. Available at: http://geomedialab.org/atlascine.html (accessed 24 December 2022).

Cosgrove D (1985) Prospect, perspective and the evolution of the landscape idea. *Transactions Institute of British Geographers* 10(1): 45–62.

Doane MA (2003) The close-up: Scale and detail in the cinema. *Difference: A Journal of Feminist Cultural Studies* 14(3): 89–111.

Doane MA (2009) Scale and the negotiation of 'real' and 'unreal' space in cinema. In: Nagib L and Mello C (eds) *Realism and the Audiovisual Media*. New York: Palgrave Macmillan, 63–81.
Doane MA (2022) *Bigger Than Life: The Close-Up and Scale in the Cinema*. Durham: Duke University Press.
Doel M and Clarke D (2007) Afterimages. *Environment and Planning D: Society and Space* 25(5): 890–910.
Edgerton S (1975) *The Renaissance Rediscovery of Linear Perspective*. New York: Basic Books.
Eisenstein S (1949) *Film Form: Essays in Film Theory*. New York: Harcourt, Brace and Company.
Eisenstein S (1974) *Au-deli des étoiles*. Paris: Union Général d'Éditions.
ESRI (2018) *Esri CityEngine Used to Create Oscar-Nominated Effects in Blade Runner 2049*. Available at: www.esri.com/about/newsroom/announcements/esri-cityengine-used-to-create-oscar-nominated-effects-in-blade-runner-2049/ (accessed 24 December 2022).
ESRI (2022) *Motion Imagery and Full Motion Video for ArcGIS*. Available at: www.esri.com/en-us/arcgis/arcgis-full-motion-video (accessed 24 December 2022).
Fletchall A, Lukinbeal C and McHugh K (2012) *Place, Television, and the Real Orange County*. Stuttgart: Franz Steiner Verlag.
Friday J (2001) Photography and the representation of vision. *The Journal of Aesthetics and Art Criticism* 59(4): 351–362.
Gleich J and Lukinbeal C (2022) A layered landscape of Western movie production: Combining geographical and historiographical methods at Old Tucson Studios. In: Stein E, Halegoua GR and Kredell B (eds) *Routledge Companion to Screen Media and the City*. London and New York: Routledge, pp. 205–216.
Haraway D (1988) Situated knowledges: The science question in feminism and the privilege of partial perspective. *Feminist Studies* 14(3): 575–599.
Harley JB (1989) Deconstructing the map. *Cartographica* 26(2): 1–20.
Kullman K (2016) The satellites' progeny digital chorography in the age of drone vision. *Forty-Five: Journal of Outside Research* 157. Available at: https://escholarship.org/uc/item/9m66j4f9 (accessed 24 December 2022).
Lukinbeal C (2010) Mobilizing the cartographic paradox: Tracing the aspect of cartography and prospect of cinema. *Digital Thematic Education* 11(2): 1–32.
Lukinbeal C (2012a) Scale: An unstable representational analogy. *Media Field Journal* 4: 1–13.
Lukinbeal C (2012b) 'On location' filming in San Diego County from 1985–2005: How a cinematic landscape is formed through incorporative tasks and represented through mapped inscriptions. *Annals of the Association of American Geographers* 102(1): 171–190.
Lukinbeal C (2015) Scale and its histories. *Yearbook of the Association of Pacific Coast Geographers* 78: 1–13.
Lukinbeal C (2018) The mapping of *500 Days of Summer:* A processual approach to cinematic cartography. *NECSUS an International Journal of the European Network for Cinema and Media Studies* Autumn 7(2): 97–120. https://necsus-ejms.org/the-mapping-of-500-days-of-summer-a-processual-approach-to-cinematic-cartography/.
Lukinbeal C (2022) Land use mapping and the topologies of a cinematic city: San Diego's Backlots from 1985–2005. In: Stein E, Halegoua GR and Kredell B (eds) *Routledge Companion to Screen Media and the City*. London and New York: Routledge, pp. 178–187.
Lukinbeal C and Sharp L (2019) Narrative formula through the geography of *Transformers: Age of Extinction*. In: Brunn S (ed.) *Handbook of the Changing World Language Map*. New York: Springer, pp. 1–23.
Mengqian Y and Caquard S (2019) Mapping *The Shawshank Redemption*: Film tourism, geography and social media. In: Lukinbeal C, Sharp L, Sommerlad E and Escher A (eds) *Media's Mapping Impulse*. Ed. Stuttgart: Franz Steiner Verlag, pp. 281–300.
Mitchell D (2003) *The Right to the City: Social Justice and the Fight for Public Space*. New York: Guilford Press.
NYC Film Maps (2020) Available at: https://nycfilmmaps.com/home (accessed 24 December 2022).
Pickles J (2004) *A History of Spaces: Cartographic Reason, Mapping and the Geocoded World*. London: Routledge.
Roberts L (2012) *Film, Mobility and Urban Space: A Cinematic Geography of Liverpool*. Liverpool: Liverpool University Press.

16

FIRING UP MAP THINKING

Music video meta-maps

Tania Rossetto

Introduction

That maps are media is a truism. The study of the communicative functions of maps as specific forms of media has a very long and vital tradition starting in the post-war era (see reviews in Kitchin et al., 2009: 4–8 and Crampton, 2010: 58–60; see Casti, 2015 for a contemporary reflection). Moreover, considering the map as a 'rhetorical *medium* for establishing various claims to truth and authority' (Cosgrove, 2008: 166, italics added) has been part of critical cartography since the 1990s. Thus, the notion of maps as media has expanded from the analytic elaboration of models of cartographic communication to the (de)constructivist take on cartographic representations. Significantly, introducing a recent collection devoted to maps and media, Lukinbeal and Sharp (2019: 9) write that cartography is not only 'one of the oldest forms of media' but also a scopic regime that mediates our socio-spatial worlds implying questions of 'meaning, ideology, and power', thus acknowledging this double communicative and critical consideration of the specific mediality of maps. Recently, then, the mediality of maps has been growingly explored in more nuanced (from critical to post-critical) ways in association with art, visual and media studies through a humanistic lens (Lo Presti, 2019). This has led to illuminating not just the mapping impulse of a variety of media but also 'mapping's media impulse', that is—far beyond the mere communicative function—the impulse of maps 'toward media' and their 'compulsion to be broadcast' (Wood, 2019: 33, 41).

Yet, as Hawkins (2021: 1710) has highlighted in the field of cultural geography, 'for some, medium is an outdated and conservative category of thought', since the 'dissolution of medium specificity' has led to a kind of 'post-medium condition'. Given the rapid confluence of cartography into a myriad of digital spatial practices, we could observe that maps today take part in this post-medium realm. Recently, for instance, mapping technologies have been considered from techno-cultural perspectives within the multifarious and wider frame of 'spatial media' (Leszczynski, 2015; Mains et al., 2015: 157–240; Kitchin et al., 2017). Significantly, Lukinbeal and Sharp (2019: 9) need to make explicit that the above-mentioned new collection on maps and media takes into consideration not only 'geospatial technologies' but also more 'traditional media forms'.

DOI: 10.4324/9781003327578-20

Despite the tendency towards a post-media condition, however, Hawkins (2021: 1713) also suggests that a persistence of a focus on the medium holds potential, since it 'is enabling cultural geographers to build more sophisticated ways of thinking and writing about and with creative practices'. In this chapter, I take up this suggestion by adopting intermediality as a creative space where cartographic imagination feeds map theory. In particular, I will elaborate on the interrelations between two different media, namely maps and music videos, as an opportunity to express map theorisation. Viewing music videos as media products that both reflect conceptions and generate thoughts on maps and mappings, I will dwell in, and play with, media borders. If, following Elleström (2021: 5), we consider that 'media are both different and similar, and intermediality must be understood as a bridge between media differences that is founded on media similarities', the encounters between maps and music videos will offer theoretical bridges to creatively perform map thinking. A selection of three case studies will show how, through the lens of cartographic theory, music videos can morph into 'meta-maps' that stimulate meditations on—as well as incarnations of—the cartographic medium.

The possibilities of cartographic intermedialities

While the 'mediality' of maps is well recognised, we could say that the specific research angle of the 'intermediality' of maps has been far less focused. To draw from Elleström's (2021: 73–82) intermedial categories, we should say, for instance, that maps may be integrated, that is, combined in a heteromedial coexistence with other media. Maps may also be subject to processes of transfer or transformation. These last may occur in the form of media representation (a medium relocated and represented in a recognisable way within another medium of a different kind) or transmediation (when the characteristics of a medium become part of another form of medium, often producing inventive versions of characteristics of both). Even if seldom referred to in terms of 'intermedial' references, the study of maps in painting, art, photography, literature, film or comics has been carried out (see respectively, among others: Hedinger, 1986; Watson, 2009; Dryansky, 2017 and Rossetto, 2017; Iacoli, 2021; Conley, 2006; Peterle, forthcoming). Often practised within extra-cartographic humanistic domains in the past, this area of interest is now at the core of cultural cartography.

Despite the fact that cartographic and map-like visuals, practices and imaginings have been populating music video cultures in varied and nuanced ways since the early 1980s, music videos have not yet been considered from a cartographic angle. Elsewhere I have proposed an historical review of such intermedial relations in the past 40 years (Rossetto, forthcoming) as well as a focus on the navigational qualities of more recent music video cartographies (Rossetto, 2024). At the base of this interest lies the idea that the peculiar intermedial audiovisuality offered by music videos, with the drastically different ways in which it materialises (Korsgaard, 2017), provides a particularly fertile ground for thinking maps through several theoretical layers. As Kelly (2019: 231) suggested, 'the ease with which pop music video can transform itself across media suggests that media borders can be blurred, transgressed or suspended, thereby locating the spaces in-between media as an emerging and fertile site for analysis'. Given their unique qualities, such as the fragmented montage, the unruly use of spatial and temporal features, the non-linear narrative structure, the frantic rhythms, the inherent mobility and the intense bodily kinaesthetic implications (Vernallis, 2004), music videos are particularly apt to reveal the fleeting qualities of maps.

In particular, in this chapter I take music videos as an opportunity for *doing* map theory, practising the reasoning on maps, searching for epiphanic moments and unexpected inspirations. The intermedial angle is crucial, since it is by working on both the similarities and differences between the two media, and on the sparks produced by the striking of them, that map theorisation may fire up. Drawing from Rajewsky (2005: 55), we could say that music videos that employ cartographic features tend to bridge the gap between the two media in the figurative mode of the 'as if', which is characteristic of intermedial references: 'an intermedial reference can only generate an illusion of another medium's specific practices'. And yet, as she suggests, it is precisely this illusion that potentially solicits, in the recipient of a medium, a sense of another medium's specific qualities. The heightened 'sense' of the cartographic medium expressed in music videos is an unmissable opportunity to continuously rethink maps from unpredictable perspectives.

Musical images and cartographic morphings: interpretive acts

In what follows, I propose three music videos as case studies, considering them distinct 'meta-maps'. I draw this idea from a seminal work by Stoichita (1996; but see also the concept of 'metapicture' in Mitchell, 1994), who wrote about the motif of 'paintings within paintings'—including the more specific motif of maps within paintings—in sixteenth- and seventeenth-century European art. Interestingly, Stoichita viewed this motif as an exercise of intertextuality (or, we could say, of intermediality): an auto-reflection on different modes of representation (or media) that produces self-aware images. These 'meta-paintings' are considered by Stoichita as theoretical objects, works on and with the image, as well as interpretive acts. Inspired by this intermedial possibility, in what follows I perform three interpretive acts considering music videos as theoretical objects with the potential to provoke meditations about cartography. None of the videos present to the audience performing artists; rather, we could say that they present performing maps, which take the stage and are broadcast while showing different configurations and modes of their existence. As we will see, these three interpretive acts of music videos' cartographic morphings engender three layers of map thinking: a representational (from technical to deconstructive) one, a post-representational (or emergent and practice-based) one and finally a speculative (phenomenological and object-oriented) one.

Act I—Representational machine: scrutinising the map in Remind Me

The 'machinic' aspect of cartography emerging from the post-war wave of the scientific study of cartographic communicative models is a perduring figure. Let's think, for instance, about the envisioning of our cartographic digital environment as a world populated by non-human, artificially intelligent, sensing cartographic machines (Mattern, 2018). Such machinic dimension of cartography is vividly performed by a music video for the song *Remind Me* by electronic Norwegian duo Röyksopp (in the remixed version by Someone Else), which was directed by Ludovic Houplan and Hervé Crécy (from graphic design and animation studio H5) and awarded as best video at the 2002 MTV Europe Music Awards. Following a model that is typical of dance/electronic music (see Gabrielli, 2010: 92), the function of the images within this video is not to paraphrase the lyrics but to create a further layer of meaning that develops independently from the verbal content of the song. The lyrics (with vocals by Kings of Convenience's singer Erlend Øye) refer to a subjective

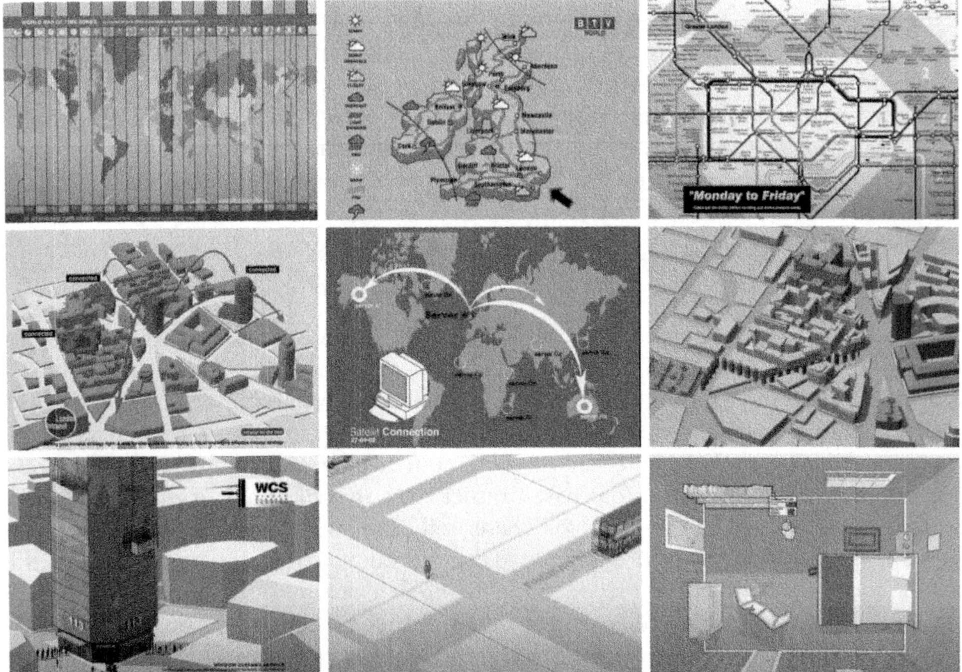

Figure 16.1 Composition of screen grabs by the author from the music video for Röyksopp's *Remind Me*.

Source: YouTube, www.youtube.com/watch?v=1Xhdy9zBEws, under Fair Use

experience of space, distance, movement ('on foot or by the help of transportation to knock on windows where/a friend no longer lived'), memory, love and silence, loss and failure ('there's always something to remind me/of another space in time/where love that travelled far had found me'). Yet the music video expands the suspended spatial evocations and sentiments expressed by the lyrics and music into a different story, that of a woman working in an IT company in London, which is entirely rendered through an animated spatial-infographic language. The video enacts a continuous transcalar movement across maps and diagrams that show the routine of an ordinary working day with a register that is analytic and 'mechanic' but simultaneously fanciful and ironic. Through an ordered and rhythmic kind of multi-spatial dance, underpinned by the electronic sound, the graphic animation comes across world, regional and urban maps, charts global networks or object life cycles, and displays exploded views, block diagrams and body anatomies, transforming the entire existence of the woman into an abstract model that exposes the 'rules' of its functioning from the global to the intimate scale (Figure 16.1).

It has been observed that 'the video *constructs and deconstructs* a day in the working life of city . . . delivering a *critique* of universal, systemic behaviour and the banal interconnectedness of human urban life' (Shane and Hanson, 2005: 71, italics added). This critique, we could say, seems directed also towards the visual regimes that visualise such schematic life, that is on the maps, charts and diagrams displayed by the animation. Of course, the dynamic qualities of the music video have the power to infuse movement and vibrancy in such representations, yet it is worth noting that the point of view remains that of a distant oblique eye that *re*presents in spatial ways the functioning of reality. The music video morphs into a complex musical cartographic machine that creatively connects multifarious spatialities. Yet, even if we are transported along such spatialities by the kinetic qualities of the musical animation, we are hopelessly kept outside the representational machine, as external observers who scrutinise the proper functioning of an abstract modelling or criticise its adoption: we *look at* the map.

Act II—Lived practice: moving with a map in The Child

The specific mediality of music videos implies an inextricable, complex relation with movement. Following Frahm (2010: 159), music videos perform not just 'movements in space'—such as rhythmic movements of performers, animated movements of figures and objects or abstract movements of lights and colours—but also 'moving spaces' that enact 'medial movements'. These medial movements, Frahm contends, are not simply formed by reference to other media (a map, in our case), rather they are reflections of the specific transformative capacity of the music video. These 'reflexive' and 'reflective', 'processual' and 'fluid', 'reshaping' and 'transforming' movements (Frahm, 2010: 160–161) performed by music videos may be found in the video for the French electronic musician Alex Gopher's song *The Child* (1999), realised by French designer Antoine Bardou-Jacquet, also from H5 studio. Developing the verbal content of the song, the video performs the wild ride through the streets of New York of a taxi which is bringing to hospital a woman who is on point of giving birth and her agitated husband. All the human and non-human, material and abstract things acting in the video (clouds, skylines, buildings, infrastructures, vehicles, people and even their feelings) are rendered in textual graphic form, generating a landscape literally made of coloured vibrant words that move following the pulsating rhythm of the music, the progress of the sonic content (which includes environmental sounds, parts spoken by the protagonists and samples from Billie Holiday's *God Bless the Child*) and the visual dynamism of the rush. As observed by Anderson (2006: 9), the tensions at the centre of the video are 'the conflict between real and constructed environments, as well as the insistent interplay of surface and depth'. In such tensions, indeed, lies the mappiness of *The Child*, which performs the navigation through the urban landscape by creatively evoking its cartographic rendering. The video creates a spatial configuration of abstract bidimensional text-imagery (which includes toponyms) that becomes three-dimensional through the movement across the configuration. In fact, what is evoked here is not a cartographic representation but a cartographic practice, a navigational action. The video adopts a shifting, fluid point of view (vertical, oblique and horizontal), but the most frequent—and salient—point of view is that of the protagonists themselves, who are anxiously crossing the city in the taxi. We actually feel the ride in that taxi, and the typographical trajectories it comes across, as if we have entered the routes of a fantastical in-car satellite navigator (Figure 16.2).

With its reflexive elaboration of a visionary and striking spatial representation, *The Child* provides a space of recognition—and transgression—of the peculiar medialities of both music videos and maps. What is emphasised here, through the moving space and medial mobility of the music video, is the practical, fluid, processual, dynamic dimension of mapping and the coexistence of maps with living subjects that live with and through them. The abstraction of the cartographic is merged with the movement of our bodies and imbued with human experience: we *perform* the map.

Figure 16.2 Composition of screen grabs by the author from the music video for Alex Gopher's *The Child*.

Source: YouTube, www.youtube.com/watch?v=URbFjz4hWMY, under Fair Use

Act III—Visceral thingness: becoming a map in Forgot My Broken Heart

The videoclip for *The Child* envisions a world that is typographically formed, which is a typical feature of the genre 'lyric videos'. However, *The Child* cannot be properly defined as a lyric video, since the featured verbal text expands the narrative rather than match the sonic/verbal content of the song. As McLaren puts it (2019: 163), 'lyric videos by definition offer a visual treatment of a song's lyrics, presented in synchronization with the recorded music'. The emergent genre of lyric video had a rapid development in the 2010s, with a dedicated category first introduced at the MTV Awards in 2014. Often released in conjunction with a single to anticipate a following official music video, lyric videos are often quickly made, simple and inexpensive. Yet, as McLaren (2019) again suggests, recently they have become more sophisticated and complex, while the aesthetic treatment of the words growingly intersects with the song the words promote in fascinating ways. There is a special relationship between lyric videos and cartographic imagery, since maps too normally embed both visuals and verbal signs. When the technique of kinetic typography, which integrates movement with texts, is used on a cartographic background, the effect is that of stimulating

Figure 16.3 Composition of screen grabs by the author from the lyric music video for Chris Cornell's *Nearly Forgot My Broken Heart.*

Source: YouTube, www.youtube.com/watch?v=zpMfZPAc1kg, under Fair Use

the video viewer to follow the movement and track words on its surface, enacting a typical practice of map reading. In a lyric video, however, the verbal text on a map is not composed of real toponyms but instead of words that generate fictional and emotional geographies, which is a common feature of map art deeply rooted in the past (see della Dora, this volume). And it is ancient cartographic devices that the official lyric video for the song *Nearly Forgot My Broken Heart* by American rock musician Chris Cornell (released together with the single in August 2015) plays with. The lyric video features a fantastic composition of details of celestial globes, celestial planispheres and armillary spheres elegantly reworked in a chromatic assemblage of black, red and gold. Following a sort of bullet, we embark in a synesthetic voyage upon the surfaces of such bi- and three-dimensional devices. This happens through a slow 'drone-like' movement—which is a trend in music videos today (Christiansen, 2022)—that transports us in very close sensorial proximity to such cartographic objects, thus provoking a 'tactile empathy' (Rossetto, 2019) towards them. The lyrics of the ballad are dynamically displayed on the ancient maps as if they were names of star constellations. The cartographic objects also include inventive alterations of their orthodox visuals to match the metaphorical imagery conveyed by the lyrics (the sky bleeding rain, the blood pouring from lips cut by little words, being like an apple bitten by the lover, or a door to be unlocked by a little key). That's how, at a certain point, we realise that at the centre of an armillary sphere, in place of the model of the Earth we expect to find because of the ancient geocentric conception of the universe, there is the gold model of a heart—the broken heart the song revolves around (Figure 16.3).

This is a moment of particular emphasis of both the song and the music video. As Gabrielli (2010: 96) states, one of the main functions of the images in music videos is to create matches with specific parts of the song through synchronisation points. Such points are moments in which a sound saliently converges with a punctual visual moment. It is in one of these salient moments, when the three-dimensional armillary sphere first appears, that

Nearly Forgot My Broken Heart takes us so near to the device that we can touch its heart. As Jacob (2006: 2, 8) suggested, although 'maps establish a new space of visibility by their distancing of the object and their replacement of it by a representational image', the map is primarily an object in itself whose effects 'result from its materiality, from the specific pragmatics of its viewer's body and gaze'. The video for *Nearly Forgot My Broken Heart* takes not only our gaze but also our body inside the material cartographic object, to explore its thingness. Once we enter its bowels, we do not observe the mechanisms of the machine. Instead, we feel the emotions it emanates: we *are* the map.

Conclusion

Advancing the idea of practice ontology, Bogost (2012: 85) thinks of 'artifacts that do philosophy'. In a similar vein, I have interpreted music videos as theoretical objects that stimulate theorisation on cartographic mediality. In fact, while music videos have constantly provided avenues for experimenting with cartographic political meaning, cultural messages, symbolic power, mobile effects, navigational features, affective atmospheres, emotional drivers, aesthetic qualities and material surfaces, they also provide an incredibly rich site for *doing* map thinking. With their peculiar mediality made of image, text and sound, music videos are excessive media that are inherently transformative. Whereas this makes them particularly elusive, as Frahm (2010: 160) suggests, the medial movements and transformations of music videos often imply a self-reflexivity on 'the epistemic question of mediality'. Here I have adopted a reflexive attitude to highlight some examples of music videos' cartographic morphings, taking the transformative qualities of music videos as a site for map theorisation. These videos thus function as meta-maps, in the sense that they could be considered reflections of/on map mediality. Taking three music videos as theoretical objects, I have placed light on three different layers of cartographic mediality that engender three layers of cartographic thinking: a representational reading that sees maps as technical devices to be decoded, or social constructions to be criticised; a post-representational take on maps as practices and events that emerge when they are lived and performed by humans; and a material and (post)phenomenological meditation on the objecthood of maps that speculates on their own emotional being. Playing with such enchanting intermedialities suggests that staying with, and ruminating on, media borders is a productive practice for map scholarship, one that makes us open to theoretical pluralism, experimental methodologies and endless empirical readings.

References

Anderson SF (2006) Aporias of the digital avant-garde. *Intelligent Agent* 6(2): 1–12.

Bogost I (2012) *Alien Phenomenology or What It's Like to Be a Thing*. Minneapolis, MN: University of Minnesota Press.

Casti E (2015) *Reflexive Cartography: A New Perspective on Mapping*. Elsevier: Amsterdam.

Christiansen ST (2022) Unruly vision, synesthetic space: Drone music videos. *The Senses and Society* 15(3): 286–298.

Conley T (2006) *Cartographic Cinema*. Minneapolis, MN: University of Minnesota Press.

Cosgrove D (2008) Cultural cartography: Maps and mapping in cultural geography. *Annales de géographie* 660(2): 159–178.

Crampton JW (2010) *Mapping: A Critical Introduction to Cartography and GIS*. Malden, MA: Wiley-Blackwell.

Dryansky L (2017) *Cartophotographies. De l'Art Conceptuel au Land Art*. Paris: INHA/CTHS.

Ellestrōm L (2021) The modalities of media II: An expanded model for understanding intermedial relations. In: Ellestrōm L (ed.) *Beyond Media Borders: Intermedial Relations among Multimodal Media*, vol. 1. Basingstoke: Palgrave Macmillan, pp. 3–91.
Frahm L (2010) Liquid cosmos: Movement and mediality in music videos. In: Keazor H and Wübbena T (eds) *Rewind, Play, Fast Forward. The Past, Present and Future of the Music Video*. Bielefeld: transcript, pp. 156–178.
Gabrielli G (2010) An analysis of the relation between music and image. In: Keazor H and Wübbena T (eds) *Rewind, Play, Fast Forward. The Past, Present and Future of the Music Video*. Bielefeld: transcript Verlag, pp. 89–109.
Hawkins, H (2021) Cultural geography I: Mediums. *Progress in Human Geography* 45(6): 1709–1729.
Hedinger B (1986): *Karten in Bildern. Zur Ikonographie der Wandkarte in holländischen Interieurgemälden des siebzehnten Jahrhunderts*. Hildesheim: Georg Olms.
Iacoli G (2021) Modus conlocandi: vicende intermediali nelle cartografie letterarie. *Moderna: semestrale di teoria e critica della letteratura* XXIII(1/2): 103–116.
Jacob C (2006) *The Sovereign Map. Theoretical Approaches in Cartography throughout History*. Chicago and London: The University of Chicago Press.
Kelly J (2019) The palimpsestic pop music video: Intermediality and hypermedia. In: Burns LA and Hawkins S (eds) *The Bloomsbury Handbook of Popular Music Video Analysis*. London: Bloomsbury Publishing, pp. 219–233.
Kitchin R, Lauriault TP and Wilson MW (eds) (2017) *Understanding Spatial Media*. London: Sage.
Kitchin R, Perkins C and Dodge M (2009) Thinking about maps. In: Dodge M, Kitchin R and Perkins C (eds) *Rethinking Maps: New Frontiers in Cartographic Theory*. London and New York: Routledge, pp. 1–25.
Korsgaard M (2017) *Music Video After MTV: Audiovisual Studies, New Media, and Popular Music*. Abingdon and New York: Routledge.
Leszczynski A (2015) Spatial media/tion. *Progress in Human Geography* 39(6): 729–751.
Lo Presti L (2019) *Cartografie (in)esauste. Rappresentazioni, visualità, estetiche nella teoria critica delle cartografie contemporanee*. Milan: Franco Angeli.
Lukinbeal C and Sharp L (2019) Introducing media's mapping impulse. In: Lukinbeal C, Sharp L, Sommerlad E and Escher A (eds) *Media's Mapping Impulse*. Stuttgart: Franz Steiner Verlag. pp. 9–29.
Mains SP, Cupples J and Lukinbeal C (eds) (2015) *Mediated Geographies and Geographies of Media*, Section II, *Transforming Geospatial Technologies and Media Cartographies*, Berlin: Springer, pp. 157–240.
Mattern S (2018) Mapping's intelligent agents. In: Bargués-Pedreny P, Chandler D and Simon E (eds) *Mapping and Politics in the Digital Age*. London and New York: Routledge, pp. 208–224.
McLaren L (2019) Katy Perry's 'Wide awake': The lyric video as genre. In: Burns LA and Hawkins S (eds) *The Bloomsbury Handbook of Popular Music Video Analysis*. London: Bloomsbury Publishing, pp. 163–180.
Mitchell WJT (1994) *Picture Theory. Essays on Verbal and Visual Representation*. Chicago and London: The University of Chicago Press.
Peterle G (forthcoming) Comics *and* maps, Comics *as* maps: Exploring intermedial connections and metaphorical interactions in comic book cartographies. In: Pedri N (ed.) *Intermediality and Comics*. Jackson: University Press of Mississippi.
Rajewsky IO (2005) Intermediality, intertextuality and remediation: A literary perspective on intermediality. *Intermédialités*, 6: 43–64.
Rossetto T (2017) Luigi Ghirri's cartographic portrayals: A review through map theory. In: Spunta M and Benci J (eds) *Luigi Ghirri and the Photography of Place: Interdisciplinary Perspectives*. Oxford: Peter Lang, pp. 121–142.
Rossetto T (2019) The skin of the map: Viewing cartography through tactile empathy. *Environment and Planning D: Society and Space* 37(1): 83–103.
Rossetto T (2024) Moving maps into musical images: Music video mobile cartographies. In: Jirsa T and Korsgaard MB (eds) *Traveling Music Videos*. New York: Bloomsbury Academic, pp. 57–71.
Rossetto T (forthcoming) Carto-clips: Forty years of map imagery in music videos. In: Springer J and Jablonowski M (eds) *Music Video Spaces*.
Shane RJW and Hanson M (2005) *Onedotzero Motion Blur. Graphic Moving Imagemakers*. London: Laurence King.

Stoichita V (1996) *The Self-Aware Image: An Insight into Early Modern Meta-Painting*. Cambridge: Cambridge University Press.
Vernallis C (2004) *Experiencing Music Video. Aesthetics and Cultural Context*. New York: Columbia University Press.
Watson R (2009) Mapping and contemporary art. *The Cartographic Journal* 46(4): 293–307.
Wood D (2019) Mapping's complicated media impulse. In: Lukinbeal C, Sharp L, Sommerlad E and Escher A (eds) *Media's Mapping Impulse*. Stuttgart: Franz Steiner Verlag, pp. 33–41.

17

WORLDS FOR SALE

Cartography in print advertisements

Davide Papotti

Introduction

The purpose of this chapter is to analyse appearances of the object 'map' in advertisements published on paper and related to services and objects other than cartographic ones.[1] This choice allows us to investigate the appeal of the cartographic object in the imaginaries shared by societies. This chapter proposes a classification of the denotative and connotative functions that the map can assume in advertisements. It is an itinerary through the cartographic world, here investigated not for its own proper functions but rather for its fascination and its evocative power. It is a widely accepted notion that the map serves certain purposes (to orient, to position and to provide information). These roles, however, do not guarantee cartographic representations the power of attraction, nor do they unveil everything about their complex nature. There are objects that are part of a repertory of frequent use in everyday life but that do not enter our emotional universe. They do not become part of our affective lexicon; they fail to attract vortexes of thought imbued with attraction and passion. The approval rating of an object changes with time according to social and individual attitudes towards certain materials and depending on the historical and cultural contexts in which it makes its appearance. Sometimes, to be able to catch the passional aspect, the hidden nature of certain objects, it is necessary to catch them by surprise and to observe them not in the ordinary terms and circumstances of apparition but in less usual places; not during their institutional work action but rather in their 'leisure time'. In any attempt to test the social approval rating of cartography, it is necessary to work in the direction of verifying certain frames in which the appearance of the object 'map' is freed from its merely functional purposes and assumes symbolic or simply expressive and decorative connotations.

In this chapter, I propose a research hypothesis that implies the crossing of the contexts in which the cartographic object is presented, with the related denotations and connotations. Study of the 'uses' of maps is a consolidated field. It suffices here to cite the work by Monmonier (1989) devoted to cartographic appearances in American journalism and the works by Rossetto (2012, 2013a, 2013b, 2015, 2019) on the materiality and physicality of maps. In an era that Carmagnola (2006) defines as the 'fiction economy', advertisements could

DOI: 10.4324/9781003327578-21

represent a promising field of inquiry. Taking into consideration the interesting suggestions provided by Muehrcke and Muehrcke (1974: 317) about the use of maps in literature:

> Such observations by authors offer the mapmaker valuable insights into the point of view of the map user, who is disassociated from the actual mapping process. Writers, as a special group of map users with an imaginative and philosophical perspective on the subject of maps, have much to say to the cartographer.

This is even more true in a context of an apparition of maps that is not mediated by the presence of a narrative text but that is rather offered, though through a printing process, to such wide audiences as those that read newspapers and magazines.

A suggestion to take this direction of research also came from the amazing collection offered in a volume by Holmes (1991), who, under the inspiring definition of 'pictorial maps', identified a rich display of cartographic *mirabilia*, map creations that were conceived and developed outside the 'normal' informative and communicative realm of cartography. Naturally, one must not investigate the appearances of maps in direct publicity, meaning advertisements for cartographic products. The evident role of advertising is to promote something.[2] Therefore, when this 'something' happens to be a map, there are very good chances that in the visual display a map would be the protagonist, clearly visible in the foreground. The language of publicity, though, works also with symbolic and evocative functions, bridging different realms, staging one thing to recall another, playing with links, pairings, references and short circuits.[3] Within the repertory of such correlations, one can exercise the maximum degree of creativity, and the pairings can be creative. The underlying rule, however, is that the object positioned next to (or even instead of) the one that must be promoted and sold must be as appealing as possible. The presence of maps in advertisements of other, non-cartographic products could be read, in this perspective, as a symptom of the potential appeal of the cartographic object and of the fascination that its image possesses in the collective imagery.

Even in the 1960s, McDermott (1969: 149) was surprised by the frequency of the appearance of maps in advertisements:

> Maps are frequently employed in advertisements in various periodicals as either the essence of the main theme or as subsidiary material to it. The frequency of application is surprising, as is the fact that maps are used to advertise a wide variety of products or services from rugs to airline passage and many types of maps, globes, and models are used to provide substance for the copy.

This dimension of a 'semantic transfer' that characterises the cartographic object when associated with the promotion of other goods gives the cartographic language an intrinsic lightness which appears freed from every duty of objectivity, reliability and likelihood.

The presence of maps in publicity for other objects or services possesses an evocative and representative value, meaning that the single cartographic object must become, so to speak, a 'champion' of the symbolic world of cartography. This metaphoric value is indeed deeply rooted in the very nature of the cartographic realm, which is based on a symbolic alphabet:

> Since the exact duplication of a geographical setting is impossible, a map is actually a metaphor. The map maker asks the map reader to believe that a mosaic of points,

> lines, and areas on a flat sheet of paper is equivalent to a multidimensional world in space and time. To 'read' a map, one needs imagination. Map symbols and their interrelationships are, in themselves, meaningless, except possibly for their aesthetic appeal.
>
> *(Muehrcke and Muehrcke, 1974: 319)*

My collection of cartographic advertisements started several years ago: I began to cut and keep in a folder images of maps that appeared in advertisements in newspapers and journals that happened to pass in front of my eyes.[4] Initially, I concentrated on proper maps—clearly identifiable cartographic objects that lured in advertisements. Then, through a progressive 'contamination', I started getting interested in weird cartographic metamorphoses, map-metaphors that appeared in the typographic spaces reserved to publicity through approximately explicit references: silhouettes of continents, contour lines sketched in a drawing, peninsulas and islands evoked in a graphic line, remnants of maps, etc. My folder titled 'Publicity maps' grew year after year through irregular supply from casual reading encounters.

The collection of samples cannot therefore be considered exhaustive, neither the sample statistically representative. Despite this, the examples can provide the basis of a possible classification in search of categories, of different typologies of appearances and of diverse kinds of use of the cartographic image. This reasoning on categories created by the multiple encounters between the creative minds of people working in the advertisement industry and the complex semantics of the cartographic language opens to this daring attempt to propose a possible 'atlas of maps in advertisements'. The cataloguing criterion is focused on the functions assumed by the cartographic appearances within the publicity frame. It is therefore a functional principle rather than a descriptive one (which would have been based on the intrinsic characteristics and typologies of the cartographic object). This kind of approach offers the opportunity to concentrate on the values and meanings of the estrangement produced by the appearance of the cartographic object in a context where the multiplicity of cross-references is potentially endless.

Functional categories of cartography in advertisements

The 'map as a map' role

The most normal appearance of the cartographic object, what could be considered the 'zero degree' of cartography in advertisement, is represented by the presence of a 'true' map: an easily recognisable one, one that can be encountered in everyday life. Inserted in a setting completed by and integrated with other objects that have the function of defining and clarifying the message for the public, the cartographic epiphany thus directly refers to 'real' and overtly recognised functions: to orient, to measure a distance, to localise. The map, often in its popular and recognisable version of the road atlas (one of the—at least till a few decades ago—most diffused and common cartographic objects), usually does not appear in the immediate forefront but rather incarnates the role of a 'non-protagonist actor', so to speak, referring to its values of objectivity and control over space. In this perspective, the map is not transformed into something else but rather appears in its 'plain' aspect (and function). In the end, it plays nothing else than itself, no matter the social consideration that it possesses among the audience at its destination.

The 'map in weird contexts'

A second degree of appearance of the cartographic object in advertisements is the display of a map in a disorienting context. This translation (in the literal meaning of the world: movement from one context to another) can be obtained by setting the cartographic appearance in an unconventional environment or through an unexpected use of the map itself. This perspective emphasises the disorientation produced by the promotional message, which relies on the creation of alternative (and generally more charming) worlds. McDermott (1969: 153), commenting on a cartographic image of Europe transformed into the shape of a dragon (used in an advertisement by the airline company Swissair back in 1968), defines as 'map caricature' the deformed use of the cartographic design. In these cases, the map does not immediately incarnate one of its primary functions but it seems to acquire an existence of its own.

Evocation through a deformed cartographic support

In other cases, the colours and features of cartography are emphasised but are accompanied or directly inserted on material supports whose shape or content are different, if not opposite, to the more common ones. The map does not appear on its usual material support (paper) but makes its appearance in other, unexpected frames (such as a shoe sole; Figure 17.1). As McDermott (2015: 19) suggests: 'Maps in advertising frequently distort familiar geographic regions into unusual shapes, often in a humorous or entertaining way, to attract viewers'.

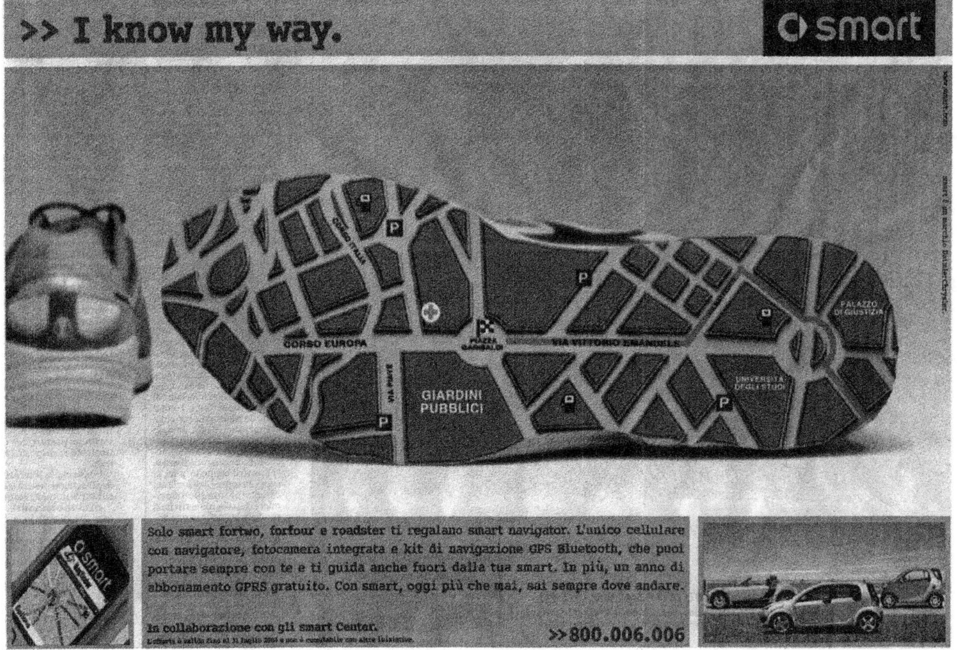

Figure 17.1 In an advertisement of the Smart automobile from the early 2000s, the map (in an urban setting and in a road atlas style) appears in an unusual context: the sole of a shoe.

These cartographic features are used as a recognisable characteristic in the interpretation of a message that aims at underlining a specific quality, at exaggerating a particular aspect and at emphasising a concentration. As Cigagna (2005: 2) states:

> Some projections are not in 2D but on rather surprising objects like a computer mouse or an apple, showing that it is possible to project a sphere on any object, with of course different levels of distortion or even inaccuracy. The global 'look' and aesthetics, and not accuracy, are normally the main concern of advertisers.

McDermott (1969: 150) uses the term 'map models' to indicate unusual representations of cartographic features on unexpected objects. This function deforms the usual cartographic expectations that can be found in more orthodox contexts of appearance. In this way, the representational authoritativeness of the map is preserved, while at the same time, elements are introduced that undermine, more or less furtively, its objective relationality in order to create a communicative short circuit. In this way, the informative and objective power of cartography remains apparently intact, saving its exterior appearance, but nevertheless a geographical *monstrum* is created, for the sake of publicity. This happens in an advertisement offered by a consortium of hotels of the island of Pantelleria, in Southern Italy, where the recognisable shape of the island is completed by a 'weird' toponymy taken from a city map of Milan (Figure 17.2).

Figure 17.2 The 'real' shape of the island of Pantelleria, in Southern Italy, is superimposed by geographical names taken from Milan's city centre to suggest the proximity between the area of departure and the tourist destination.

In this playful contamination, two 'cartographic truths' of a different nature (it is often fundamental here, the gap between the graphic language and the textual message) end up creating a cartographic lie.[5]

Evocation of a geographical form through a non-cartographic visual support

In this category, there are advertisements that offer, within the most diverse graphic and thematic environments, references to highly recognisable shapes of geographical elements such as states, continents or other elements that are well consolidated in the shared collective imagery. To be sure that the geographical references are easily recognised, it is necessary that the represented shape be part of a public domain of memory. Images related to strongly characterised geographical elements (such as islands, which have clearly marked borders) are particularly suitable for this purpose. The general profile of the geographical element assures its immediate recognisability, allowing the author of the advertisement to show its creativity in the creation of the context of appearance. This strategy, based on evocation of the geographical form—be it the state, region or city being promoted as a destination—finds its most common context of appearance in the world of tourism. This kind of representation can be supported by textual indications that favour the interpretation of the message or even by duplications (often of inferior dimensions) of the 'normal' cartographic representation.

Evocation through a single element of the cartographic language

An advertisement can also convey an original and efficient message by using only one component of cartographic language. Since cartography utilises a complex, multi-layered alphabet of signs (related to different features, such as altimetry, hydrography, administrative partitions and toponymy), it is possible to select just one field and emphasise it, making it dominant over the others. This kind of reference to the world of cartography does not take place through an explicit reproduction of a map in its visual completeness, nor through the evocative power of recognisable geographical forms, but rather through a breakdown of cartographic language into only one of its elements. The outcome, to bring the attention of the viewer to the medium rather than the content, is often an abstract form not immediately identifiable with features that populate 'real' cartography. In other cases, the profile of the cartographic design can be easily recognisable, but only because it is related to non-geographical features such as the human body, an object or an animal.[6] Sometimes, the cartographic outcome represents the very object that is promoted in the advertisement, thus creating a sort of 'cartographic good'. The cartographic system of representation of altimetry called contour, with lines indicating different layers of height, is for instance an element characterised by a high visual impact, even though it requires a minimum of 'cartographic literacy' to be properly interpreted. Contour lines are among the most specific and recognisable features of cartographic language, and thus, they can become a synecdoche of the cartographic reference.[7] Supported by the scientific appeal conveyed by a single cartographic element, the advertisement can offer the most creative short circuits.

Map as a metaphor of the world

Among the many forms, dimensions and graphic appearances that the cartographic product can assume, one of the most common and widespread elements is the three-dimensional representation of the globe. Object of cartographic appeal par excellence (its spherical shape gives the shiver of 'having the world in one's hands'), the globe preferably appears in an advertisement when a reference to a planetary dimension, to a global scale, is needed. As Cigagna (2005: 1) explains: 'For their advertising activities, companies working only nationally or locally use national or local maps. Worldwide enterprises show the wideness of their activities'. A similar function, in which the recognisability of the object 'world' plays a central role, is played by the two-dimensional planisphere. Maps of the world at a very small scale (one must remind that the larger the territory represented on a map, the smaller the scale is) are among the protagonists in the realm of 'geographical advertisements' (as in Figure 17.3, to promote a world primacy).

Their contexts of appearance are often related to travel and transportation, as McDermott suggests: 'Most common was the inclusion of a world map to emphasise the global nature of the services offered, a device frequently used in air and sea transportation' (2015: 19). Globes and planispheres can also appear in masked, transformed and unusual forms, which nonetheless maintain the easily recognisable shapes of the continents and the

Figure 17.3 An early 2000s advertisement from the Grana Padano Production Consortium (an Italian cheese protected by a denomination of origin recognised by the European Union) presents a map of the world (with the very word *mappamondo*—map of the world—in the forefront) to state the commercial primacy of the product.

conventional blue colour of cartographic oceans. The globe and the planisphere represent a sort of magic image which incarnates the possibility, or at least the illusion, of observing the world in which we live from an 'other', external perspective.

Fascination for the past: historical maps and the temporal evocation

The temporal dimension makes the power of evocation of the cartographic representation even more complex. Maps, in fact, can be classified not only in terms of scale, form or represented theme but also according to the age in which they were produced. The history of cartography thus enters the stage, closely related to the practice of collecting, which often privileges as a primary value the antiquity of the object. The connotative values linked to the appearance of 'geographical antiques' usually range from the sense of elite to the value of rarity, from the implicit communication of an innate elegance to the evocation of nobleness, from the temporal profundity of tradition to the sentimental value of the magic formula: 'once upon a time '. The patina of time deposited over a cartographic image runs against the emphasis on modernity[8] and rather assumes the reassuring connotations of tradition and appealing references to nostalgia—both usually socially diffused in societies.

Conclusion

Different forms of appearance of cartography in advertisements can often present more than one of the above-mentioned functions. Juxtapositions and contaminations are numerous. How should one classify, for instance, the appearance of an ancient globe in an advertisement? In the 'historical' category, or in the 'planetary' function? One does not exclude the other, of course. Different values, among those above listed, can coexist in the same cartographic advertisement—even integrating and reinforcing themselves. Nonetheless, it appears useful to identify the different functions that the map can assume in the promotional message in order to better understand their specific values.

The criteria of classification adopted here for the elusive and shapeshifting world of 'advertisement cartography' represent only one possible interpretative itinerary. One could offer, for instance, a classification based on the dimensions of the earth surface represented in advertisement maps or on the classic cataloguing criteria that distinguishes topographic, chorographic and geographical maps (according to the scale adopted in the representation). In my opinion, the emphasis on functions adopted here allows consolidation of a perspective that is fully coherent with the postmodern approach of deconstruction of the map proposed by Harley (1989). If the deconstructionist approach aims at unveiling all the complex 'backstages' of cartographic production, the alternance of mythologisation and demystification provided by advertisement language offers a healthy relativisation that, together with the playfulness of the promotional exercise, can present the cartographic object in unusual—and therefore potentially more appealing—forms. Maps in advertisements, thus, can contribute to postmodern abandonment of the 'vision of maps as unquestionable, univocal, objective, irrefutable scientific instrument; a vision that is strictly linked to the attribution of a positive connotation' (Spada, 2007: 57, author's translation), which informs the traditional reception of cartography.

The surprising aspect of this playful repertoire of cartographic appearances in advertisements could also have useful teaching outcomes.[9] One could teach the basic characteristics and notions of cartography by using a demystifying approach, by 'dismantling' a given image, rather than assembling the notion of cartography through a progressive explanation of theoretical principles.

Concepts such as scale, the graphic representation of three-dimensionality on a plain surface and the historical evolution of the cartographic drawing can find in these 'advertisement maps' a useful representative repertory, accompanied moreover by a pleasant sense of 'lightness', if compared to the serious scientific atmosphere that is usually associated with the teaching of cartography.[10]

The creative and artistic components of maps in advertising can also, as McDermott (2015: 22) suggests, inspire cartographers to improve the aesthetic quality of their products: 'The twenty-first century cartographer seeking ways to improve the communicative power of maps could well draw inspiration from the design of twentieth-century advertising maps'. The recreational and creative component that characterises the language of publicity can therefore permeate the cartographic realm, thus contributing to promoting the shared social appreciation of this language. This would be an important goal to achieve, especially in an epoch, like ours, which risks being characterised by the advent of a sort of 'cartographic illiteracy', mainly due to the cognitive mandate—as far as orientation and localisation is concerned—given to the ubiquitous technological tools, such as laptops and mobile phones, that we carry with us everywhere.

Notes

1 Some issues discussed in this article were also approached in a journal article published in Italian (Papotti, 2020).

2 As McDermott (2015: 18) summarises: 'Advertisements are intended to attract attention, provide information, and persuade their audience to purchase, rent, use, or subscribe to something'.

3 These aspects have been investigated within cartographic studies, especially in relation to tourist maps. The relation between maps and publicity has been approached mainly with an internal logic, observing the promotional content of the cartographic medium (Del Casino and Hanna, 2000; Makower and Bergheim, 1986).

4 Cigagna (2005: 1) confesses a similar process of collection:

> The present poster is the result of many years of informal gathering of advertisements based on cartography aspects. . . . I remember that the first time I saw an advertisement based on cartography, I kept it in order to show and discuss it with my students.

5 About the lies that can be conveyed by maps, see Monmonier (2018).

6 Sometimes the map was merged with another image evoking some quality of the advertised goods or services, a type of visual metaphor harking back to the heraldic Leo Belgicus symbolising the united Low Countries on Michael von Aitzing's map of 1538 and later versions.

(McDermott, 2015: 19)

7 Cigagna (2005: 3) emphasises the potential role of maps used in advertisements to understand level curves.

8 Which, on the contrary, gives the idea of following the latest technological discoveries and anticipating the future, as happened in the past with the use of the first satellite images made available to the public (McDermott, 1969: 151).

9 Simon Catling (1980: 15) lists 'maps in advertisements illustrating the location of a shop or store' as a category of maps that can be fruitfully used in class in junior and middle schools. For the didactic uses of maps in advertisements, see also Cigagna (2005).

10 Some thoughts on the creative and artistic use of maps can also be found in Harrison (1950).

References

Carmagnola F (2006) *Il consumo delle immagini. Estetica e beni simbolici nella fiction economy*. Milan: Bruno Mondadori.

Catling SJ (1980) For the junior and middle school. Map use and objectives for map learning. *Teaching Geography* 6(1): 15–17.

Cigagna M (2005) *Advertising and Cartography*. XXII International Cartography Conference—ICC 2005. A Coruña, Spain, 9–16 July. Available at: https://icaci.org/files/documents/ICC_proceedings/ICC2005/htm/pdf/poster/TEMA26/MARLI%20CIGAGNA.pdf

Del Casino VJ and Hanna SP (2000) Representations and identities in tourism map spaces. *Progress in Human Geography* 24(1): 23–46.

Harley JB (1989) Deconstructing the map. *Cartographica* 26(2): 1–20.

Harrison RE (1950) Cartography in art and advertising. *The Professional Geographer* 2: 12–15.

Holmes N (1991) *Pictorial Maps*. New York: Watson-Gupthill.

Makower J and Bergheim L (eds) (1986) *The Map Catalog. Every Kind of Map and Chart on Earth and Even Some Above It*. New York: Vintage Books.

McDermott P (1969) Cartography in advertising. *Canadian Cartographer* 6: 149–155.

McDermott P (2015) Advertising, Maps as. In: Monmonier M (ed) *The History of Cartography, Volume Six. Cartography in the Twentieth Century. Part 1*. Chicago: The University of Chicago Press, pp. 18–22.

Monmonier M (1989) *Maps with the News. The Development of American Journalistic Cartography*. Chicago: The University of Chicago Press.

Monmonier M (2018) *How to Lie with Maps*, 3rd edition. Chicago: The University of Chicago Press.

Muehrcke PC and Muehrcke JO (1974) Maps in literature. *The Geographical Review* 64(3): 317–338.

Papotti D (2020) Mondi di carta. La cartografia nella pubblicità a stampa. *Gnosis—Rivista Italiana di Intelligence* 26(1): 228–241.

Rossetto T (2012) Embodying the map: Tourism practices in Berlin. *Tourist Studies* 12(1): 28–51.

Rossetto T (2013a) Mapscapes on the urban surface. Notes in the form of a photo essay (Istanbul 2010). *Cartographica* 48(4): 309–324.

Rossetto T (2013b) Learning and teaching with outdoor cartographic displays: A visual approach. *J-Reading. Journal of Research and Didactics in Geography* 2(2): 69–83.

Rossetto T (2015) The map, the other and the public visual image. *Social and Cultural Geography* 16(4): 465–491.

Rossetto T (2019) The skin of the map: Viewing cartography through tactile empathy. *Environment and Planning D: Society and Space* 37(1): 83–103.

Spada A (2007) *Che cos'è una carta geografica*. Rome: Carocci.

18
MAPS AS DESIGN TOOLS
Space, time and experience

Roger Paez, Manuela Valtchanova, Ferran Larroya and Josep Perelló

Operative mapping

Maps and design have long been engaged in a fertile feedback loop of mutual affect, especially since the 1980s. Since Brian Harley famously identified the active role of maps and mapping in the generation of knowledge and, more importantly, in the generation of frameworks for knowledge (Harley, 2001), many important contributions from various fields of humanities and social sciences have delved into the multi-layered richness of maps and mapping (i.e. Christian Jacob, Denis Cosgrove, Denis Wood, Jeremy Crampton and John Pickles, just to name a few of the most relevant authors). The agency of mapping in design practices identified by Stan Allen (2000 [1989]) and James Corner (1999) spawned a renewed interest among architects and spatial designers in the operative potential of maps as design tools. More recently, contemporary approaches to design, including activist geopolitics (e.g. OMA/AMO, Bureau d'Études), planetary urbanism (e.g. Strelka, Territorial Agency), forensic architecture (e.g. Keller Easterling, Eyal Weizman), radical data-driven visualisations (e.g. Laura Kurgan, Vladan Joler) and critical practices in the fringes of art and design (e.g. Fraud, Liam Young), have widely expanded the use of maps to open new territories for design and to inform design decisions.

In the book *Operative Mapping: Maps as Design Tools* (Paez, 2019), operative mapping is conceptualised as the construction and use of maps to recognise existing opportunities in a specific environment and promote its transformation through design. We can summarise the method of operative mapping, as it is presented in this chapter, in two basic steps. The first involves mapping specific characteristics of the spatial, physical and material structure of an urban environment; events and dynamics that unfold over time; and experiences and perceptions based on subjective/collective citizen desires. This generates a concrete, specific, situated and oriented account of the context for our work. Second, these results are rearranged by performing graphic-projective operations (either analogical or algorithmic) on this mapped reality. The resulting maps show socio-urban contexts under a new light, revealing potential opportunities and, more importantly, prompting design operations to address those opportunities. Mapping is, in and of itself, a design action, as it often establishes the terrain for discussion and frames it in specific (and always value-laden)

DOI: 10.4324/9781003327578-22

ways. But besides acting as a framing device, maps can also inform design decisions through direct operations on the map to test non-actualised albeit potential realities, thus using the map as a projective device. Having said that, mapping is not a purely projective action, but one that requires a constant reality check, whether actual or virtual. Maps are generated in friction with reality—they appear as the result of testing a specific feature or condition against (known) reality, and in doing so, they expand our understanding of what is and what could be. Precisely because of this fact, we can claim that maps produce reality rather than merely represent it.

The city is a complex organism (Bettencourt, 2021), with several levels of reality (e.g. physical, social, cultural, sensitive and emotional), and needs to be visualised to understand each context in all its different dimensions and propose effective transformations. Maps can explore site conditions related not only to physical characteristics but also to flows of activation, time-related processes, relational dynamics, subjective perceptions, narrative landscapes and symbolic imaginaries. Mapping is used as a tool to detect and study the dynamics of a place and then to transform it by informing socio-spatial interventions.

This chapter addresses mapping agency in design by identifying different dimensions of urban reality that can be mapped. It also briefly shows how these different approaches to mapping can open new design aims, methods and outcomes through the documentation of a recent research project developed in the Design for City Making Research Lab (Elisava Research, UVic-UCC). It centres on three distinct approaches to operative mapping in design: spatial mapping to reveal site opportunities based on graphic operations; temporal mapping to understand dynamic behaviour based on data gathered from events; and experiential mapping to expand urban visions based on the subjective/collective desires of citizens.

Three approaches to mapping

We use a recent research project, *Civic Placemaking: Design, Public Space and Social Cohesion* (CP), to showcase three distinct approaches to operative mapping in design: *spatial mapping* reveals potential site opportunities; *temporal mapping* is used to understand dynamic behaviour displayed by data gathered from events that unfold over time; and *experiential mapping* expands urban visions drawing on subjective/collective citizen desires.

The main goal of CP, an Elisava Research project financed by 'la Caixa' Foundation, was to promote social cohesion and intercultural integration in an organic and effective way through temporary space design projects with an impact on public space. It included three main iterations (CP1, CP2 and CP3, see Paez et al., 2019, 2021 and 2022), each implemented in a different urban context in the cities of Sant Boi, Barcelona and Granollers. CP1 explored the capacities of temporary space design, in the form of a holiday lighting installation, to transform the perception of public space by the surrounding community. CP2 focused on studying how temporary space design, in the form of physical and digital micro-interventions, can promote varied interactions between young citizens to trigger novel uses of public spaces. CP3 investigated how an emerging public space could be turned into a consolidated public space, in the form of small constructions that support citizen-led temporary uses of public space, informed by a public experiment that includes GPS data collection and citizen science practices.

In CP, operative mapping tools and collaborative action methodologies are entangled with temporary space design formats to foster plural and spontaneous appropriations of public space, while at the same time generating data to inform urban transformations. This critical entanglement between mapping, relationality and ephemerality is fuelled by the logics of critical interventionism (Parry et al., 2012) and citizen science (Vohland et al., 2021), where the social dimension is particularly enhanced (Perelló, 2022) and pedestrian mobility is the central focus (Gutiérrez-Roig et al., 2016). Because of this fact, all the maps discussed in this chapter share a similar genealogy: they are high-resolution, detailed maps of specific urban settings, usually involving collaborative actions and addressing citizen behaviours and desires in order to generate innovative maps that render the city in a different light for the purpose of proliferating new urban visions and informing temporary space design decisions to transform our habitats.

Spatial mapping

Spatial mapping deals with the spatial, physical and material conditions of the city, tapping into both the built mass and the urban voids. While most conventional maps also primarily address spatiality, there are innumerable opportunities for further expanding what is being mapped and how to map it. In the context of CP, spatial mapping is used to understand the existing consolidated public space and its pertinent sociocultural activations in order to identify emergent public spaces and propose new places of opportunity for present and future design actions.

In CP1, spatial mapping depends on rigorous on-site data collection using the methods of direct observation and surveys. The data is then coded graphically to obtain three thematic readings of the site: degrees of commercial, civic and social activation; spaces of greater or lesser urban quality; and areas that the residents perceive more or less positively in terms of emotional response. These maps emphasise continuous field conditions in the form of data landscapes made up of geo-referenced numerical indexes (Figure 18.1, left). The main intention is to parametrise activation, urban quality and perception in such a way as to establish a meaningful comparison between these (usually independent) readings. In order to do that, a new series of maps are generated as 3D surfaces, where the x-y axis is consistent with the data landscape plan, while the z-axis value successively translates the indexed values of activation, quality and perception into elevations, making it possible to spatialise these three different sets of urban data in a consistent and interoperable way. As a result, Boolean operations can be performed to reveal specific urban conditions and opportunities for design actions.

In CP3, spatial mapping is used to visualise the existing conditions of consolidated urban space through the different levels of collective appropriability they present: no appropriability—built urban fabric; low appropriability—vehicular traffic spaces; medium appropriability—pedestrian traffic spaces (interior passageways, sidewalks), high appropriability—intermediate spaces (lawns, squares, undefined pedestrian spaces). This ground-floor mapping is intersected with a map of the areas of visual influence of the area's party walls: 76 windowless façades, adding up to 4,300 sq. m of blank vertical surfaces with a potential for design intervention. Both maps are graphically coded using a comparable scale of transparent grey hatches, so their superposition detects hotspots or areas of high opportunity (Figure 18.1, right). Consequently, the spatial mapping relates consolidated, existing public spaces (ground floor) with emergent, potential public spaces (party walls) to reveal eight areas of design opportunity that simultaneously have a high degree of potential ground-floor appropriability and a high degree of visual interaction with the party walls.

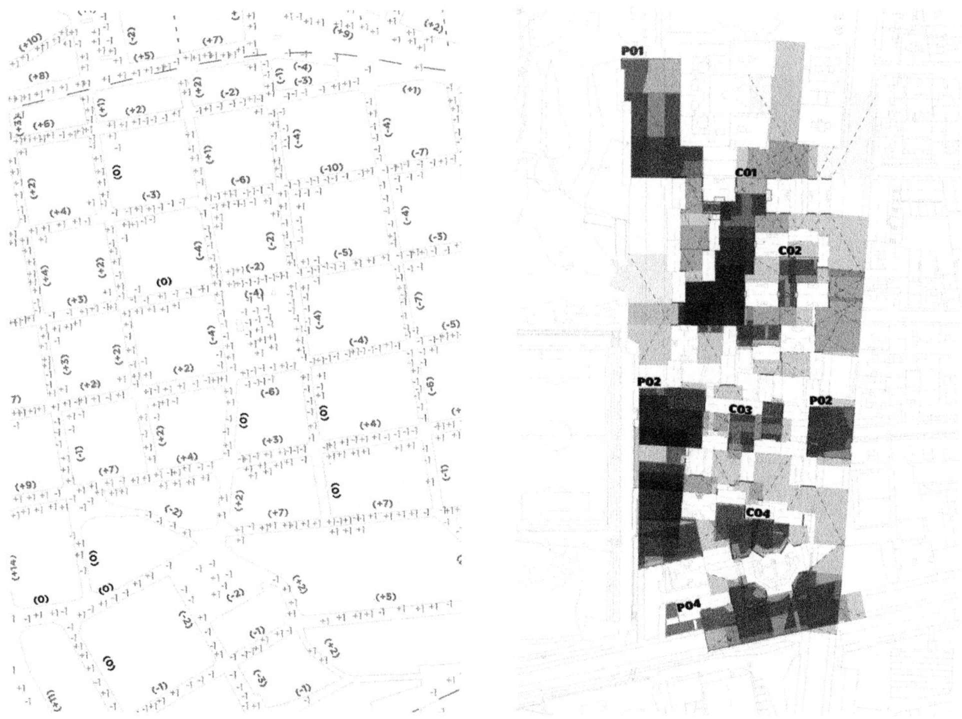

Figure 18.1 Spatial mapping. Left: Parameterisation of urban quality to inform strategic planning. Marianao, Sant Boi de Llobregat (CP1). Right: Areas of opportunity for social interaction based on party wall studies. Primer de Maig, Granollers (CP3).

Source: Image by the chapter's authors

Temporal mapping

Temporal mapping sets out to understand dynamic socio-spatial behaviour as it unfolds over time. In the case of CP, temporal mapping relies on data gathered from exceptional events that take the form of collective celebrations and shared experiences. These events allow us to study the city beyond its ordinary logics, understanding its flows and dynamics to detect opportunities and address them through design proposals that activate direct transactions between the spatial and social components of the place. In CP, the temporal mapping is articulated through three main operations: action, protocol and graphics. First, an action co-created with local communities fosters an environment of spontaneous socio-spatial interactions (e.g. a party, a game-based event, a collective celebration). Subsequently, a protocol is established for scientific GPS data collection, monitoring and processing to capture all the pertinent raw data grounded in participant's perceptions and local knowledge. Finally, a graphic code is developed to visualise the data and permit the identification of flows, clusters and tendencies associated with the time-based urban aspects being explored.

In CP1, the temporary mapping efforts registered a four-hour game-based event taking place in a neighbourhood square, which was the main setting for the project. The protocol of data gathering was based on five-second time-lapse recordings, which were

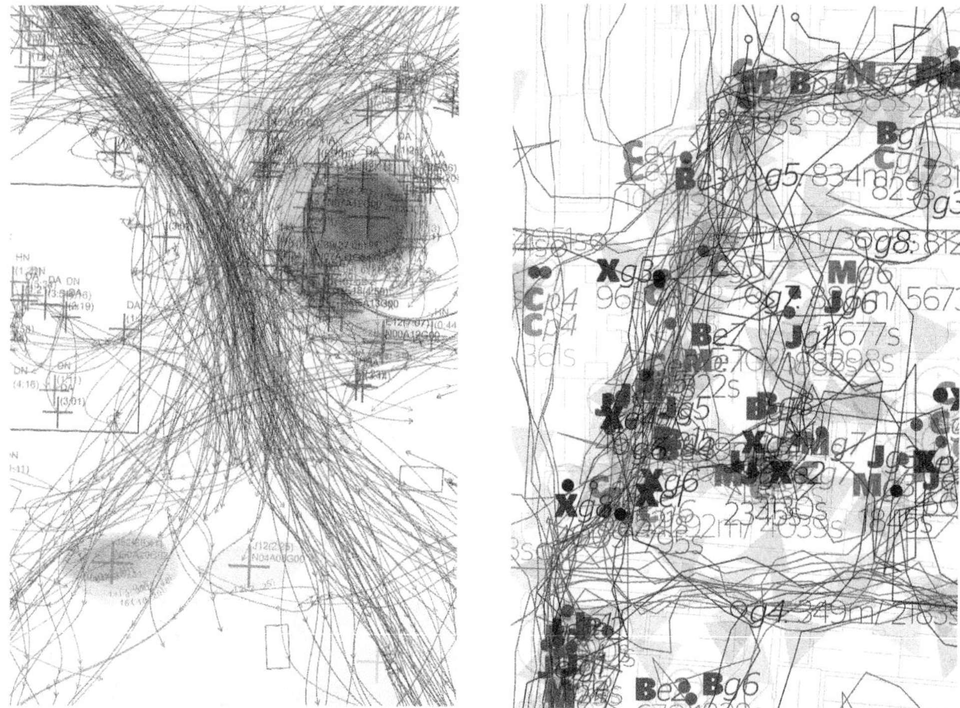

Figure 18.2 Temporal mapping. Left: Pedestrian flows and areas of gathering. Marianao, Sant Boi de Llobregat (CP1). Right: digital tracking of *Fem la Nostra Festa* action. Primer de Maig, Granollers (CP3).

Source: Image by the chapter's authors

later processed through computational analysis (using algorithms to feed the geo-tagged data into Rhinoceros and its visual scripting language add-on Grasshopper) supported by manual cross-check studies (since a fully developed machine learning algorithm could not be applied), to register and represent the movement of people and the specific clusters of activities observed. The final map (Figure 18.2, left) is based on four levels of graphic coding of specific inputs: Trajectories of directional transit movements (indicating gender and age of each path); stop-spots (indicating gender, age and time that each person stayed on the spot); clusters of activities (indicating type of activity, time of duration and quantity of people involved); and the dynamic evolution of the cluster outline every five minutes.

Whereas CP1 studied a grassroots event organised by local civic associations, CP3 designed, organised and implemented a collective celebration with the voluntary involvement of various intergenerational groups. The temporal mapping was organised in three phases: data gathering, data processing and data representation. In collaboration with the OpenSystems research group at the University of Barcelona (Perelló, 2022), a real-time tracking system was proposed that allowed for a detailed study of pedestrian mobility using the Wikiloc GPS navigation application, addressing the general absence of data available to better understand pedestrians' mobility and the inherent patterns in their movements (see previous work with the BeePath app: Larroya et al., 2023). This open-source application permits real-time anonymous tracking and recording of all the trajectories of movement. Thus, the process of data gathering does not interfere in the personal experience of participants, allowing the festive exploration of the site to be as intuitive and spontaneous as possible with minimal intervention and supervision from the research teams. Once collected, all the data is directly mapped onto the public cartography base and then processed using algorithms that classify all the movement vectors and the intentional and non-intentional stops for each group of participants (students, adults and seniors), indicating both the location and the duration of the stops (Figure 18.2, right). Based on a systematic study of all the short videos of each public space appropriation recorded by the participants themselves, each stop-spot is manually related to the festive action that happened there. Consequently, the graphic code of the temporal mapping indexes the trajectories of movement, on the one hand, and the places and actions of festive appropriation (X = to chat, M = to eat, J = to play, B = to dance, and C = to make a toast), on the other hand.

Experiential mapping

Experiential mapping deals with perception, sensation, narrative and experience. While they have an enormous influence on how we understand and navigate our cities both individually and collectively, these aspects are hardly ever considered in design-related mapping practices. This lack of factual knowledge and visual imaginaries related to these less measurable, softer, yet crucial aspects of our cities implies a double problem: on the one hand, an underrepresentation of the relational dimension that actually characterises the city; and on the other hand, an added difficulty of tapping into that dimension and addressing it in projective terms—in our case through temporary space design.

In CP2, two urban *dérives* prompted the unsupervised exploration of the neighbourhood by a group of young people (14 to 21 years old). In the first instance, participants walked through their neighbourhood and stopped in places of their choice to record a short video explaining why the place was meaningful to them. This generated a narrative-based subjective mapping that was used to prompt the second *dérive*. In it, participants were given verbatim sentences recorded by their peers and asked to locate the places they felt the given narrative applied to, generating a second map that identified matters of concern for young people and revealed specific urban areas with which they were associated most acutely (Figure 18.3, left). Finally, a third map was generated collaboratively from participants' reactions to their own and their peers' experiences by writing desires for their neighbourhood and locating them in specific spaces. This collection of maps provided a nuanced account of young people's rich perceptions and emotional responses to their city, rendered as an emotional landscape that is both revealing and actionable in terms of design and policy.

In CP3, experiential mapping focused on understanding participants' emotional responses. The raw data from the temporal mapping was clustered to identify the main areas activated and subsequently used as a graphic framework to index the activities realised. Each cluster archived all the emotional responses as expressed by the participants through photographs, videos, soundbites and written messages sent digitally to the research team. This categorised archive supports a high-resolution and situated reading of the micro-behaviours and, more importantly, the subjective and collective emotional responses of a group of citizens (Figure 18.3, right). Based on this experiential mapping, the neighbourhood was characterised through thematic activations, generating a novel understanding of

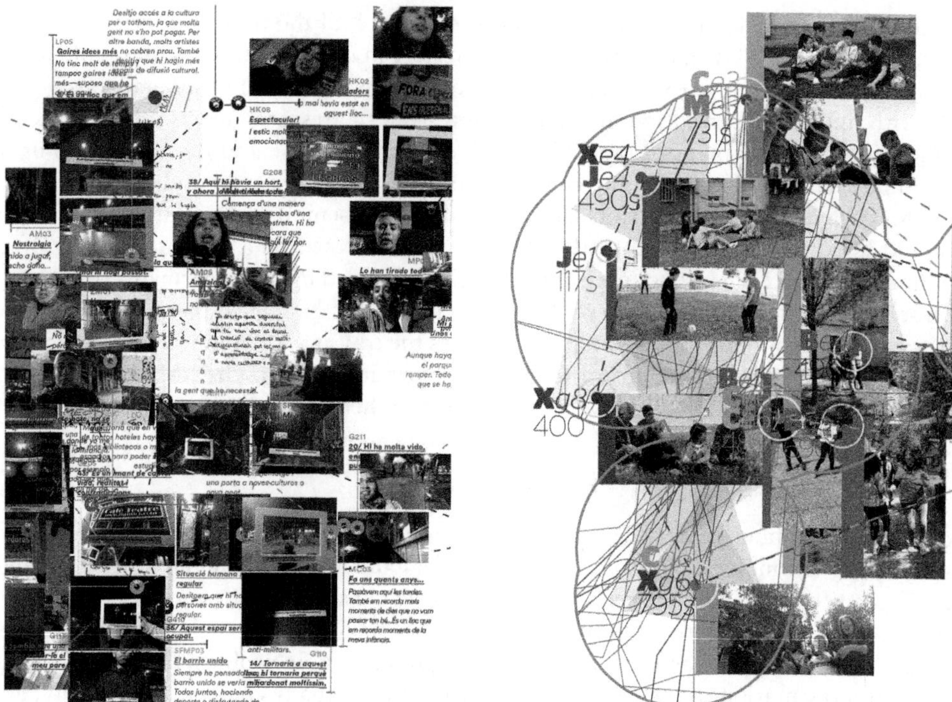

Figure 18.3 Experiential mapping. Left: Narrative-based subjective mapping. Raval, Barcelona (CP2). Right: Action-based citizen science data gathering. Primer de Maig, Granollers (CP3).

Source: Image by the chapter's authors

where certain activities (e.g. eating, dancing, playing) and the emotional responses related to them (e.g. exhilaration, boredom, happiness and calm) took place.

From map to design

The city can be understood as an ecology integrated by a complex network of things, events and meanings (Manzini et al., 2022). These three urban dimensions refer to the spatial-material structures (buildings, public spaces, infrastructures), the overlaid temporal horizons where events, actions and the very evolution of the city unfold, and the relational realities that animate it and give it specific meanings (social, cultural and linguistic).

We can no longer conceive the city from a purely spatial point of view. To do so would mean dismissing the vast richness of the urban phenomenon. And yet, that is what most architects, spatial designers and urbanists tend to do—perhaps largely because we lack access to maps to consolidate and socialise rigorous knowledge on the more supple dimensions of urban reality that could be the foundation for more nuanced, incisive and precise design approaches. These are, precisely, the approaches that are required to design the sorely needed sustainable and resilient, enriched habitats of today.

The maps we have presented humbly address all three dimensions that make up the city. Together, they become operative as they inform design decisions at many levels. Properly

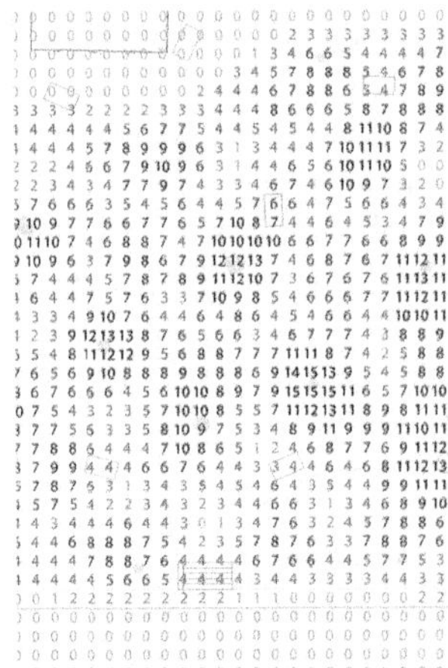

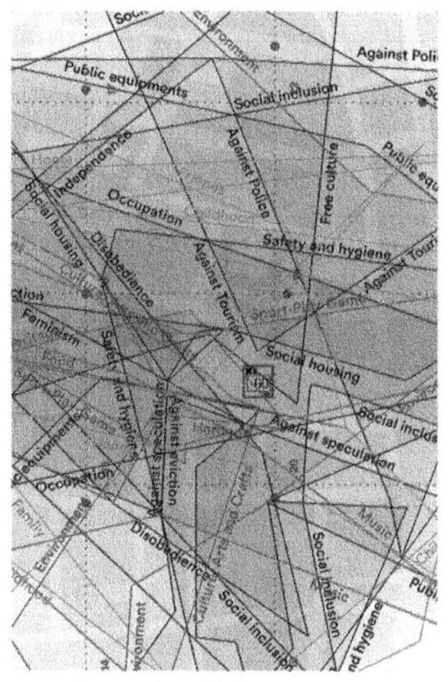

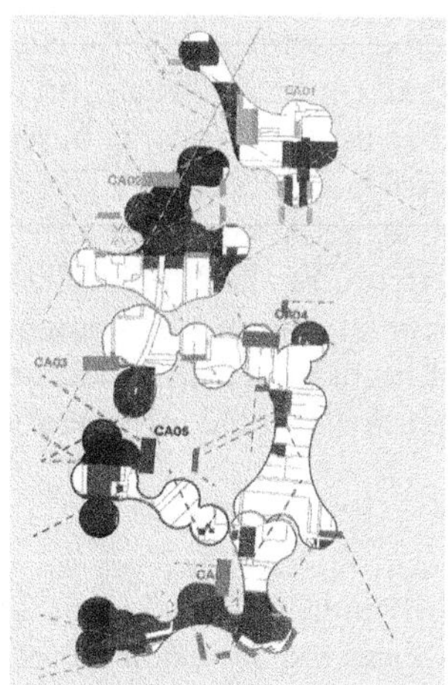

Figure 18.4 From map to design. Left: Public space parameterisation to simulate citizen behaviour. Marianao, Sant Boi de Llobregat (CP1). Centre: Narrative polygons spatialising identified matters of concern. Raval, Barcelona (CP2). Right: Clustering of citizen-activated areas. Primer de Maig, Granollers (CP3).

Source: Image by the chapter's authors

addressed, maps become *interpretive* tools to reveal opportunities that exist yet lay dormant within the city, and they generate new urban visions; *diagnostic* tools to prompt strategic planning decisions at an urban scale and to identify sites for intervention at a detailed scale; and *mediation* tools to help relate people and places in proactive and creative ways.

Specifically, in CP1, maps informed design decisions on the geometry of a collaborative holiday lighting proposal. We used the data gathered from a high-resolution mapping of a popular event on a public square (Figure 18.2, left) as inputs into an autonomous object simulation motor running on a numerical field in order to index attractors and repellers (Figure 18.4, left). This simulation, iterated several times to adjust varying site conditions, became a form-finding device that informed the final proposal or detailed information (see Paez et al., 2019).

In CP2, maps informed design briefs for infrastructures for public space interaction: a series of temporary interventions to foster intercultural cohesion and host youth-led initiatives in the public space of a particularly complex and socially challenging district. To do so, three collaborative workshops were distilled into several maps that detected matters of concern and relevant topics for the neighbourhood's youth. This task helped us not only to identify topics but also to spatialise them, revealing the specific areas where each topic was felt to be particularly relevant by the participants (Figure 18.4, centre). This map informed both the content and the location of several design briefs developed by postgraduate students for a popular initiative projector detailed information (see Paez et al., 2021).

In CP3, maps informed design decisions based on identifying six specific locations of maximum opportunity to test a new type of urban intervention to support unprogrammed, unsanctioned and spontaneous community actions. Crowdsourced data was analysed using K-means algorithms and other elaborated data science techniques to achieve a robust clusterisation method on aggregated GPS-related data. Human mobility activation with a citizen science experiment can be seen as an emergent and complex phenomenon in an urban context (Bettencourt, 2021). It is thus possible to identify key locations in terms of movement statistical features and related to stops or velocities (Gutiérrez-Roig et al., 2016; Larroya et al., 2023). The same approach and the related citizen science practices can be replicated in any other urban context since pedestrian mobility follows universal patterns (Larroya et al., 2023). When superimposed on the map to identify spaces of opportunity for intervention, the composite map of space, time and experience informed the optimal locations for a distributed system of devices to support community-led actions (Figure 18.4, right) or detailed information (see Paez et al., 2022).

Conclusions

The results obtained from this research can be translated into a series of conclusions.

First, working collaboratively with residents using mapping-informed methods of collaborative direct action and critical intervention in their own neighbourhood helps to relate different subjective narratives about what the shared living environment is, what it could be and what it should be. This facilitates the debate about the immediate future of the neighbourhood without having to reach (often superficial) consensus but instead articulating dissent in a creative and socially enriching way. Social cohesion does not come from ignoring difference, but from the radical acceptance of difference.

Second, citizen science fosters knowledge co-production by involving citizens in scientific research processes. This has two positive results: first, it produces a large amount of scientifically valuable data through the socialisation of knowledge generation and, second, it empowers citizens by making them more sensitive to the lived environment. Applied to mapping, the use of digital applications for participatory data gathering (crowdsourcing) is essential. It makes it possible to obtain a sufficient critical mass of data to generate maps that support valid conclusions. Moreover, it provides consistently formatted results that allow for a rigorous comparison and plotting of data, as well as their subsequent analysis using algorithms to generate various map formats from which to derive conclusions.

Third, mapping consolidated and emergent spaces, dynamic time-based events and experiences framed by sociocultural relations in an urban context is a valuable strategy to produce an innovative and rigorous reading of the urban milieu that helps to understand and transform the city in a more strategic and accurate way. Performing graphic-projective operations on the mapped reality helps to identify urban opportunities and places with high potential for positive change, and to inform design decision-making processes at both the conceptual and instrumental levels.

Based on these conclusions, we consider it crucial to consolidate the practical knowledge of the research model developed by *Civic Placemaking* by engaging in further research on: (1) the operative capacity of temporary space design and collaborative socio-spatial interventions as practices of critical urban interventionism; (2) citizen science and participatory data gathering for rigorous analysis and diagnosis of sites of opportunity and positive urban change; (3) collaborative mapping as a practice that works on three levels: to represent latent realities, to propose possible futures and to trigger active citizenship; and (4) novel interfaces and tools that optimise the process of data gathering, mapping and data visualisation, on the one hand, and dynamic social involvement, open dialogue and knowledge production, on the other hand.

Finally, it is relevant to clarify that the approach to mapping presented in this chapter (i.e. maps as design tools) is strongly linked with the cartographic humanities. The humanistic perspective on cartography uses technology to illuminate human experience and shed new light on how city spaces are perceived, construed and imagined, in order to inform new ways to relate to, activate, appropriate and design better human habitats. We understand mapping, both symbolically and effectively, as a worldmaking endeavour.

References

Allen S (2000) Mapping the unmappable: On notation. In: Allen S (ed.) *Practice: Architecture, Technique and Representation*. London: Routledge, pp. 31—45 [original text 1989].

Bettencourt L (2021) *Introduction to Urban Science: Evidence and Theory of Cities as Complex Systems*. Boston: The MIT Press.

Corner J (1999) The agency of mapping: Speculation, critique and invention. In: Cosgrove D (ed.) *Mappings*. London: Reaktion Books, pp. 213–252.

Gutiérrez-Roig M et al. (2016) Active and reactive behaviour in human mobility: The influence of attraction points on pedestrians. *Royal Society Open Science* 3(7): 160–177.

Harley JB (2001) *The New Natures of Maps: Essays in the History of Cartography*. Baltimore: Johns Hopkins University Press.

Larroya F et al. (2023) Home-to-school pedestrian mobility GPS data from a citizen science experiment in the Barcelona area. *Scientific Data* 10: 428.

Manzini E, Fuster A and Paez R (2022) *Plug-ins: Design for City Making in Barcelona*. New York and Barcelona: Actar Publishers.

Paez R (2019) *Operative Mapping: Maps as Design Tools*. New York and Barcelona: Actar Publishers.

Paez R et al. (2019) *Civic Placemaking 1: Disseny, Espai Públic i Cohesió Social. Marianao, Sant Boi de Llobregat*. Barcelona: Elisava.

Paez R et al. (2021) *Civic Placemaking 2: Disseny, Espai Públic i Cohesió Social. Raval, Barcelona*. Barcelona: Elisava.

Paez R et al. (2022) *Civic Placemaking 3: Disseny, Espai Públic i Cohesió Social. Primer de Maig, Granollers*. Barcelona: Elisava.

Parry B, Tahir M and Medlyn S (eds) (2012) *Cultural Hijack: Rethinking Intervention*. Liverpool: Liverpool University Press.

Perelló J (2022) New knowledge environments: On the possibility of a citizen social science. *Metode Science Studies Journal* 12: 25–31.

Vohland K et al. (2021) *The Science of Citizen Science*. Cham: Springer.

PART 4
Cultural digitalities

19

DIGITAL NARCISSISM AND GPS SELFIES

The entry of the self

Claire Reddleman

Introduction

This chapter offers a reading of the now-ubiquitous 'blue dot' of smartphone-based mapping apps as the entry of the 'self' into the map, such that the digital map now enables a range of ways in which we can visualise and surveil ourselves. I bring together a Lacanian interpretation of narcissism—in which narcissism is taken as a falsifying process of projecting an image of the self in order to sustain a fantasy of the self's integrity—with the capacity of the contemporary digital map to translate an image of the 'self's' spatial location, such that 'I' am translated into the image, or with Lacan, I am turned into the picture. I explore the practice of making 'GPS selfies' as a mode of contemporary vernacular self-portraiture, that enacts this process of narcissistic projection, and critically presents the entry of the self into the map.

'GPS selfie' is the most adequate handle I have found for those screenshots that I, and at least some other people, take to record the blue dot that locates us within the cartographic depiction offered by the mapping app as we use our smartphones. In this chapter, I offer a reading of the 'blue dot'[1] as a contemporary vernacular mode of self-portraiture. As Bridle (2019: 36) put it, 'GPS enables the blue dot in the centre of the map that folds the entire planet around the individual'. Following Bridle, the blue dot seems to have a compelling power to make the map cleave to the individual in ways that it has not previously been able to do. The idea of narcissism popularly stands for vanity and excessive self-regard, and here, I connect its aspect of vanity with the problematic of cartographic abstraction, as the map and its blue dot serve us with solipsistic self-images made using the Global Positioning System (GPS).

I am considering the blue dot as the entry of the 'self' into the map, such that the digital mapping app now enables a range of ways in which we can visualise and surveil ourselves, as well as to make self-portraits of a sort. This way of picturing is vernacular in that it responds to the map app's interface in the most primary, aesthetic way, and does not require any further knowledge or training on the part of the user. This aesthetic mode of response is perhaps a lesser function of GPS and its ability to locate (and surveil) people and objects in the world—as Dunn (2019: 13) notes in *A History of Place in the Digital*

 DOI: 10.4324/9781003327578-24

Age, 'the power of GPS on the GeoWeb lies in its ability to connect, search, synthesize, rank and, to an extent, visualize', that is, to relate data about individual locations to further data about individual locations. Appreciation of the ways in which the 'geoweb' organises data and relationships calls for some degree of specialist knowledge. The GPS selfie, by contrast, disregards these data-driven relationships and uses, to focus instead on the capacity of location data to render a depiction of the individual. Locating ourselves in the map is one of the key ways that people in general encounter and understand GPS.

GPS selfies as cartographic abstraction

In the GPS selfie, a few modes of 'cartographic abstraction'[2] are in play. By 'cartographic abstraction', I mean the ways in which cartographic techniques of depiction work by abstraction, in both senses of the word—extracting a part from a larger whole, and also positing the extracted part as a new entity, or in other words, making something new. As an example, symbolisation is a fundamental process for mapmaking. In order to create a symbol that can convey useful information in the map, the mapmaker first selects some aspect of the real to depict, such as the population size of a given city. A mark of some kind, or a graphic element, such as a star or a circle, is then used to stand for the piece of data (population size) and is deployed in the map. In this example, at least two stages of abstraction have been used in order to arrive at a useful symbol of a given city. First, a number has been chosen to represent the constantly shifting reality of the number of people who either live or exist within a chosen geographical area, and this choice of a number represents a process of abstraction from the rich and immense complexity of the reality of the city. Second, a graphic element has been chosen to depict our already-abstract number for population size. The graphic element will relate to a wider symbolic scheme in play in the particular map, enabling it to generate meaning for us as map readers. Wood and Fels (2008) have analysed how maps are able to generate these meanings in semiotic terms, as 'postings' in the 'sign plane' of the map, and have called their approach 'cognitive cartographics':

> [T]he principles underlying the graphic design of maps, far from being essentially aesthetic, are wholly at the service of the map's construction of knowledge, a construction built in *real time* by the map readers and typically validated on the spot.
>
> *(Wood and Fels, 2008: xvi, emphasis in original)*

They argue that we make meaning through using maps, often in context and '*on the fly*' (2008: xvi, emphasis in original), rather than passively consuming the information that maps appear to provide. This meaning-making is, Wood and Fels argue, often directed toward some definite purpose in action, 'toward doing something' (2008: xvi), and this is particularly applicable in the case of GPS selfies. The symbols used in our maps, then, are often read very intuitively, and we make sense of them in context and in relation to action. The abstracting steps that have been deployed in the design process are not usually in the forefront of our minds as we make use of a mapping app, and this contributes to the power of their knowledge claims.

There is a risk that these ways of considering the processes involved in making cartographic depictions makes them seem needlessly complex. However, my view is that abstract thinking is very ordinary and deeply familiar to most people, and we are accustomed to dealing with abstract entities in everyday life.

Cartographic techniques, then, necessarily involve abstracting processes and also produce abstract entities. Although this terminology can feel clunky, I prefer to say 'entities' rather than the more everyday 'things' because it is precisely how we understand the 'thingness', or the realness, of the 'things' produced by cartography that is in question. Cartography produces understandings of the world that go beyond the conceptual, or the ideal—they become real in material and social ways, and are shared among people. For example, while latitude and longitude may be seen as concepts, as particular ways of conceiving of the shape of the Earth and specifying points on its surface, they are clearly more than concepts, since they have the capacity to organise the physical disposition of people and objects on the Earth's surface, and to be understood among many people. Latitude and longitude are socially real. It is this social and material reality of abstract entities that it is useful to have a name for. In connection with mapping and its ways of depicting the world, then, I have proposed the phrase 'cartographic abstraction'. This also relates to a much larger tradition of thinking that recognises abstraction as more than thought. In particular, in the Marxian tradition, this is known as 'concrete abstraction'.[3]

To return to considering the GPS selfie, then, I suggest that creating this kind of map-based self-portrait using location data is a kind of cartographic abstraction. Indeed, the GPS as a whole uses a regime of cartographic abstraction, whereby we, in relation to our GPS receivers, are abstracted as locations in space. The GPS selfie combines the map and its rich imagery with the location data provided by the GPS. In this way, two abstract modes of creating images are combining to yield a way for us to make pictures that reinforces the concept of the isolated individual as the primary social unit in societies based on the capitalist mode of production—which is still, of course, where we live. The advent of GPS in the mapping app inaugurates this new way of individualising the experience of using the map, and this represents a renewed moment of individuation and solipsism in our experience as subjects in the society of capital.

It is the tension between the social and the individual aspects of this cartographic phenomenon that particularly interests me. In order to consider this phenomenon further, I turn to a close reading of the GPS selfie with a view to articulating a conceptual relationship between the narcissism of this type of image and the psychoanalytic and cultural concept of narcissism via Lacan and the concept of deixis. I will suggest that, while the isolated 'blue dot' individual may appear to be alone in the interface, the blue dot as a symbol misrepresents the deeply social interconnectedness of both the person depicted and the vast infrastructure via which the image is made.

Reading the GPS selfie as index

Figure 19.1 is a screenshot,[4] but not a photograph. A digital photograph still works by means of light entering a lens and being registered by a digital sensor. I think it might be helpful to think of the image as *indexical*, but not photographic. Being indexical usually means being caused by, or pointing at, the thing depicted, for example the scene or subject that exists in front of the camera lens at the moment a photograph is taken. In the case of the GPS selfie, the screenshot indexes the map app's interface.

Indexicality is one of the foundational concepts in photography theory (Frosh, 2015: 1609), drawn from Charles Sanders Pierce's sign theory, as opposed to that of Ferdinand de Saussure. Pierce's semiotic theory was tripartite, involving a sign, its object and its interpretant, and he saw this three-way relation as being irreducible to the semiotic formulation

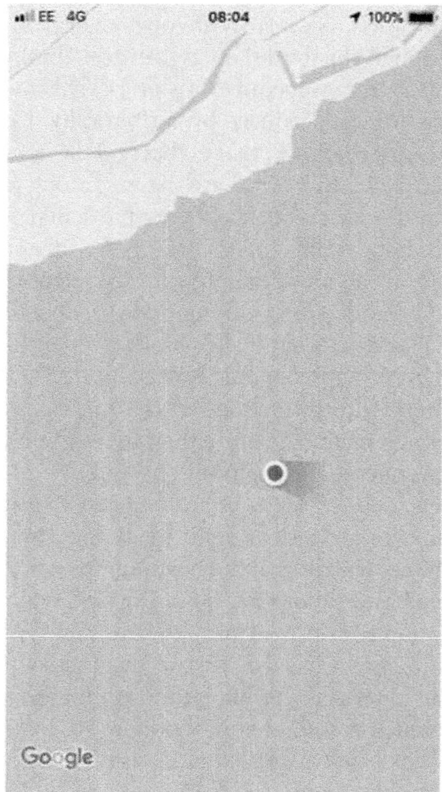

Figure 19.1 GPS selfie.

Source: Images made by the author

of pairs that Saussure popularised, involving only a signifier and a signified, to form a complete sign. There is of course a rich literature by and about both Pierce and Saussure that is beyond the scope of this short chapter, but we can take up Pierce's concept of signs functioning in three broad ways—indexing or indexicality, iconicity and symbolism. Iconicity is to do with resemblance or analogy between the object and its sign; in symbolism the relationship between object and sign is arbitrary or conventional, and so must be learned by the person interpreting the sign; and in indexicality the relationship between the object and the sign is either causal or directional, or we can say contextual—relating the sign to its spatial context. As Frosh puts it, citing Mary Ann Doane (2007), we have 'index as trace' and 'index as deixis' (Frosh, 2015: 1609). Trace is perhaps the more familiar idea, as with smoke 'indexing' fire through having been caused by that fire, being a trace of that fire, while the indexical relationship to context can be exemplified by a finger pointing at a thing. Deixis is more commonly used in linguistics and refers to terms that give contextual spatial and relational information, such as *here* and *there*, when we can only understand their referent, or object, in context. Indexicality has been foundational for photography theory in the sense of index as trace, due to the mechanism of whatever is in front of the lens being exposed to the negative or the digital sensor, as well as for cameraless photography, the photosensitive surface being exposed to the light source without a lens coming between them. The image made in this kind of way forms an index of the scene.

While indexicality has been extensively used in thinking about photography, it has not been used as much in relation to thinking about cartography because the map as sign is not caused by, so is not a trace of or an effect of, what it depicts. Instead, it has been produced through the combination of many symbolic techniques to render the physical world as spatial data (through surveying, i.e. measurement, i.e. expressing or rendering the world in the form of numerical data), and then to constitute an image of the world using, variously, scale, projection, selection, generalisation and symbolisation.[5] The map's relation to its object is often strongly iconic, achieving a form of resemblance, and always thoroughly symbolic. But the map's relation to its object has been indexical only in the form of the 'you-are-here' map,[6] such as those we encounter in town centre maps, *in situ* tourist signage, and spatial diagrams (which shade into being maps) of buildings, and shopping centres. The map's statement, 'you are here', is an instance of 'index as deixis' rather than 'index as trace', relating the map reader to their immediate surroundings, and only making sense in the correct spatial context. The blue dot's claim, or, following Wood and Fels, its argument, is of this kind—an instance of index as deixis, asserting that 'you are here' when looked at 'on the fly' in the app; when viewed later as the GPS selfie, the claim becomes retrospective, partaking of photography's condition of always depicting the past, and then asserting 'you *were* here'.

How was this image made? Without making use of the smartphone's lens, the screen has become something more like an image field or an image surface, rendering visually both the graphical user interface and the data furnished by the mapping app. It is being shown, rendered, on this surface which is transparent. The image is not registering something that exists in the world outside the smartphone, as a photograph would, but is instead recording what is rendered in visual form from within the mechanism of the smartphone. The screen becomes a field that is ready for capture by the screen-grabbing mechanism. Even though the smartphone's camera lens is not involved here, the screengrab *indexes* a way in which I have been *indexed* via the map app, as my location data. This is the sense in which it is more useful to call this image indexical instead of photographic.

What are the relationships involved in this? I stand in place, holding my networked device. The GPS receiver in my device receives the signals from at least four satellites.[7] The receiver calculates its own position in relation to the satellites, to an accuracy of approximately 0.5 metres (Kaplan and Hegarty, 2017: 3). I am co-located with the receiver, so that 'we', so to speak, are here. I am not the receiver, it is not yet implanted into me, so what is symbolised in the screenshot is the location of the GPS receiver in my phone. This location data is layered over the base map, to produce the composite image of the now-standard digital map augmented by the depiction of my personal location. In this way, we arrive at the newly self-referential form of digital mapping, in which the 'self' has entered the cartographic image by means of being abstracted as location data, and then symbolised, in the usual way, in the map. The user is symbolised here as a stand-alone collection of location data.

The narcissism of the GPS selfie: 'I am turned into the picture'

The mapping app has become a deeply normalised way of depicting ourselves to ourselves, for purposes of wayfinding (navigating from one place to another), as well as to simply identify one's exact or approximate location. As users, we have taken up a place in the map app, and we are now accustomed to being able to make the map 'return' to us when we have moved the depiction away from 'ourselves', to put ourselves back at the centre, via the W3C geolocation standard.[8] The location in which we presently exist becomes a dynamic symbolic element in the interactive digital map experience.

The selfie is popularly considered to be an expression of excessive self-regard: 'the accusation of narcissism is one of the most common themes in public discourse about selfies' (Frosh, 2015: 1621). Indeed, 'narcissism has become a defining feature of the modern era' (Yakeley, 2018: n.p.) and the idea of a wide-ranging 'culture of narcissism' was popularised by Christopher Lasch in 1978. Lasch (1979) was influential in bringing the concept of narcissism from specialist psychoanalytical use into mainstream discourse, where it has taken on the sense of excessive self-regard or self-obsession. The myth of Narcissus, a young man who is cursed to fall in love with his own reflection, and consequently dies, is best known from Ovid's *Metamorphoses*. The idea first came into psychiatric thought via the work of Havelock Ellis in 1898 (Yakeley, 2018: n.p.) and has since been elaborated into an official personality disorder. In his essay *On Narcissism: An Introduction* (1914/1925),[9] Sigmund Freud 'described a process in which we become subjects by taking our "self" as our first love object' (Tyler, 2007: 344), through incorporating 'an "ego-ideal"' (Tyler, 2007: 344) that combines an idea of a parental or societal gaze as well as an ideal of what the self should be. In one sense, then, narcissism is a way of internalising ideals and norms that are important in the subject's society and this is a vital way in which the subject carries out self-management and navigates the fraught process of developing object relations, the ability to relate to others (Holmes, 2001: 6). This is a foundational process that 'inaugurates subjectivity' (Tyler, 2007: 344). Freud proposed that "A power of this kind, watching, discovering and criticizing all our intentions, does really exist. Indeed, it exists in every one of us in normal life" (Freud, 1925/1981: 95). This is the surveillant aspect of the narcissism process, which, for Freud, is to some extent necessary, but in the popularised usage of narcissism is seen as excessive and antisocial.

In the original Narcissus myth, the role of the reflection is central. The image of the self is what occasions Narcissus' excessive self-love and ultimate death. Lacan has developed the relationship between self and image further, seeing 'narcissistic identification as an "imaginary" process of taking on an image and "appropriating" it as if it represents the self' (Gaitanidis and Curk, 2007: 9). For Lacan, this process presents crucial doubts about the legitimacy of the ego itself, exposing it as 'a false acceptance of an image as real' (Gaitanidis and Curk, 2007: 9). Lacan proposes that the gaze is what turns 'me' into a picture: 'What determines me, at the most profound level, in the visible, is the gaze that is outside' (Lacan, 1981: 106). In the context of the GPS selfie, we can read the gaze as the radio signals which are being transmitted to my GPS receiver, which is the location of the process of calculation which ascertains my location, *and* then the process of symbolisation of that location data, and its expression in the cartographic field of my mapping app—through which 'I' am translated into the image, or with Lacan, I am turned into the picture.

We can debate whether we should attribute the satellites themselves the position of the 'other' who originates the gaze, or whether it would more properly be the GPS as a whole, or perhaps the institution which operates it, the U.S. Space Force. All are involved. But the initiator of the operation is still me, controlling the smartphone, choosing to instigate this network of viewing relations among myself, my GPS receiver, the satellites, the U.S. Space Force, the internet and the map app company, and through doing so to create a 'selfie' based only on location data, I create a self-abstraction, a depiction including only my spatial existence in relationship to my device. I have selected out of consideration other characteristics of my existence in this place, most notably any kind of photographic depiction of myself. There is a mode of narcissism in play here which is less to do with the popularised 'cultural narcissism' version and more to do with the Lacanian rendering of the self as picture.

Conclusion

In conclusion, after thinking of my GPS selfie as indexical in the sense of deixis rather than trace, in its capacity to attest 'I am here' and now, later, 'I was there'; and after thinking of it as a result of processes of abstraction in which my location is abstracted away from all other aspects of me and symbolised in the map; and after thinking of the relation I have sought out with the 'other' of the GPS/U.S. Space Force in order to symbolise myself as a picture; the reason I am still interested in this set of ideas is that it seems to me to render a picture of aloneness, via this mechanism's emphasis on creating a symbol of my GPS receiver on its own. Whereas conventional photographic selfies often involve pairs and groups of people, the GPS selfie so far appears to be a way of making pictures of the self alone—in which the presence of other people and the relationships and wider social relations and *context* involved are de-selected for depiction, not at the level of the maker of the picture directly, but at the level of the image-making mechanism. Thus, the entry of the self into the map appears as a mode of digital narcissism.

Notes

1 The scientist Carl Sagan coined the phrase 'pale blue dot' in his 1994 book 'Pale Blue Dot: A vision of the human future in space', reflecting on the appearance of the Earth as photographed by Voyager 1 in 1990 as it left our solar system. The photograph can be viewed at https://solarsystem.nasa.gov/resources/536/voyager-1s-pale-blue-dot/ [accessed 31 August 2023]. The first referent of the phrase, therefore, was the Earth as a specific location within the solar system, which has been further adapted and specified in the contemporary blue dot's ability to symbolise a specific location on the Earth's surface.
2 I have offered a fuller, theoretical account of this concept in Reddleman, 2018.
3 For a more in-depth theoretical account of this tradition, see Reddleman, 2018.
4 For a contemporary account of the screenshot as a cultural and artistic form, see Nešović, 2022.
5 Mark Monmonnier gives a useful account of these fundamental cartographic processes. See Monmonier, 1996.
6 See Montello, 2015.
7 See www.gps.gov/systems/gps/space, accessed 12 April 2023.
8 See https://w3c.github.io/geolocation-api/, accessed 12 April 2023.
9 *On Narcissism: An Introduction* was first published in German in 1914, and first published in English translation in 1925. See the editorial introduction for an account of the development of the ideas and the essay, in Freud, S. et al. (1981) *The Standard Edition of the Complete Psychological Works of Sigmund Freud: Volume XIV*. London: Hogarth Press and the Institute of Psychoanalysis, pp. 69–71.

References

Bridle J (2019) *New Dark Age: Technology, Knowledge and the End of the Future*. London and New York: Verso.
Doane MA (2007) The indexical and the concept of medium specificity. *Differences* 18(1): 128–152.
Dunn SE (2019) *A History of Place in the Digital Age*. Abingdon and New York: Routledge.

Freud S (1925/1981) On narcissism: An introduction. In: Freud S et al. (1981) *The Standard Edition of the Complete Psychological Works of Sigmund Freud: Volume XIV*. London: Hogarth Press and the Institute of Psycho-analysis.
Frosh P (2015) The gestural image: The selfie, photography theory, and kinesthetic sociability. *International Journal of Communication* 9: 1607–1628.
Gaitanidis A and Curk P (2007) *Narcissism: A Critical Reader*. London: Karnac.
Holmes J (2001) *Narcissism*. Cambridge: ICON.
Kaplan ED and Hegarty C (eds) (2017) *Understanding GPS/GNSS: Principles and Applications*, 3rd edition. Boston and London: Artech House (GNSS Technology and Applications Series).
Lacan J (1981) *Book XI the Four Fundamental Concepts of Psychoanalysis*, Sheridan A (trans.). New York and London: W. W. Norton and Co. (The Seminar of Jacques Lacan).
Lasch C (1979) *The Culture of Narcissism: American Life in An Age of Diminishing Expectations*, reissue edition. New York: W. W. Norton and Company.
Monmonier MS (1996) *How to Lie with Maps*, 2nd edition. Chicago: University of Chicago Press.
Montello DR (2015) Experimental Studies in Psychology. In: Monmonier M (ed.) *Cartography in the Twentieth Century*. Chicago and London: University of Chicago Press (The History of Cartography), pp. 1080–1082.
Nešović D (ed.) (2022) *PrtScn: The Lazy Art of Screenshot*. Amsterdam: Institute of Network Cultures. Available at: https://networkcultures.org/blog/publication/prtscn-the-lazy-art-of-screenshot/ (accessed 12 April 2023).
Reddleman C (2018) *Cartographic Abstraction in Contemporary Art: Seeing with Maps*. London and New York: Routledge.
Tyler I (2007) From 'The Me Decade' to 'The Me Millennium': The cultural history of narcissism. *International Journal of Cultural Studies* 10(3): 343–363.
Wood D and Fels J (2008) *The Natures of Maps: Cartographic Constructions of the Natural World*. Chicago: University of Chicago Press.
Yakeley J (2018) Current understanding of narcissism and narcissistic personality disorder. *BJPsych Advances* 24(5): 305–315.

20
AUTOMATED MAPPING CULTURES

Sam Hind

Introduction

The automation of cartography has come in various forms, contexts and waves. A short history of the UK's largest motoring organisation—the Automobile Association (AA)—offers an insight into automation writ large in the post-WWII era. By 1965, the organisation was processing over 1.25 million individually tailored route requests for journeys up and down the country (AA, 2023). AA employees called 'route compilers' would manually select relevant route sections and 'piece them together to form the completed route' (Calder, 2012) before posting to the AA member. Struggling to keep up with the sheer volume of requests, the AA designed so-called 'Throughroute' maps in 1968, showing AA-certified routes from 50 different towns across the UK to over 500 destinations (AA, 2023). The maps were intended to reduce the volume of individual requests, as members could simply consult the certified routes, rather than commit a route compiler to manually produce them. In 1984, the AA's 'Home Routes Service' was computerised, so routes could be 'generated automatically without the need for staff to "pick" and collate each route' (AA, 2023). In 1990, a dedicated 'Routes Processing Unit' was formed, charged with collecting data used for generating navigational routes, from road layouts to signposting. In 1999, the AA's online 'Route Planner' service was launched, with members now able to generate routes by themselves, on the web. In 2010, the service registered its one billionth route calculation (AA, 2023).

The history of the AA in the modern era, then, can be considered a history of cartographic automation: a gradual, piecemeal, process in which different parts of the making, compiling and using of navigational commands slowly became automated. While the launch of an online route planning service might represent the apex of automated route-calculation, the history above demonstrates that many other moments are key to this story too. In this context, while the compilation of routes in the 1960s might be understood as a distinctly manual activity, it is telling that the work involved compiling ready-made route segments, rather than generating them from scratch. Likewise, while the Homes Route Service computerised route collation at the backend, ordinary members still had to request them manually.

 DOI: 10.4324/9781003327578-25

In this chapter, I will examine the role of automation in the making, distribution and use of maps. Using 2010 as a rough start point—following on from when the AA's Route Planner recorded its one billionth route calculation—I will consider the various 'mini-automations' that have taken place at the level of the mobile, web-enabled, map app, namely: 'auto-complete', automatic geolocation and automatic route-calculation. Now ubiquitous, the map app can be considered the successor to desktop, web-based mapping services such as AA's Route Planner and, most famously, MapQuest, acquired by internet giant AOL in 1999. I begin by introducing three of these mini-automations, before discussing the role of 'machinic entities' in cultivating navigational experiences. I then move on to examine the rise of the map 'user' before discussing two costs of cartographic automation: algorithmic control and environmental destruction.

Mini-automations

First, users of typical map app services such as Google Maps can take advantage of 'auto-complete' features that predict locations and addresses when users begin to type. Originally integrated into web browsers by (then) Google (now Alphabet), the feature largely removed the need for users to type out requests in full, or to correctly spell the object, person or phenomenon the user sought information on. Upon integrating the feature into Google Maps, users were likewise spared from having to type out full addresses, strangely named streets, or the jumble of numbers (and sometimes letters) that comprise that most modern of spatial technologies: the postal code.

The second mini-automation is the automatic geolocation of users themselves—or at least, their mobile devices. Freed from their desktop computers, early mobile users found themselves in a predicament: where exactly in the world were they? Able to surf the early mobile web—think WAP and the Opera Mini browser—users still struggled to navigate out in the world with their mobile devices. Subject to small screens, accessibility issues and bandwidth problems, those desiring to navigate their way using such a device would have been hard-pressed to locate themselves in the first place. GPS devices existed but were largely the preserve of recreationalists like hikers and sailors, with specialist navigational requirements. The inclusion of GPS receivers in smartphones meant these new mobile, multifunctional devices opened up the opportunity for navigating in the wild, in addition to web browsing and messaging on-the-go. Critically, GPS receivers or chips were able to do the work to locate the mobile device they were embedded within without much effort at all on behalf of the user. Ordinarily this could take just two button taps: one to turn the receiver on, and another to open the map app.

The third mini-automation is the automatic calculation of routes. Here, things got trickier in a different way. Rather than having to move heaven and earth to put satellites in the sky, the automatic calculation of routes required the operationalisation of Dijkstra's algorithm: a technique for determining the shortest path from one point to another. Certainly not a new technique—Dutch computer scientist Edsger W. Dijkstra first designed it in 1956—emergent mapping firms still had to find a way to apply shortest-path algorithms to relevant cartographic datasets. More specifically, to develop functional automated route-calculation systems that could efficiently route users from A to B—without error. While automatic route-calculation has a longer history to text auto-completion or a comparable history to GPS (see the AA example), in terms of mobile device navigation, automated route-calculation makes little sense (certainly *less* sense) without these other mini-automations.

Indeed in many other instances, neither auto-complete nor geolocation is present, arguably making automatic route-calculation that little more onerous.

Collectively, what I have called here 'mini-automations' might not be so small at all. Arguably, the integration of GPS receivers into mobile devices, or the operationalisation of shortest-path algorithms, cannot ordinarily be considered 'small' advances. Indeed, their comparative developments yielded colossal technological, social, cultural and often political consequences: like the members of the Israeli Defence Force who mistakenly entered Palestinian territory thanks to Waze (Tarantola, 2016; Carraro, 2021), or the football fans who erroneously drove to Frankfurt (Oder) in eastern Germany rather than Frankfurt (Main) in the west, 600 km away (O'Brien, 2019). Yet for the user of such services, phenomenologically, these features are only ever experienced momentarily, fleetingly, in the tapping of an icon, or the parsing of a destination. The technological advances that make such mini-automations possible lie largely underneath the surface, like the bulk of an iceberg beneath the water.

Machinic entities

What these specific developments—auto-complete, geolocation, automatic route-calculation—represent is the gifting of greater forms of agency to machinic entities. Yet the kind, depth, scale and form of machinic agency granted in these cases is arguably different in each. Auto-complete ensured that the search box became a critical technical object in the mobile map interface, completing its migration from the search engine website, to the desktop map interface, to the mobile map app. Subsequent updates refined auto-completion further, with location suggestions based on a user's previous search history. With this, the search engine began to operate less like a retrieval tool and more like a recommendation machine, less 'here's what you're looking for', and more 'here's what we think you might like'.

Automatic geolocation offered a similar tale: rather than requiring the navigator to deduce their position from where they had been, retracing their steps or locating local landmarks, mobile map app users could rid themselves of having to determine their location entirely. In letting their device do the work, users were not only handing control over to a tiny GPS chip (and much larger satellites) but also granting permission for map app providers to locate them. While mobile network providers had long been able to triangulate their customers based on cell tower location, this was always a fuzzier estimate. With GPS chips, devices (and therefore users) could be located down to just a few metres—with all the privacy issues this shift entailed. Consequently, the GPS chip became an invaluable part of the mobile smartphone, underpinning all subsequent navigational requests made by the user.

Automatic route-calculation was perhaps the most visible of the mini-automations, through which users were able to witness—if not entirely understand—the computational work being performed. Most interestingly, automatic route-calculation yielded a new kind of relation between user and map, with the former able to assess the desirability, or even viability, of routes generated. While easy enough to accept—talk to the IDF officers or the football fans—the automatic generation of routes could also be seen as suggestions. Able to toggle different modes of transport (bus, train, underground, tram, car, bicycle), select a departure time, or choose between a 'best' route or 'less walking' on the Google Maps app, automatic route-calculation can clearly be seen as an assistive tool, more-or-less helpful, more-or-less of the time. While the mobile navigational experience is enabled by all of the aforementioned features and more, the mobile map app would be far less useful if users were forced to wholly determine routes themselves—even if we have come to recognise the idealised world they operate in.

But what marked these developments as significant was not that each of them granted greater agency to singular technical objects independent of anything else or each other. Instead, what was notable was the extent to which each of them *inter-acted, inter-connected* and *inter-operated* with each other, constituting a hybrid, complex, multi-agential navigational experience. With these developments it became clearer that the machinic agencies being generated were no longer solely locatable at the geographic location of the user. While they might have previously relied on the manual route compiler (think again of the AA) or paper map—both of which would have drawn their own machinic connections—the entanglements generated by the mini-automations meant these networks would be ever present and ever-changing. Every navigational moment, in every location, for every user, would rely on, and re-draw the entangled connections between auto-complete, geolocation, automatic route-calculation and more.

The user

Yet rather than seeing this process as necessarily taking away the 'art' of human navigation, these developments produced a suite of 'ironic' (Bainbridge, 1983) cultural effects, ushering in new ways to navigate. If one is to understand navigation (digital or otherwise) as a hybrid activity, composed of both ostensibly 'human' and 'non-human' components, of extant knowledge on behalf of the navigator, and forms of so-called tertiary memory on behalf of the technology, then the art, craft and skill of the mapmaker or map user is always already wrapped up with machinic entities. What happens with each technological shift, however big or however small, is a rearrangement of these components, rather than a wholesale replacement of one (human) with another (non-human). This is not to downplay or ignore the very real effects that are wrought on different kinds of human labour related to the production and use of maps, only to re-iterate that the terrain over which the battles are fought is a hybrid one, where human skill and non-human agency are entwined.

Those opposed to mini-automations understand this very well. Each mini-automation was integrated to smooth and improve the navigational experience for one specific type of person: the fabled 'user'. Although it may seem like an ahistorical, and certainly apolitical term, the user—as synonymous with the human or a 'person who uses our service'—has a history to it. In the annals of computing, the user is a relatively new phenomenon, born of the personal computer age. The user is what Microsoft or Apple saw, not IBM. IBM dealt in mainframe computers and centralised computing systems for large organisations (after all, IBM stands for International Business Machines). Centralised computing systems were not designed to be used (or even really accessed) by single, distributed, non-expert 'users'. In contrast, mainframe computers were used, maintained and upgraded by specialists within firms who relied upon computing infrastructure for performing administrative tasks like database management.

Yet it was a revolution in *personal* computers that created the user—if not identical to a living, breathing human, then at least a nominal representation of one, or however a person identified themselves when logging into *their* computer. At least in theory, if not always in practice, desktop computers were not meant to be used by just a single person, only that each person could log into them as a separate user and access personal files tied to their (user) account. The birth of mobile computing, mobile devices and mobile apps precipitated a shift from the user as a nominal representation of an individual to a (near) one-to-one mapping of person onto user—at least at the level of the mobile device, or the smartphone. Here it became more obvious that the person using the device or app was the *only* user and, most importantly, that their use of the device or app could be determined not from an *a priori* label ('user X or Y') or login credentials but from their *use* of the device or app itself. If operators of these devices, apps and of course, maps, could now determine the profile of a user based on their user activity, users had an increased *value* to firms who developed, delivered, maintained and updated the platforms themselves, infinitely determined by their user activity—what they typed into auto-complete, where they activated their GPS chip, the route they selected from A to B. Doing so repeatedly, iteratively, daily, or whenever, offered mapping companies incredible access to the daily lives of their users and, consequently, to build up an incredibly finely detailed picture of who their users were, and where they might want to go next.

Automation costs

So what has been the result of all these mini-automations? Have 1,000 mobile map app mash-ups bloomed? A million map API requests blossomed? A billion customised user experiences flowered? Maybe, but the overriding belief is that the mini-automations ushered in by big tech have only led to their consolidation. Rather than merely passively or subtly collecting data on their users, mapping companies long decided to actively shape the activities their users participated in, where they wanted to go, and what modes of transportation or routes they wanted to take. For general leisure users using Google Maps to go to a new restaurant on the other side of town, this appears genuinely useful. For Uber drivers forced to take routes instructed to them by an invisible algorithm dictating where they drive, this appears like a form of workplace control. In this penultimate section it is worth sketching out how these mini-automations have ushered in forms of 'algorithmic control', seeking to either 'nudge' behaviour in more subtle ways or fully shape and restrict activities through more explicit means.

Uber is perhaps the most well-known example in the latter category, dependent upon the design of a 'routing engine' to plan 'efficient' routes for their drivers to take. As Nguyen (2015) has written about, use of Dijsktra's search algorithm is ordinarily too slow for the kind of real-time route calculation required by the platform. Using something called 'contraction hierarchies', Uber's routing engine allows them to 'pre-process' a routing graph, in part by weighting each road segment according to their suitability. By also employing an 'A* search algorithm', a fancy version of Dijkstra's algorithm, Uber's routing engine is ordinarily able to respond to dynamic road conditions—factors like traffic that invariably affect the viability of a given road segment or route. Factors also that AA route compilers in the 1960s could only dream of considering! Recent employment of machine learning (ML) has further refined Uber's routing engine, intended to generate more accurate vehicle ETAs (estimated time of arrival) (Hu et al., 2022). Alongside more subtle nudges to drivers displayed through the Uber app (e.g. what music to play or how to keep their car tidy), automatic route-calculation pressures drivers to accept certain jobs and routes, in order to deliver optimised ETAs for both customer and platform. A more efficient route, determined algorithmically, means a quicker journey for the customer and a maximisation of profit for the platform. For the driver it ultimately means greater stress, forcing them to drive faster and work longer, while still maintaining the level of service determined by Uber themselves—save for losing precious stars when rated by customers at the end of their ride.

The other deleterious aspect of embedding these mini-automations in the wild is the environmental costs of building, running and updating the digital infrastructure that supports them. Although experienced as minor product updates, offering fleeting visibility to users, the infrastructures that support these mini-automations, from auto-complete to routing engines, are huge. Although calculating the specific cost of every navigational request or action reliant upon these mini-automations is complex, the scale of the problem is easy enough to understand. For example, there were 24.7 million downloads of the Google Maps app in the United States in 2022, with a further 9.1 million downloads of fellow Google/Alphabet-owned navigation app Waze (Statista, 2023). In 2020, the average data use of the Google Maps app was calculated at 0.73 MB per 20-minute period, with Waze using an average of 0.23 MB (Statista, 2020). Each navigational request made by a user relies on a global network of data centres from Dallas to Delhi and cloud-based services to manage requests from every user and every device. Each routing engine calculation, as the Uber example above suggests, now equally relies on computationally intensive ML methods. Each cutting-edge semiconductor chip and battery within each mobile device used for ever-more demanding navigational requests incorporating these mini-automations and more, requires increasing quantities of rare earth minerals, chemicals and water to manufacture. The costs of these mini-automations are manifold, and only set to proliferate with each new automated feature or addition.

Conclusion

Many of the developments may appear far removed from the practice of making or using maps. They may also appear as resolutely 'a-cultural' or at least thoroughly technical—rooted in the established practice of designing software but not much else. Yet this chapter has tried to tease apart the navigational reliance on such systems, and the necessary nature of their involvement in facilitating navigational requests made by all kinds of users on a daily basis. From finding a restaurant in an unfamiliar city, to picking up a passenger from the hospital, daily life around the globe is determined by the power of automated technologies, shaping what this part of this handbook refers to as 'cultural digitalities'. While this chapter has primarily focused on those 'mini-automations' just about visible to this daily user, there are myriad further automated processes that have changed how navigational requests are made at the so-called backend. Again, while these might appear to be lacking a cultural dimension, it is clear that cultural demands, expectations, norms and beliefs shape the very conditions of mobile navigation, and that new forms of 'automated culture' (Andrejevic et al., 2023) consequently emerge. In choosing to embed mini-automations such as auto-complete, automatic geolocation, and automatic route-calculation into map apps, firms like Google/Alphabet and Uber have fundamentally lowered the work required by human users to navigate. The effect of this in some cases might well be an easier, smoother, more hassle-free experience, but in many other cases, it has led to forms of algorithmic supervision, management and control. In addition, the environmental costs of building and supporting the development of such features risks increasing the burden on the earth to an even greater, impossible, largely foreseeable, but ultimately destructive, degree. The challenge—both now and in the near future—is how to cultivate fewer psychically and ecologically destructive navigational experiences in an age of automation. Whether this is possible remains to be seen.

References

AA (2023) *The AA Timeline: From the Early Years to the Present Day*. Available at: www.theaa.com/about-us/aa-history/timeline (accessed 17 February 2023).

Andrejevic M, Fordyce R, Li L and Trott V (2023) Automated culture: Introduction. *Cultural Studies* 37(1): 1–19.

Bainbridge L (1983) Ironies of automation. *Automatica* 19(6): 775–779.

Calder S (2012) *How to Get from A to B (20 Million Times a Month)*. Available at: www.independent.co.uk/travel/news-and-advice/simon-calder-how-to-get-from-a-to-b-20-million-times-a-month-8298282.html (accessed 17 February 2023).

Carraro V (2021) *Jerusalem Online: Critical Cartography for the Digital Age*. Singapore: Palgrave Macmillan.

Hu X, Cirit O, Binaykiya, T and Hora R (2022) *Deep ETA: How Uber Predicts Arrival Times Using Deep Learning*. Available at: www.uber.com/en-GB/blog/deepeta-how-uber-predicts-arrival-times/ (accessed 17 February 2023).

Nguyen T (2015) *ETA Phone Home: How Uber Engineers an Efficient Route*. Available at: www.uber.com/en-GB/blog/engineering-routing-engine/ (accessed 17 February 2023).

O'Brien S (2019) *Benfica Fan Travels to the Wrong Frankfurt Ahead of Europa League Clash*. Available at: https://talksport.com/football/529808/benfica-fan-wrong-frankfurt-europa-league/ (accessed 17 February 2023).

Statista (2020) *Average Data Use of Leading Map and Navigation Apps in the United States as of October 2020*. Available at: www.statista.com/statistics/1186009/data-use-leading-us-navigation-apps/ (accessed 17 February 2023).

Statista (2023) *Leading Mapping Apps in the United States in 2022, by Downloads*. Available at: www.statista.com/statistics/865413/most-popular-us-mapping-apps-ranked-by-audience/ (accessed 17 February 2023).

Tarantola A (2016) *Waze App Leads IDF Soldiers into Palestine, Conflict Erupts*. Available at: www.engadget.com/2016-03-01-waze-app-leads-idf-soldiers-into-palestine-conflict-erupts.html (accessed 17 February 2023).

21
MAP FETISHISM AND THE POWER OF MAPS
A feminist-technoscience perspective

Valentina Carraro

Introduction

The idea that maps have power is a tenet of critical cartography, to the point of sounding cliché. In this chapter, I revisit this truism, drawing on the scholarship on feminist technoscience, a transdisciplinary field investigating how gender and other markers of identity are entangled in technoscientific knowledge and practice. This perspective, I suggest, can help clarify what it means to assume that maps—as objects, images or technologies—can inform and persuade, create and transform, surveil and discriminate. My starting point is the work of Donna Haraway. Certainly, Haraway cannot be made to stand for the whole field of feminist technoscience, but her work has been remarkably influential in critical cartography and critical GIS, most famously through her critique of the god-trick, the cartographic gaze that purports to see the 'everything from nowhere'. Here, map scholars have largely focused on Haraway's epistemological writings (1985, 1991 [1989]), paying less attention to the empirical analyses of technologically mediated vision that she developed in subsequent years (Haraway, 1997, 2013)—a trend that, according to Åsberg and Lykke (2010: 300), characterises engagements with Haraway in the social sciences and humanities more broadly. Hopefully, then, this chapter can also offer new insight to readers who are not familiar with this side of Haraway's work or who are put off by her intentionally meandering prose.

In the following section, I start by discussing how Haraway's ideas about situated knowledges have influenced critical cartography, and highlight differences and overlaps with other critical approaches to maps. I then turn to Haraway's work on technologies of vision, a term that underscores how ways of seeing and understanding the world emerge from historically specific 'infoldings' (some would say, assemblages) of narratives, material objects, humans and non-humans. Specifically, I consider Haraway's examination of gene mapping and the associated discussion of map fetishism. In the conclusion, I link these discussions to recent anthropological scholarship that points to fetishes as a means to creating new social relations. Overall, my argument is that fetishism is a productive conceptual tool to think about maps, especially contemporary digital mappings, because it allows us to push against technological determinism while recognising the (very real) power of maps.

DOI: 10.4324/9781003327578-26

The god-trick of modern cartography

In her essay *Situated Knowledges*, Haraway (1991) draws on science and technology studies (STS) and feminist science studies to develop a notion of objectivity that is neither universalising nor relativistic. Put simply, the problem at stake is how to build reliable, truthful accounts of the world once we recognise that all knowledge is historically specific and socially shaped. For Haraway, modern Western science has relied on a ruse, or *god-trick*, to establish its legitimacy and trustworthiness. The god-trick consists in giving the illusion that science allows us to view 'everything from nowhere', a disembodied view that is free from perspective distortions and thus *seems* to coincide with the cartographic gaze of modern maps.[1] The trick conceals the social positioning and subjectivity of the viewer—the cartographer, the scientist, Man—generating the illusion that *his* perspective is not a perspective at all, but indeed 'life itself'.

To see the god-trick 'in action', we can turn to the work of critical cartographers, examining how modern cartography is implicated in the emergence of nation states, European colonialism and capitalism. As a well-known example, modern surveying techniques were devised and perfected to facilitate land transactions, and thereby the transfer of power, from local landlords to colonial elites. Mapping the land transformed the relation between mapmakers and mapped populations, but also land itself, consolidating its status as a commodity and legitimising its conquest (Harley, 1988; Edney, 1997). Looking at contemporary cadastral maps, it is easy to forget this history and thus also the fact that different ways of seeing (land) could have led to different maps and different politico-economic arrangements. This type of constructivist cartographic analysis started to emerge in the late 1980s, more or less at the same time as Haraway's work on situated knowledges, but it was inspired by Foucault (2007) and Said (1979) more than feminist technoscience. In the words of Brian Harley, critical cartography's putative father, the aim was to 'deconstruct the map', leveraging insight from social theory to 'search for the social forces that have structured cartography and to locate the presence of power—and its effects—in all map knowledge' (Harley, 1989: 2). At least in this respect, Haraway's approach is almost antithetical and prefigures the more recent turn to processual cartography; rather than drawing on critical epistemologies to critique maps, she takes mapping technologies as a starting point for the development of a distinctive epistemological stance. Her aim is to move beyond the critique of 'fixed appearances' and 'end products', investigating instead 'the apparatuses of visual productions' that make specific ways of seeing possible (Haraway, 2008: 196). The discussion of gene mapping in the next section provides a clear example of this approach.

For Haraway, the antidote to the god-trick lies in the recognition that all knowledge is situated, and that this partiality does not undermine objectivity. On the contrary, it strengthens it because it requires knowledge-producers to take responsibility for their claims and their claims' consequences, for the 'monsters' they create:

> The moral is simple: only partial perspective promises objective vision. This is an objective vision that initiates, rather than closes off, the problem of responsibility for the generativity of all visual practices. Partial perspective can be held accountable for both its promising and its destructive monsters.
>
> *(Haraway, 1991: 190)*

Situated knowledges do not mean that all perspectives are equally useful, a position that Haraway disavows as the 'mirror-twin' of totalising science, and as equally deceitful. Rather, producing situated knowledges entails forging connections between different ways of seeing, making them commensurable or, as Haraway (1991: 196) puts it, generating 'power-sensitive' conversations between them.

Fittingly, in her subsequent work, Haraway (1997) uses a cartographic example to clarify these arguments, drawing on research about maps and land ownership by Helen Verran and David Turnbull (1995). In the 1990s, after years of Aboriginal activism, the High Court of Australia recognised that indigenous people had owned the land before 1770, when the British laid claim to it. The courts ruled that native titles were valid in all cases where control had not been explicitly relinquished. Thus, changing power relations between Aboriginal and European Australians opened up an epistemological question: how do we know who owns a given piece of land? The ensuing legal negotiations forced both Aboriginals (the discussion focused on the Yolngu group) and Europeans to lay bare their knowledge systems in relation to land, making explicit their spatialisation practices. Through the encounter/friction between knowledge systems, it becomes possible to see land differently, and to produce a better, more objective map.

In the intervening decades, numerous map scholars and practitioners have built on the notion of situated knowledges to advance alternative mapping practices that privilege 'local' or 'indigenous' perspectives, framing them as juxtaposed to Western scientific knowledge (see Sletto, this volume). For anyone interested in cartographic humanities, these are fascinating and valuable endeavours. Despite the semantic similarities, however, *situated* knowledges and *local* knowledge are vastly different concepts. Indeed, feminist-technoscience and STS scholars take issue with the notion that local and scientific knowledge are qualitatively different, highlighting how scientific developments are driven by local action (Latour, 1993), and so-called local knowledge systems have their own means to travel through standards and technical devices (Verran and Turnbull, 1995). Western science *seems* universal purely because of the god-trick: its standards and technical devices are so powerful that they appear to us as 'natural'.

On map fetishism, or *Who is playing the god-trick?*

I find this to be an incredibly useful frame for thinking through how science works, and how it can be critiqued and nurtured at the same time. The god-trick trope, however, raises a question, namely: *who is playing the trick?* Scientists are arguably the first to be fooled; I am not keen on personifying science, and blaming a literal god lets humans off the hook too easily. To answer, it is helpful to turn to the notion of fetish, a term coined by Portuguese merchants to refer to the objects used in religious practices by West Africans. In more general terms, fetishes are material objects that are transformed by becoming objects of what is considered disproportionate or irrational desire or value. The transformation lies in the fact that a fetishised object *really does* exert power over those who worship it.

To explore this idea, I turn to Haraway's (1997: 131–148) discussion of gene mapping. In this chapter, Haraway takes issue with the scientific and popular discourses surrounding the rapid advancements in genetic biology that would lead to the mapping of the complete human genome in 2003. Sarah Franklin (1995)—whose work inspired Haraway's reflection—characterised this quest as a 'postmodern romantic adventure' with 'medieval resonance'. In many ways, these tones are those long used to celebrate the masculinist subjugation of nature at the hands of scientists, but what is distinctive here is that the start and end of it all is not nature, but the gene. Put differently, genes—the DNA strings that specify the makeup of the proteins of which organisms are made of—come to be seen as 'life-itself', as reality free from troping. The notion of map fetishism refers to this misrecognition and suggests that, because of how we relate to maps (and other scientific artefacts), they *really do* have power over us.

The map as commodity

In Haraway's reading, fetishism has a political-economy, psychoanalytic and cognitive dimension. The first dimension explains the power of the *map-as-commodity*. For Marx, commodity fetishism is one instance of alienation, the broader phenomenon by which even though social systems are human creations, people often fail to understand their workings, let alone control them. Commodity fetishism refers to how, under capitalism, consumers have come to see value as intrinsic to the objects themselves, rather than as a social relation. This misrecognition is made possible by the fact that the social relations between producers and consumers have been completely obscured by the relationship between things, specifically between money and products. To use a borrowed example (Harvey, 2018: 40), the price we pay for a head of lettuce at the supermarket expresses a social relation; it is the monetary representation that, as a society, we have attributed to the work that goes into producing it. This market exchange disguises the social relations that link consumers to the farmers, truck drivers and supermarket employees involved in making lettuce available to us. The system of production is so complicated that we cannot possibly see all of it, and so we mistake the part of it that we can see, the head of lettuce, for its totality. Nor can commodity fetishism be 'un-done' by recognising its illusory character. The exchange-value of lettuce is *truly* expressed by its supermarket price and cannot be changed by insisting on paying a different sum. Thus, commodification turns objects into mysterious things, with metaphysical properties, as if they had agency (see Hind, this volume).

This perspective is extremely helpful for making sense of the 'mapping revolution' that is said to have been underway since the early 2000s. Granted, strictly speaking, maps have long been commodities (see Brückner, this volume). Indeed, the moment maps became commodities roughly coincides with the moment they became scientific, as waged surveyors, cartographers and printing workers replaced craftsmen. On the other hand, the commodification of maps has expanded and intensified with the rise of an exceptionally lucrative market for geospatial data and the code used for its circulation and analysis. The social relations through which geographic information is produced are extremely complex and opaque. They involve highly paid software developers in Silicon Valley (and, at different points on the wage scale, Shenzhen, Chennai or Lagos), subcontracted workers tasked with more menial tasks that AI is too dumb to perform, users 'volunteering' information, factory workers producing the necessary circuits and chips, miners extracting the rare earth metals essential to the manufacturing of electronic devices, and, in keeping with Haraway's sensitivity for non-humans, the environmental costs involved in running the whole apparatus. Platform companies selling geographic information can capitalise on this labour, extracting value from spheres of social life that were previously relatively uncommodified (think of activities such as running in the park, or keeping in touch with friends). This trend is so significant that it has been identified as a new stage of capitalism (Srnicek and De Sutter,

2016), one in which (geospatial) data has become an end in itself, a means to general value, that is, a form of capital (Sadowski, 2019).

The map as masterplan

The psychoanalytical dimension of map fetishism enables Haraway to scrutinise the discourses that surround the gene as a masterplan for life in its totality. At odds with accepted biological knowledge, these accounts characterise the gene as a 'selfish' autonomous subject intent on replicating itself and, in the process, determining the characteristics of individuals, and, by extension, driving history. Just as Freud's archetypical young man is haunted by the knowledge that the fetish is only a surrogate (for his mother's phallus), so too we know on some level that the gene is 'just' a sequence of nucleotides with a crucial but defined biological function, and is only part of a complex cellular machinery. Yet, in a balancing act of knowledge and faith, 'DNA as information bearer' comes to be seen first as blueprint, then as masterplan, and then as master molecule (Lewontin, 1992: 240).

To understand how these ideas relate to contemporary mappings, it is useful to turn briefly to the work of another feminist-technoscience scholar, Wendy H. K. Chun. Chun (2008) observes a similar process of fetishisation in how media studies scholars regard source code[2] as 'sourcery', the essence that holds the key to understanding all digital objects and moments. As with genetic code, source code comes to be viewed as a language that transcends signification, that carries instructions while being free of meaning and thus existing outside the familiar structures of power-knowledge. The fetishism consists in conflating the process through which human-readable commands are turned into action with the source code, attributing to it almost-magical properties. This way of conceiving of hardware and software is historically specific and gendered, originating in the devaluing of the work performed by women operators in computations during World War II (Chun, 2005). Here too the information bearer is turned into the masterplan, and 'thinking' is valued above the 'physical labour' necessary to execute commands.

Psychoanalytical fetishism illuminates an iteration of the god-trick that is distinctive to contemporary digital mappings, namely our collective anxiety about algorithms and fetishistic belief/disbelief in geospatial software's power to 'do things'. Depending on one's attitudes towards technology, these 'things' include analysing data, finding patterns, managing and optimising flows, planning cities or economies, surveilling, discriminating and controlling people. On some level, we do know that it is not *really* code that performs these actions, but a combination of software, people and machines. As with Haraway's genetic scientists, however, knowledge and faith coexist. In fact, ethnographic observations suggest that even (or maybe especially) expert developers are prone to code fetishism (Thomas et al., 2018). And such is the working of fetishes that misrecognition has real consequences and confers actual power to code.

The map as truth

The third and last dimension of map fetishism originates in what Whitehead (1948) calls 'the fallacy of misplaced concreteness'. Whitehead's philosophy, which has profoundly influenced Haraway (as well as 'companion-thinkers' Bruno Latour and Isabelle Stengers), puts forward a processual ontology whereby reality is a process of becoming and everything takes the shape of an ongoing event. Whitehead's conception stands in contrast to the modern scientific ontology that instead imagines the world as made of entities (matters) located in space, and whose properties can undergo change but whose essence remains essentially stable. Incidentally, we can see a clear manifestation of this scientific ontology in geographic information systems, where spatial entities (point, lines and polygons) are associated with empirical properties (attribute data). Whitehead considers the scientific ontological model a useful abstraction and readily acknowledges that locating and measuring 'things' has produced real knowledge about the world. The fallacy, or, as Haraway would have it, the fetishistic misrecognition, lies in mistaking this abstraction for the concrete, for 'reality as it really is'.

Feminist technoscience's contribution to this onto-epistemological discussion has been highlighting the analytical and political importance of dwelling on things, people and situations that do not fit existing categories or measurement methods, that 'resist being enrolled' (Star, 1990). It is an argument for valuing marginality as a privileged perspective, where marginality should be understood in sociological terms, as the condition of inhabiting multiple categories and thus defying categorisation (Star, 2015). This condition does not necessarily coincide with subjugated or subaltern identities; indeed, as Haraway (1991: 192) notes, the very idea of identity presupposes stable subjects, at odds with a processual ontology. It can be transient and mundane, like having a rare allergy (Star, 1990), or ticking the 'None of the above' option when responding to a survey. Such situated perspectives are not purer, or more innocent, but they are likely to recognise the abstraction entailed in the process of categorisation and to realise their view is partial—in short, to call out the god-trick.

Conclusion

In classic anthropology and common speech alike, the word 'fetish' has strongly negative connotations, which rest on the assumptions made by Europeans[3] about the original African fetishists and about themselves. First, that African fetishists were entirely incapable of abstract thinking, that is, of being on some level aware that objects of worship stand in for something else: a god, or, in Graeber's (2005) interpretation, social bonds. Second, that the value system of gold-thirsty European merchants was more rational and objective than those of their African counterparts (Pietz, 1985). As it turns out, neither assumption is accurate.

More recent anthropological work has emphasised the important social functions played by fetishes. For example, as already mentioned, Graeber views fetishes as the material manifestation of social bonds. In this view, then, the admittedly irrational belief in the fetish is a generative leap of faith that is necessary to create new bonds and thus, ultimately, new social arrangements and institutions. From this perspective, the fetishists are right, in as far as our actions and the objects we create really do have power over us. The power of maps cannot be undone by revealing its existence, and, arguably, that is not even something we should wish for. As Graeber (2005: 431) puts it, in Haraway-esque terms:

> The key factor would appear to be, not whether one sees things as a bit topsy-turvy from one's immediate perspective . . . but rather, whether one has the capacity to at least occasionally step into some overarching perspective from which the machinery is visible, and one can see that all these apparently fixed objects are really part of an ongoing process of construction.

Maps can be insightful, useful, profound and transformative, as long as we regularly remind ourselves that they do not provide us with the full picture, indeed, that a full picture does not exist, and all knowledge is partial and situated. The danger comes from believing in the gods we create in a blind and dogmatic manner.

Notes

1 Technically speaking, some common map projections are perspective projections, meaning that they result from viewing the globe from a specific vantage point, usually located at a relatively short distance from the Earth's surface. It is also worth mentioning that good map-makers are all too aware that perspective distortions are unavoidable, in line with Haraway's insight that scientific practices do not need to fall into the god-trick.

2 Source code is written in a programming language using words and symbols. A compiler (a piece of software) translates source code into machine code, that is, long streams of noughts and ones that can be 'understood' by chips.

3 A notable example is Friedrich Hegel (1956: 93–94).

References

Åsberg C and Lykke N (2010) Feminist technoscience studies. *European Journal of Women's Studies* 17(4): 299–305.

Chun WHK (2005) On software, or the persistence of visual knowledge. *Grey Room* 18: 26–51.

Chun WHK (2008) On 'sourcery,' or code as fetish. *Configurations* 16(3): 299–324.

Edney MH (1997) *Mapping an Empire: The Geographical Construction of British India, 1765–1843*. Chicago: University of Chicago Press.

Foucault M (2007) *Security, Territory, Population: Lectures at the Collège de France, 1977–78*. Berlin: Springer.

Franklin S (1995) Romancing the helix: Nature and scientific discovery. In: Stacey J and Pearce I (eds) *Romance Revisited*. London: Falmer Press, pp. 63–77.

Graeber D (2005) Fetishism as social creativity: Or, Fetishes are gods in the process of construction. *Anthropological Theory* 5(4): 407–438.

Haraway D (1985) Manifesto for cyborgs: Science, technology, and socialist feminism in the 1980s. *Socialist Review* 80: 65–108.

Haraway D (1991) Situated knowledges: The science question in feminism and the privilege of partial perspective. In: *Simians, Cyborgs, and Women: The Reinvention of Nature*. New York: Routledge, pp. 83–202.

Haraway D (1997) *Modest_Witness@Second_Millennium. FemaleMan_Meets_OncoMouse: Feminism and Technoscience*. New York: Routledge.

Haraway D (2008) Crittercam: Compounding eyes in nanoculture. In: *When Species Meet*. Minneapolis: University of Minnesota Press (Posthumanities, 3), pp. 249–264.

Haraway D (2013) *When Species Meet*. Minneapolis: University of Minnesota Press.

Harley JB (1988) Silences and secrecy: The hidden agenda of cartography in early modern Europe. *Imago Mundi* 40(1): 57–76.

Harley JB (1989) Deconstructing the map. *Cartographica* 26(2): 1–20.

Harvey D (2018) *A Companion to Marx's Capital*, Complete edition. London: Verso.

Hegel GWF (1956) *The Philosophy of History*, Sibree J (ed.). New York: Dover Publications.

Latour B (1993) *The Pasteurization of France*, First Harvard University Press paperback edition. Cambridge, Massachusetts: Harvard University Press.

Lewontin R (1992) The dream of the human genome. In: Lewontin R and Levins R (eds) *Biology Under the Influence: Dialectical Essays on Ecology, Agriculture, and Health*. New York: Monthly Review Press, pp. 235–265.

Pietz W (1985) The problem of the fetish, I. *RES: Anthropology and Aesthetics* (9): 5–17.

Sadowski J (2019) When data is capital: Datafication, accumulation, and extraction. *Big Data & Society* 6(1). DOI:10.1177/205395171882054.

Said EW (1979) *Orientalism*, 1st Vintage Books edition. New York: Vintage Books.
Srnicek N and De Sutter L (2016) *Platform Capitalism*. Cambridge, UK and Malden, Massachusetts: Polity Press (Theory redux).
Star SL (1990) Power, technology and the phenomenology of conventions: On being allergic to onions. *The Sociological Review* 38(1_suppl): 26–56.
Star SL (2015) Misplaced concretism and concrete situations: Feminism, method, and information technology. In: Bowker GC, Timmermans S, Clarke AE and Balka E (eds) *Boundary Objects and Beyond: Working with Leigh Star*. Cambridge, Massachusetts: MIT Press, pp. 143–167.
Thomas SL, Nafus D and Sherman J (2018) Algorithms as fetish: Faith and possibility in algorithmic work. *Big Data & Society* 5(1). DOI:10.1177/2053951717751552.
Verran H and Turnbull D (1995) Science and other indigenous knowledge systems. In: Felt U, Fouché R, Miller CA and Smith-Doerr L (eds) *Handbook of Science and Technology Studies*. London, Thousand Oaks and New Delhi: SAGE Publications, pp. 115–139.
Whitehead AN (1948) *Science and the Modern World: Lowell Lectures, 1925*. New York: The New American Library (Pelican Mentor Books).

22

ETHNOGRAPHY AND MAPS IN THE DIGITAL AGE

Mike Duggan

Ethnographic mappings and ethnographic maps

Ethnographers have long used 'ethnographic mapping' to develop and convey an understanding of people and practices, alongside other representational apparatus including sketches, writing, photography and film. At the same time, 'ethnographic maps' have been popular devices used to represent different ethnicities for a range of social and political actors. This chapter is mainly about the former, with a focus on how digital mapping technologies have led ethnographers to find new ways to understand people through maps, and to develop new mapping practices to engage participants and audiences.

The history of ethnographic mapping can be traced back to the anthropological handbooks (see Garson and Read, 1892) and the so-called godfathers of the ethnographic method in social and cultural anthropology, dating from the late 19th century.[1] Franz Boas famously mapped the Inuit cultural practices on Nunatsiarmiut by geo-locating knowledge, stories, rituals and toponyms, as well as recruiting Inuit people(s) to produce maps for him in the hope that they would shed light on cultural practices and land-relations. Similarly, but without direct involvement from participants (as far as I know), Bronislaw Malinowski (1922) produced maps of the Kula Ring kinship system in New Guinea to highlight significant places, trade routes and the flows of intra-cultural rituals. With a focus on urban environments, Robert Park and Ernest Burgess (1925) created the now well-known concentric city models of Chicago, which went on to shape the Chicago School's approach to urban ethnography and a fair amount of urban geography for much of the 20th century.

Since then, ethnographic mapping has come in all shapes and sizes and has been defined as a specific form of anthropological mapping and a key element of visual ethnography (Pink, 2021). It differs from topographical or political mapping, in that it focuses on representing how people interact with spaces and places, and used to show how spaces shape social and cultural practices (Martin, 2018; Pelto, 2016).[2] Maps have become indispensable for many ethnographers, used for preliminary enquiries and constructed 'in the field', as well as used during the analysis and dissemination of their work. These have ranged from

DOI: 10.4324/9781003327578-27

quickly drawn sketch maps of local geographies, to detailed and carefully mapped representations of daily rhythms and practices (Gell, 2001), participatory and counter-maps produced with participants (Sletto, 2009; Sletto et al., 2020), to diagrammatic and systems maps laying out kinship relations, family trees and trade-exchange networks (Partridge, 2013). Then there are the maps that have been used to contextualise field sites and findings among broader social, economic and environmental processes. These include maps produced by and with professional cartographic and statistical agencies, governmental authorities and other researchers. Above all else, maps have been devices to *think with*, to guide the ethnographer through the processual nature of their practice (Becker, 2007).

The history of ethnographers using maps should therefore be recognised as experimental and process driven, with evidence of researchers trialling different theoretical and practical approaches to mapping. This is unlike the history of other methodological approaches where maps and mapping practices have tended to support research outputs, but have done so by following the conventions of cartography, which is to locate things on the archetypical geographical map (see e.g. anglo-centric geography during the 'quantitative revolution' of the 1950s and 1960s). This either reflects the adaptable and reflexive nature of ethnographic study, whereby researchers are forced to change and reflect on their practices based on the situations they find themselves in, or shines a light on the lack of published reflexivity in positivist methodologies.[3]

The history of ethnographic mapping is also a history of ethnography, tied as it is to colonial, racist and patriarchal projects to understand, control and coerce 'the other'. It is not simply a method, but rather a praxis that has caused a great deal of damage along the way. To map *as* an ethnographer is a privileged position, and with that comes the power to shape how someone else's life might be affected by your map. It is no surprise to find that the history of ethnographic mapping is one of affluent white men in privileged positions.

Today, ethnographic mapping is a practice decoupled from the traditions of longitudinal ethnographic research and is now widely used to describe research that maps human experiences and social practices, or participatory research practices that use maps as form of people-writing to represent a range of localised social issues. King et al. (2022), for example, have recently used 'ethnographic mapping' to describe a single participatory mapping workshop designed to draw out the local concerns and responses to waste management in fast growing cities in the Global South. We have also seen the 'ethnographic mapping lab' emerge as a learning space for understanding what maps can do to represent social and cultural practices and empower marginalised communities.[4] These trends represent an altogether different approach to ethnographers that immerse themselves in another culture for lengthy periods of time. They align more with the traditions and histories of participatory mapping and action research (see Guldi, 2017). As ethnography has gained acceptance and traction in other social and humanities disciplines as part of various disciplinary 'ethnographic turns' in the 1990s and 2000s, the grip that anthropology once had on the method, and especially its longitudinal form, has lessened. Like much of ethnographically informed research today, what counts as 'ethnographic mapping' has very different interpretations depending on where you look.

The 'ethnographic map' on the other hand is a wider and more popular category of map used to locate ethnic or racial groups.[5] Ethnographers do, of course, produce and inform ethnographic maps, but so do others with an interest in this form of representation. This is an important distinction to make as the practice of *doing ethnography with* maps is not the same as *designing* a map using data on ethnicity or race. As above, doing ethnography with maps involves using maps to help shape the process as well as disseminate the findings, whereas designing a map with such data is a cartographic practice.

There is a contentious history to the ethnographic map as it makes use of cartographic conventions including borders, toponyms and colour to lay claim to territory, shape national identities and exert power over marginalised groups. It is a history tied to exertions of colonial power, nationalism and multiple discriminations. This can be seen on maps such as the 'Ethnographic map of the world showing the present distribution of the leading races of man' (Cornelius S. Cartee, 1856),[6] which readers will recognise as inaccurate and ignorant of the world's people. Culcasi and Asch (2015) argue that ethnographic maps were especially popular in the inter-war years of 20th-century Europe, as nation states aimed to establish themselves along ethno-political lines. And they remain in use today, used to convey simple messaging during geopolitical conflicts such as wars and migration crises. At the same time, ethnographic maps have been beneficial in fights for recognition used to support claims to land and jurisdiction (see Vaughn, 2018). Decolonial mapping projects are a case in point. These ethnographic maps are intended to put ethnicities back on the maps they have historically been removed from.[7] Elsewhere, ethnographic maps can be found widely as cultural objects including in books and atlases, found in museum gift shops, symbolised in tourist paraphilia and depicted regularly on film and TV.

Ethnography and digital mapping practices

A key feature of ethnographic mapping over the past two decades has been the digital technologies used to make and disseminate maps. Most notably, geographic information systems (GIS) and internet mapping platforms have given ethnographers novel ways to gather and represent spatial information about their participants, whether it be layering information over a base map, broadening participatory mapping approaches through free-to-use web technologies, or easily disseminating their maps through the internet.

These trends continue to this day, when it is common to find ethnographers adopting GIS and digital mapping platforms into their work. 'Geo-ethnography' has been one such field to emerge in recent years to explicitly frame and theorise the coupling between GIS and ethnography (MacNell, 2018; Matthews et al., 2005). To a lesser extent, 'spatial anthropology', the 'spatial humanities', 'deep mapping' and 'citizen science' methodologies are other fields that have found valuable benefits in combing digital maps with ethnographic enquiry (Brennan-Horley et al., 2010; Low, 2016; Roberts, 2016). 'Cartographic storytelling' is one such area that has recently come to the fore in this area. Digital mapping platforms such as ESRI's popular StoryMaps tool and the UK based Humap have given ethnographers the opportunity to tell participant stories—a key feature of ethnographic work—through augmenting maps with audio, visual and textual material (see e.g. Saxinger et al.'s 2022 cartographic stories of life along railway lines in Serbia). From a dissemination point of view, there is little doubt that these tools provide a simple and engaging way for ethnographers to tell and share spatial stories beyond the academic monograph or article. Nevertheless, as Lo Presti (2022) rightly notes, storytelling with maps does not have to end with putting stories on maps. Instead, ethnographers could take a broader approach to telling stories with maps by designing methodologies that utilise the power of maps as objects, practices and tools to encourage conversation, elicit debate and inspire deeper reflective narratives from their participants and field sites (see also Rossetto, 2019). In this sense, *storymapping* may be a more apt term, as following Cosgrove, 'the mapping's record is not confined to the archival: it includes the remembered, the imagined, the contemplated' (1999: 2).

Alongside these ethnographically defined studies, there are many studies that have combined qualitative interviews, observation and content analysis with GIS in what's come to be known as 'qualitative GIS'. This is a burgeoning field in itself, often encompassing ethnography, and attracting the attention of academics, activists, artists, governments, NGOs, planners and policy makers alike (for much more on qualitative GIS, see Jung and Elwood, 2019; Kwan and Knigge, 2006).

The focus of ethnographic mapping has tended to be on what is shown on the map, rather than on the socially and materially situated ways that maps are being produced, used and circulated in the world. Adjacent to the digital developments in ethnographic mapmaking, there have been calls and responses for ethnographic studies into those that produce and use digital maps, including amateur and professional cartographers, and everyday users. Although John Fels and Denis Wood recognised the need for what they called an 'anthropology of cartography' in 1986 (Wood and Fels, 1986), it was only later, during the post-representational turn in critical cartography that this strand of ethnographic enquiry properly took off. Key to this was Suchan and Brewer's (2000) and Del Casino and Hanna's (2005) work that urged us to move beyond the binaries that separated maps as spaces of representation from the spatial practices that involved them. Further interventions by Perkins (2004) and Dodge et al. (2009) clearly positioned ethnographic enquiry as a unique way to study the nuances of what maps do in the world, and what kinds of social and cultural activity they support. Grasseni (2004) offered a concrete example of this with her ethnographic study of Valtaleggio, North Italy, which examined the processes that went into producing a map designed to attract tourists to the area. Through observing the (sometimes tense) negotiations and participation of the local stakeholders involved, Grasseni produced a 'cartographic memoir' as her ethnographic account of how the map came to be.

This is where I see the greatest opportunity for ethnographic engagement with maps today, especially when digital and mobile maps have come to feature so heavily in much of what people do on a daily basis. This is not simply about the map. There is much value in these approaches for understanding social and cultural life and how it unfolds with and through the digital map, and in how economic trends are fast becoming established with new mapping technologies. Here, I am specifically talking about the trends in location-based social networking and gaming apps (e.g. foursquare, Pokémon Go and 'augmented reality' gaming), location-based design features (e.g. location sharing and tracking), and the significant economic value that has been placed on spatial and location-based data (Wilken, 2020), all of which maps have become a primary interface for.

Brown and Laurier's (2005, 2012) studies about how navigating with maps and GPS devices is a social as well as spatial activity were pioneering examples of ethnographic approaches to understand map user behaviour. By unpacking the minutia of navigational practices and experimenting with video and speech analysis, they created an ethnomethodological framework for navigational studies that to this day are rarely surpassed in terms of attention paid to the process of map reading. Following this, other ethnographic studies of digital navigation, including Wilmott's (2017) study of mobile navigation in post-colonial urban environments, and Duggan's (2017) study of multi-modal navigational encounters with maps, have underlined the importance of context in understanding how maps are used.

Away from ethnographies of navigation have been studies of digital cartographic studios and professional agencies. Gekker's (2016) study of digital mapmaking in the commercial sector offers a good example of how long-form ethnographies can reveal the nuances in how social relations and technical interfaces affect the production of maps. Similarly, Boria and Rossetto's (2017) study of a professional cartographer in Italy is notable for merging an ethnographic perspective with other textual and qualitative methods with the aim to understand how maps emerge in the world. Digital mapping platforms and open sourced mapping communities have also encouraged ethnographic studies of participatory mapping communities (Duggan, 2019; Forero, 2021). Gerlach (2015), for example, studied how OpenStreetMap was used in Lima to form participatory and creative communities around the act of world building with maps. Lin (2011) utilised ethnographic methods to explore how one becomes and instils the values of an 'OpenStreetMapper' in a community group through repeat collaboration with people and what she calls the 'boundary object' of a map itself.

Finally, autoethnographic studies of digital mapping have recently started to gain traction. David Garcia (aka @mapmakerdavid), for example, has documented the process of becoming and unbecoming a GIS and open-source mapper. Others have looked to their own experience in cities to study the glitches that unfold in everyday uses of locative media in urban environments (Leszczynski, 2020). Going further still, some have experimented with 'carto-fiction' as a way to document the autoethnographic study of lived urban experiences (see Peterle, 2019, and her chapter in this volume). While ethnographers of maps have long taken a self-reflexive approach to their research, these studies highlight how personal experiences with maps can reveal nuanced details in how they shape everyday practices that would be difficult to gather from research participants.

Conclusion

Maps and mapping methodologies are popular among ethnographers in many fields. We can expect this to continue for as long as maps remain useful for developing an understanding of people, places and practices. Researchers across the social sciences and humanities still place great value in maps for doing so (Rucks-Ahidiana and Bierbaum, 2015). Nevertheless, there is more to be done here in *deconstructing ethnographic mapping* (cf. critical cartography) and reflecting on how maps fix people-writing (ethnography) in place. The danger is that ethnographic mappings are not seen for what they are, which is always-biased representations of the world, and instead used as a research output rather than a processual tool for understanding human experiences. Much like ethnographic writing, photography, video and sounds, maps encounter the same impossible problem of truly conveying ongoing and dynamic life stories, and this is something to be reckoned with. This does not make them useless because they do add another viewpoint from which the ethnographer can tell the stories of others, which is valuable if it engages with the reader, but this form of reflection should become part of the reflexivity central to the ethnographic method.

Ethnographers have, ever since Clifford Geertz's (1973) influential writings at least, operated on the epistemological grounds that interpretation and representation are recognised as iterative and selective processes. It follows then that maps should be folded into this approach, not just representing the findings of a study, but instead used along the way to shape research questions and directions. Understanding mapping as a performative and processual—encouraging dynamic behaviours, attitudes and practices to emerge—is key here, because it situates ethnographic maps and mapping as objects and practices that are never finished, but rather always already a part of an ongoing processes of understanding people and what they do. I urge ethnographers to continue to reflect on how and why mapping is the appropriate method for them. If ethnographic mapping is an object and practice that is taken for granted, then its powerful effects may go unchecked. One only has to look at the effects that ethnographic maps have had throughout history to see this. In a world where digital technologies continue to make producing and disseminating maps easier than ever, ethnographers have a duty to reflect on this.

As we have seen, the digital era offers exciting possibilities for mapping ethnographers and ethnographers of mapping practices. We find ourselves seemingly overrun with ways to digitally map and tell the stories of people. Many are already taking full advantage of this, across academia and beyond. Nevertheless, the role of interpretation and good storytelling remains central to effective ethnographic scholarship, and we should not let the alluring veneer of digital mapping technologies distract us from this. The key question of how to produce a reflexive account of culture(s) that audiences can relate to, emphasise with and react to, should remain our primary concern. In some circumstances, mapping technologies will be an excellent medium for storytelling, but in others they will not. Therefore, ethnographers should begin not with the mapping technologies they would like to use, even encouraged to use, but instead consider the story they want to tell and the audience they want to share it with. This follows the tradition of reflexivity in ethnographic research, which asks researchers to carefully consider their mode of representation and their positionality as storytellers.

Notes

1 All of which remain problematic when we consider their roles in re(producing) racist ideologies, disseminating Indigenous knowledge without permission and erasing indigeneity among other ills.
2 See also 'ethnograms', https://anthropologyoffthegrid.wordpress.com/ethnograms/
3 Clearly it would be untrue to suggest that positivist researchers do not reflect on the processes of doing research with maps, but owing to the conventions of published scientific work, it is rare to read the kinds of reflexive accounts about mapping that you will come across often in ethnographic, and more broadly qualitative literature.
4 See, for example, the ethnographic mapping lab at the University of Victoria, Canada.
5 Linked to this are 'ethno-linguistic' maps, dialogue maps (see Ormeling, 1983).
6 www.davidrumsey.com/luna/servlet/detail/RUMSEY~8~1~288330~90059867:Ethnographical-map-showing-the-dist
7 See, for example, https://decolonialatlas.wordpress.com and https://native-land.ca

References

Becker HS (2007) *Telling About Society*. Chicago: Chicago University Press.

Boas F (1888) *The Central Eskimo*. Washington: Annual Report 6 of the Bureau of Ethnology to the Secretary of the Smithsonian Institution, 1884–1885.

Boria E and Rossetto T (2017) The practice of mapmaking: Bridging the gap between critical/textual and ethnographical research methods. *Cartographica* 52(1): 32–48.

Brennan-Horley C, Luckman S, Gibson C and Willoughby-Smith J (2010) GIS, ethnography, and cultural research: Putting maps back into ethnographic mapping. *The Information Society* 26(2): 92–103.

Brown B and Laurier E (2005) Maps and journeys: An ethno-methodological investigation. *Cartographica* 40: 17–33.

Brown B and Laurier E (2012) The normal, natural troubles of driving with GPS. *Proceedings of the SIGCHI Conference on Human Factors in Computing Systems*: 1621–1630.

Cosgrove D (1999) *Mappings*. London: Reaktion Books.

Culcasi K and Asch CR (2015) Ethnographic map. In: Monmonier M (ed.) *History of Cartography Project Volume 6*. Chicago: Chicago University Press, pp. 409–413.
Del Casino Jr VJ and Hanna SP (2005) Beyond the 'binaries': A methodological intervention for interrogating maps as representational practices. *ACME: An International Journal for Critical Geographies* 4(1): 34–56.
Dodge M, Kitchin R and Perkins C (eds) (2009) *Rethinking Maps: New Frontiers in Cartographic Theory*. Abingdon: Routledge.
Duggan M (2017) *Mapping Interfaces: An Ethnography of Everyday Digital Mapping Practices*. PhD Thesis, Royal Holloway University of London.
Duggan M (2019) Cultures of enthusiasm: An ethnographic study of amateur map-maker communities. *Cartographica* 54(3): 217–229.
Forero J (2021) Composition is also coordination: Multi-actor digital mapping in Madrid. *Livingmaps Review* 11.
Garson JG and Read CH (1892) *Notes and Queries on Anthropology*. London: Anthropological Institute.
Geertz C (1973) *The Interpretation of Cultures*. London: Fontana Press.
Gekker A (2016) *Ubiquitous Cartography: Casual Power in Digital Maps*. PhD Thesis, Utrecht University.
Gell A (2001) *The Anthropology of Time: Cultural Constructions of Temporal Maps and Images*. Abingdon: Routledge.
Gerlach J (2015) Editing worlds: Participatory mapping and a minor geopolitics. *Transactions of the Institute of British Geographers* 40(2): 273–286.
Grasseni C (2004). Skilled landscapes: Mapping practices of locality. *Environment and Planning D: Society and Space* 22(5): 699–717.
Guldi J (2017) A history of the participatory map. *History Faculty Publications* 10.
Jung J and Elwood S (2019) Qualitative GIS and spatial research. In: Atkinson P, Delamont S, Cernat A, Sakshaug JW and Williams RA (eds) *Sage Research Methods Foundations*. London: Sage.
King J, Sajjad F and Ahmed S (2022) *Ethnographic Mapping as a Critical Pedagogical Tool in Planning and Architecture*. LSE Cities Working Paper. Available at: www.lse.ac.uk/Cities/Assets/Documents/RRR/LSE-Cities-WP-Ethnographic-mapping.pdf (accessed 20 January 2023).
Kwan M-P and Knigge L (2006) Doing qualitative research using GIS: An oxymoronic endeavor? *Environment and Planning A: Economy and Space* 38(11): 1999–2002.
Leszczynski A (2020) Glitchy vignettes of platform urbanism. *Environment and Planning D: Society and Space* 38(2): 189–208.
Lin Y-W (2011) A qualitative enquiry into OpenStreetMap making. *New Review of Hypermedia and Multimedia* 17(1): 53–71.
Lo Presti L (2022) Leaving or rescuing the (story) map? Commentary to Saxinger, Sancho Reinoso and Wentzel. *Fennia* 200: 68–71.
Low S (2016) *Spatializing Culture: The Ethnography of Space and Place*. Abingdon: Routledge.
MacNell L (2018) A geo-ethnographic analysis of low-income rural and urban women's food shopping behaviors. *Appetite* 128: 311–320.
Malinowski B (1922) *Argonauts of the Western Pacific*. London: G. Routledge & Sons.
Martin T (2018) Ethnographic mapping. *CUNY Ethnography Made Simple Series*. Available at: https://cuny.manifoldapp.org/read/untitled-fefc096b-ef1c-4e20-9b1f-cce4e33d7bae/section/38541549-7b00-4307-a5b3-c361e5dc6f2b (accessed 20 January 2023).
Matthews SA, Detwiler JE and Burton LM (2005) Geo-ethnography: Coupling geographic information analysis techniques with ethnographic methods in urban research. *Cartographica* 40(4): 75–90.
Ormeling FJ (1983) Minority toponyms on maps: The rendering of linguistic minority toponyms on topographic maps of Western Europe. In: *Volume 30 of Utrechtse Geografische Studies*. Utrecht: University of Utrecht.
Park R and Burgess E (1925) *The City*. Chicago: The University of Chicago Press.
Partridge T (2013) Diagrams in anthropology: Lines and interactions. *Anthropology of the Grid*. Available at: https://anthropologyoffthegrid.wordpress.com/ethnograms/diagrams-in-anthropology/ (accessed 31 January 2023).
Pelto PJ (2016) *Applied Ethnography: Guidelines for Field Research*. Abingdon: Routledge.

Perkins C (2004) Cartography-cultures of mapping: Power in practice. *Progress in Human Geography* 28(3): 381–391.
Peterle G (2019) Carto-fiction: Narrativising maps through creative writing. *Social and Cultural Geography* 20(8): 1070–1093.
Pink S (2021) *Doing Visual Ethnography*, 4th edition. London: Sage.
Roberts L (2016) Deep mapping and spatial anthropology. *Humanities* 5(1): 5.
Rossetto T (2019) *Object-Oriented Cartography: Maps as Things*. London and New York: Routledge.
Rucks-Ahidiana Z and Bierbaum AH (2015). Qualitative spaces: Integrating spatial analysis for a mixed methods approach. *International Journal of Qualitative Methods* 14(2): 92–103.
Saxinger G, Sancho Reinoso A and Wentzel SI (2022) Cartographic storytelling: Reflecting on maps through an ethnographic application in Siberia. *Fennia—International Journal of Geography* 199(2): 242–259.
Sletto BI (2009) "We drew what we imagined": Participatory mapping, performance, and the arts of landscape making. *Current Anthropology* 50(4): 443–476.
Sletto BI, Bryan J, Agner A and Hale C (2020) *Radical Cartographies: Participatory Map Making from Latin America*. Austin, TX: Texas University Press.
Suchan TA and Brewer CA (2000) Qualitative methods for research on mapmaking and map use. *The Professional Geographer* 52(1): 145–154.
Vaughn L (2018) *Mapping Society*. London: UCL Press.
Wilken R (2020) *Cultural Economies of Locative Media*. Oxford: Oxford University Press.
Wilmott C (2017) *Living the Map: Mobile Practices in Postcolonial Cities*. PhD Thesis, University of Manchester.
Wood D and Fels J (1986) Designs on signs: Myth and meaning in maps. *Cartographica* 23(3): 54–103.

23
A HUMANISTIC REWIRE OF GISCIENCE

Bo Zhao

Introduction

Geographical Information System (GIS), as used in this chapter, describes the system or technology that creates, processes, displays and analyses geographical data, and GIScience refers to the scientific field that studies these systems and the data they engage with. Over the past two decades, GIS has intertwined increasingly with our daily lives than ever before. The emergence of GIS instantiations, such as Location-Based Services, Geospatial Artificial Intelligence, Internet of Things, Metaverse and Digital Twin, has accumulated vast amounts of geospatial data that have significantly transformed state governance, business operations and our everyday lives. In response to the constant proliferation of geospatial data in the Anthropocene, computer scientists have advocated for incorporating humanism into the contemporary development of AI, such as 'human-centred AI' (Li and Etchemendy, 2018) and 'making AI that is good for people' (Li, 2018). Drawing inspiration from these endeavours, this chapter casts a critical gaze on how GIScience, as a scientific field, can incorporate a humanistic viewpoint to better navigate today's data-intensive society. This chapter encourages cartographers, GIScientists and geographers to embed essential humanistic values in their research and to extend humanistic care to underserved communities. Subsequently, the remainder of this chapter delves into the evolution of humanism and its impacts on modern geographies and the phenomenology of technology. It then analyses how a humanistic lens can rewire GIScience and can provide a caring stance when creating or using GIS instantiations. This chapter concludes with a summary and suggestions for potential research directions.

Humanism and its impacts on geographies

Humanism is a philosophical stance that emphasises the value and agency of human beings (Penman and Adams, 1982). J.K. Wright (1947), in his AAG presidential address during the mid-20th century, urged geographers to incorporate a humanistic perspective in their research and focus on the terra incognita—the subjective geographical understanding of the world. Building upon this call, Humanistic Geography emerged as a prominent research perspective in the 1970s, drawing inspirations from humanism tradition. Yi-Fu Tuan (1976)

DOI: 10.4324/9781003327578-28

coined the term 'Humanistic Geography' to describe a list of research themes that aimed to provide a comprehensive understanding of various human experiences in the geographical world, building on earlier discussions in the field. For example, Tuan (1971) argued that the notion of 'human-being-in-the-world' was a critical research subject for geographers because it integrated the environmental, geographical and place aspects of human experiences. Ann Buttimer (1974) examined the notion of the 'taken-for-granted world' and how individual, unselfconscious understandings shaped geographical knowledge. In her work, Buttimer (1976) also interpreted the idea of 'lifeworld' as the 'taken-for-granted world' of everyday living. These conceptualisations have greatly advanced humanistic exploration of social space, place and time-space rhythms. Later, David Ley and Marwyn Samuels (1978) compiled various early individual and collective endeavours into a book titled *Humanistic Geography: Prospects and Problems*. This edited collection illustrated various theoretical aspects of humanistic geography in the 1970s, including philosophical foundations, epistemological orientations and methodological implications. It demonstrated how a humanistic perspective expanded the research scope of modern geography and envisioned potential research directions.

Humanistic geography faced criticism from critical geographers after 1978, particularly on issues like essentialism, masculinist bias and an uncritical interpretation of social life without deep consideration of power relations. This criticism prompted an introspective examination of humanistic geography's core principles (Seamon and Lundberg, 2017). Indeed, humanistic work recognised human diversity, evident in David Ley's early work (1972) on exploring the lives of African Americans in Philadelphia's inner city. The 'being' in humanistic geography does not overrule the existence of various individuals. Similarly, the human experience is not an abstract concept but includes various experiences stemming from differing social-economic and cultural statuses, such as sexual orientation, gender identity, race/ethnicity and age range. Humanistic geography did accept Marxist criticism that it inadequately addressed social or power relations. Facing criticism, many humanistic geographers turned towards a social constructive perspective on human experience and praxis by studying philosophers like Michel Foucault, Jacques Derrida, Gilles Deleuze, Bruno Latour and other critical social theorists (Cresswell, 2014). This trend in humanistic geography has been well documented in the edited collection *Texture of Place* (Adams et al., 2001). The authors also called for a reconsideration of the research themes of humanistic geography and urged for an engagement with critical social theories.

Phenomenological perspectives on technology

In addition to the influence on humanistic geography, phenomenology also shed light on a humanistic perspective on the use of modern technology in today's data-intensive society. Edmund Husserl (2012 [1931]), the founder of phenomenology, sought to distil the essence of a human to their interaction with the lifeworld. Similarly, Martin Heidegger (2010 [1927]) focused on the essence of being, which refers to the philosophical idea of a human being from the first-person perspective, rather than the overused biological status denoted by the term human. As Heidegger noted, the essence of being is being-in-the-world (*Dasein*), signifying that humans are defined by their interactions with the world. Heidegger also discussed the essence of technology in his later work, arguing that technology is not merely a collection of tools, machines or advanced methods; instead, technology is a way of revealing—it shapes how we perceive and interact with the world. Nevertheless, as modern technology advanced significantly, Heidegger observed an unsettling shift in this way of revealing. He noted that humans and nature were increasingly perceived and treated as mere resources ripe for exploitation. This shift, Heidegger argued, was changing the fundamental essence of technology. Technology was no longer just a way of revealing the world; it was now enframing our understanding of the world, putting a restrictive frame around the way we perceive nature and humans. This shift in the essence of technology also invites us to re-evaluate our relationship with the world and reconsider the potential consequences of our technological pursuits.

Building on Heidegger's view, Don Ihde (1990) brings attention to the mediating function of technology within human practices in the world. He builds up his theory under the overarching frame of post-phenomenology. It innovatively describes the mediating role of technology and further notes that technology can be embodied, read, interacted with or exist merely in the background of human experience. Ihde also insightfully observes that technology can enhance and simultaneously diminish human experience. The field of post-phenomenology has laid the foundation for a systematic examination of technology in its relationships with human practices and the surrounding environments. Don Ihde's post-phenomenology (1990) has also been further developed by his follower Peter Verbeek (2020) and geographers, such as Ash and Simpson (2016) and Lea (2020). This perspective transcends a focus on human experience and instead acknowledges the creative forces that shape the world and involve both human and non-human entities. This creative force extends experience trans-humanly and spreads across bodies and places. It does not just occur within the human body but can also take in all manners of things, including both human and non-human entities (e.g. animals, bots, algorithms).

Humanistic rewire

By adopting a post-phenomenological perspective, GIS can move beyond a narrow theoretical focus on human experience and encompass a broader range of human or non-human entities in its interplay. For instance, when a Facebook user checks into a restaurant using a GPS-enabled smartphone, the user (a human entity) initiates a creative force directed at the restaurant (a non-human entity), which is then interpreted as a check-in—a piece of geographical data on the smartphone. This geographical data on smartphone mediates the creative force between the user and the restaurant. In addition, the creative force can exist without human actors, such as when two self-driving cars interact with each other while running on the same road (see Hind, this volume). Each car models its surrounding environment and identifies other vehicles on the road, allowing them to navigate within the flow of traffic and interact with each other. In this instance, either of the two self-driving cars is a non-human actor.

Furthermore, I examined different creative forces that spread through GIS and different types of human or non-human entities (Zhao, 2023). By analysing the proximity between the GIS and the entities involved, I identified four major types of GIS, including Embodiment GIS, Hermeneutic GIS, Autonomous GIS and Background GIS (Zhao, 2022). In the case of Embodiment GIS, the GIS becomes an integral part of the actor's experience. An example of this would be a wearable device such as a Virtual Reality (VR) goggle, where the wearer and the device become a symbiotic unit, and the wearer's visual sensory input considers the virtual environment in the VR system as a real place. Hermeneutic GIS, on the other hand, interprets geographical environments into representations such as maps or geospatial data. Navigation maps are a good example of this type of GIS, as they interpret traffic patterns in a way that drivers can easily understand. Autonomous GIS behaves as an independent entity, such as a self-driving vehicle that is capable of navigating itself within the flow of traffic. The fourth type, Background GIS, seamlessly merges into geographical environments and is not perceivable by users. However, it still exerts influences. An example of this type of GIS is a natural park management platform that collects environmental parameters using sensors or monitoring cameras. Although visitors may not be aware of these devices, the collected data about their presence informs the management of human flows within the park. It's worth highlighting that a specific GIS might not fit into a single type of GIS category mentioned earlier. Instead, it can embody a combination of types simultaneously. For instance, a metaverse can be an embodiment GIS by integrating with its user when they wear VR goggles, and it can also function as a hermeneutic GIS when it interprets a virtual environment on its screen. This new classification system is useful for illustrating the various ways in which GIS enables creative interactions between different human and non-human entities.

Some may argue that the post-phenomenological perspective broadens the definition of GIS to include anything that momentarily interacts with geographic data. While I agree that this perspective does expand the concept, I also see it as a key contribution of the post-phenomenological approach. It recognises that a thing should not be defined solely by its function but must take into account the creative forces that shape its relationship with GIS and other human or non-human entities. For instance, a self-navigating food-delivery robot may be considered a GIS when it constructs the geographic environment using its internal LiDAR, even though it is not typically viewed as a GIS when it is performing its primary delivery function. This viewpoint challenges the notion of a universal GIS model and encourages a more comprehensive and dynamic understanding of GIS. Consequently, GIScientists can examine those instances when things interact with geographic data.

Humanistic care

The constant flow of creative forces has a profound impact on the involved actors and can even influence the entire society as a whole when these forces converge. Certain influences are enhanced while others may diminish, with the amplified influences being more powerful and noticeable. Although some suggest completely avoiding GIS due to its potentially negative effects, this is impractical given that even the most remote rural areas are now covered by GPS or other positioning signals. On the other hand, blindly relying on GIS is also problematic, as the dynamic creative forces it mediates can lead to unforeseen consequences. Therefore, neither extreme scepticism nor blind dependence is an appropriate approach to the complex consequences of GIS.

To navigate these challenges, it is essential to learn from humanistic values that shed light on how GIS mediates and is influenced by creative forces. For instance, the ethics of care is a moral and philosophical theory that emphasises the importance of relationships, empathy and compassion. It is based on the idea that caring is a fundamental human value, and it should be an essential aspect of all ethical decision-making (Schuurman and Pratt, 2002). In addition, the concept of harmony is a core principle of Chinese philosophy, which has evolved over thousands of years. Harmony is often seen as a social and ethical value, highlighting the importance of social cohesion, cooperation and respect for others (Fan, 2022). Moreover, the idea of Ubuntu is a traditional African philosophy that underscores the interconnectedness of all things and the importance of community and relationships. Ubuntu is often translated as 'humanity towards others' and is based on the idea that one's humanity is defined by their relationships with others (Sambala et al., 2020). Each of these humanistic values emphasises the importance of relationships, empathy and interconnectedness in different ways.

Humanistic GIS places the importance of being responsible to both human and even non-human entities and demonstrates care through the mediation of GIS. This care extends not only to humankind but also to our planet and companions such as animals and even robots. This idea aligns well with Morrill's (1984) earlier call for geographers to take responsibility for the Earth, and Lawson's (2007) advocacy for the geographies of care. For example, multiple locative apps launched an ownership-labelling campaign to increase the visibility of black-owned businesses. It is also critical to note that historically oppressed peoples, such as women, people of colour, LGBTQ+ individuals and Indigenous peoples, face various ongoing struggles today, and they can use GIS to amplify their voices (Elwood, 2022). For example, the Standing Rock Sioux Tribe utilised a range of tactics to resist the development of the Dakota Access Pipeline. Among these methods was the use of geo-tagging in their social media posts, enabling the protest's location to be marked and broadcasted to a wider audience. This creative use of geographic data garnered substantial backing from social media users, who demonstrated their support for the indigenous community by virtually checking in at Standing Rock (Zhang et al., 2021). This caring stance challenges GIS analyses towards a population of undifferentiated individuals and highlights the differences between each other and the power relations in shaping bodies and worlds (Ash and Simpson, 2016; Kinkaid, 2021). Indeed, the traditional use of GIS reduces the individual to a mere abstract representation, like a geo-tag, space-time path, or a cell in cellular *automaton*, while this caring stance recognises that GIS must be examined in relation to its users and their practices. After all, GIS is always in use as geospatial data, algorithms and platforms. By acknowledging the significance of human practice, the proposed humanistic GIS provides an opportunity to balance the demands of human beings and the proliferation of GIS artefacts. This proposed humanistic rewire of GIScience will empathise with diverse lived experiences and care for the well-being of others and ourselves.

Conclusion

This chapter presented a novel research agenda that seeks to integrate humanism into GIS. Beginning with an examination of humanism and its relevance to geographies, it then elaborated on a humanistic rewire of GIScience. By analysing how GIS mediates the bodies, places and other sentient beings, this agenda identified four major types of GIS, including Embodiment GIS, Hermeneutic GIS, Autonomous GIS and Background GIS. It is crucial to understand that GIS is a constantly evolving technology that can influence society and ourselves differently. It is possible that the use of GIS can become negative forces, such as a deadly GPS-guided missile (Smith, 1992). To prevent such consequences, a humanistic rewire of GIS offers a feasible pathway that can produce better awareness of the diverse lived experiences and other forms of creative forces. Through this humanistic approach, GIScientists are encouraged to explore newly emerging and under-examined GIS applications, to develop a comprehensive understanding of the social implications of GIS, including both immediate and potential risks and more importantly, to adopt a caring stance towards others, the earth and ourselves. Overall, the integration of humanistic principles into GIS can potentially lead to improvements in this technology by considering the needs of human users and the impact of GIS on the world.

Therefore, the future GIScience should not only prioritise technological advancements but also embrace a humanistic rewire. While this humanistic rewire holds promise, it would greatly benefit from a more critical examination of the development and use of GIS (Ward and Wasserman, 2010), as well as a deeper reflection on the relationships between GIS and non-human entities within the context of more-than-human or posthuman geographies (Panelli, 2010; Greenhough, 2014). Although there is still room for improvement in the current agenda, it can serve as a foundation for future assessments and criticisms of this humanistic rewire. Furthermore, it presents an opportunity to foster collaboration among cartographers, GIScientists and human geographers to imagine new pathways for GIScience.

References

Adams PC, Hoelscher SD and Till KE (2001) *Textures of Place: Exploring Humanist Geographies.* Minneapolis: University of Minnesota Press.

Ash J and Simpson P (2016) Geography and post-phenomenology. *Progress in Human Geography* 40(1): 48–66.

Buttimer A (1974) *Values in Geography.* Washington: Commission on College Geography.

Buttimer A (1976) Grasping the dynamism of lifeworld. *Annals of the Association of American Geographers* 66(2): 277–292.

Cresswell T (2014) *Place: An Introduction.* Chichester and Malden, MA: John Wiley & Sons.

Elwood S (2022) Toward a fourth generation critical GIS: Extraordinary politics. *ACME: An International Journal for Critical Geographies* 21(4): 436–447.

Fan J (2022) A century of integrated research on the human-environment system in Chinese human geography. *Progress in Human Geography* 46(4): 988–1008.

Greenhough B (2014) More-than-human geographies. In: Lee R et al. (eds) *The Sage Handbook of Human Geography*, vol. 1. London: Sage, pp. 94–119.

Heidegger M (2010 [1927]) *Being and Time.* Albany, NY: Suny Press.

Husserl E (2012 [1931]) *Ideas: General Introduction to Pure Phenomenology.* Oxford: Macmillan.

Ihde D (1990) *Technology and the Lifeworld: From Garden to Earth.* Bloomington, IN: Indiana University Press.

Kinkaid E (2021) Is post-phenomenology a critical geography? Subjectivity and difference in post-phenomenological geographies. *Progress in Human Geography* 45(2): 298–316.

Lawson V (2007) Geographies of care and responsibility. *Annals of the Association of American Geographers* 97(1): 1–11.

Lea JJ (2020) Post-phenomenology/post-phenomenological geographies. In: Kitchin R and Thrift N (eds) *International Encyclopedia of Human Geography.* Oxford: Elsevier, pp. 373–378.

Ley DF (1972) *The Black Inner City as Frontier Outpost: Images and Behavior of a Philadelphia Neighborhood.* Philadelphia: The Pennsylvania State University.

Ley DF and Samuels M (1978) *Humanistic Geography: Prospects and Problems.* Chicago: Maaroufa Press.

Li FF (2018) *How to Make A.I. That's Good for People.* Available at: www.nytimes.com/2018/03/07/opinion/artificial-intelligence-human.html (accessed 24 October 2018).

Li FF and Etchemendy J (2018) *Introducing Stanford's Human-Centered AI Initiative.* Stanford. Available at: http://hai.stanford.edu/news/introducing_stanfords_human_centered_ai_initiative/ (accessed 24 October 2018).

Morrill RL (1984) The responsibility of geography. *Annals of the Association of American Geographers* 74(1): 1–8.

Panelli R (2010) More-than-human social geographies: Posthuman and other possibilities. *Progress in Human Geography* 34(1): 79–87.

Penman KA and Adams SH (1982) Humane, humanities, humanitarian, humanism. *The Clearing House* 55(7): 308–310.

Sambala EZ, Cooper S and Manderson L (2020) Ubuntu as a framework for ethical decision making in Africa: Responding to epidemics. *Ethics & Behavior* 30(1): 1–13.
Schuurman N and Pratt G (2002). Care of the subject: Feminism and critiques of GIS. *Gender, Place and Culture: A Journal of Feminist Geography* 9(3): 291–299.
Seamon D and Lundberg A (2017) Humanistic geography. *International Encyclopedia of Geography: People, the Earth, Environment and Technology* 6: 1.
Smith N (1992) History and philosophy of geography: Real wars, theory wars. *Progress in Human Geography* 16(2): 257–271.
Tuan YF (1971) Geography, phenomenology, and the study of human nature. *Canadian Geographer/ Le Géographe canadien* 15(3): 181–192.
Tuan YF (1976) Humanistic geography. *Annals of the Association of American Geographers* 66(2): 266–276.
Verbeek PP (2020) Politicizing Postphenomenology. In: Miller G and Shew A (eds) *Reimagining Philosophy and Technology, Reinventing Ihde*. Berlin; Springer, pp. 141–155.
Ward SJ and Wasserman H (2010) Towards an open ethics: Implications of new media platforms for global ethics discourse. *Journal of Mass Media Ethics* 25(4): 275–292.
Wright JK (1947) Terrae incognitae: The place of the imagination in geography. *Annals of the Association of American Geographers* 37(1): 1–15.
Zhang S, Zhao B, Tian Y and Chen S (2021) Stand with# StandingRock: Envisioning an epistemological shift in understanding geospatial big data in the "post-truth" era. *Annals of the American Association of Geographers* 111(4): 1025–1045.
Zhao B (2022) Humanistic GIS: Toward a research agenda. *Annals of the American Association of Geographers* 112(6): 1576–1592.
Zhao B (2023) Reorienting GIScience for a data-intensive society. *Dialogues in Human Geography*. Epub ahead of print. 30 May 2023. DOI:10.1177/204382062311792.

24

THE CINE-TOURIST'S ONLINE CARTOGRAPHIC CURIOSITY CABINET

Tadas Bugnevicius

Introduction

In the documentary *Be Natural: The Untold Story of Alice Guy-Blaché* (Green, 2018), a scholar takes us on a walking tour in the 19th arrondissement of Paris.[1] His goal is to have us visit the shooting locations of films like *The Drunken Mattress* (1906) and *The Rolling Bed* (1907) by the cinema pioneer Alice Guy. The documentary shows present-day streets while superimposing footage from these early films of action occurring in the same places. The result is layered images that juxtapose past locations in black and white with their present-day appearance. This juxtaposition emblematises the working method of our tour guide, film and literary scholar Roland-François Lack (1960–2021). A self-described 'cine-tourist', Lack investigated film's relation to place. This chapter offers an intellectual history of this Cine-Tourist's queries and discusses the role of cartography in them.

In a self-reflexive moment of *Be Natural*, Lack's finger can be seen pointing at the walking tour itinerary digitally marked on an old analogue map (Figure 24.1). The combination of the early twentieth-century map with digital compositing evokes his method: what Lack describes as 'old-school location hunting, with an old map in hand' (2017c: 95) and, eventually, with the aid of Google Maps and Google Street View. By the 2010s and until his illness in 2020, Lack was also giving public lectures and warm-up acts on how to identify locations and map films, producing image-essays and offering tours such as the one documented in *Be Natural*. Crucial to this public engagement was his Cine-Tourist website which he initially created as a repository for his research.[2] Brimming with images and possessing its own quirky organisational principles, this is the spectator's cabinet of curiosity. By far the largest collection of images on the website is Lack's 'Maps in Films' blog, tallying 823 entries. I therefore focus on these map shots and also on his own maps, what he liked to call, respectively, the 'map in the film' and the 'map of the film'. But before that, some words on the emergence of the Cine-Tourist website itself.

Intellectual origins of the Cine-Tourist website

Lack's early research on French literature and critical theory was a precursor of many of his later cinematic preoccupations. As co-editor and co-translator of the *Tel Quel Reader*,

 DOI: 10.4324/9781003327578-29

Figure 24.1 Lack's finger pointing at the walking tour itinerary for a location of an Alice Guy film in *Be Natural*.

Source: Screenshot by the author, under Fair Use

he delved into the collective of the *Tel Quel* journal (1960–1982) that reformulated writing and reading in terms of discontinuity, difference and intertextuality, as opposed to representation and truth ('Division of the Assembly', 1998: 22–23). Lack's monograph *Poetics of the Pretext: Reading Lautréamont* (1998), appearing in the same year as the *Reader*, expanded the theory of intertextuality developed by Kristeva, according to whom 'any text is constructed as a mosaic of quotations; any text is the absorption and transformation of another' (1986: 37). Kristeva soon expanded this 1966 gloss of Bakhtin's dialogism to affirm that 'the literary text inserts itself into the set of all texts' (1998: 29). In short, the modernity of a text depends on the extent to which it acknowledges the existence of other texts enmeshed in it. This conception of the layered text is something Lack carried over to his film scholarship.

Poetics of the Pretext trumpets a concern with tracking sources that predicts Lack's future Cine-Tourism. The double sense of the word 'pretext'—an excuse as well as a precedent—is crucial to Lack's reframing of Kristeva's 'other texts' as so many alibis. Lack's book takes these alibis seriously, referring to his task as 'source-hunting' and his role as a 'source-critic' and a 'searcher after sources' (1998: 4, 6) much as he would later describe himself as a 'source hunter' exploring intertextuality in Godard's films (2004a: 208), and even later still, a Cine-Tourist going on 'location hunting'. Alibi means 'elsewhere' in Latin, so Lack's notion of intertextuality translated easily into the eminently spatial medium of film. His earliest efforts focused on 'pretextual materials' such as actor, character and mode of narration in his reading of *Le Petit Soldat* (Godard, 1963) as a pretext of *Beau Travail* (Denis, 1999; Lack, 2004b: 36). Later on, pretexts were common locations—hotel rooms, streets and neighbourhoods—shared between different New Wave films (Lack, 2017c, 2018b). He also called these overlapping locations 'cine-memories' (Lack, 2017b: 12). Whether applied to literature or film, pretextuality was for Lack a rigorous way of talking about influence and adaptation.

Up until around the year 2000, Lack was still primarily a literary scholar. He had been a cinephile for a long time and he was familiar with film theory, so it made sense to reconfigure his own brand of intertextuality for the study of film. Lack's transition gained a new development when in 1999 the Harvard feminist historian Jann Matlock joined the Department of French at University College London, where Lack was teaching. Shortly thereafter she introduced him to her colleague Tom Conley, at that time on a visit to London, who reinforced Lack's long-standing interest in maps. Conley's latest book at the time was *The Self-Made Map: Cartographic Writing in Early Modern France* (1996) and he was already researching what became *Cartographic Cinema* (2007). A few years after the meeting, Conley's and Matlock's colleague Giuliana Bruno published *Atlas of Emotion: Journeys in Art, Architecture, and Film* (2002), which became a crucial assigned reading in Matlock and Lack's burgeoning interdisciplinary film programme at UCL. Both poured over old maps while they were writing their first articles on cinema's relation to place, Matlock on Chicago (2005), Lack on London (2007). Both contributed to the project of critical human geography *Moving Pictures/Stopping Places: Hotels and Motels on Film* (2009), whose editors were asking why filmmakers had been so attracted to hotels and motels, described as 'fixed place[s] dedicated to movement' (Clarke et al., 2009: 3). From then on, combining pretextuality and cinematic cartography, Lack could envisage Cine-Tourism.

At the 2010 *Cities Methodologies* event, Lack presented 'How To Map a Film', a triptych of image-essays each plotting a film on the Google Map of one of three cities on which he would come to specialise: *Last Tango in Paris* (Bertolucci, 1972) for Paris; *Blow-Up* (Antonioni, 1966) for London; *Le Petit soldat* for Geneva. The event was part of UCL's Urban Laboratory founded by cultural geographer Matthew Gandy. Encouraged by the results, Lack launched in April 2011 the Cine-Tourist website, posting a map a day for a year and engaging hundreds of other films in the next decade, often with multiple still shots and maps (Figure 24.2). Another catalyst was the Autopsies Research Group convened in 2009 by Matlock, Lack and their graduate students as an inquiry into modernity and obsolescence. Asking 'how objects die', each member researched up to four 'dead objects', such as the typewriter, the tape recorder or the jukebox, and wrote up their 'obituaries'.[3] Lack's collection of maps was an outgrowth of the Autopsies Group website and shared with the latter an attention to objects and their histories.

Map shots

Be Natural contains a map and would therefore be a welcome addition to the Cine-Tourist's collection of maps in films. Currently this collection (accessible through the 'blogs' section) stands at 823 films, of which 523 are included in the chronological index. Numerous posts, however, mention two or more films, effectively raising the number of films with maps to about 900. The daily map blog even contains a separate set of 31 entries of maps in books. This disparity suggests that while maps are rarely integral to printed fiction, they are very much at home in fiction films. Indeed, as Conley reminds us, 'maps appear in most of the movies we see' (2007: 1). Geographers Sébastien Caquard and D.R. Fraser Taylor echo this sentiment when they affirm that 'maps are ubiquitous in movies' (2009: 5). We only need to

Figure 24.2 Last entry in the 'Maps in films' blog on the *Cine-Tourist* website.

Source: Composition of screenshots by the author, under Fair Use and permission by Verene Lack

think of the logos of production companies, such as the familiar Universal and RKO globes or the lesser-known globes of British International Pictures, Francinex and Nordisk Film, that mark the credits of so many films. But beyond the logos, maps are prevalent as part of the set: as props, as backdrops and even as accidental objects in the profilmic reality. There are also 'cine-maps', a term Lack often reserves for maps produced through montage or animation. To my knowledge, Lack's collection is the best empirical proof the public has of the ubiquity of maps in cinema. His collection also suggests that the list could continue indefinitely.

Hard evidence is far from the only merit of the Cine-Tourist site. The collection indicates the extraordinary diversity of function and appearance that maps can take on in films. One thing we learn from perusing this evidence is that maps are particularly present in places that are associated with institutions and the state: classrooms, police stations and military headquarters. Coming-of-age films, crime films and war films are therefore among the genres where maps can be expected. A quintessential French New Wave film like *The 400 Blows* (Truffaut, 1959) is ordinary in its inclusion of maps in the classroom and police station scenes. Documentaries, especially educational, ethnographic and propaganda films, thirst for cine-maps. In the first episode of Frank Capra's *Why We Fight* series (1942–1945), the ominous Disney-animated blob of black ink spreading over maps conveys efficiently the threat of the Axis powers and contrasts poignantly with figurative drawings by unsuspecting American children. Finally, films using urban locations are likely to capture street plans and transit maps.

Lack's map shots remind us that cinema has incorporated every imaginable cartographic form of its time and even held some in store for the future: as Caquard explains, cine-maps in films like Lang's *M* (1931) anticipated digital cartography by several decades (2009: 49–50). Lack was particularly keen on collecting images of globes and hand-drawn maps. But not every map is as central to the meaning of the film as the one in *M*. The globe signifies more in Chaplin's *The Great Dictator* (1940), where the fascist dictator Hynkel dances with an oversized balloon representing Earth, than it does in Hawks's *The Big Sleep* (1946), where globes blend with the settings of the bookstore and the detective's office. Discreet globes such as those in Hawks's film serve, according to Lack, 'two basic purposes: 1) to connote intellectual curiosity (like books); 2) to modify subtly the composition by introducing soft round lines into angular spaces' (*Cine-Tourist*). However discreet, a map is never just like any other prop. To borrow from Markus Nornes's theorisation of calligraphy in cinema, a map is at the same time something to read and something to look at. Like brushed characters, maps 'demand accounting for, whether they are legible or not' (Nornes, 2021: 101). They thematise the very act of interpretation. The irony of course, as Lack shows it in multiple cases, is that paying attention to the map often reveals contradictions in the fabric of the film. The large wall map behind the Berlin police chief in Lang's *The Thousand Eyes of Mabuse* (1960) appears not to match any maps of Berlin from that time. But films also know how to flaunt these contradictions, as in what Lack playfully calls the 'saga of "Bugs Bunny misreads the map" films' (*Cine-Tourist*).

In addition to showing the sheer numerical presence of maps in films and what they do, Lack's collection also teaches the visitors of his website how to respond to them. While a good chunk of his posts contains bare images, many do accompany screenshots with a short commentary that draws attention to thematic, narrative, or cinematic uses of the map in question. Among the highlights, for example, is Lack's discussion of Cissé's *Finye* (1982) where he points out the contrast between the maps in the student protesters' clandestine print shop and the map in the military governor's office. In several posts, Lack checks the film map against the real map. A hand-drawn map in *Le Chevalier de Maison Rouge* (Capellani, 1914) set during the French Revolution compares well with a 1797 map, but a hand-drawn map in *Les Vampires* (Feuillade, 1916) does not map the place it claims it does. Perhaps the most interesting critical accompaniment comes in the form of citation. The very first entry, on *The Red Desert* (Antonioni, 1964), inaugurates this strategy by accompanying a screenshot of Monica Vitti's character looking at a map with quotations from Bruno and Gandy, and no other commentary. Most often the citation bears directly on the film or the map in question, but occasionally the relationship between the image and the text is indirect, as in the juxtaposition of a map from *King Kong Escapes* (Honda, 1967) and an excerpt from Jameson on globalisation. Here Lack draws inspiration from one of his favourite films, Godard's *Histoire(s) du cinéma* (1988–1998), made almost entirely of visual and textual citations. Bricolage in general characterises Lack's posts which, like Godard's film, are meant to generate ideas. One idea that comes to mind perusing his collection is that maps, which normally impose an identity of space and text, lose some of their representational rigidity once incorporated in film. In film, maps become evocative, as opposed to closed circuits.

Plotting film locations

Maps are so important for moving pictures that in the case of urban, topographically oriented films such as *Paris Belongs to Us* (Rivette, 1961) and *Blow-Up*, the absence of maps is significant in itself. Scholars have contended, however, that films also map in a deeper sense, in which case the inclusion of a concrete map, itself not compulsory, is a symptom of that deeper mapping (see Avezzù, this volume). For Conley, 'a film can be understood in a broad sense to be a "map" that plots and colonizes the imagination of the public it is said to "invent" and, as a result, to seek to control' (2007: 1). Bruno speaks of cartographic techniques in cinema (such as the establishing shot) that approximate the practice of map reading (2002: 271). Similarly, focusing on techniques like pan shots, panoramic shots and aerial shots, Teresa Castro (2009: 9) identifies cinema with what art historians and geographers have known as the 'mapping impulse', suggesting that 'the coupling of eye and instrument that distinguishes cartography's representation of space is in many ways very similar to cinema's coding and scaling of the world'. Lack shared this agenda, but he also shifted to other questions. In the 'how to map a film' section of his website, he lists the familiar questions such as 'how to read a map in a film', 'how to read a film, as a map', and 'how to read a map as a film' (including their 'why' variants), but he also asks: 'How to read a film with a map (in your hand)?' 'Why read a film with a map (in your hand)?'

The map-in-hand motif recurs in Lack's writings (2007: 168, 2017c: 95) and indicates a desire to locate places on an actual map. For him, spectatorship is never confined to the viewing experience—it is a prelude to identifying and, ideally, travelling to places seen in the films. He expresses his research agenda in the following way:

> For the ciné-tourist—a film spectator who is a traveller arriving in a city—the space in which he is to be secured is not that of the narrative, to which he is indifferent. What establishes his position is not a shot that maps him, but one that can be mapped.
>
> *(2007: 168)*

Here, Lack engages not only with Bruno and Conley but also with the influential 1970s film-theoretical concern with the spectator's placement in what Stephen Heath (1986) calls 'narrative space'. In that canonical article, Heath reiterates that cinema turns space into place (1986: 392, 394, 397 and passim). By this he means that the conventions of continuity, the effect of suture and the centrality of narrative endow artificial, fictional space with all the trappings of a real place. But there is an obverse side to this mechanism, in which Heath is less interested, but which I think grounds Lack's concern with concrete place—which is that cinema also turns place into space. It uses real places and reconfigures them into narrative space. Film narrative, even when not constrained to the studio, tends to obscure place, which is why 'location hunting' is a 'hunt' and not a walk in a park. Lack's Cine-Tourist proclaims indifference to narrative and reads for the place, scrutinising shots for topographical cues—terrain, landmarks, signage and building façades—in order to be able to put a scene on a map.

Plotting film locations can be seen as a way to wrestle place from space. It opens new avenues for thinking about the film historically and formally. Lack started mapping films in the early 2000s as a teaching tool, beginning with such works as *Blow-Up* and *The Passenger* (Antonioni, 1975), initially on analogue maps and then gradually using Google Maps. In his posts on *Rififi* (Dassin, 1955) and *Le Pont du Nord* (Rivette, 1981), we can see how the meticulous identification of the quasi-totality of all locations enables him to plot the points and trace an itinerary. The *Pont du Nord* locations appear as an image-essay in Kino Lorber's 2015 DVD release of the film, but without Lack's perhaps most evocative map which traces the chaotic itinerary of Marie and Baptiste after they meet each other (Figure 24.3). A close look at any cinematic itinerary reveals that films often imply illogical,

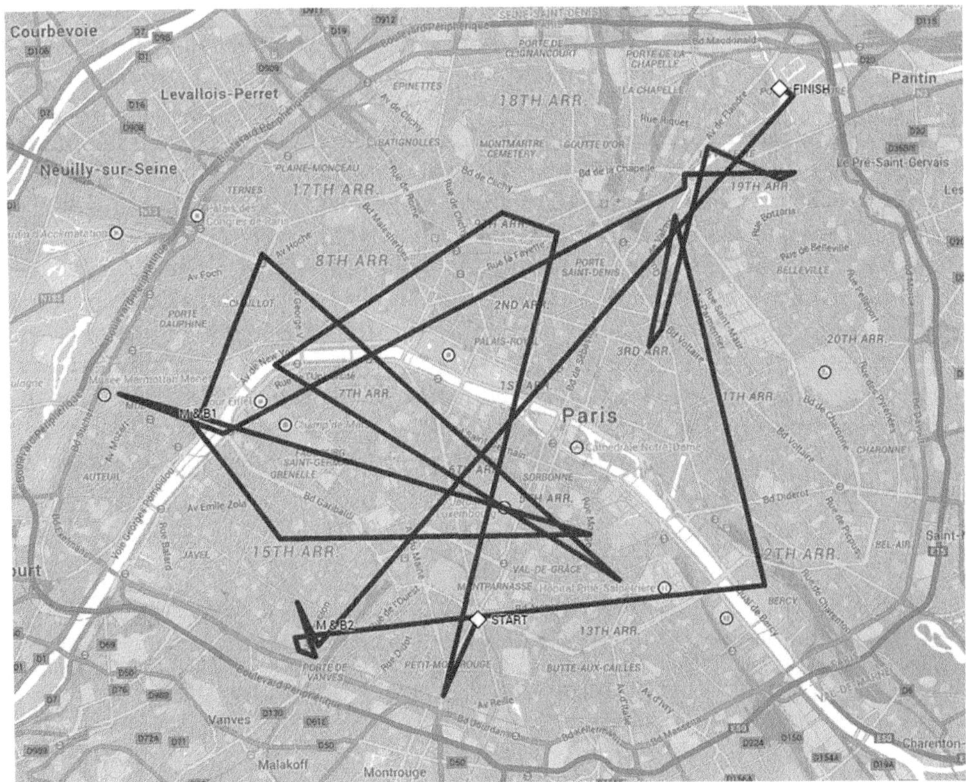

Figure 24.3 Lack's plotting of Marie and Baptiste's chaotic itinerary after they meet near the Place Denfert-Rochereau in Rivette's *Le Pont du Nord*.

Source: Screenshot by the author, under Fair Use and permission by Verene Lack

even impossible character routes, as in the case of *Blow-Up*. But in *Le Pont du Nord* the playful, unruly movement of the two women across Paris functions as an acknowledgement of cinema's well-known secret. They draw the 'Game of the Goose' on top of their map of Paris, laying fiction over the representation of the real place, thus offering a rare instance of a map in a film that tells us how to understand the map of the film.

Conclusion

For Lack, recognising a place had value in itself. He was very proud of taking weeks to solve what for most of us would appear the most trivial question: where exactly was the 30-second scene of Irma Vep posting a letter in *Les Vampires* filmed? Using the 1911 edition of the *Paris-Hachette* directory, old postcards, photographs and site plans of the 18th arrondissement, in addition to his own pictures of the neighbourhood today, Lack was able to determine that the fictional Irma Vep was at a real postbox opposite 46 rue Leibniz (*Cine-Tourist*; Lack, 2013). The place has changed a lot since then. Trees grow where there was the postbox; one of the two buildings in the scene has been demolished and replaced, and the narrow passage between them has been filled.

Lack was invested in what he variously called 'visual evidence', 'visual record' or 'visual documentation' (2018a: 24) of ordinary places transformed over time. Elsewhere he opined that 'it is a function, sometimes even a purpose, of cinema to record its object for posterity' (2007: 156; see also 2017a: 24–25). The irony is, as I mentioned, that films obscure at the same time as they record these ordinary places. This meant that his research prioritised films that actively engage with real places: the bulk of published film-related material deals with the French New Wave, while his latest book proposal was on early cinema. That book project calls attention to a shift from generic spaces to recognisable locales in the narrative film of the 1900s and 1910s, and treats this 'topographical shift' as a significant contribution to the narrative film form as we know it today. In Lack's approach, there remains something of Kristeva's conception of the double text with its representational surface and its infinite intertext: the idea that a limited narrative space is entangled with the vast and rich profilmic reality. As Lack's work cannot stress enough, this entanglement of space and place has a history that can be written with the help of maps. The Cine-Tourist invites other spectators like ourselves to haunt the backroads of the cinematic imaginary, each armed with a map, in the pursuit of that history.

Notes

1 I thank Roland-François Lack's colleague and friend Jann Matlock for providing suggestions on this chapter. I am also grateful to Verene Lack for reading and responding to it.
2 www.thecinetourist.net. The website was archived in archive.org (https://web.archive.org/details/http://thecinetourist.net) and deep-saved by the Bibliothèque nationale de France.
3 www.autopsiesgroup.com

References

Bruno G (2002) *Atlas of Emotion: Journey in Art, Architecture, and Film.* New York: Verso.

Castro T (2009) Cinema's mapping impulse: Questioning visual culture. *The Cartographic Journal* 46(1): 9–15.

Caquard S (2009) Foreshadowing contemporary digital cartography: A historical review of cinematic maps in films. *The Cartographic Journal* 46(1): 46–55.

Caquard S and Taylor DRF (2009) What is cinematic cartography? *The Cartographic Journal* 46(1): 5–8.

Clarke DB, Pfannhauser VC and Doel MA (2009) Checking in. In: Clarke DB, Pfannhauser VC and Doel MA (eds) *Moving Pictures/Stopping Places: Hotels and Motels on Film.* Lanham: Lexington Books, pp. 1–12.

Conley T (1996) *The Self-Made Map: Cartographic Writing in Early Modern France.* Minneapolis and London: University of Minnesota Press.

Conley T (2007) *Cartographic Cinema.* Minneapolis and London: University of Minnesota Press.

Heath S (1986) Narrative space. In: Rosen P (ed.) *Narrative, Apparatus, Ideology: A Film Theory Reader.* New York: Columbia University Press, pp. 379–420.

Kristeva J (1986) Word, dialogue and novel. In: Moi T (ed.) *The Kristeva Reader.* New York: Columbia University Press, pp. 34–61.

Kristeva J (1998) Towards a semiology of paragrams. In: Ffrench P and Lack RF (eds) *The Tel Quel Reader.* London and New York: Routledge, pp. 25–49.

Lack RF (1998) *Poetics of the Pretext: Reading Lautréamont.* Exeter: Exeter University Press.

Lack RF (2004a) *A bout de souffle*: The film of the book. *Literature/Film Quarterly* 32(3): 207–212.

Lack RF (2004b) Good work, little soldier: Text and pretext. *Journal of European Studies* 34(1–2): 34–43.

Lack RF (2007) London circa sixty-six: The map of the film. In: Cunningham G and Barber S (eds) *London Eyes: Reflections in Text and Image.* New York and Oxford: Berghahn Books, pp. 149–176.

Lack RF (2013) Irma Vep poste une lettre. *Le Tigre* 34: 36–40.

Lack RF (2017a) 'Local film subjects': Suburban cinema, 1895–1910. In: Hirsch P and O'Rourke C (eds) *London on Film*. Cham: Springer International Publishing, pp. 15–26.
Lack RF (2017b) Mapping *Out 1*: Thirteen cartographic footnotes. *The Cine-Files* 12. Available at: www.thecine-files.com/mapping-out-1/ (accessed 25 March 2023).
Lack RF (2017c) The cine-tourist's map of new wave Paris. In: Penz F and Koeck R (eds) *Cinematic Urban Geographies*. New York: Palgrave Macmillan, pp. 95–111.
Lack RF (2018a) Lumière, Méliès, Pathé and Gaumont: French filmmaking in the suburbs, 1896–1920. In: Met P and Schilling D (eds) *Screening the Paris Suburbs: From the Silent Era to the 1990s*. Manchester: Manchester University Press, pp. 23–36.
Lack RF (2018b) The new wave hotel. In: Phillips A and Vincendeau G (eds) *Paris in the Cinema: Beyond the Flâneur*. London: BFI Palgrave, pp. 66–75.
Matlock J (2005) Chicago. In: Jousse T and Paquot T (eds) *La ville au cinéma*. Paris: Éditions des Cahiers du cinéma, pp. 368–379.
Nornes AM (2021) *Brushed in Light: Calligraphy in East Asian Cinema*. Ann Arbor: University of Michigan Press.
Tel Quel (1998) Division of the assembly. In: Ffrench P and Lack RF (eds) *The Tel Quel Reader*. London and New York: Routledge, pp. 21–24.

Filmography

Beau travail (1999) Directed by Claire Denis. [DVD]. France: La Sept-Arte, Pathé Télévision, S. M. Films and Tanaïs Productions.
Be Natural: The Untold Story of Alice Guy-Blaché (2018) Directed by Pamela B. Green. [DVD]. United States: Be Natural Productions.
The Big Sleep (1946) Directed by Howard Hawks. [DVD]. United States: Warner Bros. Pictures.
Blow-Up (1966) Directed by Michelangelo Antonioni. [DVD]. United Kingdom, Italy, United States: Carlo Ponti Production.
The Drunken Mattress (1906) Directed by Alice Guy-Blaché. [DVD]. France: Gaumont.
Finye (1982) Directed by Souleymane Cissé. [DVD]. Mali: Les Films Cissé.
The Great Dictator (1940) Directed by Charles Chaplin. [DVD]. United States: Charles Chaplin Productions.
Histoires(s) du cinéma (1988–1998) Directed by Jean-Luc Godard. [DVD]. France: Canal+, CNC, France 3, Gaumont, La Sept, Télévision Suisse Romande, Vega Films.
King Kong Escapes (1967) Directed by Ishiro Honda. [DVD]. Japan, United States: Toho and Rankin/Bass.
Last Tango in Paris (1972) Directed by Bernardo Bertolucci. [DVD]. Italy, France: Produzioni Europee Associati and United Artists.
Le Chevalier de Maison Rouge (1914) Directed by Albert Capellani. [DVD]. France: Pathé Frères.
Le Petit soldat (1963) Directed by Jean-Luc Godard. [DVD]. France: La Société nouvelle de cinématographie.
Le Pont du Nord (1981) Directed by Jacques Rivette. [DVD]. France: Les Films du Losange, Lyric International and La Cecilia.
Les Vampires (1916) Directed by Louis Feuillade. [DVD]. France: Gaumont.
M (1931) Directed by Fritz Lang. [DVD]. Germany: Nero-Film AG.
Paris Belongs to Us (1961) Directed by Jacques Rivette. [DVD]. France: Ajym Films and Les Films du Carrosse.
The Passenger (1975) Directed by Michelangelo Antonioni. [DVD]. Italy, Spain, France: MGM, Compagnia Cinematografica Champion, Les Films Concordia, and CIPI Cinematografica.
The Red Desert (1964) Directed by Michelangelo Antonioni. [DVD]. Italy, France: Film Duemila and Francoriz Production.
Rififi (1955) Directed by Jules Dassin. [DVD]. France: Pathé Cinéma, Indus Films and Prima Films.
The Rolling Bed (1907) Directed by Alice Guy-Blaché. [DVD]. France: Gaumont.
The Thousand Eyes of Mabuse (1960) Directed by Fritz Lang. [DVD]. West Germany, Italy, France: CCC Film, CEI Incom, and Critérion Film.
Why We Fight (1942–1945) Directed by Frank Capra. [DVD]. United States: U.S. War Department.
The 400 Blows (1959) Directed by François Truffaut. [DVD]. France: Les Films du Carrosse and SEDIF Productions.

PART 5

Troubles and disruptions

25
EMPTYING AND FILLING. MAPS OF INLAND AFRICA

Andrea Pase

Introduction

In a well-known passage from Conrad's (1990) short novel *Heart of Darkness*, the young Marlow is portrayed eagerly observing all the blank spaces that still existed in the Atlases of the second half of the nineteenth century, in particular at the Poles and along the Equator. He places his finger on them and says, 'I will go there'. The largest among them is in Africa. However, by the time Marlow was an adult, that 'blank space of delightful mystery' had in the meantime become 'a place of darkness'. On his journey up the Congo River, Marlow faces at least three dimensions of this 'darkness': the darkness of the predatory and violent practices of the European colonisers; the darkness of the equatorial forest as a threshold that conjures up a 'prehistoric earth'; and the darkness of Mr. Kurtz's mind—the ivory trader—as he progresses into madness. All three reflect an ambivalence implicit in Conrad, recognised by various authors who have spoken of 'schizophrenic contradiction' (Brantingler, 1985), of 'contradictions' (Ngugi, 2017) or of 'double vision' and 'contradiction in terms' (Šešlak, 2020). On the one hand, this novel certainly has passages that denounce the cruelty of colonialism, which was expressed—at the time when Conrad was writing—with particular ferocity precisely in the so-called Congo Free State. However, on the other hand, he has been accused of being 'a thoroughgoing racist', as did the Nigerian writer Achebe in 1975, considered 'the father of African literature in English language' (Achebe, 2016; Phillips and Achebe, 2007). I will return to this topic in the conclusion of this chapter.

What matters now is to observe the finger that lands on the largest blank on the map. Three elements capture our attention: the desire, the gesture and the object. The map and even more the blank spaces inside it create desire, that desire that is aroused by any 'delightful mystery'. Desire triggers action: first, it is that of the gesture, of pointing the finger; then the action of the journey towards and within the unknown space that is ahead. Finally, there is the object—the map—in its specific tactile relationship with the body (Rossetto, 2019a, 2019b). A map is certainly one of those landscape materials-objects capable of enchanting, of arousing emotions, as della Dora (2009: 334) states: on a map every adventure is potentially possible. Following this thought a bit further, if then inside fascinating objects such as maps there is a white space, fantasies are activated and projected: everything

DOI: 10.4324/9781003327578-31

that is desirable can be hidden in that void. Hence, the first impulse is movement, pointing the finger while touching the surface and proclaiming: 'I will be there'. Touching the map and the map as an object have been investigated by Rossetto (2019a, 2019b), reiterating the value of the 'cartographic surface'. The many fingers touching a 'You-Are-Here' map leave a sign of wear on the surface (Rossetto, 2019b: 92–93), testifying to the importance of the tactile dimension, which also manifests itself when a finger marks a path, the itinerary of a journey, almost caressing the surface of the map, its skin (Rossetto, 2019b). While touching the blank spaces, young Marlow fantasises on the 'You-Will-Be-There' maps. The rest, both the story and Marlow's journey, are in some way a consequence of that first desire, that first movement and that first tactile sensation.

This chapter intends to reflect on the transformation of blank spaces into mapped places, within the process of colonial appropriation. With respect to an area defined precisely by geographic coordinates (i.e. the part of the earth's surface between the 6th and 18th parallel N and between the 22nd and 27th meridian E), two cartographic documents—created almost a century apart: 1827 and 1924—enable us to observe the transition from blank space to a topographic survey. We are in what Close (1925: 349) defined as 'the most remote part of Africa, along that strip in the centre of the continent which is more than 1,000 miles from the coast'. The coast was the most frequented and best-known African area in European cartography. The consequences of this blank-filling process will be fatal.

Blank spaces, just in the middle

The first cartographic document consists of two sheets from the *Atlas Universel de Géographie Physique, Politique, Statistique et Minéralogique*, vol. 5, *Africa*, published by Philippe Vandermaelen in 1827: sheet 24, *Darfour et Kordofan*, and sheet 31, *Donga*. Vandermaelen (Brussels, 1795–1869) has been defined both as 'a geographic fanatic' and as 'the Mercator of the young Belgium' (Silvestre, 2016). Self-taught, in 1825, he began to publish his *Atlas Universel*, a colossal and hitherto never attempted representation of the whole planet on the same large scale (1:1.641.836). The Atlas is made up of 400 lithographed sheets, collected in six volumes in-folio, that 'could be assembled so as to form one gigantic globe of 7.75 m diameter' (Silvestre, 2014: 248; Bracke, 2021). His Atlas had great critical and commercial success (Silvestre, 2014; Bracke, 2021). The *Carte d'assemblage* of the volume dedicated to Africa was already quite noteworthy (Figure 25.1): there are several blank areas, but 'the biggest' is precisely at the Equator and includes the whole area that would later become the Congo Free State. The inscription reads 'Country unknown to Europeans'. It is suggestive that even before the formation of the Belgian state, an Atlas produced in Brussels highlighted this void—this blank space—over which Leopold II would, within a half century, perpetrate his power and around which Conrad will write.

The two sheets of interest to us (Figure 25.2) are also poor in details, in comparison with what happens in particular of the sheets of the regions along the coast. The large blank spaces inside the two sheets offer room for the insertion of long written notes. The two sheets contain meagre toponymical and geomorphological indications: regions such as Darfour, Kordofan, Donga and Ferti; populations (e.g. *Arabes Cubbabees*); towns, connected by tracks (sheet 24); deserts and mountain ranges; hypotheses of river courses, with the Niger and the Congo converging near *Mont Deir el Tougalla* (sheet 31); an explorer's itinerary (*Route de Brown en 1793 et 1796*, sheet 24). Further notes include information on many localities, provided by explorers, as well as supposed paths of the rivers or even

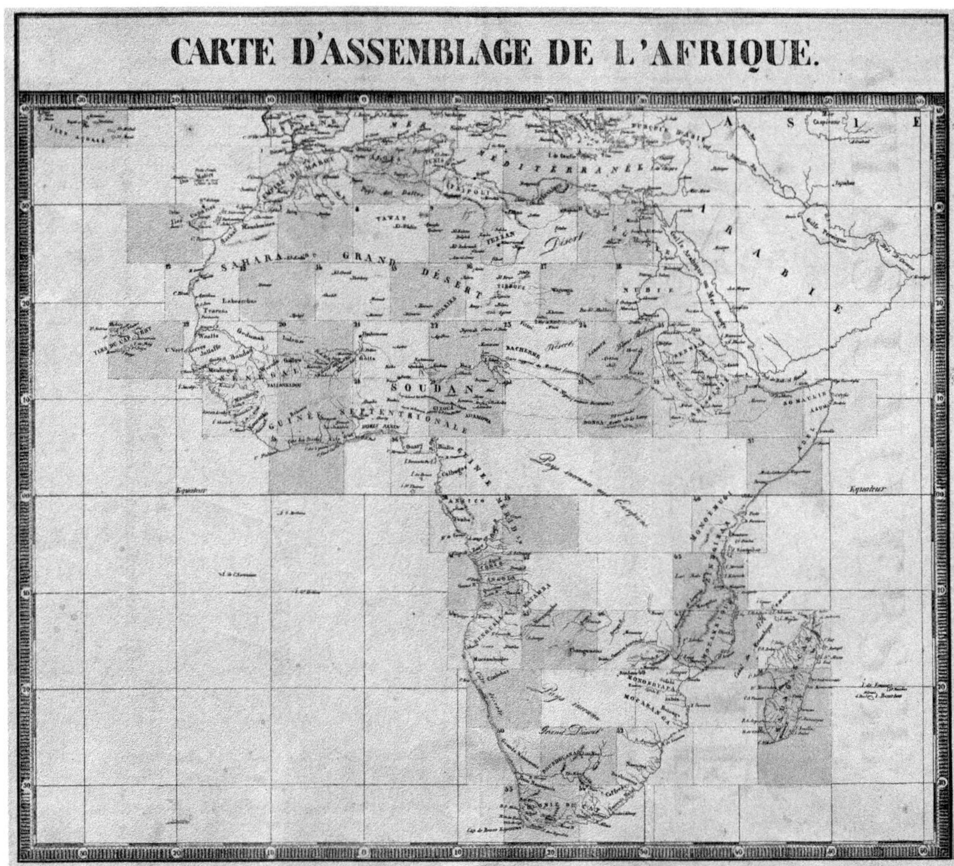

Figure 25.1 Vandermaelen, *Atlas Universel*, *Africa* (1827). Carte d'assemblage.
Source: Courtesy of Museum of Geography, University of Padua

ancient sources such as Ptolemy or Idrisi, whose names are placed in brackets next to the toponyms. The Mountains of the Moon, an essential landmark for ancient cartography in Africa, is also annotated in this way (sheet 31). The presence of information drawn from ancient sources bears witness to the fact that the Atlas marks a transition between the old and the new cartography of the continent, a phase that had already begun with Jean B.B. D'Anville 'whose 1749 map of Africa is famous for its extensive blank spaces' (Bassett, 1994: 322). The intent of the French cartographer was to report only verified information, thus expelling from the interior of the African continent 'animals, imaginary mountains, large and flamboyant lettering, and descriptive texts' (Bassett, 1994: 322). On the one hand, this choice testifies to the desire for 'scientificity' of the cartographic description: 'a scientific approach leads to the removal of many legends and assumptions', as Stone (1988: 58) put it. However, on the other hand, the 'use of blank spaces also resulted from disregard of indigenous geographical knowledge and from attempts by Europeans to keep their knowledge secret' (Bassett, 1994: 322–323). Taking up the line of thought followed by the South Korean-German philosopher Han (2018) regarding the disappearance of the Other in the

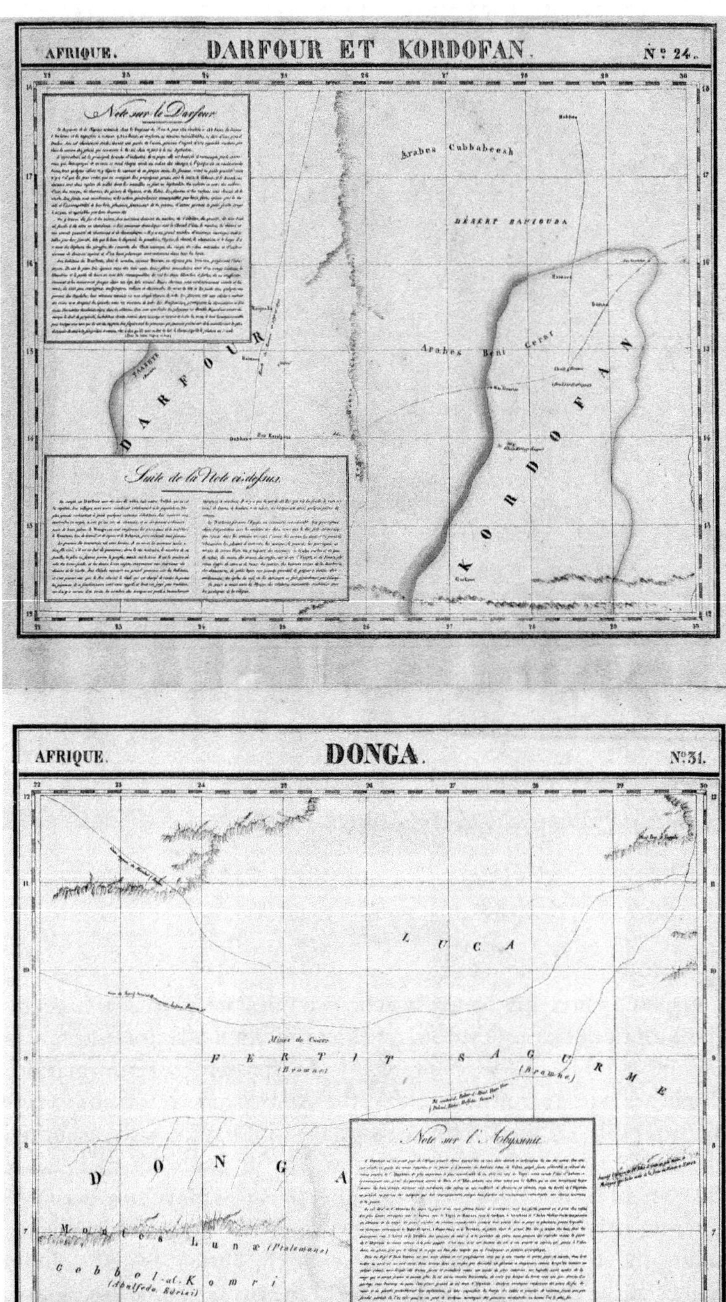

Figure 25.2 Vandermaelen, *Atlas Universel*, *Africa* (1827). Carte n° 24 and Carte n° 31, juxtaposed.

contemporary world, it is this very possibility of an authentic, autonomous Other that vanishes with the cancellation of the imaginative world that populated inland Africa up to that time. Only by accepting its being 'enigma or secret' can the Other be approached without deleting it: 'the *Other as an enigma* . . . eludes all exploitation' (Han, 2018: 68, italics by the author). In fact, blank spaces are anything but 'passive' but rather actively construct a hierarchy among knowledge, giving voice to some and silencing others (Lobo-Guerrero et al., 2021: 6; dos Reis, 2021: 108). In addition to that, the 'empty spaces' can be thought of as a 'no man's land' and therefore become the 'legitimate' prey of European imperial aspirations (dos Reis, 2021: 116–118). However, the path started by d'Anville was not yet concluded in Vandermaelen's Atlas, as also testified by the presence of the information sheets, which recorded—among other things—a devaluation of the African populations. The inhabitants of Darfur, 'this half-wild country', are described in these terms in sheet 24: 'they are not brave, and filthy, thieves and underhanded'. Furthermore, 'an extreme dissolution reigns in the relations of one sex to the other. Polygamy is unlimited'. These very negative statements of Vandermaelen have their roots in the deterministic prejudices against the people of the hot regions, widespread at the time.

Tracking an imaginary line in an alien land

Let us imagine a century has passed. Same area. From the late nineteenth and early twentieth centuries, the opposing imperialistic designs of France and England spread as far as the enormous forest and desert spaces lying between Lake Chad and the Nile. The mapping of these hitherto 'unknown' lands (by Europeans, of course) took place as part of the colonisation process itself: advancing military columns against the local sultanates and 'rebels' prepared the first cartographic surveys, while the Commissions in charge of establishing the border between French Equatorial Africa and Anglo-Egyptian Sudan carried out the detailed topographic surveys. Utilising the most modern means of deployment available at that time, often preceded by the silence imposed by machine guns, cartography attempts to 'define' the territories of the colonies: it succeeds, but only on maps. Its failure to understand the deep dynamics of the spaces it was redesigning—measurable just by observing the gaps and inconsistencies between the expectations of the cartographers and the uncontainable otherness of these different spaces, ironically noted by the colonisers themselves—provide an insight into cartography's role in the violent reality of colonialism.

From 1 February 1922 until the first days of May 1923, for 15 months, two Commissions—one English and the other French—worked jointly to define the colonial border (Pase, 2011: 188–222). The region covered extends from deep in the desert of the Sahara to the tropical forests of the watershed between the Nile and the Congo. Actually, it was a question of following on the ground and drawing on the map a theoretical line—decided thousands of kilometres away by the European governments—inside an alien land, of which very little was known. The reasons for this uncomfortable mission can only be understood by placing it in the context of the 'scramble for Africa'. As well described by one of the same protagonists, General Mangin, 'an inevitable law' pushed European penetration into the 'barbarian continent' to advance from one coast to the other sea or until it met the limits of another power, another 'civilized country' (Grossard, 1925: I–VI). This is how France and England met in the heart of the continent. The primary task of the Joint Commissions was to determine the boundary line, describing it in detail in a protocol, and to draw the boundary from the 29°30′ parallel to the 5° parallel (Figure 25.3).

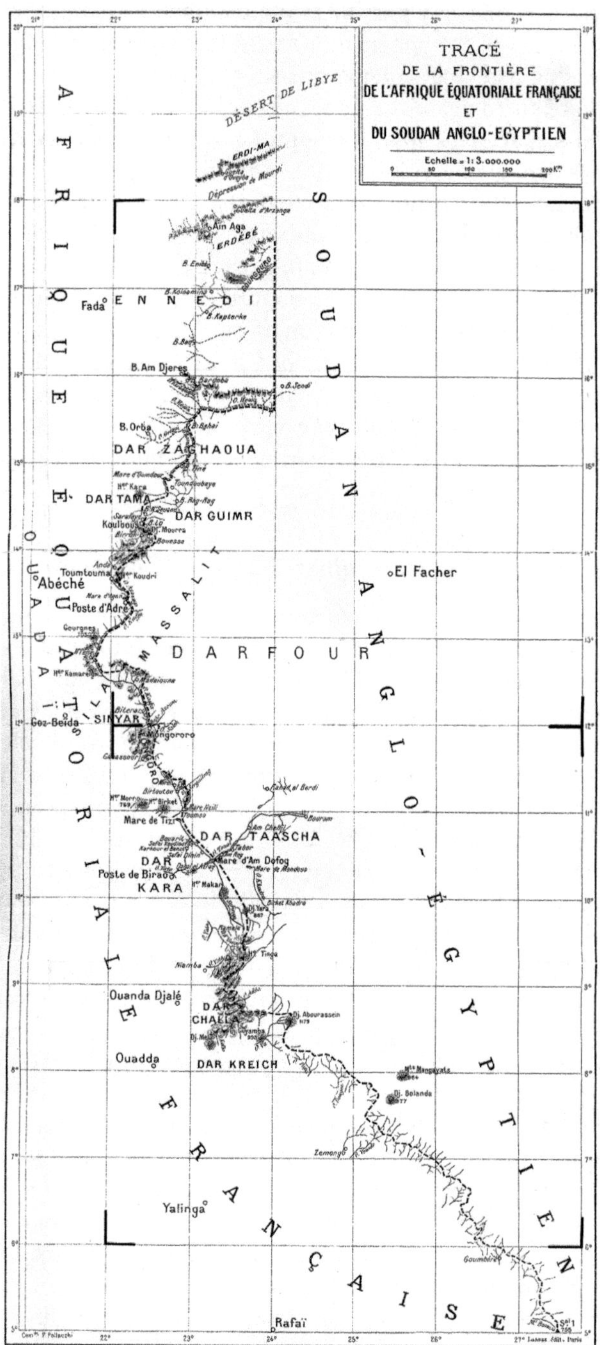

Figure 25.3 *Tracé de la frontière de l'Afrique Equatoriale Française et du Soudan Anglo-Egyptien.* The signs drawn on the map identify the overlapping areas with sheets 24 (above) and 31 (below) of Vandermaelen's Atlas.

Source: Grossard (1925), attached map

It had to be 'a rigorously exact document', to allow the two Commissioners to define the boundary line 'in relation to the significant points of the terrain'. The precision had to be such that the Commissioners could identify 'the current rights of the tribes on the map, after investigations made in the area', as stated by Lieutenant Grossard, head of the French expedition, in his detailed final report (1925: 44). The Commissioners were charged with a second task: collecting the highest volume of scientific expertise on the territory crossed. The results were 2,290 km of topographic survey and the description of an 'entirely new geographical landscape', with observations involving the various scientific branches: geology (with collection of samples brought back to France), botany, zoology, anthropology and ethnography. In short, the result was hailed as a comprehensive compendium, a 'definitive document for the geography of Central Africa' (ND, 1924: 17). The map produced, in ten sheets, five at a scale of 1:500,000 and five at a scale of 1:200,000, included a strip of land about 20 km wide located on both sides of the theoretical frontier. The protocol agreed upon by the Commissioners was signed on 10 January 1924. A paradoxical declaration closed the report: 'Everything leads one to believe that it will give satisfaction to the tribes of the frontier region' (Grossard, 1925: 191). Indeed, the Commissioners had often tried to contact the 'indigenous authorities'. For this, the sultans and the chiefs of the tribes of the regions concerned were summoned and consulted: some collaborated, others were indifferent, still others hostile, even if not openly. However, these 'authorities' only played the role of informants: they never became actors in the process, they were not asked for an opinion or even just advice. The Commissioners were personally directed 'to study the current rights of the tribes mentioned in the Convention and to defend their interests'. The testimony of sultans and notables was simply gathered and even with great difficulty (Grossard, 1925: 188–189):

> It was useless to look for written documents relating to the history and borders of these provinces; the sultans always took care that none were ever established, lest they be communicated to foreigners. They thought that by remaining unknown to the rest of the world, they had a better chance of keeping their independence intact.

Often the information gathered turned out to be nothing more than 'a fabric of childishness and inconsistency'. As Grossard (1925: 189) further argued: 'For most indigenous people, the history of their tribe boils down to the genealogy of their sultans' and 'their memory is short'. The unreliability of the 'authorities' also extended to knowledge relating to the territory (Grossard, 1925: 90):

> Overwhelmed by his zeal, the topographer is sometimes tempted, in geographically little-known countries, to extend the surface of his survey by reporting the location of points obtained from indigenous information. This is one of the causes of the sometimes surprising errors that our maps of Africa contain: one is never sure that a question, although exactly translated to a black, has been well understood.

For this reason, any reported information was excluded from the maps. Indeed, it was potentially dangerous, capable of damaging the quality of the cartographic result. The territory was the one surveyed by topographers, not the one described by the inhabitants. The Commissioners and topographers never even dealt with the fact that they were facing another way of conceiving space, political relations and social structures. What mattered was the quality of the topographic survey, ensured by the precision of the triangulations carried out. The clear satisfaction with the technical result achieved in such difficult conditions did not however hide perplexities inherent in the territory, particularly in the description of hydrography, slopes and villages. For example, rivers were represented by continuous blue strokes while interrupted blue strokes identify all other watercourses. 'But some have water for five or six weeks, in the rainy season . . . while others are always dry, whatever the season' (Grossard, 1925: 88): it was not possible 'to distinguish one from the other', just as the topographers failed to 'differentiate with a special sign, the water points or permanent wells from those that are often dry'. The map was not sufficient and therefore it was accompanied by an 'annex relating to hydrography' which contained 'all the information collected on the wadis or water points whose names are shown on the map' (Grossard, 1925: 276). The same was true for the layout of the slopes which 'almost everywhere in the rainy season . . . disappear under the tall grass'. As a result, the lines which on the map 'connect the villages one to the other simply indicate that no insurmountable obstacle is encountered on the itineraries thus traced' Grossard, 1925: 88). Also in this case, the differences of the territory required completing the map with a list of the main itineraries, indicating the condition of the routes and the necessary stops. The greatest discrepancy, however, related to the settlements: 'The villages and camps were indicated on the map at the points where the topographers and Commissioners found them as they passed' (Grossard, 1925: 88–89). But 'the natives, especially in the northern regions, have the habit of often changing the location of their huts and their crops, according to needs' (Grossard, 1925: 89). The villages move so much that it is necessary to specify that 'when mention is made of a village in the text [of the Final Protocol], it is understood that it is the position in which that village was found by the Commissioners, or in the position shown by the map' (Grossard, 1925: 210). The Commissioners understood that their maps filled the blank spaces but were not able to describe the functioning of the territory in a satisfactory way, not even for the purposes of colonial control, and were forced to expand the cartographic representation with descriptions and classification schemes. Africa slips out of the hands of western cartographers, as sand between fingers.

The sedentary paradigm and the logic of moving sands

The territory that European cartographers ponder is based on a sedentary paradigm. The French geographer Retaillé (1998, 2007) contrasts western geography which represents a world firmly fixed, anchored to the ground, precisely the result of a sedentary culture, in which the social contract is transformed into topographical space (Retaillé, 2007: 172), with a different geography that is identified by the author in nomadism, mainly referring to the Saharan and Sahelian area. In the nomadic world, the organisation of space is profoundly crossed by the anthropological values linked to movement (Retaillé, 2007: 176–178). The first characteristic of this nomadic space is its constellation structure, where distance is not an obstacle but a link between the points of the space. The limit of nomadic space is not a line but a horizon, determined by time which allows you to connect the points to make a single place. The number of places is indefinite and depends on the possible combinations of points in space. Each constellation is activated by paths in an almost provisional way. What matters, therefore, is the movement that creates the places. In this way, political identity is not anchored to a delimited space, but to the movement. The Cameroonian philosopher Mbembe reaches similar conclusions, citing once again the example of the region we are talking about—between Lake Chad and the NIle—when he notes how the new economic geography in Africa 'is not unlike the geography that prevailed in the nineteenth century . . . just prior to colonial conquest and partition' (Mbembe, 2021: 181). At that time, the diverse, vast multiethnic regional spaces

> were not characterized by stable and precise borders, or by clear figures of sovereignty, but rather by a complex series of vertical corridors, lateral axes, and networks that were often mutually imbricated according to the *principle of intertwining and multiplicity*.
>
> (*Mbembe, 2021: 181–182, italics by the author*)

The central act of the colonial disarticulation of this space

> was not so much the arbitrary division of previously united entities. . . . It was, rather, the attempt to shape pseudostates on the basis of what fundamentally was a *federation of networks* and a *multinational space* made up not of 'peoples' or 'nations' as such, but rather of *networks*. It was an attempt to set rigid borders in what was structurally a *space of circulation* and negotiation . . . flexible and with a changeable geography.
>
> (*Mbembe, 2021: 183, italics by the author*)

This changing and mobile geography was founded on commerce, predation and war, mainly in the form of raids. What prevailed was a 'translocal' dimension, which 'obeyed what could be called the *logic of moving sands*' (Mbembe, 2021: 183, italics by the author).

In these considerations, we can find an echo of the distinction proposed by Deleuze and Guattari (1980: 434–527) between the dynamics of a mobile, nomadic, reticular, 'smooth' space, as opposed to the 'striated' one of the State. Cartography's need to establish borders played a central role—first cognitive and then practical—in the process of imposition of the sedentary paradigm and the 'striated space' on the African continent.

Conclusion. At the dawn of a long, dark night

Blank spaces turn into darkness, in the long night of colonisation. Let us go back for a moment to Marlow's finger that rests on them, in the maps of the Atlas: his skin is white. White are the cartographers who drew those atlases and the maps of the colonial borders. Precisely because of his 'admiration of the white skin' and his adherence to the stereotype of 'African barbarity', Achebe accuses Conrad of 'the recycling of racist notions of the "dark" continent and her people' (Phillips and Achebe, 2007). Conrad's obsession with 'the physicality of the negro' (Phillips and Achebe, 2007: 59) can be better understood through Fanon's reflections in *Black Skin, White Masks* where the Caribbean psychiatrist well expresses the point of view of someone who feels, in the gaze of whites, 'the whole weight of his blackness' (1986: 150). Without history, with a short memory and unreliable even in the knowledge of their territory, as stated by the Commissioners in charge of drawing the colonial borders, blacks are dispossessed of their being Other, as well as of the control of their lands: 'racism . . . is fundamentally a technology of dispossession' (Mbembe, 2021: 53). With reference to Fanon's thought, Mbembe proposes an exit from this 'dark night' hypothesising 'a praxis of self-defense' that leads to the formation of a 'decolonized community', 'a community of walkers', 'a vast, universal caravan' that moves leaving a provincialised Europe, looking upon it 'with solicitude and compassion, and breathing back into it the supplement of humanity it had lost' (Mbembe, 2021: 224–225). The long night of darkness is giving way to dawn, ushering in a new era. An era in which the carto-humanistic approach is breaking away from the 'whitestream' (Daley and Murrey, 2022: 172) and the 'colonial knowledge chain' (Mbembe, 2021: 56), thus coming to terms with 'the racism embedded within colonial cartographic dissection of the continent' (Daley and Murrey, 2022: 161). One hopes that the skin of the world and its cartographic transpositions will then reflect all the colours of each Otherness.

References

Achebe C (2016) An image of Africa: Racism in Conrad's *Heart of Darkness*. *The Massachusetts Review* 57(1): 14–27.

Bassett TJ (1994) Cartography and empire building in nineteenth-century West Africa. *Geographical Review* 84(3): 316–335.

Bracke W (2021) From the Atlas de l'Europe by Philippe Vandermaelen (1828–1833) to the Weiss Map by Thomas Best Jervis (1854). The role of the Établissement géographique de Bruxelles in the map production of European Turkey. *Proceedings of the International Cartographic Association* 3(4).

Brantingler P (1985) Heart of darkness: Imperialism, racism, or impressionism? *Criticism* 27(4): 363–385.

Close CF (1925) The western frontier of the Sudan. *The Geographical Journal* LXVI(4): 349–352.

Conrad J (1990) *Heart of Darkness*. New York: Dover.

Daley PO and Murrey A (2022) Defiant scholarship: Dismantling coloniality in contemporary African geographies. *Singapore Journal of Tropical Geography* 43: 159–176.

Deleuze G and Guattari F (1980) *Mille plateaux: capitalisme et schizophrénie*. Paris: Les éditions de minuit.

della Dora V (2009) Travelling landscape-objects. *Progress in Human Geography* 33(3): 334–354.

dos Reis F (2021) Empires of science, science of empires: Mapping, centres of calculation and the making of imperial spaces in nineteenth-century Germany. In: Lobo-Guerrero L, Lo Presti L and dos Reis F (eds) *Mapping, Connectivity and the Making of European Empires*. Lanham, Boulder, New York and London: Rowan & Littlefield, pp. 105–137.

Fanon F (1986) *Black Skin, White Masks*. London: Pluto Press.

Grossard JH (1925) *Mission de délimitation de l'Afrique Équatoriale Française du Soudan Anglo–Égyptien*. Paris: Larose.

Han B-C (2018) *The Expulsion of the Other. Society, Perception and Communication Today*. Cambridge and Medford: Polity Press.

Lobo-Guerrero L, Lo Presti L and dos Reis F (2021) Mapping and the making of imperial European connectivity. In: Lobo-Guerrero L, Lo Presti L and dos Reis F (eds) *Mapping, Connectivity and the Making of European Empires*. Lanham, Boulder, New York and London: Rowan & Littlefield, pp. 1–18.

Mbembe A (2021) *Out of the Dark Night*. New York and Chichester: Columbia University Press.

ND (1924) La Mission Ouadaï–Darfour (mission Grossard): Délimitation de l'Afrique equatoriale Française et du Soudan Anglo-Egyptien. *L'Afrique française* XXXIV(1): 15–17.

Ngugi WT (2017) The contradictions of Joseph Conrad. *Sunday Book Review, The New York Times*, 16 November: 13.

Pase A (2011) *Linee sulla terra. Confini politici e limiti fondiari in Africa subsahariana*. Roma: Carocci.

Phillips C and Achebe C (2007) Was Joseph Conrad really a racist? *Philosophia Africana* 10(1): 59–66.

Retaillé D (1998) L'espace nomade. *Revue de Géographie de Lyon* 73(1): 71–82.

Retaillé D (2007) Quel est l'impact de la mondialisation sur le développement local? Les échelles paradoxales du développement. *Cahiers d'OutreMer* 238: 167–183.

Rossetto T (2019a) *Object-Oriented Cartography: Maps as Things*. London and New York: Routledge.
Rossetto T (2019b) The skin of the map: Viewing cartography through tactile empathy. *Environment and Planning D: Society and Space* 37(1): 83–103.
Šešlak MŽ (2020) The double vision of Joseph Conrad: *Heart of Darkness*, a contradiction in terms. *Lipar/Journal for Literature, Language, Art and Culture* 73: 113–131.
Silvestre M (2014) L'Atlas universel de Philippe Vandermaelen (1825–1827): Histoire d'un succès commercial. *In Monte Artium, Journal of the Royal Library of Belgium*: 247–266.
Silvestre M (2016) Philippe Vandermaelen, the Mercator of young Belgium. *Brussels Studies* [Online], General Collection, 106. Available at: https://journals.openedition.org/brussels/1438 (accessed 3 September 2023).
Stone JC (1988) Colonialism and cartography. *Transactions of the Institute of British Geographers* 13(1): 57–64.

26
CARTOGRAPHY CONTRA COLONIALISM

Clancy Wilmott

'Before us lay the trackless immeasurable desert, in awful silence.'

Watkin Tench, A Complete Account of the Settlement at Port Jackson, 1793

Introduction

This chapter explores the limits of colonial cartography, as both an epistemological and representational structure, and the way in which these limits operate in producing oppositional knowledges of colonised landscapes which still resonate today. It focuses on what scholars might understand as cartography in the most 'classic' sense, as a tradition of scientific representation that emerges from the Enlightenment, and reaches prominence, with the term 'cartography' emerging during the nineteenth century (Crampton, 2010). I argue that although cartography and associated cartographic sciences are often considered to be agents of colonial processes (Akerman, 2017), whose power may only be tackled via concerted counter-mapping practices (Lucchesi, 2018), their limitations in representing the immeasurable, unclassifiable and heterogeneous results in inadvertent anti-colonial events: cartography contra colonialism.

Cartography is a scientific endeavour. Mapping, as a more general term, might encompass the stick-and-stones 'rebbelib' charts made by the Marshall Islanders to navigate the currents of the Pacific Ocean, or it might include the tones and timbres of the songlines sung by Aboriginal Australians to traverse the desert. Equally, it might also be a way of describing the myriad translations that occur between people and spaces in everyday life and how we make sense of unsettled worlds (Wilmott, 2020). It is based in both the principles of the mathematical measurement of space, as well as what Farinelli (1998) has described as 'cartographic reason'—the principle that (a) the world is a stable, rational system and (b) once this system is described, that description can be used to reason and build. Joseph (2011: xiii) argues that 'the concept of mathematics found outside the Graeco-European praxis was very different. The aim was not to build an imposing edifice on a few self-evident axioms but to validate a result by any suitable method'.

DOI: 10.4324/9781003327578-32

This difference became particularly evident with the development of Cartesian algebraic geometry—the geometry upon which the cartographic coordinate system is developed. The grid enables spatial problems to be solved as algebraic equations: all places become points in sets of x and y, and theoretically equal. Furthermore, in the European tradition of mathematical-based cartography, the world becomes theoretical—a singular, universal, deductive system upon which cartographers can aspire to bring the earth into order. Importantly, Joseph (2011) argues that European mathematics was both *idealist*—in that mathematics was free of values or bias—and *elitist*—because its pursuit was limited to an exclusive, intellectual strata, rather than a grounded everyday practice. This mathematical praxis was folded into mapmaking practices—through modes of cartographic measurement and calculation, the development of cartography as a scientific practice, designed for navigation rather than description, bringing the world into a single, universal, scalable representational system.

Cartographic lines and slippery landscapes

This is the cartography that settler-colonists used to conquer the territories of the new world. Across Africa, North America, the Caribbean, South America, Asia and the Pacific, cartography preceded colonialism: the first process of territorialisation was the sketching of a coastline, the measurement of the depths of the bay and the river, the identification of shoals, sandbars, sightlines and declination, the details of entry into harbours and bays, the survey of hills and flatlands. For example, Lt. Cook's 1770 map of Stingray Bay, now Botany Bay, shows the tentative outline of the shoreline, with a few dozen sounding depths sketched in pencil through Cruwee Cove at the mouth of the Bay. As cartographic efforts moved inward, coastlines turned into cadastral tracts. The straight-edge cartographic outline of parcels of land, again, preceded the survey of the geography of the landscape itself. De Brahm's 1757 'A Map of South Carolina and a Part of Georgia' shows the eastern coastline of North America with inlets, sandbars and rivers stretching up into the interior. Some swamps, marshes and oak forests are marked but for the most part, the parishes and individual land parcels remain bare except for numbers of lots linked to each 'proprietor' or property owner. The 20 or so tribes of Native people who lived and owned the land along the coast of the Carolinas, and the distinct landscapes they inhabited, are nearly completely absent.

In *The Black Shoals* (2019), Tiffany Lethabo King (referring to poet Kei Miller) writes:

> In the poem, 'What the Mapmaker Ought to Know', Miller notices and writes that islands and landscapes 'fidget'. Speaking back to the colonial violence of mapmaking, Miller warns the cartographer the 'landmarks shift' and 'slip' from the grip of those intending to fix and dominate people and the earth.
>
> *(King, 2019: 76)*

As King describes it, the desire to conquer the earth through cartography is ultimately futile. King points to the shoal, or the shore, as a space of counter-cartographic analysis—one which exposes the limits of what she terms 'White cartography' and its inability to complete the process of conquest, colonisation and settlement. Specifically, she argues: 'At the shoal (and shore)—or the in-between space—the White cartographic subject is a by-product of Black and Indigenous pressure or friction from the outside' (King, 2019: 78).

King points to the way in which cartography—specifically de Brahm's map—creates orders of interior and exterior, with Black and Indigenous liveliness on the outside, cast off and erased from the map. Furthermore, the White cartographic subject resides inside the map through the lines drawn on and into landscapes—the boundaries of properties, the fixed line of the coast and the toponyms from remote conquistadors afar.

However, while King sees the map as an agent of the White cartographer, against Black and Indigenous 'pressures and frictions', I want to argue that colonised landscapes too are exterior others which fail to be both understood and contained by the map, and thus creates their own enduring pressures and frictions. This argument is not to detract from the social and political resistance—the people—who continuously fight structures of settler colonialism and the white possessive (see Moreton-Robinson, 2015). Instead, it is an attempt to expand and explain what I have described in *Mobile Mapping* (2020)—the way in which even in the spaces where the violence of colonialism has completely dispossessed Native people of their land, the White cartographic subject struggles to wholly conquer and control the fidgety, slippery, absconding landscape through its own limitations in knowledge and representation. Cartographic frictions emerge in a myriad of forms—from cadastral boundaries that cut apart rivers, hills and watersheds, to mythological formations of fertile pastures, endless resources and lost lakes, to morphological processes made still. Such lines, once drawn, are hard to remove from the colonial and territorial imaginations of settlement, and continue to reroute, destabilise and sabotage those who follow them. The map becomes, thus, counterproductive: it promises an authority that it does not have, and in doing so, subverts its own power.

The flood lands of Dyarubbin

A particular example can be found in the case of Pitt Town Bottoms in Darug country in NSW, Australia. Pitt Town Bottoms was a small farming colony established by freed convicts along the banks of Dyarubbin, known by settlers as the Hawkesbury-Nepean River, near Pitt Town in what is now north-western Sydney. Following the establishment of the Colony of New South Wales in Port Jackson in 1788, a detailed survey and mapping of the coasts, inlets and rivers directly to the north and south of the harbour began. Karskens (2016: 319) writes: 'The Hawkesbury-Nepean is an unusual river. It flows out of mountains in the south and back into mountains in the north'. Since the landscape was scrubby and dense with trees, water provided one of the easiest modes of travel for surveying and exploring the interior of the Australian continent. As early as 1788/1789, William Bradley, naval officer and cartographer of many maps of the colonial period made immediately after settlement in NSW, had produced a detailed survey of Broken Bay (the inlet immediately to the north of Port Jackson) with one of its tributaries, the Hawkesbury River. The river is exquisitely detailed, stretching out westward into the empty spaces of the map, dotted with continuous depth measurements along its meander.[1] To the south, westward expeditions from Prospect Hill near the farthest reaches of Port Jackson had happened upon the Nepean River by 1789, and by the time J. Walker's map was published in 1793 in Watkin Tench's 'A complete account of the settlement at Port Jackson, in New South Wales', they had been connected together as the Hawkesbury-Nepean. South Creek also appears on the map as it flows northwards as another of the seven tributaries.

This was perhaps the first folly of cartography—the overreliance on a view from the exterior and on accessible entryways into the interior which once traversed were filled with information, but if not, left bare. The emptiness of these spaces beleaguered more ambiguity: does the emptiness mean that what exists in that space is not known, or not worth knowing? (see Pase, this volume). The early colonial surveys into the rocky, silty, loamy landscapes around the basin were generally in aid of finding arable land for farming. But the surveys themselves do not seem to comprehend—nor depict—the complex realities of water in a land known for intense drought and flooding rains, non-equilibrium rivers and that: 'the flow rates of Australian rivers, including the Hawkesbury-Nepean, are twice as variable at rivers in the rest of the world' (Karskens, 2016: 319).

Walker's map in Tench's journal draws its information from survey expeditions between 1788 and 1793. Across the map, it describes the different qualities of land and soil—'sandy', 'wretched and bushy', 'miserable', 'stony and barren', 'poor', 'mossy and sour', 'very bad', 'bad', 'better' and the faintest of praise, 'tolerably good'. Flooding is noted in this early expedition—along the 'supposed' course of the Nepean, it is noted that 'In floods, the water rises to heights of 50 ft perpendicular, leaving Reeds &c. in the Trees'. But further north, in the flat basin described by Karskens as being subject to flooding, the only note of the land where Pitt Town would eventually lay is a note on the creek that runs from the Hawkesbury to Prospect Mount 'very dreadful Country the whole that we saw upon this Creek the ground covered with large stones as if paved'.

By 1794, freed convicts and soldiers had begun to set up farms along the Hawkesbury at the bottoms of Pitt Town (Karskens, 2016). While earlier maps showed the river as a smooth and fixed path of water—if rocky—the tendency of flood meant that the settlement on these banks were not fixed by colonial farming and agriculture, but rather were 'zones of flux' (Karskens, 2016: 319). Importantly, however, it was precisely this geomorphological flux that produced such fertile soil at its banks:

> The receding waters leave behind layers of sediments, but what kinds is a lottery: rich, black, silty mud from the Paleozoic rocks of the Cos and Wollondilly, sterile sand from the Warragamba, Nepean, and Grose Gorges, or a mixture of both.
>
> *(Karskens, 2016: 319)*

The result was a suite of farms that traced the riverbank at Pitt Town Bottoms, run by ex-convicts and soldiers living off the fertile soil in conditions designed to be quickly packed up or rebuilt. This mode of farming—as flux rather than fixity—defies cartographic expectation in a way often seen on the Australian continent. Paul Carter writes:

> There is no sharp line dividing the line in theory from the line in practice; one always bleeds into the other—and, if we accept the figurative basis of geographical reasoning, then there is no line of thought that does not find its precedent in drawing the line. Perhaps the first to be aware of this were practical geographers themselves, the field surveyors charged with organizing the lie of the land into a linear design. They not only found that different environments resisted surveying in different ways, but that the act of surveying demanded its own environment, one with novel human and professional protocols.
>
> *(Carter, 2008: 54)*

In general, landscapes do not conform to the cartographic expectations of fixity laid out upon them. In the basins of Hawkesbury-Nepean River, however, they explode cartographic expectations, and the river bursts out of its lines as

> Seasonal rains send floodwaters rushing down . . . a catchment of some seventy thousand square kilometres. . . . When the waters reach the bottleneck at Sackville, they back up and drown the valley floor like one vast bath.
>
> *(Karskens, 2016: 319)*

The evidence of this bath is all but washed away on the maps described earlier, as the Hawkesbury-Nepean sits as a contained, orderly flow weaving northward into Broken Bay. No inscriptions of rapids or falls, of swamps or signs of fluvial swells are indicative of the way in which the river waxes and wanes according to the season. At the same time, there is substantial documentation of contact with Darug people by early surveyors and the *Dyarubbin* mapping project claims that, in the specific bend of the river where Pitt Town Bottoms was colonised, the local Darug place name was *Werriling* or *Wiriliny*, which the project tentatively cites as meaning 'Either an ill-omened place or a name describing flat lowlands here' (Karskens et al., n.d.).

Inundation and the enduring blue line

The marks of these floods across the landscape were evident—and *legible*—to the surveying missions from which these maps were made. Watkin Tench, who was in the first crew of settlers to encounter the Nepean, and also in the party who established that the Hawkesbury and the Nepean were two parts of the same river, writes in his journal as his party traced the paths of the river:

> We proceeded upwards, by a slow pace, through reeds, thickets, and a thousand other obstacles, which impeded our progress, over coarse sandy ground, which had been recently inundated, though full forty feet above the present level of the river.
>
> *(Tench, 1793: 28)*

It was clear that the floods in the basin were terrific—and that the river had played a fundamental role in not only the geomorphology of the landscape but also its ecology. As the water flows down from the Blue Mountains and the Southern Highlands plateaus, moving into the easy meanders of what is now north-west Sydney it can, during flood seasons, reach choke points at in the narrow sandstone gorges between Sackville and Brooklyn, which sends the flood waters rushing back and inundating the valley. The surge of flood waters in the region is so fast that it created palpable tension in those who witnessed it:

> we proceeded from there up to Richmond Hill by the river side; mounted it; slept at its foot; and on the following day penetrated some miles westward or inland of it until we were stopped by a mountainous country, which our scarcity of provisions, joined to the terror of a river at our back, whose sudden rising is almost beyond computation, hindered us from exploring.
>
> *(Tench, 1793: 127)*

Flooding continued to prevent further exploration by Tench and others, specifically at the gorges, partially out of fear and uncertainty of what would precipitate a flood blocking their return.

> The difficulty of penetrating this country, joined to the dread of a sudden rise of the Hawkesbury, forbidding all return, has hitherto prevented our reaching Carmarthen mountains.
>
> (*Tench, 1793: 161*)

Here, the limitations of Tench's survey—and the maps by Walker which accompanied his *Accounts*—are revealed: the river-as-fixed is very legible as a cartographic object measured or drawn by eye; but the river as a space of inundation and indeterminacy is far more unpredictable and inexpressible by contemporaneous cartographic techniques. As Carter elaborates, 'One begins to see why Australian explorers were constantly anxious about the progress of their work' (Carter, 2008: 54).

This disjuncture between colonial cartography and the experience of the flooding Australian landscape along Dyarubbin is long reaching throughout the nineteenth and into the twentieth centuries, along with the nonchalance of the colonial survey towards hydrological shifts in the environment. Karskens (2016: 327) writes that '[b]etween 1795 and 1821, the Hawkesbury-Nepean rose in freshes and floods at least thirty-seven times', and emphatically makes the point that there was continuous lived experience of flooding in the region to the point where it established a *mythos* within local settler wisdom. Yet 'A Chart Shewing the Inudundation at the Hawkesbury' between 22 and 24 March 1806 again depicts the river as a bounded space—a stable meander, with the flooded landscape coloured over. In certain parts, it depicts the flood waters reaching above 8 ft, and swallowing an area calculated at 42,000 acres. By 1817, a map by Scott[2] depicts a small tract of land along the north-west corner of the Hawkesbury River, with a small patch of darkened land plots, labelled with 'Creek', and the description 'Fine rich Soil' to the east of Richmond Hill. A cadastral map from 1826 'Pitt Town' by J. Whatman shows parcels of farmland pushing up against the river—yet again fixed and bounded by its banks. While there are panoramic and landscape views of the flooding—from 1816, for instance—most maps of the region still depict Dyarubbinn/the Hawkesbury-Nepean as a continuous, contained feature. And still, the farmers kept returning—to the despair of Governors and planners alike.

Over 130,000 people now live in the flood plains of Dyarubbin. A map from 2011 released by Hawkesbury City Council depicts the flood risk of the northern region. A thin dark blue line follows the same meanders sketched by Walker in 1791, albeit with better positioning and more accuracy. And like the 1806 maps, the river is encompassed by the extent of a flood. This is a hypothetical flood—a light blue wash of the extent of a 1:100 or once in 100-year flood, with a 1% chance per year. It stretches out along the lowlands—Windsor, Pitt Town Bottoms, Cornwallis, Clarendon, Mulgrave—and is then bounded by a darker blue to show the Probable Maximum Flood (PMF). The land in this map is a land of flood, but at the same time, a land of *rare* flood. And yet, since 1790, there have been over 130 floods recorded along Dyarubbin which have reached ten metres or more. In 2022, that predicted once in 100-year flood occurred *twice*, with thousands evacuated each time.

Conclusion

The discourses set out by colonial cartography—of a fixed and contained river, surrounded by fertile soil—resonate in flood and cadastral maps, in evacuations, and in lived experiences. Although the farms at Pitt Town Bottoms have by and large moved, though the river still swells, the constant throughout the invasion and settlement by colonisers along Dyarubbin is the thin blue line of the river on the map. It persists, a fixed object, despite hundreds of experiences and observations to the contrary. The epistemological insistence by cartographic technique on the stable, and the fixed, results in the illegibility of landscapes which shift and fidget. This is not just the shoals, as King describes, but all landscapes—and most especially those that have been bound up by colonial processes of surveying and territorialisation. The desire for 'fertile soils' results in an inability for the map to see, the difficulty in expressing land as both floodplain and farm, to show scalar connections across both space and time. The line of the river becomes folded into cadastral maps of the plains, and then into hydrological maps, and then into digital maps. To this end, we see continued floods which cause evacuations and destruction of places that perhaps should not have been built. In its own limits, colonial cartography becomes accidentally adverse to its own purpose, creating maps that wipe away traces of experience and drawing what a place is (a river) rather than what it does (a place that floods), hindering the progression of total colonisation for centuries: cartography contra colonialism.

Notes

1 Harbours in the County of Cumberland New South Wales [cartographic material]: The harbours survey'd by Capn. Hunter, 1788, 1789. Those parts not shaded with Indian ink on the land side is eye sketch/W. Bradley.

2 This map is titled 'A chart of the three harbours of Botany Bay, Port Jackson and Brocken [*sic*] Bay showing the ground cultivated by the colonists with the courses of the rivers, Hawkesbury, Nepean'.

References

Akerman JR (ed.) (2017) *Decolonizing the Map: Cartography from Colony to Nation*. Chicago: The University of Chicago Press.

Carter P (2008) *Dark Writing: Geography, Performance, Design*. Honolulu: University of Hawaii Press.

Crampton JW (2010) *Mapping: A Critical Introduction to Cartography and GIS*. Chichester: Wiley-Blackwell.

Farinelli F (1998) Did Anaximander ever say (or write) any words? The nature of cartographical reason. *Philosophy and Geography* 1(2): 135–144.

Joseph GG (2011) *The Crest of the Peacock: Non-European Roots of Mathematics*, 3rd edition. Princeton: Princeton University Press.

Karskens G (2016) Floods and flood-mindedness in early colonial Australia. *Environmental History* 21(2): 315–342.

Karskens G, Watson LM, Wilkins E, Seymour J and Wright R (n.d.) *The Dyarubbin Project: Aboriginal History, Culture and Places on the Hawkesbury River*. Available at: https://portal.spatial.nsw.gov.au/portal/apps/MapSeries/index.html?appid=82ae77e1d24140e48a1bc06f70f74269# (accessed 7 July 2023).

King TL (2019) *The Black Shoals: Offshore Formations of Black and Native Studies*. Durham: Duke University Press.

Lucchesi AH (2018) "Indians don't make maps": Indigenous cartographic traditions and innovations. *American Indian Culture and Research Journal* 42(3): 11–26.

Moreton-Robinson A (2015) *The White Possessive: Property, Power, and Indigenous Sovereignty.* Minneapolis: University of Minnesota Press.

Tench W (1793) *A Complete Account of the Settlement at Port Jackson in New South Wales, Including an Accurate Description of the Situation of the Colony; of the Natives; and of Its Natural Productions: Taken on the Spot.* London: Nichol and Sewell.

Wilmott C (2020) *Mobile Mapping: Space, Cartography and the Digital.* Amsterdam: Amsterdam University Press.

27
INDIGENOUS CARTOGRAPHIES

Davi Pereira Junior and Bjørn Sletto

Introduction

Indigenous and Afrodiasporic[1] cartographies challenge nation-state control over cartographic production and displace the traditional meaning of map production. Unlike state-sponsored mapmaking, Indigenous cartography is not concerned with defining territorial limits, excluding groups that do not possess territorial rights, or protecting private property under capitalist land market regimes. Instead, by exercising the power to map, Indigenous peoples demonstrate that cartography can be used to protect their rights to existence, territory and identity. To the extent that Indigenous peoples are able to produce their own cartography based on their own, autonomous spatial and social criteria, they challenge the erasure of subordinated, territorial epistemologies inherent to state-led cartographic production.

Because Indigenous cartographies emerge from specific territorial, social and cultural realities and are shaped by Indigenous epistemologies and ontologies, they reflect the diverse, lived experiences of Indigenous peoples, embody situated forms of storytelling and reveal the symbolic connections that Indigenous peoples maintain with their territories. This sets Indigenous mapping fundamentally apart from state-led, western ways of thinking about space and mapmaking (see Willmott, this volume). Indigenous cartographic processes are informed by collective political mobilisation and territorial struggles, thus serving as a response to neoliberal policies, the expansion of capital, and the commodification and extraction of Indigenous land and natural resources. Since Indigenous cartographies emerge from autonomous mapping processes, allowing Indigenous people to exercise their agency and act as protagonists in the mapping process, Indigenous mapmaking serves as a political instrument of self-defence against various forms of genocide, silencing, exploitation and extraction.

Politics of Indigenous cartographies

Since Indigenous cartography emerges from situated processes of mapmaking that take place in spaces beyond state control, it serves as a powerful political instrument (Farias Júnior, 2010; Brown and Raymond, 2013) to reinforce traditional knowledges and identities while

DOI: 10.4324/9781003327578-33

mobilising Indigenous people in struggles for territory, political authority and distribution and protection of rights (de Almeida, 2013; Bryan, 2011; Sletto et al., 2020). By using appropriate, community-based methods (Louis et al., 2012; Pearce and Louis, 2008), Indigenous mapping processes seek to maintain the group's autonomy while allowing communities to control the process of knowledge production during map production. The mapping process can thus be understood as a form of political intervention, allowing marginalised groups to give visibility to their struggles through the symbolic appropriation of physical spaces (Farias Júnior, 2010). For Indigenous people, maps are never individual expressions. Instead, mapping processes reflect and produce political actions (Crampton, 2001: 16) and must therefore be understood as plural and collective manifestations (Farias Júnior, 2010).

Because Indigenous maps are instruments of political mobilisation, Indigenous mapmakers eschew generic classifications and homogenisation of people, cultures and landscapes. This is because geographical borders and spatial nomenclature imposed by the nation-state are insufficient for understanding the profound processes of Indigenous territorial construction. Instead, Indigenous maps embrace specific political contexts and local realities, presenting situational claims according to the needs and conflicts experienced by the group. Indigenous cartographies are thus never frozen or static (Kitchin et al., 2013). Indigenous cartographies assume that the symbolic dimensions of space, including ancestral memories and sacred places, can always be reinterpreted and serve to defend claims to territories. While Indigenous maps capture specific moments, their meanings are always reinvented depending on who interprets the maps and how they are used in storytelling (Caquard and Cartwright, 2014). As Crampton (2001) asserts, cartographic knowledge production is always situated within a given, political and social context and thus normalised by particular relations of power (Radcliffe, 2012).

When Indigenous people thus map social relations within their territories, they develop a deeper understanding of the boundaries that distinguish and demarcate their territorialities as well as the collective identities that correspond with this territoriality. Indigenous territories are conceived through awareness of the self and collective identity, and Indigenous cartographies bring forth specific, local lexica used by Indigenous peoples to (un) name the social spaces where they live. By naming their own territorial domains through cartography, Indigenous people symbolically express their territorial epistemologies and the ways in which these give meaning to their social practices and symbolic relations with the landscape.

In one case in 2007, the Brazilian government authorised the binational company Alcântara Cyclone Space[2] (ACS) to illegally invade the farming and fishing areas of the Quilombola communities of Alcântara in Maranhão, Brazil, in order to build a platform to launch spacecraft and rockets. Faced with this threat, community leaders decided to initiate a community-based mapping process to protect their territories, and in doing so, mobilised Quilombola residents who took to building roadblocks to stave off the invasion. The maps resulting from the community-based process were later incorporated into the broader territorial rights struggles, providing support for lawsuits filed by Quilombola leaders both through the Brazilian legal system and in international human rights courts. The Quilombolas also started to use mapping to monitor the territorial integrity of their lands. Today, when their territories are threatened, they resort to mapping as a way of mobilising communities and producing knowledge to support their territorial demands.

By teaching participants about the landscape, social relations and community rules for appropriation of territory and natural resources, Indigenous cartographies provide pedagogical opportunities and contribute to the reproduction of Indigenous identities. Indigenous cartographies serve to record epistemologies, cultural memories, political struggles and histories of resistance, tracing the construction of Indigenous identities over time in ways that bridge different generations. By providing a means to access ancestral memory, cartography allows Indigenous people to connect their ancestral past with their present while also envisioning their own futures. Cartography makes it possible for Indigenous people to describe and georeference their material landscapes but also their symbolic landscapes, building visual narratives to transmit their epistemologies to future generations.

Mapping projects are thus always immersed in intergenerational pedagogical experiences as they bring together the knowledge of elders and youth. The mapping experience provides a means for elders to share knowledge accumulated throughout life based on collective memory, personal experiences and knowledge present in the territories and in peoples' bodies. For young people, the mapping experience also provides opportunities to engage with technologies such as Global Positioning Systems (GPS), digital photography and Geographic Information Systems (GIS) to design maps and construct symbols used in mapping. Most importantly, the Indigenous mapping process allows young people to learn about their own community from elders who have experienced the same reality as them.

When I—Davi Pereira Junior—conducted my first mapping project with Quilombola community of Alcântara where I was born, I looked for elders to guide me through our territory and take me to the boundaries established by our ancestors (Pereira Junior, 2020). In Indigenous and Afrodiasporic landscapes such as the Quilombola territories, not everyone has permission to cross sacred territories and walk freely in certain spaces, requiring elders to accompany younger community members on their first journey (Sletto et al., 2021). As we proceeded with the process of georeferencing our territorial boundaries, other young Quilombola leaders asked if they could join us on our walks. For many young Quilombolas like me, this was our first opportunity to visit the places that are of fundamental importance to our community and to learn the location of our territorial boundaries from our elders. If it weren't for this mapping project, maybe I and many other young people from Itamatatiua would never have had the opportunity to visit these important sites.

Since Indigenous cartographies are intimately connected with the defence of Indigenous territories, Indigenous mapping begins by collectively discussing the goal of the mapping process, what should appear on maps, and what should not be mapped. As a pedagogical tool, the construction of Indigenous maps is integral to the social, cultural, economic, religious, ontological and epistemological reproduction of the community, but on the other hand, maps can also be used as external political tools. Since Indigenous maps may eventually be made public and used in ways that are beyond their control, communities need to decide if the mapping process is designed to primarily serve internal needs of knowledge reproduction or if it is intended to meet political objectives.

Indigenous cartography thus emerges from a participatory trend in international planning and development, whereby local communities are invited to shape the maps that impact their lives. However, not all mapping projects serve to challenge the hegemony of the state. Although Indigenous maps have indeed been used as counter-hegemonic representations to further local struggles, they may also embody contradictions. Depending on who leads the mapping process and how it is implemented, the resulting map could serve both as an instrument of domination or as a tool of resistance and empowerment of subaltern groups. Since maps grant authority to their creators (Huggan, 2011), Indigenous mapping projects might also provide opportunities for political manipulation, create new, internal power hierarchies and fuel political fragmentation, and weaken processes of identity formation.

Representational strategies in Indigenous cartographies

Indigenous cartographies feature symbolic elements that give meaning to communities' ways of life, allowing Indigenous maps to express epistemologies, anthologies and identities in ways that escape the logic of conventional Cartesian cartography (Sletto, 2015). The counter-hegemonic potentials of Indigenous cartographic representations can in part be explained by the ways in which they challenge dominant, western understandings of borders, bodies and territory. From an Indigenous perspective, bodies and land should not be thought of in binary terms (Zaragocin and Caretta, 2020) but rather as intimately connected and contingent. Indigenous cosmology and ontology hold that being and existing is shared and experienced by all as a community, and Indigenous maps thus emerge from a collective form of existence. In doing so, Indigenous cartographies challenge the conventional and orthodox conceptualisations of maps that are intrinsic to western cartography and instead further decolonise understandings of the intimate connections between the sacred, the body and territory. Since Indigenous people understand landscapes as spaces that embody both spirituality and humanity rather than as simply 'natural' and non-human terrains, Indigenous cartographic representations evoke feelings of love, fear, courage, struggle, memory and the sacred. When Indigenous maps are experienced collectively by Indigenous communities, they reproduce a sense of common identity and belonging that sustains community life.

Mapmaking by Indigenous people thus reflects epistemological self-awareness and serves to define social relationships, sustain social rules and strengthen social values (Harley, 1988; Chambers, 2006; Sparke, 1998). In Indigenous cartographies, the awareness of Indigenous ways of existing in the world is expressed through an etymology of classification based on their deep understanding of territoriality, identity and people's relationships with their ancestors and their landscapes. That is to say, the true purpose of Indigenous mapmaking is not to produce realistic representations of space but rather to give symbolic meaning to things, objects and places in ways that make sense in communities' social world. This leads to a mapmaking process that follows its own rules in contradiction to orthodox and hegemonic ways of producing maps, the better to foster fluid and creative cartographic processes that serve political needs in anti-colonial struggles (Kitchin et al., 2013).

Since the symbolic relationships of indigenous peoples with topography and places are fundamental to understanding their social world, Indigenous maps need to be as dynamic as the realities of Indigenous peoples, expressing criteria such as identity, gender, race and ethnicity while reflecting spatial elements that give meaning to their existence. Because Indigenous cartographic representations are not constrained by the dominant logics of colonial authority, Indigenous maps are fraught with material and symbolic complexity, making them sometimes difficult to understand for those who do not share the same social world.

In order to reflect the deep meanings of topography and places in the lives of Indigenous people, the representational regime of Indigenous maps departs from the traditional conventions of established mapping systems through different approaches to lines and polygons, colour choices and forms and styles of map symbology. For example, because Indigenous conceptions of borders differ from those of the modern nation-state, Indigenous cartographies call for thinking of 'borders' without the fixity of lines and polygons, and even imagining representing borders without lines or polygons. The understanding of borders in Indigenous epistemologies is based on relationships with places, landscapes and ancestral memory (Larsen and Johnson, 2012; Novoa, 2022), which means that borders are inherently and constantly shifting and permeable (see Pase, this volume). This leads to experimentation with irregular lines and other symbology that reflect the fundamental dynamism of borders, thus revealing a symbolic economy inspired by the natural, material and spiritual elements of the territory.

To represent the complex lifeworld of Indigenous peoples, Indigenous cartographers seek to portray rivers, oceans, swamps and lakes in ways that reflect how these natural resources are lived and perceived by the communities. In doing so, Indigenous cartographers eschew Cartesian conventions that seek to preassign standardised representational forms to complex socio-material spatialities. Representations of forests, toponymy, topography, natural features and landscapes thus go beyond the concreteness of western cartography to instead deploy symbology that reflects complex meanings and socio-natural relations. In the case of the Amazon, for example, the colour of the water depends on the natural chemical processes associated with the dissolution of vegetation and the presence of algae, which leads Indigenous people to use colours and symbols that reflect their intimate, situated and symbolic relationship with water.

In ancestral Indigenous cosmovision, territory is constituted by topographies and places that are sacred and often secret, and therefore can only be accessed—and represented—by people authorised by their deities and spiritual leaders and who are familiar with the language and epistemology of their ancestors. Such sacred places are central to performance of rituals and therefore fundamental for their existence (Sletto et al., 2021). To access sacred places, rules of movement and access determined by the ancestral spirit, 'owner' of the territory, must be followed, suggesting that the circulation of bodies within indigenous territories is controlled or mediated by symbolic as well as political realities (Larsen and Johnson, 2012; Radcliffe, 2012). A good example are the sources of rivers and ancient water wells, which are generally sacred places that belong to enchanted or sacred entities. Unwanted visitors to these sacred sites are subject to punishment ranging from fever or illness to death, and the visit may change the enchantment in the place and the entity may disappear. Only spiritual leaders have the power and knowledge of the rituals required to free a person from these punishments.

To Indigenous people, these spiritual relations give meaning to their existence and help make sense of their relationship with territory. Indigenous cartographers are thus tasked with spatialising the complex interplay of joy, knowledge and religiosity that characterise communities' social and symbolic relationships with territory, which leads to experimentation with innovative strategies to represent intangible yet profoundly meaningful socio-spatial relations. To Indigenous people, the forest is not simply a natural resource that can be monetised or a place where animals live, for example. Instead, the forest is understood as fundamental to the epistemological reproduction and cosmological existence of the group. Similarly, a hill may not be simply a topographic formation but rather a sacred dwelling for ancestors or an enchanted being. It is these symbolic relationships that Indigenous and Afrodiasporic people establish with nature that are at stake in the process of representation, and it is also these symbolic relationships that prompt their desire to protect their territory and thus safeguard their existence. The concern of Indigenous cartographers, then, is to properly represent spatial elements as they exist in the symbolic world of indigenous peoples and the meaning they play in their lives, rather than replicating the rules of conventional cartography.

Conclusion

Indigenous cartographies provide new ways of understanding maps and mapping processes. Indigenous cartographies have contributed to a deeper understanding of the political dimensions of mapmaking, revealing the ways in which maps and mapping processes may serve as empowering tools for politically underrepresented groups. Practitioners in the field of Indigenous cartographies are now seeking to develop methodologies that can better respond to the political demands of traditional groups, as they fight for collective rights and protection of territories in order to guarantee our right to physical, cultural, identity and religious reproduction. At the same time, Indigenous cartographies have also fostered creative innovations by incorporating ethnic factors, identity, cosmology and alternative ontologies and epistemologies into cartographic representations. Through Indigenous mapmaking, complex social relationships and notions of collectivity have been brought to light, thus fostering a decolonisation of cartographic thought and epistemology.

Indigenous maps are not produced for purely instrumental reasons intended to meet specific demands vis-à-vis the nation-state. Instead, for Indigenous peoples, maps are an important tool to transmit their knowledge and cosmology to the next generation. Indigenous cartographies thus break with the paradigm of orthodox cartography, where maps are understood primarily as instruments to make war or to support the work of an administrative bureaucracy as it seeks to impose the logic of state governance. By challenging the paradigmatic role of state maps, Indigenous cartographies open the possibility for maps to represent borders in new ways instead of reproducing the traditional symmetrical geometric lines of the nation-state. Indigenous maps reveal emotional geographies such as fear, love, longing, loneliness and sadness, thus speaking to readers in a different register than traditional state maps. Indigenous maps open the door to imagination, fostering a connection between readers and mapmakers in ways that challenge the orthodox conceptualisation of technical cartography.

Notes

1 In the following, we use the terms Indigenous cartographies and Indigenous maps to also refer to map-making in Afrodiasporic communities.

2 Alcântara Cyclone Space was a binational public company funded with Brazilian and Ukrainian capital established on 31 August 2006 with the objective of commercialising and launching satellites using the Ukrainian Cyclone-4 space rocket from the Alcântara Launch Center. Due to limited investments, the company closed in 2018 without achieving the aims expected by the governments of Brazil and Ukraine.

References

Brown G and Raymond CM (2013) Methods for identifying land use conflict potential using participatory mapping. *Landscape and Urban Planning* 122: 198–208.

Bryan J (2011) Walking the line: Participatory mapping, Indigenous rights, and neoliberalism. *Geoforum* 42(1): 40–50.

Caquard S and Cartwright W (2014) Narrative cartography: From mapping stories to the narrative of maps and mapping. *The Cartographic Journal* 51(2): 101–106.

Chambers R (2006) Participatory mapping and geographic information systems: Whose map? Who is empowered and who disempowered? Who gains and who loses. *The Electronic Journal of Information Systems in Developing Countries* 25(1): 1–11.
Crampton JW (2001) Maps as social constructions: Power, communication, and visualization. *Progress in Human Geography* 25(2): 235–252.
de Almeida AWB (2013) *People and Traditional Communities: New Cartography Editions*. Manaus: UEA Editions.
Farias Júnior EdeA (2010) Social cartography and traditional knowledge associated with the claim of specific territorialities. In: Almeida A, Dourado BS, Menezes ESde, Farias Júnior EdeA, Nakazono E and Barauna GMQ (orgs) *Discussion Papers New Social Cartography*. Manaus: Casa8, pp. 90–105.
Harley JB (1988) Maps, knowledge, and power. In: Cosgrove D and Daniels S (eds) *The Iconography of Landscape: Essays on Symbolic Representation, Design and Use of Past Environments*. Cambridge: Cambridge University Press, pp. 277–312.
Huggan G (2011) First principles of a literary cartography, from territorial disputes: Maps and mapping strategies in contemporary Canadian and Australian fiction. In: Dodge M, Kitchin R and Perkins C (orgs) *The Map Reader: Theories of Mapping Practice and Cartographic Representation*. Oxford: John Wiley & Sons, pp. 412–421.
Kitchin R, Gleeson J and Dodge M (2013) Unfolding mapping practices: A new epistemology for cartography. *Transactions of the Institute of British Geographers* 38(3): 480–496.
Larsen S and Johnson JT (2012) In between worlds: Place, experience, and research in Indigenous geography. *Journal of Cultural Geography* 9(1): 1–13.
Louis RP, Johnson JT and Pramono AH (2012) Introduction: Indigenous cartographies and counter-mapping. *Cartographica* 47(2): 77–79.
Novoa M (2022) Insurgent heritage: Mobilizing memory, place-based care and cultural citizenships. *International Journal of Urban and Regional Research* 46(6): 1016–1034.
Pearce MW and Louis RP (2008) Mapping Indigenous depth of place. *American Indian Culture and Research Journal* 32(3): 107–126.
Pereira Junior D (2020) Political appropriation of social cartography in defense of Quilombola territories in Alcântara, Maranhão, Brasil. In Sletto B, Bryan J, Wagner A and Hale C (eds) *Radical Cartographies: Participatory Mapmaking in Latin America*. Austin: University of Texas Press, pp. 183–202.
Radcliffe SA (2012) Relating to the land: Multiple geographical imaginations and lived-in landscapes. *Transactions of the Institute of British Geographers* 37(3): 359–364.
Sletto B (2015) Inclusions, erasures, and emergences in an indigenous landscape: Participatory cartographies and the makings of affective place in the Sierra de Perija, Venezuela. *Environment and Planning D: Society and Space* 33(5): 925–944.
Sletto B, De la Torre GB, Lamina Luguana AM and Pereira Júnior D (2021) Walking, knowing, and the limits of the map: Performing participatory cartographies in Indigenous landscapes. *Cultural Geographies* 28(4): 611–627.
Sletto B, Wagner A, Bryan J and Hale C (eds) (2020) *Radical Cartographies: Participatory Mapmaking from Latin America*. Austin: University of Texas Press.
Sparke M (1998) A map that roared and an original atlas: Canada, cartography, and the narration of nation. *Annals of the Association of American Geographers* 88(3): 463–495.
Zaragocin S and Carreta MA (2020) Cuerpo territorio: A decolonial feminist geography method for the study of embodiment. *Annals of American Association of Geography* 111(5): 1513–1518.

28
BLACK CARTOGRAPHY AS MEMORY WORK

Stephen P. Hanna

Introduction

In 1873, John M. Washington wrote his memoir, *Memorys of the Past*. Eleven years had passed since he emancipated himself during the United States' Civil War by leaving Fredericksburg, Virginia, crossing the Rappahannock River and entering a Union army camp. At the time he wrote his memoir, he and his wife were raising five sons in Washington, DC. John worked as a house painter and served as the superintendent of Sunday schools at a church founded by once-enslaved people from Fredericksburg (Blight, 2007: 94).

Historian David Blight suggests that Washington wrote a memoir 'to leave his story to that brood of sons who otherwise might never really know where they came from' (2007: 104). While this is a logical rationale for the memoir's existence, it is unlikely we will ever fully know why Washington included a hand-drawn map of Fredericksburg in its pages—a decision that makes this memoir unique among autobiographies written by once-enslaved people in North America.

Washington's map accompanies the portion of his memoir where he recounts the events of 18 April 1862, the day he escaped enslavement (see Figure 28.1). Taking up most of a page, the map displays a stretch of the Rappahannock River between Fredericksburg, represented as a grid of streets in the southeast corner, and Falmouth, a smaller rectangle to the north-west. Washington carefully printed labels, complete with serifs, for the river, two railroads, a canal and the roads connecting the two towns. The Union camp, symbolised with a stippled pattern, lies to the northeast of Falmouth. Washington numbered important places from both that most significant of days and from earlier life events included in *Memorys of the Past*. He gave meaning to those numbers in a legend to the left of the map. The highest number, '16', labels a line crossing the Rappahannock and is defined in the legend as 'where I crossed the River'.

In his edited version of Washington's memoir, Crandell Shifflett argues that the numbered points on the map are 'coordinates of freedom—places where his life intersected with events, people, or moments when the thoughts of liberty touched his memory' (2008: 81). In other words, by mapping his Fredericksburg, Washington pinned key places and events from a Black man's memories to spaces that the antebellum White leaders of Fredericksburg had regulated to deny freedom to Black people.

 DOI: 10.4324/9781003327578-34

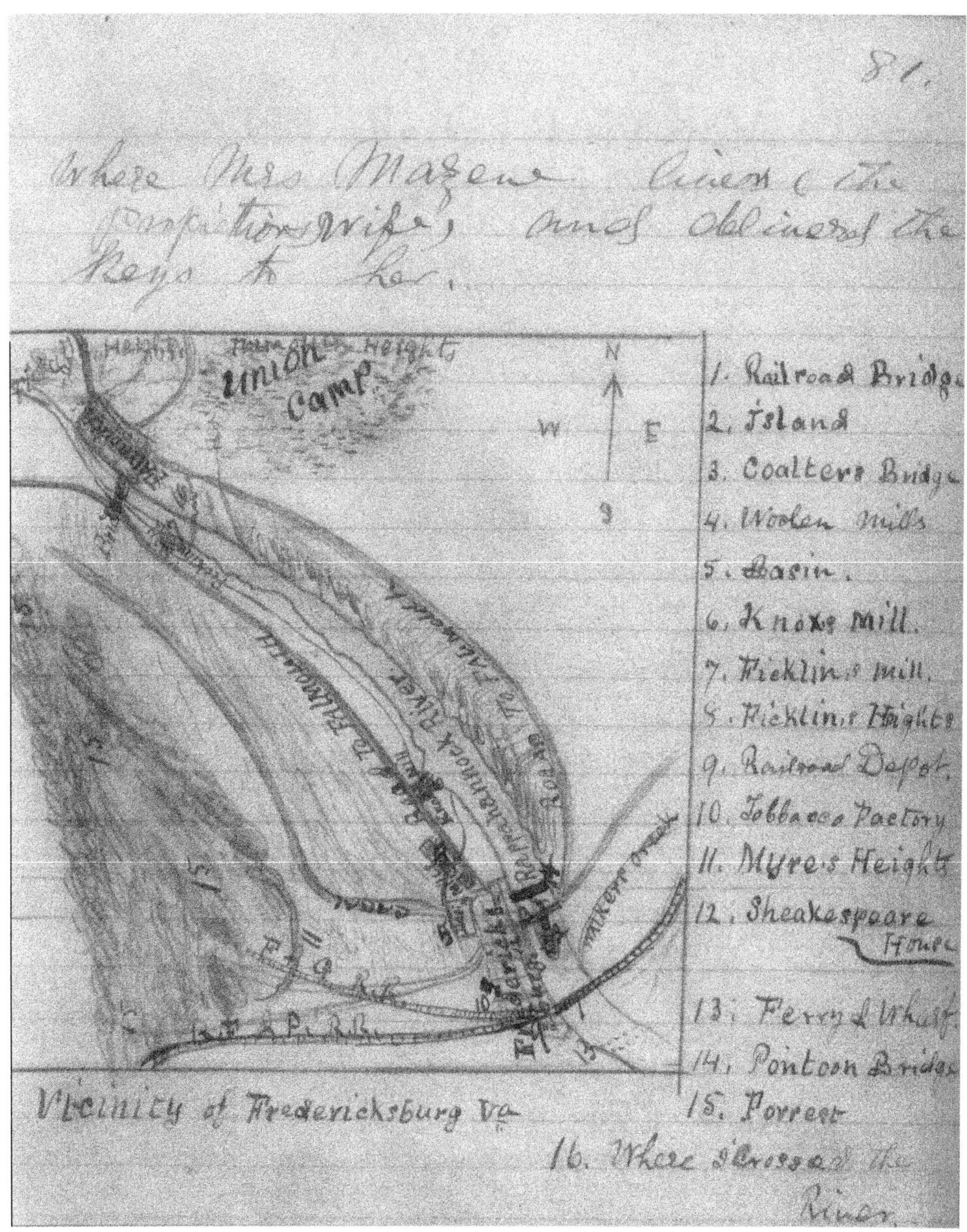

Figure 28.1 John Washington's mapping of Fredericksburg, Virginia. John Washington Papers, 1858–1865, 1982, Accession #15000, Special Collections, University of Virginia Library, Charlottesville, VA.

Washington's map is an exemplar of counter-mapping as memory work in both the past and the present. During the moments of its initial creation in 1873, Washington's physical act of using a pencil to lay out Fredericksburg helped him access his past as an enslaved Black child and young man. As Blight (2007: 41) argues, in 'the detail of the map, it is as if Washington is declaring his need to never let this memory recede from his mind'. As such, it is a nineteenth-century example of Black cartographies (Alderman et al., 2021) that repurposed mainstream mapping practices to counter the erasure of enslavement from the nation's memory and the intensifying efforts of White southerners to crush 'the rights of blacks (*sic*) by intimidation, murder, and disfranchisement' (Blight, 2007: 98). Because Washington's descendants preserved his memoir, his map has re-emerged in a present marked by struggles to acknowledge and repair the legacies of race-based slavery. After more than a century of erasing or marginalising Black experiences from Fredericksburg's official histories, city officials and local activists now practise Washington's cartographic memories by placing historical markers and adding his story to walking tours through the city's commemorative landscape.

Mapping Washington's memories

The emergence of critical cartography (Harley, 1988; Wood, 1992; Crampton, 2001) in the 1980s and 1990s allowed maps, such as Washington's, to be understood as representations produced within specific historical contexts and shaped by deeply unequal power relations. From this perspective, Washington's map can be understood as a 'subaltern mapping that stands in opposition to the top-down representational texts of professional cartographers, which erase' the lived experiences and memories of Black people (Hanna, 2012: 49; see also Kelley, 2021; Rankin, 2020). As such, the map represents a Black geography of nineteenth-century Fredericksburg authored by a Black cartographer.

To understand Washington's map as the memory work of a free Black man living in the nineteenth century, however, it is necessary to also examine it as a more-than-representational practice (Kitchin and Dodge, 2007; Hanna and Del Casino, 2017). From this perspective, the meanings of Washington's map are not fixed, but emerge when the map is used. While the moments when Washington himself first mapped the city of his childhood cannot be reconstructed, geographers today ask research participants to draw maps in order to elicit the memories, perspectives and emotions they attach to particular places (Argent and Walmsley, 2009; Potter, 2015). Given the efficacy of such methods, it seems reasonable that the act of mapping the events of 18 April 1862 helped Washington recover his actions and emotions from that day as well as from earlier events of his childhood.

Del Casino and Hanna (2000) argue that maps are not separable from the spaces they represent and Washington crafted this map to represent the Fredericksburg described in his memoir. Therefore, an interpretation of Washington's mapping should not be limited to the symbols found on this single page of his journal. The memoir is replete with precise locations of towns, plantations, buildings and other places he deemed important to his story. He makes narrative space to describe, using cardinal directions and street names, the intersections where important buildings were located, including the African Baptist Church and the bank building where he lived and worked as an enslaved child. Recognising the entirety of the memoir as part of Washington's mapping follows other theorists who work to enlarge the language of cartography (della Dora, 2009; Nurmi, 2016).

Washington's mapping of Fredericksburg 11 years after his escape was more than an act of recovering his own memory. By adding serifs to labels and appropriating symbols found on contemporary maps of the town, Washington appears to have wanted other people to use his mapping to understand the spaces where his experiences of enslavement and emancipation unfolded. Perhaps most significantly, this mapping of his memories is not drawn from the perspective of a 24-year-old fugitive slave (Nurmi, 2016), but one that positions a 35-year-old map author above and outside of the Fredericksburg of his memories.

In other words, Washington's memories of enslavement and emancipation in Fredericksburg emerged through his appropriation of a western cartographic imagination which this self-taught man could only have gained by working with published maps (Hanna, 2012). Even while enslaved, Washington fed his appetite for learning by stealing moments to read *Harper's New Monthly Magazine*—a publication that included maps among its illustrations. During the Civil War, northern newspapers and magazines published battle maps and, after the war's end, included maps in illustrated histories of the war. In fact, a map of Fredericksburg published in *Harper's Pictorial History of the War of Rebellion* (1866) uses symbols for railroads and the city itself that resemble those Washington used on his map. So, part of Washington's memory work almost certainly involved practising other mappings and appropriating both their symbology and the god's-eye perspective of White, western cartography (Pickles, 2004).

This effort to learn and practise formal mapping techniques of the nineteenth century reflects Washington's role as an educator and, more generally, Black Americans' embrace of learning as what a Mississippi freeman called 'the next best thing to liberty' (quoted in Foner, 2005: 88). In the wake of the Civil War, Black people built schoolhouses and created educational associations because they believed that education was a form of empowerment (Foner, 2005). Washington embraced this mission becoming superintendent of his church's Sunday school. Sunday schools, which taught adults and children, were part of Black people's collective educational efforts and their curricula often included maps to teach students geography (Pierce, 1971; Blight, 2007). Thus, it is reasonable to think that the techniques Washington used to map his own past were, at least in part, inspired by his belief in education.

Recognising that Washington's memory mapping reflected the importance of education to Black people's efforts to realise freedom and equality helps place his mapping within the racial contexts of the 1870s. By the time he mapped Fredericksburg, Black people's economic and political freedoms were under assault as White southerners employed violence and terror to regain control of their society (Blight, 2007: 98). In addition, a White supremacist remembering of the Civil War had already begun—one that would ultimately make the 'Lost Cause' mythology, complete with its trivialisation of enslavement, the dominant way White Americans would position the Civil War within the nation's historical narrative (Blight, 2001).

Given this context, Washington's mapping of Fredericksburg from the remembered perspective of a formerly enslaved Black man places this work in the category of countermapping (Alderman et al., 2021). While not made public, like subsequent Black mappings of lynchings or W. E. B. Dubois's masterful use of maps to explain the centrality of slavery to the antebellum Atlantic economy (Battle-Baptiste and Rusert, 2018), Washington's mapping (re)places slavery and emancipation as the central stories of Civil War Fredericksburg.

Through the map's creation, Washington remembers stolen moments of freedom while enslaved and those hours of 18 April 1862 when he was a fugitive. The fact that this mapping is a work of memory does not make it less performative or embodied than the in-the-moment Black material cartographies performed by fugitives while escaping bondage (Nurmi, 2016: 125). Washington, as a self-emancipated and self-educated Black man, can be imagined sitting at a desk in his home and pencilling his memories onto paper using symbolic practices learned through his engagements with other mappings of Fredericksburg. Seen in this way, his performative mapping reflects not only his temporal and spatial distance from the geography of his escape from bondage but also his desire to formalise his memories in the language of nineteenth-century cartographic practice. As such, his mapping was 'an empowering . . . device' that allowed Washington to 'orient himself in the physical and cultural landscape' (Rossetto, 2015: 479) of Black emancipation and post-war struggles to realise the full promise of freedom and citizenship.

Practising Washington's map in the present

John Washington's counter-mapping was preserved by his descendants and made public by professional historians, thereby bringing his remembered geographic experiences in Fredericksburg into our present. Yet, traditional and exclusionary perspectives on what maps 'should be' can interfere with how it is used today. Crandall Shifflett provides an exemplar of this when he initially dismissed Washington's map as a 'crude drawing . . . done in pencil with relative locations' (2007: 81). If public historians and city officials practise the map with that perspective, they might view it as an unreliable guide to mid-nineteenth-century Fredericksburg and, therefore, not use it while making the history of the enslavement and emancipation more visible in the city's commemorative landscape. Fortunately, the work of Harley (1988, 1990), Wood (1992) and others opened up approaches to seeing Washington's map, and the memoir it ties to Fredericksburg's landscape, as an alternative or counter representation of a nineteenth-century Black man's experiences.

Yet, the critical cartographic perspective merely leaves us with a valuable artefact representing a Black man's memories. It does not help us recognise that this map's meanings emerge through practice—that the context within which this mapping is performed, as well as the nature of that performance determines whether and how Washington's memories are made present in Fredericksburg's public spaces today. As Rossetto reminds us, this mapping 'will be encountered by different people in diverse ways' (2015: 487). The ways people engage with Washington's Black cartographic memories enable and limit the nature of its use.

If the existence of Washington's memoir had been widely known prior to 2000, White city officials and many others working in heritage tourism might have worried that making his mapping present in public spaces would be too negative or controversial. At that time, narratives of Black history were all but absent in public mappings of Fredericksburg's past. The two exceptions were the Slave Auction Block, an inadequately interpreted relic sitting at the town's primary site for slave auctions, and the self-guided African-American History Tour. Between the end of the Civil War and 2000, no sustained effort had been made to establish respectful commemorative practices at the auction block and many in Fredericksburg's Black community viewed it as a sign of continuing oppression. And, while the African-American History Tour's brochure was available at the City's visitor centre, there were no *in situ* interpretations of the places marked on the brochure's map.

After 2000 and due to the efforts of a Black employee in the city's planning department and of influential congregants from the city's historical Black churches, the City installed wayside panels interpreting these churches' involvement in the struggle for civil rights.

Figure 28.2 The 'Bound for Freedom' wayside panel marking where Washington crossed the Rappahannock River to freedom. Photograph by author.

Erected between 2000 and 2010, these signs mapped portions of the African-American History Tour onto the city's commemorative landscape. Stories of enslavement and emancipation, however, remained mostly invisible.

It took the independent efforts of David Blight (2007) and Crandell Shifflett (2008) to bring Washington's mapping of Fredericksburg to the attention of local public history practitioners. By publishing their edited versions of the memoir and placing Washington's work in historical context, Blight and Shifflett ensured that public historians, city officials and other memory workers now trust the memoir's authenticity and are willing to inscribe this Black man's memories into Fredericksburg's landscape in the form of guided tours and historical markers.

By 2017, the city, working in partnership with National Park Service historians and Black business owners, had made Washington's mapped memories physically present in the city's landscape in three places. The first is marked on his map as #16, 'Where I crossed the River'. In 2010, the city placed a wayside panel titled, 'Bound for Freedom', on the banks of the Rappahannock, very close to where Washington mapped his crossing (see Figure 28.2). The panel describes Washington's walk from Fredericksburg to this location where he saw Union soldiers on the river's opposite shore. The soldiers rowed a boat over to where Washington stood and, to quote the panel, he 'took the fateful step of crossing the river to freedom'. A photographic portrait of Washington as a mature and successful man makes him present in this space. The panel also includes a quote mapping his memoir to the riverbank, 'I could not begin to express my new born hopes for I felt already like I was certain of my freedom now'.

The second location, the Farmers Bank building where Washington remembers being forced to live apart from his family while still a child, is not symbolised on Washington's map, but is mapped precisely within the memoir as on the 'N.W. Cor. Of George and Princess Ann Streets' (Washington, 2007: 170). The bank building is extant on the landscape today, but until recently, wayside panels and guided tours ascribed the building's significance as a site where Abraham Lincoln delivered a speech. After Blight and Shifflett enabled more people to engage with Washington's mapped memories, the building is increasingly associated with this Black man's experiences as an enslaved child. When the first floor of the bank was converted into a restaurant in 2016, the owners commissioned a portrait of Washington that hangs above one of the restaurant's fireplaces. Washington's face is comprised of words from his memoir recalling the sadness he experienced as an enslaved child.

During the renovation, the original stairs at the bank's side entrance were moved one block away to the grounds of the Fredericksburg Area Museum where a wayside panel, titled 'Historic Footsteps', notes that both Abraham Lincoln and John Washington set their feet on these stone steps. On this third mapping of Washington's memories onto the city's present-day landscape, his portrait is placed between a painting of Lincoln addressing Union troops and a photograph of the steps in their original location. A quote detailing his first night of freedom maps his emotional realisation of freedom to this location, 'The Soldiers assured me that I was now a free man . . . [that] I never would be a Slave no more. . . . Life had a new Joy awaiting me'.

For many members of Fredericksburg's Black community, however, the wayside panels dedicated to Washington's experiences and those located near historically Black churches were not sufficient. In the wake of the 2017 White Supremacist riots in nearby Charlottesville, Virginia, Black residents and their allies focused the anger stemming from their continuing marginalisation in official public histories on the Slave Auction Block, a relic that had been subjected to decades of neglect and disrespect. After 18 months of public discussion facilitated by the International Coalition of Sites of Conscience, City Council voted to move the Block to the Fredericksburg Area Museum and heeded the Coalition's call to make Black experiences more fully a part of the histories mapped to public spaces through organised tours and physical markers.

To this end, the city's Economic Development and Tourism Department partnered with the University of Mary Washington and the Black community to create 'Freedom, A Work in Progress', a Civil Rights trail marking sites prominent in the local Black struggle for equality and justice from the nineteenth century to the present. Since John Washington's self-emancipation represents an early victory in this struggle, this mapping once again makes his memories present at the Farmer's Bank Building. A 2022 City grant allowed the Fredericksburg Area Museum to hire a curator for African-American history who, in addition to creating new exhibits within the museum, is working with city residents, other museums and city government to create a tour mapping the history of Black entrepreneurship and to find other ways to make the city's public history more diverse and inclusive. A new tour on Black resistance to enslavement featuring the mapped memories of John Washington, as well as those of others once-enslaved in the Fredericksburg area, is planned as well.

Conclusion

The significance of Washington's memories as an enslaved child to Fredericksburg's mapped public history was remapped again in 2023 when Virginia's Department of Historic Resources installed a new state historical marker, 'Great Exodus from Bondage' near the Farmer's Bank building. The marker designates the building as Washington's home and workplace and identifies him as one of the first of over 10,000 enslaved people from the region to free themselves eight months before Lincoln issued the Emancipation Proclamation. Given that Washington last saw his enslaver in this building on the day of his escape, the new marker could become the first stop on a trail people could use to remap the path to freedom he traced with his pencil in 1873. Such a path could feature Shiloh Baptist Church (Old Site), a historically Black church occupying the same site as the African Baptist Church where Washington experienced his spiritual awakening, the site of the tobacco factory where, while still enslaved, Washington earned his own wages, the Shakespeare Hotel where Washington first heard of the arrival of Union troops on 18 April 1862 and, of course, point #16 on his map—'where I crossed the river'.

Ideally, such a trail's brochure, website and wayside panels would prominently feature an image of Washington's map of Fredericksburg and interpret it in ways that help people understand the significance of this unique example of nineteenth-century Black cartographic practice. Yet such representational re-mappings would not be sufficient. This mapping would have to be practised year after year by people setting their feet on city streets and walking to places where they could imagine the world Washington and others worked to change. Only through such performative mapping could Fredericksburg emerge as a place where Black people, such as John M. Washington, survived enslavement, secured emancipation and began the long and continuing struggle for equality and justice.

References

Alderman DH, Inwood J and Bottone E (2021) The mapping behind the movement: On recovering the critical cartographies of the African American freedom struggle. *Geoforum* 120: 67–78.

Argent NM and Walmsley DJ (2009) From the inside looking out and the outside looking in: Whatever happened to "behavioural geography"? *Geographical Research* 47(2): 192–203.

Battle-Baptiste W and Rusert B (eds) (2018) *W.E.B. Du Bois's Data Portraits*. Princeton: Princeton Architectural Press.

Blight D (2001) *Race and Reunion: The Civil War in American Memory*. Cambridge: Harvard University Press.

Blight D (2007) *A Slave No More: Two Men Who Escaped to Freedom, Including Their Own Narratives of Emancipation*. Orlando: Harcourt.

Crampton J (2001) Maps as social constructions: Power, communication and visualization. *Progress in Human Geography* 25(2): 235–252.

Del Casino Jr VJ and SP Hanna (2000) Representations and identities in tourism map spaces. *Progress in Human Geography* 24(1): 23–46.

della Dora V (2009) Performative atlases: Memory, materiality, and (co-)authorship. *Cartographica* 44(4): 240–255.

Foner E (2005) *Forever Free: The Story of Emancipation and Reconstruction*. New York: Vintage.

Hanna SP (2012) Cartographic memories of slavery and freedom: Examining John Washington's map and mapping of Fredericksburg, Virginia. *Cartographica* 47(1): 50–63.

Hanna SP and Del Casino VJ (2017). Representation and presentation. In: Richardson D, Castree N, Goodchild MF, Kobayashi A, Liu W and Marston RA (eds) *International Encyclopedia of Geography: People, the Earth, Environment and Technology*. Malden: John Wiley & Sons, Ad Vocem.

Harley JB (1988) Maps, knowledge, and power. In: Cosgrove D and Daniels S (eds) *The Iconography of Landscape: Essays on Symbolic Representation, Design, and Use of Past Environments*. Cambridge: Cambridge University Press, pp. 277–312.

Harley JB (1990) Introduction: Text and contexts in the interpretation of early maps. In: Buisseret D (ed.) *From Sea Charts to Satellite Images: Interpreting North American History Through Maps*. Chicago: University of Chicago Press, pp. 277–312.

Kelley E (2021) 'Follow the tree flowers': Fugitive mapping in *beloved*. *Antipode* 53(1): 181–199.

Kitchin R and Dodge M (2007) Rethinking maps. *Progress in Human Geography* 31(3): 331–344.

Nurmi T (2016) Shackle, sycamore, shibboleth: Material geographies of the underground railroad. In: Bishop K (ed.) *Cartographies of Exile*. New York: Routledge, pp. 111–132.
Peirce P (1971) *The Freedmen's Bureau: A Chapter in the History of Reconstruction*. New York: Haskill House.
Pickles J (2004) *A History of Spaces: Cartographic Reason, Mapping, and the Geo-Coded World*. London: Routledge.
Potter A (2015) The commons as a tourist commodity: Mapping memories and changing sense of place on the Island of Barbuda. In: Hanna SP, Potter AE, Modlin Jr EA, Carter P and Butler DL (eds) *Social Memory and Heritage Tourism Methodologies*. New York: Routledge, pp. 109–128.
Rankin W (2020) Race and the territorial imaginary: Reckoning with the demographic cartography of the United States. *Modern American History* 3: 199–230.
Rossetto T (2015) The map, the other and the public visual image. *Social & Cultural Geography* 16(4): 465–491.
Shifflet C (ed.) (2008) *John Washington's Civil War: A Slave Narrative*. Baton Rouge: Louisiana State University Press.
Washington JM (2007) Memory's of the past. In: Blight D (ed.) *A Slave No More*. New York: Harcourt, pp. 165–212.
Wood D (1992) *The Power of Maps*. New York: Guilford.

29
GENDER AND MAPPING CULTURE

Christina E. Dando

Introduction

In 1929, the Hibbing Branch of the American Association of University Women (AAUW) created and published *The Arrowhead*, a pictorial map of Hibbing's Minnesota region (Figure 29.1). Insets in the lower right identify its creators: 'Research Committee Dorothy Hurlbert, Iris Walker' and 'Map Irene Anderson, Blocks & Border Kathryn Walker'. On this women-created map, history, culture and art are brought together to present a visually appealing take on their home landscape, the creators utilising their educations to craft a product to raise funds for women's higher education scholarships. With this map, they are both mapping their culture and engaging in their own mapping culture.

When we think of maps, we often envision graphic depictions facilitating spatial understandings, abstractions of reality perceived as 'gender-less' or neutral. In western cultures over the last 1,000 years or more, maps have been made predominantly by men, for men, with cartography at times called the 'the science of princes' (Harley, 2001: 88). With this provenance, maps represent masculine knowledge and understanding and so are gendered. But not all maps are scientific: some are created to entertain, to educate and/or be a thing of beauty, such as *The Arrowhead*. Pictorial maps are 'artistic renderings of places, regions, and countries . . . [that] combine map, image, and text, frequently for the purposes of telling a visual story or to capture a sense of place' (Hornsby, 2017: 2). The earliest pictorial maps date to the European Medieval period or perhaps even earlier (Hornsby, 2017: 4; Bird, 2016: 53). In the early twentieth century, new advances in lithography combined with burgeoning American popular culture resulted in a 'Golden Age' of pictorial maps (Hornsby, 2017). Pictorial maps were popular with the American public: they 'instantly summarized a sense of place, delighting the eye and stimulating the viewer's imagination' (Hornsby, 2017: 4). Major map publishers produced and marketed pictorial maps. Women's magazines ran articles on home decorating with maps. It is with pictorial maps that we see the greatest visible presence of women in American mapping, with Ruth Taylor White and Louise Jefferson being well-known mistresses of the form (Bird, 2016: 55).

DOI: 10.4324/9781003327578-35

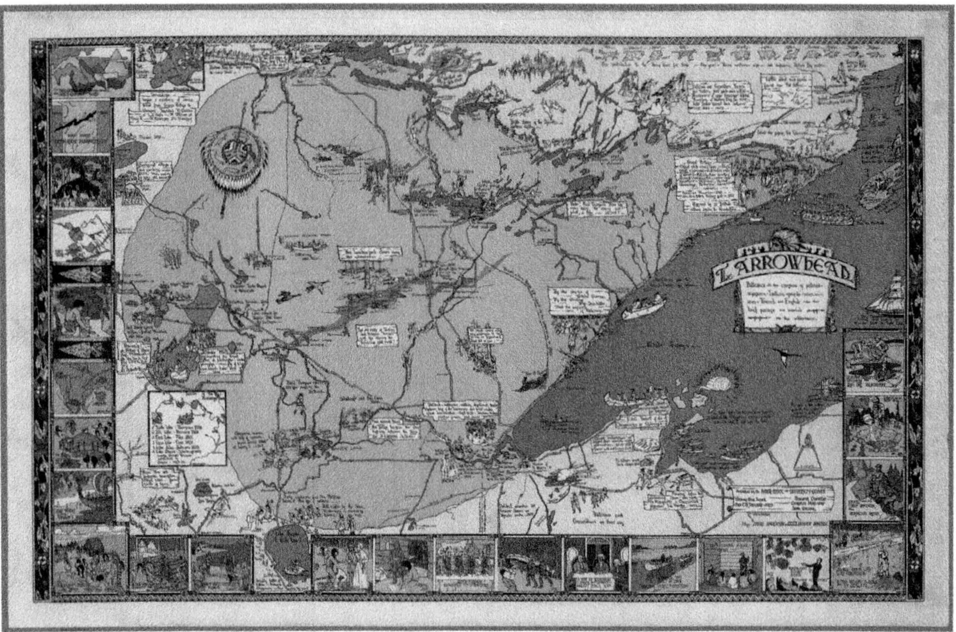

Figure 29.1 *The Arrowhead*, 1929. Created and published by the Hibbing Branch of the American Association of University Women, Hibbing MN. This full-colour, richly illustrated map focuses on the history and culture of the 'Arrowhead' region of northern Minnesota, enlivened with scenes and text of its life, history and mythology. Just north of Hibbing, at the centre of the map, there is a camp scene with a lumberjack chopping down a tree, a label above reading 'The lumberjack clears away the wilderness'. To the east (north is at the top of the map), by the shores of Lake Superior, there is a scene with a Native American woman and child outside a wigwam, with the opening lines of Henry Wadsworth Longfellow's poem *Hiawatha*: 'By the Shores of Gitchee Gumee, By the Shining Big-Sea-Water, Stood the Wigwam of Nokomis' (1855). Around the left, bottom and right of the map is a vignette border, illustrating significant points in Minnesota history, beginning in the upper left and moving counterclockwise, with geologic history, early Native peoples, European explorers, early Minnesota history, as well as folklore (Paul Bunyan and Babe, the Blue Ox). On the left and right of the map is a beautiful floral border. Size of the original: 54 cm × 87 cm.

Source: Image courtesy of the David Rumsey Map Collection, David Rumsey Map Center, Stanford Libraries. By permission of the Hibbing Branch of the American Association University Women, Hibbing, Minnesota

Hibbing AAUW's foray into map production represents a new chapter in women and cartography: utilising their educations to create their own maps that would sell. Fundraising maps were created by the Hibbing chapter, by other AAUW chapters and by the national organisation (Table 29.1), as well as by other women's organisations. AAUW chapters have promoted their work with the slogan 'changing the world, one woman at a time', referring to their support for women's education, but they were also changing their worlds through their creative use of their educations. I begin with an overview of the gendering of cartography and mapping before considering *The Arrowhead*. I argue that the AAUW maps are an example of women's mapping culture and an early practice of feminist mapping.

Table 29.1 Maps produced by American Association of University Women chapters

Date	*Title*	*Author(s)*	*Publisher*	*Notes*
1929	*The Arrowhead* [Minnesota]	'Map: Irene Anderson & Blocks Borders Kathryn Arnquist'; 'Research Committee: Dorothy Hurlbert. Irma Walker'	American Association of University Women	'Copyright 1929 Mrs. C. H. Stewart, Pres.'
1929	*Orange County: A Historical Map* [California]	Jean Goodwin Ames	American Association of University Women	'Printed in the Santa Ana High School Print Shop'
1929	*Oregon*	Doris Wildman	American Association of University Women	'Copyright by the American Association of University Women'
1929	*Los Angeles as It Appeared in 1871*	Gores	Women's University Club, Los Angeles	Possible AAUP affiliate/branch?
1930	*Pictorial Map of Kansas*	At the bottom of the map in small print: 'Art Work by Herschel C. Logan'	American Association of University Women, Salina Branch	'Copyrighted' AAUW
1930	*Elgin Past and Present: 1835–1930*	'Animated historical map by Robert Brightman and Robert Gatechair. Produced by the Elgin Branch of the American Association of University Women'		
1931	*Historical Map of the State of Minnesota: (The Land of Cloud-tinted Water)*	Clara Searle Painter, Barbara Bell	Minneapolis Branch of the American Association of University Women	'Copyright 1931, Minneapolis College Women's Club'
1931	*Historical Map of the State of Washington*	Bertha Ballou	'Published and copyrighted by the American Association of University Women, Spokane Branch, 1931'	Copyrighted by American Association of University Women, Spokane Branch.
1933	*The Conquest of the Continent*	'Designed and copyrighted by August Kaiser. Historian Clara Searle Painter, B.A. Mt. Holyoke College. Checked for historical accuracy by Agnes Larson, B.A. St. Olaf College, M.A. Columbia University, M.A. Radcliffe College'	'Sponsored by National Fellowship Appeal Committee, American Association of University Women'	
1940	*Founded in 1839 as the City of Bloomington, Named Changed to Muscatine in 1849* [Iowa]	Art Department of the American Association of University Women	Art Department of the American Association of University Women	
1941	*Cartographic Map of Winnebago County* [Illinois]	Ted Mackechinie; 'produced by The Legislative Study Group, Rockford Branch, AAUW'		

Gender and mapping culture

The Arrowhead superficially appears to be an attractive, mass-produced map from the early twentieth century. But *The Arrowhead* presents an opportunity to consider gender and mapping: to consider not only who was mapping and their audience but also the gendering of landscapes and depictions of gender on maps.

Cartography as a practice traces its roots back at least to the Renaissance (Edney, 2019: 5). In the fifteenth century, cartography was shifting from an art towards an empirical science, practised by a variety of disciplines: explorers and surveyors sketched in the field, compiled data and created maps; businesses or government agencies compiled data and made maps; artisans working in printing shops crafted and reproduced maps (Rees, 1980: 63; Raisz, 1937). These maps were wielded by kings, presidents and governments, and deposited and preserved in map collections and archives (Harley, 2001: 88). Maps were part of the 'masculine-dominated spheres of influence', have long been a form of knowledge essential to controlling space and resources and 'professes to capture the truth about a place in pure, scientific form' (Varanka, 2009: 216; Dell'Agnese, 2007: 439; McClintock, 1995: 27). Much of the canon of the history of cartography has focused on maps viewed as important, largely confining our view of cartography to the 'science of princes' (Harley, 2001: 88).

In this construction of masculine knowledge, there is a western tradition of explorers explicitly feminising lands. Annette Kolodny's *The Lay of the Land: Metaphor as Experience and History in American Life and Letters* (1975) examines how explorers and settlers of North America presented the lands they were encountering in feminine terms. Colonist John Hammond wrote in 1656, 'Having for 19 year served Virginia the elder sister, I casting my eye on Mary-land the younger, grew in amoured on her beauty, resolving like Jacob when he had first served for Leah, to begin a fresh service for Rachell' (quoted in Kolodny, 1975: 13). To 'know' a place, to have 'knowledge' of it, was likened to 'knowing' a woman in an intimate way (McClintock, 1995: 23–24). Not only were new lands gendered as feminine but also cartographers included images of scantily clad women in their maps and atlases, at times personifications of continents, catering to the predominantly male gaze of their audiences (Harley, 2001: 74–76).

Only a handful of scholars have considered the gendered construction of cartography. J. Brian Harley famously wrote of cartography's 'silences and secrets', the intentional and/or unintentional suppression of knowledge in maps (Harley, 2001: 56). Nikolas Huffman suggests there is the irony in Harley's silences and secrets, as Harley himself 'remains silent about one of the most pervasive categories of social relations: gender', and takes historians of cartography to task for 'excluding women cartographers from the canons of cartographic history' (Huffman, 1995: 1623–1624). Dalia Varanka's essay on 'The Manly Map' begins to address the implications of '"masculine" cartographic style', paralleling the rise of modern cartographic design with the rise of masculine ideals (Varanka, 2009: 206).

Despite the gendered construction of cartography, women have always been involved (Hudson, 1989; Ritzlin, 1986; Tyner, 2015, 2020). Scholarship has uncovered women involved in cartography—as draftsmen, publishers, map sellers, engravers, globemakers and colourists—but their roles were often obscured, their identities erased, in what Tyner terms the 'hidden cartographers' (Tyner, 1997). In most of these instances, the women involved with cartography were facilitating masculine practices of cartography; this would not change until women had the agency to produce their own maps.

But there is a difference between cartography and mapping, with cartography focused on the art and science of making maps and mapping focused on the process of creating a graphic of sorts to facilitate a spatial understanding. All cultures map and have their own mapping culture. Cartography, with its focus on the creation of an 'ideal' form, narrows the consideration of mapping and maps down to a relatively small number of maps created by kings, governments, businesses (Edney, 2019).

In the United States, in the nineteenth century, girls were trained, just as boys were, in the basics of mapmaking in primary school as part of their geography lessons. Geography was integral to American education, beginning not long after the founding of the country, with mapping part of the pedagogy. As women's higher education developed with women's seminaries and colleges, a background in geography, maps and mapping was expected for admission and was often a core curriculum component (Dando, 2018: 3–5).

In the 1890s, women took on greater public roles, largely associated with home and domestic issues, working within women's organisations to bring about social change. Women's clubs and organisations advocated for women to investigate for themselves problems in their communities then work towards solutions. Geographic knowledge and maps were an important element of their work. Jane Addams and the Residents of Hull House in Chicago were at the forefront with their mapping projects and their advocacy for mapping as a tool in community improvement (Dando, 2018: 119–172). The suffrage movement crafted persuasive maps and used them extensively as part of their campaign for the right to vote (Dando, 2018: 173–205). With these mapping practices in the late nineteenth and early twentieth centuries, we have women producing maps from data they gathered themselves, drafted themselves and at times reproduced themselves, to advance work in their own interests.

Was this women's mapping culture feminist? Dalia Varanka in her consideration of 'The Manly Map' concludes with a reflection on feminine-style cartography, suggesting that it involves: 'the importance of personal narratives and influence of interpersonal relationships within community' (Varanka, 2009: 216). Feminist cartography today is said to centre the subject, their embodiment and their experiences in the production of cartographic media, while acknowledging different experiences as well as different modes of production (Gollaz Moran, 2022; Huffman, 1995). With the geospatial revolution of the last 30 years, mapping is now within the reach of many peoples, including many different women and women's groups. Meghan Kelly and Amber Bosse write: 'Feminist digital geographies, design justice, and data feminism have reinvigorated attention to mapping contexts by grounding feminist principles such as power through situated knowledge and intersectionality—particularly in relationship to systems of oppression—in maps and mapping' (Kelly and Bosse, 2022: 400). So, yes, we might consider the Progressive Era mapping feminist.

The AAUW's fundraising with maps

The AAUW has its roots in a women's college alumnae group, formed in 1881, united in an interest in 'practical education work' (Talbot and Rosenberry, 1931: 12). The women were looking for opportunities to utilise their own educations while encouraging other women to pursue them. Towards these ends, AAUW created women's fellowships for graduate training in the United States and Europe. AAUW branches were established across the country, providing opportunities for their members to engage in intellectual activities, such as study groups on International Relations. Initially, scholarship monies came from donations and voluntary dues. Branches fundraised to create their own local scholarships. In 1927, the

AAUW undertook a nation-wide endowment campaign and by 1956 they had created a $1.6 million endowment while awarding more than $2 million in fellowships and grants (Tryon, 1957). It is in this context that the Minnesota map can be situated.

The Arrowhead was created by two librarians and three high school art teachers of the Hibbing branch, 'to raise money for its Junior College Scholarship Fund' (Brown, 2021; Arrowhead Map, 1929a, 1929b). An accompanying booklet further explains the region's history and provides suggestions for further reading for both adults and children (Walker, 1929). Map and booklet present a colourful synopses of physical geography (glaciation), popular culture (Longfellow's poem *Hiawatha*; folklore giants Paul Bunyon and Babe the Blue Ox), exploration, and development, with particular attention to the region's iron ore. There are also two references to mapmaking on the map, in a vignette 'Ochagach [Native American] draws map for Verendrye' and on the map a figure of a frontiersman with 'David Thompson returns in 1798 after mapping the Northwest'. The artists used rich colours and dynamic human and animal figures to craft a highly appealing map. The map's publication was announced in Minnesota and Wisconsin library newsletters (Arrowhead Map, 1929a, 1929b).

While much of the history on the map could be described as male dominated, there are glimpses of women's history and culture, albeit from the perspective of educated, white American women. In the north-west corner is a scene of two native women processing animal skins with the notation '700 women scraping skins', which the booklet explains as 'seven hundred Indian women were retained by the company [North West Company, a fur trading business] to scrape and clean the skins, and to make up the packages of pelts' (Walker, 1929: 9). Native women are depicted gathering blueberries on the west central portion of the map, simply labelled 'Blueberrying'. And below the lines from *Hiawatha* to the east is an image of a native woman standing with a little boy, likely a depiction of Nokomis, Hiawatha's grandmother.

One of the most interesting depictions of gender on the map is of a nude man, in a vignette adjacent to the names of the women mapmakers (Figure 29.2). The vignette depicts a young James J. Hill (viewed from behind) fording a stream, his clothes held above his head to keep them dry, to meet with Lord Strathcona, where the two men supposedly came up with the ideas of the 'twin roads', the Great Northern and Canadian Pacific Railways (Walker, 1929: 13). This cheeky image is the only nudity on the map.

We are left to wonder what prompted them to undertake this map as a fundraiser. It seems likely something triggered this project because there were other maps created the same year for the same purpose (Table 29.1). A 1950s AAUW publication mentions a branch bulletin with 'ninety-two different kinds of successful money-raising project that have been reported' (Tryon, 1957: 182). Might one be pictorial maps? Outside of the AAUW network, one of the most famous pictorial map creators, Jo Mora, published a very similar map of San Diego in 1928. Could the AAUW maps have been inspired by Mora? Further research will hopefully identify the prompt for these fundraising map projects. There is, however, a tradition of women creating goods to raise funds for organisations or causes that are important to them. In the mid-nineteenth century, anti-slavery organisations in the northern U.S. states produced and sold cakes and cookies, hats and mittens at anti-slavery fairs to raise money for the cause (Van Broekhoven, 1998: 40). American suffragists produced all sorts of goods to raise funds for the movement (Finnegan, 1999). Fundraising through maps appears to be a variation on fundraising practices women have used to fund their causes in the United States.

Figure 29.2 Detail of *The Arrowhead* (1929). This detail from the lower right corner captures the identification of the map creators as well as examples of the vignettes and the floral border. The corner vignette includes the only nudity on the map: the naked back of James J. Hill as he fords a stream, clothes balanced on his head, for a fateful meeting with Lord Strathcona.

Source: Image courtesy of the David Rumsey Map Collection, David Rumsey Map Center, Stanford Library. By permission of the Hibbing Branch of the American Association University Women, Hibbing, Minnesota

Producing maps for AAUW fundraising continued after 1929. In 1933, the AAUW produced a national map, 'The Conquest of the Continent', focused on 'the western progress of the pioneer'. With this 1933 map, the AAUW used artist August Kaiser for the imagery but, as the cartouche makes clear, women were clearly steering the project: 'Designed and copyrighted by August Kaiser. Historian Clara Searle Painter, B.A. Mt. Holyoke College. Checked for historical accuracy by Agnes Larson, B.A. St. Olaf College, M.A. Columbia University, M.A. Radcliffe College'. The listing of their credentials is an assertion of their authority as scholars.

Other women's organisations similarly created, published and sold maps to raise funds for their initiatives. *The Fox River Valley of Wisconsin 1634–1880* was published in 1931 by the Oshkosh Museum Auxiliary, likely a fundraiser for the museum. Its creators are identified on the map: 'Research—Mary A. O'Keefe, Map—Nile Behncke, Border—Mary G. Rogers, Pictures—Emma P. Comstock'. *The Fox River Valley* follows the same design layout of *The Arrowhead*: a wildflower border, historical vignettes in a border in a counter-clockwise progression, beginning with the geology, with scenes and text on the map about the region's history and culture. It also has an accompanying booklet. In 1937, the Garden Club of North Carolina commissioned *A Map of North Carolina for Nature Lovers* as a fundraiser. An inset reads:

> This map was designed by Mabel Pugh, head of the Art Department of Peace, a junior college for women in Raleigh, N.C. for Mrs. R. L. McMillan, Chairman of the Map

> Committee and First Vice-President of the Garden Club of North Carolina in the spring of 1937.

It too has a booklet.

In considering gender and mapping culture with *The Arrowhead*, we have a clear example of white American women's mapping culture. Women were the drivers of the research, design and execution of the map. The audience is still being considered but, thus far, announcements about the map have been found in regional library and educational newsletters, which at the time would have been a largely female audience. With its subject matter focused on the history and culture of the region, the map was not intended for navigation or use in making decisions, but rather for educational purposes: the vignettes and map images are designed to stimulate curiosity and encourage further reading. The depictions of gender on the map are largely traditional, mostly male explorers and leaders important to regional development. But there are women depicted on the map (and all clothed) engaging in the everyday activities that helped build the region—gathering food, raising children and engaging in the production of skins, an often overlooked labour task essential to the fur industry. The only naked figure is of a famous 'empire-builder' whose nude bottom is offered to sharp-eyed map readers in a vignette that, while it captures a historic moment, also suggests that the women mappers may have been offering a little bit of humorous nudity for their largely female audience. With *The Arrowhead*, the Hibbing Branch of AAUW were mapping their culture (the culture of their region) while engaging in their own mapping culture. They were not 'cartographers', but were women applying and employing their educations to craft a product interesting and appealing to a wide audience, to advance their agenda of raising scholarship monies.

AAUW maps as mapping culture and feminist mapping

The Hibbing AAUW Branch's map was created with an understanding of what maps are, was designed to look like a 'proper' map, and reflects the history of both mapping and the region. But it also represents a practice of mapping that can be described as feminist: created by women, reflecting to some extent their own worldview, and circulated for their own purposes. It is an example of educated white American women of the Progressive Era mapping their culture and engaging in their own mapping culture.

As an example of a pictorial map, to my knowledge, *The Arrowhead* has not been studied from a 'history of cartography' perspective: it is not a 'scientific' map, it is not groundbreaking in design. Tyner (2016: 12) observes:

> [S]cientific maps tend to be the primary focus of most histories of cartography. Most of the studies about women and maps have focused on the same kinds of maps studied in classic works and often unconventional maps are dismissed because they are not legitimate-seeming maps or because they do not appear to contribute to the canon of cartography.

Pictorial maps in general are not viewed as 'scientific' but have proven to be extremely popular and highly desirable. Stephen Hornsby's (2017) book on pictorial maps attracted broad public attention because of their great visual appeal. While Hornsby does not investigate in-depth women and pictorial maps, they are quite present in his book, with over 30 women named as map creators. Despite women being significant creators of pictorial maps, their works as a particular practice of mapping have largely been overlooked.

The creators of maps such as *The Arrowhead* are not 'hidden cartographers' (Tyner, 1997). They are quite publicly claiming authorship, with the names not only visible on the map but often highlighted in the cartouche or in their own inset box (Figure 29.2). As the Hibbing women crafted their map, they were wielding the 'language of science', exerting their own control over spaces, territories, taking on the hegemonic masculine practices of map creation. But they were clearly using what could be described as 'less threatening', more female-acceptable aspects of higher education, the humanities, to craft maps that did not have to be scientific, did not have to be judged in terms of 'accuracy' but were focused on appealing to the public and it could be argued, reaching a wider audience.

Changing the world, one woman at a time

The Arrowhead map, now approaching 100 years old, is still in circulation. Today, on Etsy.com (an online marketplace), you can purchase contemporary reprints of the map, as well as PDFs of its booklet. Some reprints are being sold by vendors specialising in historic map reproductions, but consumers can also purchase a copy from the Hibbing AAUW, who are again using it to raise funds to benefit Iron Range women's educational opportunities. There is some irony in the fact that a women-created pictorial map has likely been reproduced and hangs on more walls today than many maps in the canon of the history of cartography.

This foray into considering historic women's map creation is but scratching the surface on the subject of gender and mapping culture. Just with this study of *The Arrowhead*, there is more work to be done: what prompted or inspired the creation of this map? Did the women involved have any backgrounds in geography or cartography? Are there more AAUW maps out there? And do the other fundraising maps created by women's organisations also reflect what could be described as feminist mapping? And there are many outstanding questions regarding gender and mapping culture: are there other examples of women's mapping culture that have not been previously considered because it was not recognised as part of the canon and preserved in an archive? Could, through further research on women's mapping cultures, scholars begin to identify the ways women's mapping culture are, on one hand, similar to and, on the other hand, distinct from the masculine canon of cartography, from the subject matter of their mapping to the techniques and symbols employed? And might we begin to identify multiple women's mapping cultures, shaped by their ethnicity, social class, education and other crucial life experiences? What might the Arrowhead map look like if created by the indigenous women from north-eastern Minnesota? All cultures have their mapping traditions and mapping cultures: women are no different. Scholars have studied the mapping traditions of many different global cultures, while overlooking half the world's population (female). It is time we changed this world view.

References

Arrowhead Map (1929a) *Library Notes and News (Dept. of Education, State of Minnesota)* 9(7): 143. Archives.

Arrowhead Map (1929b) *Library Bulletin Wisconsin* 25(December): 394. Archives.

Bird C (2016) Pictorial cartography: Its American expressions. *IMCOS Journal* 147(Winter): 53–59.

Brown C (2021) Female mapmakers detailed Minnesota's Arrowhead back in 1929. *Star Tribune* (Minneapolis, MN), 13 June.

Dando C (2018) *Women and Cartography in the Progressive Era*. New York and London: Routledge.
Dell'Agnese E (2007) Geo-graphing: Writing worlds. In: Cox K, Low M and Robinson J (eds) *The Sage Handbook of Political Geography*. Thousand Oaks, CA: Sage, pp. 439–453.
Edney M (2019) *Cartography: The Ideal and Its History*. Chicago: University of Chicago Press.
Finnegan M (1999) *Selling Suffrage: Consumer Culture and Votes for Women*. New York: Columbia University Press.
Gollaz Morán A (2022) Embodied urban cartographies: Women's daily trajectories on public transportation in Guadalajara, Mexico. In: Harcourt W, van den Berg K, Dupuis C and Gaybor J (eds) *Feminist Methodologies: Experiments, Collaborations and Reflections*. Cham: Springer International Publishing, pp. 189–209.
Harley J (2001) *The New Nature of Maps: Essays in the History of Cartography*. Edited by Paul Laxton. Baltimore and London: The Johns Hopkins University Press.
Hornsby S (2017) *Picturing America: The Golden Age of Pictorial Maps*. Chicago, IL: University of Chicago Press.
Hudson A (1989) Pre-twentieth century women mapmakers. *Meridian* 1: 29–32.
Huffman N (1995) Silences and secrecy in the history of cartography: J.B. Harley, science and gender. In: *Proceedings of the 7th International Cartographic Conference, Barcelona*, vol. 2. Barcelona: Institut Cartogràfic de Catalunya, pp. 1622–1630.
Kelly M and Bosse A (2022) Pressing pause, "doing" feminist mapping. *ACME: An International Journal for Critical Geographies* 21(4): 399–415.
Kolodny A (1975) *The Lay of the Land: Metaphor as Experience and History in American Life and Letters*. Chapel Hill, NC: The University of North Carolina Press.
McClintock A (1995) *Imperial Leather: Race, Gender, and Sexuality in the Colonial Contest*. New York and London: Routledge.
Raisz E (1937) Outline of the history of American cartography. *Isis* 26(2): 373–391.
Rees R (1980) Historical links between cartography and art. *Geographical Review* 70(1): 60–78.
Ritzlin A (1986) The role of women in the development of cartography. *AB Bookmans Weekly* 77(23): 2709–2713.
Talbot M and Rosenberry L (1931) *The History of American Association of University Women 1881–1931*. Boston and New York: Houghton Mifflin Company.
Tryon R (1957) *Investment in Creative Scholarship: A History of the Fellowship Program of the American Association of University Women 1890–1956*. Washington, DC: American Association of University Women.
Tyner J (1997) The hidden cartographers: Women in mapmaking. *Mercator's World* 2(6): 46–51.
Tyner J (2015) Women in cartography. In: Monmonier M (ed.) *History of Cartography, Volume 6: Cartography in the Twentieth Century*. Chicago, IL: University of Chicago Press, pp. 1758–1761.
Tyner J (2016) Mapping women: Scholarship of women in the history of cartography. *Terrae Incognitae* 48(1): 7–14.
Tyner J (2020) *Women in American Cartography: An Invisible Social History*. Lanham, MD: Rowman & Littlefield.
Van Broekhoven D (1998) 'Better than a clay club': The organization of anti-slavery fairs, 1835–60. *Slavery and Abolition* 19(1): 24–45.
Varanka D (2009) The manly map: The English construction of gender in early modern cartography. In: Dowler L, Carubia J and Szczgiel B (eds) *Gender and Landscape: Renegotiating the Moral Landscape*. New York and London: Routledge, pp. 206–220.
Walker I (1929) *The Story of the Arrowhead Country from the Age of Stone to the Age of Steel*. Hibbing, MN: Hibbing Branch of the American Association of University Women.

30
MAPPING AS A MODE OF GOVERNANCE IN THE ANTHROPOCENE

David Chandler

Introduction

This chapter analyses the development of mapping as a mode of governance in the Anthropocene, starting with a critique of modernist assumptions of the human subject and of rationalist claims to knowledge and the development of this framework of thinking in the field of systems ecology and philosophically through the approach of assemblage theory. The first point that should be established is that mapping has little in common with traditional cartography, in fact, it is an explicit critique of this approach to the conceptualisation of both time and space; viewing these as outcomes of relational processes rather than as the containers in which they operate. John Brian Harley, in his influential work, which helped establish the field of critical cartography, accurately describes the limited nature of mapping in modernity with its emphasis on the 'true' map (Harley, 1989): 'Its central bastions were measurement and standardisation and beyond there was a "not cartography" land where lurked an army of inaccurate, heretical, subjective, valuative, and ideologically distorted images' (Harley, 1989: 5). For critical approaches to cartography, 'the map is an authoritarian image', removing life and context, and thus facilitating managerial, bureaucratic and autocratic modes of governance, distanced from the complex reality on the ground (Harley, 1989: 14). For Latour (1986), the map as an 'immutable mobile' was key to the homogenising and universalising drive of modernity, enabling the abstraction of knowledge and the creation of 'objective' space:

> Even the very notion of scale is impossible to understand without an inscription or a map in mind. The 'great man' is a little man looking at a good map. In Mercator's frontispiece Atlas is transformed from a god who carries the world into a scientist who holds it in his hand!
>
> *(Latour, 1986: 27)*

Mapping in modernity constituted space as an empty container filled with distinct autonomous parts, side by side as separate entities, without context or relation: it created a fictional world amenable to subject-centred human rule (Latour, 2016: 7). As Benjamin Bratton states:

DOI: 10.4324/9781003327578-36

> Lines that are linked, folded, and looped become a frame, keeping things in or out. . . . The modern nation-state is itself also [a] function of a cartographic projection that conceives the Earth as a horizontal plane filled with various allotments of land.

Thus, there is 'no stable geopolitical order without an underlying architecture of spatial subdivision' (Bratton, 2015: 24). Mapping was thereby a process of 'emptying out' space of its constitutional relational dynamics and its replacement by a 'universal spatial order based on mathematical formalisation and geographic interchangeability': a 'groundless materialism' of 'false equivalences' that could be 'divided up like an algebraic equation' (Bratton, 2015: 30).

Mapping conceptualised as a mode of governance is very different to this modernist understanding. Rather than the two-dimensional flat or universal space of modernity, mapping develops an understanding of space as a product of inter-relationality. Therefore, as Doreen Massey noted: 'we understand space as the sphere of the possibility of the existence of multiplicity in the sense of contemporaneous plurality; as the sphere in which distinct trajectories coexist; as the sphere therefore of coexisting heterogeneity' (Massey, 2005: 9). For mapping as a mode of governance, space, actively produced through plural interaction, is understood as a relational outcome, which can be mapped only through seeking to concretise it as a specific or unique set of contingent relations.

In this sense, following from Kitchin and Dodge (2007), mapping is not seen as a fixed or objective representation of the world but as an iterative and processual attempt to visualise a particular set of relationships to facilitate problem-solving. Rather than abstracting from the world—as if a single map could be 'viewed as a universal and essential solution to a range of questions (that there can be a "best" or "most accurate" map that all people understand and use in the same way to address a range of problems)' (Kitchin and Dodge, 2007: 12)—mapping is seen as deconstructing universal forms of representation and causality, bringing knowing closer to reality. Mapping is thus conceived here as processual rather than as representing fixed points or relations. Where this framing differs from that of Kitchin and Dodge (2007) is that the relational problems addressed in mapping are not merely spatial but also temporal. Mapping as narrative tracing can deal with relations in time as well as space (see e.g. on mapping and narrative, Caquard, 2013). Mapping becomes central in the Anthropocene precisely in recognition of the appreciation of relationality and differentiation.

Mapping is therefore an important governance mode to grasp as it reverses the imaginary of classical liberal assumptions of universal or flat 'Newtonian' space and instead emphasises the spatial, contextual and relational development of social, economic and political institutions, ecosystems and, in its logical development, all entities and assemblages (see further in Massey, 2005). It is opposed to reductionist understandings of the world, which empty it of relationality and historical specificity. Mapping has thus powerfully entered social and political theorising to explain difference and non-linear outcomes, for example, why the introduction of markets or democracy might make problems of conflict or development more intractable, or how social and political institutions affect the impact of social and environmental changes. Rather than taking appearances for granted and assuming the power of agential forces (either human or non-human) or properties of fixed entities, pragmatic, institutional and mapping approaches see these appearances as concealing the work of relation and translation.

This chapter analyses mapping as a governance mode of the Anthropocene as it develops through a clear and well-articulated critique of modernist or linear understandings of governance and provides a field for negotiation and experimentation in terms of what it might mean to govern without the clear goals and top-down mechanisms of 'command-and-control' or assumptions of universal or linear frameworks of cause and effect. Key to mapping is the notion of 'emergence': the understanding that causality is not the unfolding of fixed essences or relationships but a process of complex interaction in which outcomes are non-linear. Non-linearity means that outcomes are mediated, that is, that they depend not merely on 'inputs' into the system but rather how these inputs, in terms of information, interaction, system disturbances, etc., are perceived, understood and responded to. Mapping thereby develops as the study of the internal relations and interactions of the object of policy intervention (in the sphere of international relations this object would be states or societies subject to policy interventions in the diverse spheres of economic, political, social and environmental problems). Mapping bears the clear legacies of its modernist heritage and is very much premised on assumptions of governmental agency. However, this agency is no longer exercised from the 'top-down' but the 'bottom-up'. In neoliberal or neo-institutionalist thought, the mapping of these relations and interactions is often termed 'process-tracing' or the charting of 'path-dependencies', to reveal the contingent nature of interactive processes and the possibilities for intervening to adjust or manipulate these.

Mapping the assemblage

Mapping as a set of immanent or endogenous understandings of non-linear outcomes perhaps receives its clearest conceptualisation in the approach of assemblage theory, which is seen to move 'away from the anthropocentrism' of modernist thought (Acuto and Curtis, 2014: 2). Acuto and Curtis (2014) argue that assemblage theory can be read to cover a variety of different approaches with multiple intellectual roots. In international theorising, assemblage theory has become increasingly discussed as a critical methodology focusing upon immanent and emergent understandings of causation (see e.g. Anderson et al., 2012). Philosopher Manuel DeLanda argues that a theory of interactive assemblages (and thus of mapping as a mode of governance) can be derived from ideas dispersed across the work of Gilles Deleuze and Felix Guattari (Delanda, 2006: 3). Assemblage theory is understood as overcoming scientific framings, which tend to focus on either the parts (reductionism) or the wholes (structuralism) rather than interactions and relations. As Ben Anderson et al. state:

> By beginning from the claim that 'relations are exterior to their terms', assemblage thinking allows us to: foreground ongoing processes of composition across and through different human and non-human actants; rethink social formations as complex wholes composed through a diversity of parts that do not necessarily cohere into seamless organic wholes; and attend to the expressive powers of entities.
>
> *(Anderson et al., 2012: 172)*

The conceptualisation that 'relations are exterior to their terms', derived from Deleuze, is conceptually central as relations are not reduced to mere intermediaries between autonomous entities but neither are entities constituted or fully determined by their relations. Relations are the key to the understanding of the contingent emergent effects of interaction. DeLanda (2006), the leading theorist taking Deleuze's work further in this area, argues that the key to non-linear understandings (and thereby the precondition for mapping) is the ontopolitical assumption that the internal organisation of an assemblage or entity is more important that the external or extrinsic factors, which 'are efficient solely to the extent to which they take a grip on the proper nature and inner processes of things' (DeLanda, 2006: 20). One and the same external set of policies or causal actions 'may produce very different effects' (DeLanda, 2006: 20).

This point is the cornerstone for mapping as a mode of governance grounded on the ontopolitics of the Anthropocene. If developments, whether societal, economic, or environmental, were the outcomes of complex interplays between different and overlapping processes of interaction and contingency, policy interventions would make little sense as the process would be too complex to make meaningful intervention possible. Turning mapping from a framework of knowledge scepticism into a policy approach of governance intervention methodologically required the construction of a conceptual division between the internal processes within the assemblage and the external stimuli or information, which were necessarily processed and adapted to in contingent ways. It would necessarily be the internal relations, which were open to mapping as a methodology of understanding and for intervention to shape outcomes. For this reason, some form of assemblage thinking was necessary for actually existing neoliberalism to develop as a coherent mode of governmental reason.

Ecological resilience

The framing of assemblage thinking which has been perhaps most instrumental in facilitating and cohering mapping approaches, especially in the terminology of resilience, has been ecologist Holling's work on complex interaction in the field of ecology. Jeremy Walker and Melinda Cooper (2011), in a well-cited article on the genealogy of resilience, emphasise the interconnections between the neoliberal thought of Friedrich Hayek and the assemblage approach of ecological systems theory developed by Holling, noting that 'Holling and Hayek, writing in the early 1970s, were simultaneously preoccupied by questions of epistemic limits to prediction and assertions of ecological limits to growth' (Walker and Cooper, 2011: 144). Thus, mapping as a mode of governance develops and gains coherence through parallel discussions in ecology, considering how ecosystem's relations of interaction enable them to withstand or adapt to external pressures.

Holling, in his path-breaking 1973 paper, 'Resilience and Stability of Ecological Systems', distinguished 'ecological resilience' from traditional understandings of 'engineering resilience' (Holling, 1973). Whereas 'engineering resilience' was concerned with quantitative measuring, derived from classical physics understandings of fixed relations, 'ecological resilience' was more concerned with the qualitative nature of changing relationships within a system. The assumption being that systems are likely to be 'transient' or changing all the time, rather than operating around a fixed or 'natural' equilibrium that would be 'returned' to. 'An equilibrium-centred view is essentially static and provides little insight into the transient behaviour of systems that are not near the equilibrium. Natural, undisturbed systems are likely to be continually in a transient state; they will be equally so under the influence of man' (Holling, 1973: 2; see also Holling, 1986: 76).

Not only did Holling question the idea of a natural equilibrium, he fundamentally challenged the dominant idea that space was a pre-existent universal container, in which causal relations operated in a linear fashion. Here, the distinction between 'Newtonian' conceptions of space as homogenous and alternative 'topological' conceptions of space as agentially constituted through interaction is important (see e.g. Bryant, 2014: 143–147; Morton, 2013: 55–68). Holling emphasised that it was vital to 'recognize that the natural world is not very homogeneous over space, as well, but consists of a mosaic of spatial elements with distinct biological, physical, and chemical characteristics that are linked by mechanisms of biological and physical transport' (Holling, 1973: 16). Thus, for understandings of ecological resilience, the differences of context make all the difference: there could be no universal understanding of causal interaction nor of the outcomes of policy intervention, as the internal relations of the system were key to the responses, which would necessarily be non-linear: ecological resilience 'emphasizes variability, spatial heterogeneity and nonlinear causation' (Holling, 1986: 72).

Mapping was therefore an essential precondition for any form of governance intervention seeking to make a system more sustainable or resilient, as there could be no 'reductionist' view that external actors could assume a linear causal outcome, or transfer lessons from one system to another (Folke, 2006). It was the internal relations that were decisive. In language very similar to those deploying neoliberal or neo-institutional forms of intervention in international relations, Holling saw reductionist approaches, which assumed linear causality, to have counterproductive and sometimes disastrous effects:

> crises and surprises . . . are the inevitable consequences of a command-and-control approach to renewable resource management, where it is (implicitly or explicitly) believed that humans can select one component of a self-sustaining natural system and change it to a fundamentally different configuration in which the adjusted system remains in that new configuration indefinitely without other, related, changes in the larger system.
>
> *(Holling and Meffe, 1996: 330)*

Ecological resilience exhibited very similar concerns about mapping and non-linearity as those exercised in neoliberalism as a framework of intervention. Intervention could not be direct, or goal directed with a modernist telos, but only indirect, enabling systems based on their own internal relations and interactions:

> A management approach based on resilience, on the other hand, would emphasise the need to keep options open, the need to view events in a regional rather than a local context, and the need to emphasize heterogeneity. Flowing from this would be not the presumption of sufficient knowledge, but the recognition of our ignorance; not the assumption that future events are expected, but that they will be unexpected. The resilience framework can accommodate this shift of perspective, for it does not require a precise capacity to predict the future, but only a qualitative capacity to devise systems that can absorb and accommodate future events in whatever unexpected form they may take.
>
> *(Holling, 1973: 21)*

Ecological resilience requires the mapping of internal system relations to understand how the system copes with and responds to external pressures. One aspect of internal system relations which was vital was a mapping of the 'adaptive cycle': composed of a 'front loop'—a period of production and accumulation but of increasing rigidities—and a 'back loop' of experimentation and reorganisation. The stage of the cycle influences responsiveness:

> The reshuffling in the back loop of the cycle allows the possibility of new system configurations and opportunities utilising the exotic and entirely novel entrants that had accumulated in earlier phases. The adaptive cycle opens transient windows of opportunity so that novel assortments can be generated.
>
> *(Holling, 2001: 397)*

Thus, Deleuze can be read to have captured well the processes of the adaptive cycle where there are tendencies towards rigidity and organisation as well as towards fluidity and disorganisation, which he analysed in the terminology of 'territorialization' and 'deterritorialization'. As already discussed in terms of assemblage theory, ecological approaches to resilience also view systems as nested scales of interaction:

> These scales represent ecosystems, which are defined here as communities of organisms in which internal interactions between organisms determine behaviour more than do external biological events. External abiotic events do have a major impact on ecosystems, but are mediated through strong biological interactions within the ecosystems. It is through such external links that ecosystems become part of the global system.
>
> *(Holling, 1986: 77)*

Thus, the process of mapping is a nested one, where interaction takes place both within systems, or assemblages, and between them. In ecological resilience thinking, mapping becomes generalised as a way of tracing cross-scalar interactions in what is often conceptualised as a 'panarchy' (Holling, 2001; Gunderson and Holling, 2002). Panarchy offers a vision of complex social-ecological systems as nested sets of adaptive cycles. Thus, as a mode of governance, mapping reconfigured sustainability from a linear problem of optimising scarce resources into a non-linear problem of systemic adaptation to emergent social, economic and environmental conditions. Mapping resilience works to enable system governance on the ontopolitical assumption of a nested hierarchy of interactive relations in which assemblages are primarily self-organising as autopoietic systems at the same time as being impacted by external stimuli.

Work in the complexity sciences and in ecology, which heavily influenced Deleuze, was important for the extension and development of mapping, taking governance beyond the initial neoliberal focus in understanding bounded rationality. In this shift, mapping moves from being a question of epistemology (the focus on subjectivity and socially constructed understandings) to one of ontology (the material entanglements of interactive life). It is important to clarify the nature of this shift. As we saw in the section above, for neoliberal and neo-institutionalist thought, the problem is epistemological, in that sub-optimal decisions are taken due to different and distinct path-dependencies that create barriers to adaptation. For neo-institutionalist thought, these lie at the level of formal institutions, for example, property laws or barriers to trade, and at the informal or cultural level. Thus interventions, as in those of peacebuilding or disaster risk management, seek to adapt or alter these institutions to enable better or more efficient outcomes. The solution is seen to be indirect work on shaping the choice-making environment (for more detail, see Chandler, 2010). This adaptation takes place on the assumption that there is a hierarchy of institutional 'fits', in this case that democracy and markets work best but that they depend on lower institutions having the right 'fit' in terms of cultural and religious values, etc.

Mapping as neoliberal or neo-institutional problem-solving seeks to shape or influence these institutions according to preconceived (Eurocentric or western) ideas of 'what works'.

Conclusion

This chapter has elucidated the ontopolitical assumptions underlying the emergence of mapping as a mode of governance, seeking to map societal interactions in order to intervene and redirect the processes and outcomes and as a method of locating the productive source of potential policy solutions. Assemblage theory, with a clear distinction made between the entity (and its internal relations) and the external perturbation or disturbance, provokes non-linear responses or outcomes and thus reveals the processes of interaction. The governance framework expands from a consideration of shaping the 'choice-making environment' of individuals to a broader, ontological framing of the processual becoming of life itself. It is not only thought that is contextually shaped and 'bounded' in its rationality but also the processes within which thought itself operates.

Mapping, as developed through the generalisation of ecosystems theory of resilience or as assemblage theory, is concerned with ontology more than epistemology, or, to put this another way, seeks to drill down in much more detail to the reality of the world in its plurality, flux and difference; rather than seeking to align maladjustments or 'sub-optimal' systems to an ideal model. While for neoliberal or neo-institutionalist models—for 'actually existing neoliberalism' policies of governance intervention—there is an assumption of what the correct outcome is, in terms of an efficient economy and social, economic, or democratic 'progress', this is less the case for mapping understood as an ontological theory of interaction. For assemblage theory, differentiations lead to further differentiations but there can be no external goal of 'progress' to pre-set goals, only the careful management or modulation of interactions to attempt to balance and ease the strains of adaptation as an ongoing process.

References

Acuto M and Curtis S (2014) Assemblage thinking and international relations. In: Acuto M and Curtis S (eds) *Reassembling International Theory: Assemblage Thinking and International Relations*. Basingstoke: Palgrave Macmillan, pp. 1–15.

Anderson B, Kearnes M, McFarlane C and Swanton D (2012) On assemblages and geography. *Dialogues in Human Geography* 2(2): 171–189.

Bratton B (2015) *The Stack: On Software and Sovereignty*. Cambridge, MA: MIT Press.

Bryant L (2014) *Onto-Cartography: An Ontology of Machines and Media*. Edinburgh: Edinburgh University Press.

Caquard S (2013) Cartography I: Mapping narrative cartography. *Progress in Human Geography* 37(1): 135–144.

Chandler D (2010) Neither international nor global: Rethinking the problematic subject of security. *Journal of Critical Globalisation Studies* 3: 89–101.

DeLanda M (2006) *A New Philosophy of Society: Assemblage Theory and Social Complexity*. London: Continuum.

Folke C (2006) Resilience: The emergence of a perspective for social-ecological systems analyses. *Global Environmental Change* 16: 253–267.

Gunderson L and Holling CS (eds) (2002) *Panarchy: Understanding Transformations in Human and Natural Systems*. Washington, DC: Island Press.

Harley JB (1989) Deconstructing the map. *Cartographica* 26(2): 1–20.

Holling CS (1973) Resilience and stability of ecological system. *Annual Review of Ecological Systems* 4: 1–23.

Holling CS (1986) The resilience of terrestrial ecosystems: local surprise and global change. In: Clark W and Munn R (eds) *Sustainable Development of the Biosphere.* Cambridge: Cambridge University Press, pp. 292–320.
Holling CS (2001) Understanding the complexity of economic, ecological, and social systems. *Ecosystems* 4: 390–405.
Holling CS and Meffe G (1996) Command and control and the pathology of natural resource management. *Conservation Biology* 10(2): 328–337.
Kitchin R and Dodge M (2007) Rethinking maps. *Progress in Human Geography* 31(3): 1–14.
Latour B (1986) Visualisation and cognition: Drawing things together. In: Kuklick H (ed.) *Knowledge and Society Studies in the Sociology of Culture Past and Present*, vol. 6. Amsterdam: Jai Press, pp. 1–40.
Latour B (2016) *Does the Body Politic Need a New Body?* Yusko Ward-Phillips Lecture, University of Notre Dame, 3 November. Available at: www.bru no-latour.fr/sites/default/files/151-NOTRE-DAME-2016.pdf.
Massey D (2005) *For Space.* London: Sage.
Morton T (2013) *Hyperobjects: Philosophy and Ecology After the End of the World.* Minneapolis: University of Minnesota Press.
Walker J and Cooper M (2011) Genealogies of resilience: From systems ecology to the political economy of crisis adaptation. *Security Dialogue* 42(2): 143–160.

PART 6

Elicitations and co-creations

31

CO-CREATIVE MAPPING OF MEMORIES

Élise Olmedo, Emmanuelle Kayiganwa and Sébastien Caquard

'I share this fragile story for anyone who will read it.
I share this unbearable story because the words are hard to find.
I share it with the 'I' that has become 'we', a protective crutch.
I share my intimate conviction that life is beautiful and that it must be protected from the ugliness that fogs the brain.
I share the sacredness of love that transcends pain and illuminates the world.
I share the sorrow of our people who can't accept what they have gone through.
I share, as many have before me, even though they remain in shock (Agahoma munwa).
I share, at the beginning of the beautiful Fall season, the loveliest season in the lovely province of Quebec.
I share before the snow covers this lovely nature.
I share words of comfort (Mpore) to people who are wounded in their flesh and in their heart.
I share the pain with all the people who sorely regret their missing family.
I share my conviction of eternal life, we are them, they are us, love does not die and remains our sacred heritage.
I share my prayers to find peace in life's miracles.'

Emmanuelle Kayiganwa, September 2022

Introduction

Emmanuelle Kayiganwa is a survivor of the 1994 genocide against the Tutsi in Rwanda. She shared her life story in 2009 in a three-hour interview as part of the 'Montreal Life Stories' research project in which approximately 500 people, who fled their home countries to come to Montreal to escape various forms of violence and persecution, shared their life experiences (High, 2014). Twenty-one of these stories were mapped in an experimental digital atlas, the *Atlas of the Rwandan Life Stories*. Emmanuelle Kayiganwa's story was one of them. It also became the subject of a second atlas: A *Subjective Atlas* made of sensibility maps (Olmedo et al., 2023). The maps which comprised these two atlases were regularly presented and discussed with Emmanuelle Kayiganwa. This exchange intensified from 2020 to 2023, along with an increasing back-and-forth between the maps and the story,

 DOI: 10.4324/9781003327578-38

between the visual and the oral, between memory and its interpretation, between the past and the present, between the one who holds the memory and those who intend to map it. This back-and-forth process has left various tangible traces such as cartographic sketches, recordings and photographs, but also less tangible ones such as memories, informal discussions around food and tea, and a certain habit to work collaboratively. Although the initial project was to map emotions associated with places in life stories, our intentions shifted towards developing collaborative mapping practices with Emmanuelle herself. Indeed, the mapping process turned into a co-creation process of recalling, sharing and listening to memories. By engaging in this collaborative process, Emmanuelle Kayiganwa identified a major vacancy in the first version of her life story shared in 2009: her childhood memories. Uncovering this omission brought along a strong need to fill this gap and complete her life story. This is when she decided to be actively involved in the mapping process. In this chapter, we present this process, along with the concept of co-creative mapping and its outcome.

Defining co-creative mapping

Co-creative mapping can be considered as a form of collaborative mapping in which participants agree to share time, skills, or experiences to design and produce maps. Co-creative mapping is generally based on the idea that the entire mapping process can serve to achieve common or individual goals. It can involve sharing the act of mapping, which means the act of taking a pencil, a mouse, a needle, or a brush to sketch the map through cartographic symbols, words or images to elicit spatial knowledge, memories, desires, or projects. But co-creative mapping can also take place through a more specific division of tasks such as 'I tell', 'you map', 'I comment', 'you modify', as well as through discussions and reflections that can sometimes lead to the enunciation of a 'we': 'we tell', 'we map', we comment' and 'we modify'. In this co-creative process, roles are neither fixed nor assigned a priori. They take shape over the development of the collaboration and evolve over time; the one who knows can become the one who does and vice versa. Likewise, the steps of the mapping process are not predefined; they also evolve according to the conversations and the listening of the other(s). Such an approach is situated at the crossroads of different movements in contemporary cartography. It is inspired by post-representational cartography with its interest in the process, its consequences and its possible instrumentalisations (Kitchin et al., 2013). Just like in post-representational cartography, in co-creative mapping the map is always in the process of becoming (Del Casino and Hanna, 2006). It also resonates with participatory mapping approaches that emphasise the intimate and visceral relationships we have with places, which may involve traumatic memories (Sletto et al., 2023). It is also influenced by work done in non-Euclidean cartography and alternative cartographies for the unrestricted creativity they offer in terms of design (Olmedo, 2018; Westerveld and Knowles, 2020). Co-creative mapping draws on narrative cartography in its willingness to mobilise the map not only as a storytelling medium but also as a means of making visible and as transparent as possible the various stages of the mapping process (Caquard and Cartwright, 2014; Alavez, 2022). Finally, it draws on work in cartographic mediation that envisions the map as a point of contact that fosters exchange, interaction and listening to one another (Martouzet et al., 2010; Mekdjian et al., 2014). Co-creative mapping is thus a collaborative approach designed to bring personal or collective experiences to the surface of the map while focusing on the process. As a process, it must always be mindful of potential ramifications this type of engagement might imply for the storyteller and knowledge holder, particularly when sharing traumatic memories. It recognises the importance of cartographic forms that are not defined a priori, but that emerge from this process. This approach potentially offers an interesting framework for projects that address memory and issues that may be difficult to express, especially when they are associated with trauma.

Co-creative mapping and sensibility mapping

As their name indicates, sensibility maps are particularly interested in revealing the worlds of sensibility such as perceptions, emotions and intimacies associated with places. They strive to represent what is evoked, what is sketched, what is left unsaid and what emotions are withheld, as well as the disjointed and fragmentary nature of any story. By mobilising context-specific forms of expression, sensibility maps seek to reveal parts of the personal and intimate dimensions of our relationships with the 'guts of the story' (Olmedo and Caquard, 2022), or what Sletto and colleagues (2023) called 'visceral geographies'. Sensibility maps are usually designed for people who share their stories and are appreciated by anyone for whom these forms of representation resonate because they share similar experiences, or because they are interested in the academic, pedagogical, or creative dimension of these maps. While Euclidean maps primarily engage viewers on a rational, intellectual level, sensibility maps engage them viscerally. They mobilise their empathy and reactivate their personal experiences. To do so, they involve traces of the past. These traces can be intangible, such as memories and stories, or material, such as notes, photographs, drawings or documents. Starting from these traces, sensibility mapping borrows from conventional cartography's practices such as data collection protocol, graphic semiology and the use of legends as well as from creative practices. Throughout this process and its outcomes, they expose the audience and contributors to a personal interpretation of the intimate aspects of this material. Sensibility maps offer a space of expression of the interpretation of stories and memories based on close and careful listening.

The foundations of the collaboration

The collaborative work with Emmanuelle Kayiganwa started in 2009 when she shared her life story in a three-hour interview as part of the 'Montreal Life Stories' project. In this co-constructed story with the interviewer Sandra Gasana, another Rwandan exile living in Montreal who also shared her life story, Emmanuelle talks about her exile from Rwanda in 1973 and her migration journey. Following the segregation she experienced at school as a Tutsi, she was forced to leave her family at the age of 11 and flee her country for Burundi at the age of 20. The 1994 genocide occurred while she was in the city of Bukavu in the Democratic Republic of Congo, right on the border with Rwanda. At that point, she decided to leave the region with her children and go to Canada, where her husband joined them later.

This life story is one of the first ten stories selected to be mapped in 2014 as part of the 'Mapping Life Stories' project. The initial goal of this project was to map emotions associated with places using the open-source software Atlascine. This mapping approach required to rigorously define and identify the different types of places that structure the stories and the themes associated with them. The first maps were produced in 2016 and were shared with all the storytellers who were invited to offer their perspectives on these maps. One of the most striking comments from Emmanuelle Kayiganwa at that point was that these maps, with their abstract symbols, were able to communicate unspeakable experiences

(Caquard et al., 2019). On the other hand, this mapping process failed to identify emotions in a consistent way throughout these stories. It became clear that mapping emotions might require a more dedicated approach.

Élise Olmedo joined the project in 2019 as a postdoctoral fellow, with the intention of applying a sensibility mapping approach to map the more intimate dimensions of these stories, including emotions. After a close listening of Emmanuelle's story and extensive note taking, she identified key emotional moments in this story and started to map them. These maps along with some of the notes were compiled in a *Subjective Atlas* that was given to Emmanuelle in the Spring of 2020. This atlas marked a turning point in our collaborative relationship. Between September 2020 and March 2023, a dozen of interviews and conversations took place, first remotely by phone (due to the Covid pandemic) and then, in person. Progressively in this series of exchanges, Emmanuelle identified certain important moments in her life that she didn't share in her original life story, including her childhood in Kabgayi, Kaduha and Nyanza in the 1960s. This period is essential for her, since it corresponds to a stable and happy time in her life preceding a much more difficult period with the disappearance of her father during the waves of violence perpetrated against the Tutsi in the 1960s which triggered her exile in 1973. Between January 2022 and March 2023, Emmanuelle wrote the missing part of her childhood story and started to build up a personal archive of this period by gathering, in a blue binder, texts, collages of photographs and an annotated map of Rwanda describing her exile journey. At that point, she asked Élise if they could design a sensibility map of her childhood (Figure 31.1).

From collaboration to co-creative mapping: the childhood map

The design of the childhood map started with a photocopied map of Rwanda on which Emmanuelle located the significant places of her childhood using coloured dots associated with small cards on which she described the events and emotions connected to these places. She describes moments of carefree innocence as well as painful ones, such as the separation from her parents which foreshadows her exile. These documents inspired Élise's design of a hand-drawn map using pencils and watercolours. More memories and documents were slowly enriching this map. As Emmanuelle Kayiganwa recalled her childhood, she (re)discovered traces of her past such as forgotten photographs. The map was then drawn and redrawn several times shedding multiple iterations along the way (Figure 31.2).

The first and second sketches (at the top) emphasise the structure of the story. The third one (bottom left) deepens the meaning and the unfolding of the different events while in the fourth one (bottom right) Emmanuelle associates them with specific colours. Here, the co-creative mapping process encourages the design of the map based on the contributions and traces produced by both participants. It thus creates a space for verbal and non-verbal expression that accumulates the cartographic trials more than it erases them. In this way, the co-created map reveals different stages of the process of which it is the result. This palimpsest also reveals the layering of memories and how they are remembered and shared in different phases, requiring a constant adaptation of the map. In that sense, this palimpsest map somehow reflects the complex layering of our relationships with memory. It became obvious to use the palimpsest map as a base to host the unfolding of Emmanuelle's memories.

From one version to the next, this cartographic palimpsest reveals the importance of tracing each contribution in order to better chart the evolution of the memory and the

Figure 31.1 Collaborative map representing Emmanuelle's childhood, a missing part of her initial story, Emmanuelle Kayiganwa, Élise Olmedo, 2023.

Figure 31.2 Successive versions of the collaborative map representing Emmanuelle's childhood, Emmanuelle Kayiganwa, Élise Olmedo, 2023.

evolution of the map, even though these traces may sometimes seem unfinished or erratic. For instance, the first version of the map aimed to represent the two main places—Kaduha and Nyanza—as two juxtaposed bubbles (Figure 31.2, top left). But a more thorough study of the story helped elucidate the significance of the links between these two places for Emmanuelle, which led to their representation as two interlocking circles. In the finalised map, the childhood story begins in Kaduha when her mother tells Emmanuelle and her sister (in the centre of the map) that their father, who was close to the royal family and the Tutsi government, has disappeared in 1959 (represented by a faded character next to the previous one). This moment is represented through a protective family bubble which breaks (first pink delimited circle in the centre of the map), submerged by the (navy-blue) wave of this announcement. The family bubble is then reformed (second turquoise-blue circle around the first one) but in a more diffuse way (vaporous circle), when Emmanuelle is exfiltrated with her sister and their cousin and taken to Nyanza. Two distinct bubbles then appear at the opposite edges of the map, representing her parents when they were deported to Nyamata in 1968 (crimson circle at the bottom right) and Emmanuelle attending high school in Kigali afterward (turquoise-blue circle at the top left). Traces of each of the three first mapping attempts remained in the fourth one, revealing the transformation of the map under the influence of the co-creative mapping process.

From 'I' to 'we', sharing the story

Through this mapping process, the personal story became a plural narrative because it includes other people such as family members, friends, or people met along the way even if it does not pretend to tell their story in their place. Beyond the people that are part of these stories, the story keepers also involved the researchers who contributed to 'carrying it' in some way:

> The keepers of our story (it has become ours because the 'I' is no longer appropriate, since some of the events I experienced do not belong to me personally) were empathetic and compassionate while facing the description of highly traumatic events. Collaborating on this project has become a real commitment, a frightening endeavor since revisiting the memory is painful. How to select and sort memories? How to tell some of them? Should we tell horrors? The collaboration therefore allowed us to say 'we'. The process was less painful thanks to this supportive relationship. The lightness of the meetings, the communication (even remotely due to the pandemic) made the feeling of anxiety disappear.
>
> *(Emmanuelle Kayiganwa, 2022, personal writing)*

This transition from 'I' to 'we' is another evidence of the co-creative mapping process at work. Here, the 'we' refers not only to all genocide survivors but also to the three authors of this chapter who built together an evolving version of this story through the cartographic process. It is also a way for the memory holder to share the responsibility of taking care of the story with the researchers; both the storyteller and the researchers become responsible for the story. The story as it appears in the different co-produced maps is no longer a pre-existing material waiting to be mapped, but something that formed throughout the collaboration and that evolved with it.

Processing the story

The mapping process performed a real transformation of the story. Indeed, working with memory means exploring a fluctuant material. Maps tend to fix memories, just like any other types of data, admitting the difficulty of approaching this material in a way that respects and reflects its fluctuations. The goal of the sensibility map was to address this issue. Although a form of coherence seemed to emerge from the whole process with the last version of the childhood map being more detailed and refined than the first one, we renounced any attempts at a stable and definitive representation of the memory. This long iterative process of drawing and redrawing the map contributed to the elicitation of a traumatic experience.

> This collaborative mapping work is an attempt to share the emotions that I carry with me, to share these words that will shape—subjective—sentences. Subjectivity allows me to respect my inability to write. Indeed, my emotions do not freeze the story. Telling a story is an opportunity (a gift from heaven) that was offered to me with grace. . . . Even—subjective—, there are real parts of the story which are very painful and unreal for those who have not experienced it. Working together on this story, however, has created a serenity to continue. Thanks to you and thanks to all those who are committed to provide a space to the denied humanity that led to the Genocide of the Tutsi.
>
> *(Emmanuelle Kayiganwa, 2022, personal writing)*

The visual result of this process is very far from cartographic canons. While conventional cartography insists on the visual efficiency of the design to convey a message as clearly as possible, co-creative mapping primarily focuses on creating and adapting a cartographic language that corresponds to what the storyteller wants to express. It is only at a later stage that these collaborative maps can be revisited to reach a wider audience, for instance by including some texts, legends or other forms of mediation such as videos. This is the paradox of memory co-construction which oscillates between personal representations that aim to be as close as possible to the vision of those who hold the memories and more conventional representations that are formatted to be autonomous and accessible to a broader audience (Caquard et al., 2019). However, Emmanuelle emphasises the importance of the freedom this co-creative mapping process offers: 'I did not feel the pressure to share the story in a fixed way limited to academic expectations. I allowed myself to change, skip, or modify certain steps in the story' (Emmanuelle Kayiganwa, 2023, personal writing). This sense of freedom in terms of format, outcome and forms of collaboration is probably one of the major assets of co-creative sensibility mapping.

Conclusion

Through this collaborative work, we have identified some essential aspects of co-creative mapping. First of all, it needs a clear commitment from the people who hold the knowledge to be mapped. Although this may seem obvious, participatory mapping projects are often mainly carried out by researchers, as it was the case at the beginning of this project. But, with time and the constant sharing of the researcher's results with the storyteller, the need to develop the childhood map clearly came from Emmanuelle Kayiganwa and was largely carried by her. Co-creative mapping then takes on a slow mapping process (see Seemann, this volume); it requires time: time to know each other and to build trust, as well as time for the storyteller to remember and organise their memories and time to be ready to share them; time to understand what mapping could be, what form it could take and how it could serve a particular purpose; and time to understand that mapping could be liberating and inviting, rather than intimidating and restrictive when one doesn't master its conventions, codes and techniques. Co-creative mapping is then a powerful process to expose anyone to the relevance and benefits of creative forms of spatial expressions, such as palimpsest maps that resonate with both the process of remembering and the process of mapping memory. This cartographic palimpsest also resonates with the blurry and fragmented structure of memory that we need to navigate through when we try to remember, furthermore when we try to remember traumatic memories.

Finally, the unfinished dimension of the cartographic results of this process and its detachment from cartographic conventions expands the possibilities of representation. While post-representational cartography envisions maps as always in the state of becoming (Del Casino and Hanna, 2006), here we turned this vision into a mapping practice. In co-creative mapping, the map is never stable or finished. The map of childhood keeps on evolving. More importantly, the co-created map presented here embraces this unfinishedness by looking fragmented, blurry, imprecise, disconnected and confusing; all kinds of characteristics that could also be used to describe memory. It is probably because of all these *imperfections* that this co-creative mapping process has been so easily adopted by Emmanuelle Kayiganwa.

Acknowledgment

Emmanuelle Kayiganwa would like to express special gratitude to Tania Rossetto and Laura Lo Presti for their encouragement and support in facilitating the publication of this chapter. The interest shown and the dedicated space provided for this work honor the memory and offer a source of comfort for a humanity deeply affected by the genocide of the Tutsi in Rwanda.

The authors would also like to thank Léa Denieul-Pinsky for text editing. This project was supported by the Social Sciences and Humanities Research Council of Canada (SSHRC) [435–2016–1178] and a Banting Postdoctoral Fellowship.

References

Alavez J (2022) Mapping intimate geographies of grief and loss. *Cartographica* 57(4): 270–280.

Atlas of Rwandan Life Stories. Available at: https://rs-atlascine.concordia.ca/rwanda/index.html.

Caquard S and Cartwright W (2014) Narrative cartography: From mapping stories to the narrative of maps and mapping. *The Cartographic Journal* 51(2): 101–106.

Caquard S, Shaw E, Alavez J et al. (2019) Mapping memories of exiles: Combining conventional and alternative cartographic approaches. In: De Nardi S et al. (eds) *The Routledge Handbook of Memory and Place*. London: Routledge, pp. 52–66.

Del Casino Jr VJ and Hanna SP (2006) Beyond the 'binaries': A methodological intervention for interrogating maps as representational practices. *Acme: An International E-Journal for Critical Geographies* 4(1): 34–56.

High S (2014) *Oral History at the Crossroads. Sharing Life Stories of Survival and Displacement*. Vancouver: UBC Press.

Kitchin R, Gleeson J and Dodge M (2013) Unfolding mapping practices: A new epistemology for cartography. *Transactions of the Institute of British Geographers* 38(3): 480–496.

Martouzet D, Bailleul H, Feildel B and Gaignard L (2010) La carte: Fonctionnalité transitionnelle et dépassement du récit de vie. *Natures Sciences Sociétés* 18(2): 158–170.

Mekdjian S, Amilhat-Szary AL, Moreau M, Nasruddin G, Deme M, Houbey L and Guillemin C (2014) Figurer les entre-deux migratoires: Pratiques cartographiques expérimentales entre chercheurs, artistes et voyageurs. *Carnets de Géographes* 7.

Olmedo E (2018) Textile maps: Using sensitive mapping for crossovers between academic and vernacular worlds in the Sidi Yusf working-class neighbourhood in Marrakech. In: Orangotango K (ed.) *This Is Not an Atlas*. Bielefeld: Transcript Verlag, pp. 264–269.

Olmedo E and Caquard S (2022) Mapping the skin and the guts of stories: A dialogue between geolocated and dislocated cartographies. *Cartographica* 57(2): 127–146.

Olmedo E, Kayiganwa E et al. (2023) *Atlas Subjectif; Les Entrailles d'un Récit de Vie Rwandais*. Montréal: Auto-Édition.

Sletto B, Novoa M and Vasudevan R (2023) 'History can't be written without us in the center': Colonial trauma, the cartographic body, and decolonizing methodologies in urban planning. *Environment and Planning D: Society and Space* 41(1): 148–169.

Westerveld L and Knowles AK (2020) Loosening the grid: Topology as the basis for a more inclusive GIS. *International Journal of Geographical Information Science* 35(10): 2108–2127.

32

MAPPING AS THE ART OF LISTENING TO JEWISH MEDITERRANEAN MIGRATIONS

Piera Rossetto

Introduction

A passage from the autobiography of Shimon Ballas, an Israeli author born in Baghdad in 1922, is the first memoir by a Jew from an Arab-Muslim country which drew my attention to the deep connection between memory and space/place in the context of contemporary Jewish diasporas across the Mediterranean and beyond:

> 'Maybe the two houses where I grew up still exist today!' A young Iraqi friend I met in Paris showed me a typical tourist map of the city [Baghdad], I found myself looking at streets, gardens, markets, bridges and building blocks being built on the outskirts of town. I have another map, I told him, a map showing paths that turn and intersect. I could sketch it on a piece of paper; every curve, every niche, every vault, every window, and the projecting walls of the houses, sharp angles where men stop at night and urinate. I remember each of them. 'A lot of areas have been destroyed, yours could be one of them', my friend replied. Destroyed or not, my neighbourhood was still there, in its place.
>
> *(Ballas, 2009: 38)*

Ballas is one of the about 700,000–800,000 Jews from Arab-Muslim countries who, between the 1940s and 1970s, decided or were forced to leave their country of birth and resettle, mainly in Israel but also elsewhere.[1] As Miccoli (2020) noted, writing the history of such diverse (also internally) Jewish populations, from Morocco to Iran and including Yemen, their progressively shrinking presence and eventually disappearance from these regions, is not an easy task; even more so if the aim—as is the case in my research agenda—is to include the subjective points of view of the migrants themselves on the how and why this all happened.

On the one hand, argues Miccoli, 'if one were to search for the exact year, day, or event that caused the end of this world and therefore the departure of the Jews from North Africa and Egypt [and other countries], one would not find it' (2022: 2). Indeed, we are dealing with a complex cluster of migratory phenomena that goes beyond any clear-cut

 DOI: 10.4324/9781003327578-39

categorisation and could only benefit from an interdisciplinary analysis (Katz et al., 2017; Meir-Glizenstein, 2018; Moreno, 2020). On the other hand, continues the literary scholar, one would find a plethora of stories 'of contrasted feelings for a vanished world whose echoes many can still hear, as if it were a phantasmic absence and the shadow of a past that becomes also present and future' (Miccoli, 2022: 1–2). Echoes of such contrasted feelings are enshrined in a vast literary production by Jewish authors of Middle Eastern and North African descent, as well as in personal recollections performed in in-depth, semi-structured interviews.

By focusing on this latter form of memorialisation (Rossetto and Spadaro, 2014; Rossetto, 2022), in the last few years, I have been investigating how, if at all, individual trajectories can shed light on this complex socio-historical phenomenon. In order to address this question, I followed implicit invitations by many interviewees and adopted space and place as key categories of interpretation of the recollections I was entrusted with. This perspective prompted an interest in deep and creative mapping practices and eventually the development of a research-creation collaboration[2] to explore 'to which extent the form of the map (whether creative, deep, or thick) fits the knowledge (complex, diverse or contradictory) embedded in the recollections by and about North African and Middle Eastern Jews' (Rossetto, 2024). In this chapter, I recall the initial coordinates of this personal and collaborative journey and then explore in more detail one of the six creative maps produced in the collaboration.

The memories

In my effort to make sense of the complex processes Middle Eastern and North African Jews underwent in their (forced) migrations, I conceive my ethnography as a 'diary of the streets I walk',[3] the streets being the recollections, emotions and perceptions of my interlocutors. The interviewees themselves induced me to do so, as the following vignettes taken from my fieldwork will exemplify. Interviewees often spontaneously adopted a spatial language as well as some kind of cartographic practice (sketching a map onto a piece of paper, displaying objects on the table to reconstruct the former urban environment).

Vignette one: Rome, late spring 2012. Victor sits in his office and I in front of him.[4] Born in Tripoli, Libya in 1957, Victor is willing to share with me his personal recollections of a playful youth spent on the shores of the Mediterranean Sea, between the notes of Italian pop singers, who used to tour Tripoli in the 1950s and 1960s (David, 2014), and the unique melody of his family language, Ladino (or Judeo-Spanish) which bore the traces of a long Jewish diaspora across the Mediterranean. 'No, I've never been to Tripoli', I admit. Victor takes a pen and a piece of paper and draws the shape of the city harbour. Through the quick stroke of the pen, he invites me to virtually stroll at the seaside as I follow the lines he draws. While his recollections unfold, a real, intimate cartography takes shape (Papotti, 2011).

Vignette two: Milan, late autumn 2018. Yoram welcomes me to his apartment.[5] Born in Tripoli in 1953 to a Jewish family of Italian origin, Yoram starts our conversation by retrieving memories of the riots that broke out in Libya on 5 June 1967: a direct consequence of the Six-Day War between Israel and its neighbouring countries that began that very day. The narration is a detailed account of the events of the first day of violence, and how he risked his life to cycle home from school. After 12 days locked up at home, Yoram and his family, as well as the majority of the approximately 6,000 Jews living in the country at the time, were escorted by a military convoy to the airport and flown to Rome. But flying over the city, safe and relieved, they had not imagined that this flight would be their last chance to see their city, their home and their country. The interview lasted about three hours. When I thought it was time to go, Yoram invited me to sit at the computer and follow him while he searched Google Maps for the exact route he had taken that day. As he virtually rides his bike more than 50 years later, Yoram clearly remembers every street, corner and building as well as the fear, distress and anxiety he had felt that day. The online mapping tool allowed him to 'come back' to his city and, in a way, to take me there too.

As mentioned, I did not elicit such cartographic practices or languages. I interpret them as an invitation made by my interlocutors to embark on a virtual journey through the streets of their past, as a way to try and 'take me there', *là-bas*, with them. Quite interestingly, the word 'there', *sham* in Hebrew, is used in contemporary Israeli literature, starting with David Grossmann, to signify the places of the Shoah, the places of the utmost suffering, without naming them directly. Indeed, many of the places I am taken to during the interviews are places of suffering: these are stories difficult to tell for those who directly experienced displacement, loss and hardship. Moreover, they are difficult to 're-tell' for me as I measure my distance from the interlocutors' experiences and try 'to walk in their shoes' (Jones and Ficklin, 2012: 103): 'No, I've never experienced a forced departure', I must admit to them and to myself.

In the emotionally charged context of the 'inter/view'—'*uno scambio di sguardi*', an 'exchange of looks' in Portelli's words (2007: 78)—I turned to mapping because the spatial perspective allowed me to grasp, beyond the mere movement of migrants across space, the deep existential implications of leaving, encountering and inhabiting places for individuals and groups (Tuan, 1977). This approach elicited my empathy, understood as 'a complex imaginative process involving both cognition and emotion', the ability to take up another person's psychological perspective and imaginative experience while maintaining 'a clear sense of my own separate identity' (Coplan, 2004: 143). In this sense, mapping memory meant 'charting empathy' (Rossetto and Melilli, 2021): locating the positionality of each participant in the space of betweenness, which constitutes the anthropological dialogue (Tedlock, 1983).

The map

When I started approaching maps and mapping in relation to recollections by and about Jewish migrants from the Middle East and North Africa, what I lacked was not enthusiasm about maps but perhaps clarity, a concrete idea of 'what makes a map a map' (Wood, 1993) in my research context. Thanks to the sensitivity and expertise of all those engaged in the research-creation collaboration I initiated,[6] mapping memories evolved progressively from a metaphorical level into a much more concrete form, as the maps created show. The tension towards mapping memories generated six creative products, which I consider forms of creative maps and mapping practices: a storyboard on a Forex (PVC) support; a digital storyboard; an artistic map embroidered on a handkerchief; a digital visualisation of about 100 migration routes of Jews from North Africa and the Middle East to Milan, Italy; a foldable map&letter print on paper; and finally, a 'map of words', that is, the narrative podcast *Shamailang*.[7]

As it would be impossible to comment here on each map, I suggest dwelling on one, the map&letter *Ze haya be-leil Shabbat, The Eve of the Shabbat* (Figure 32.1).

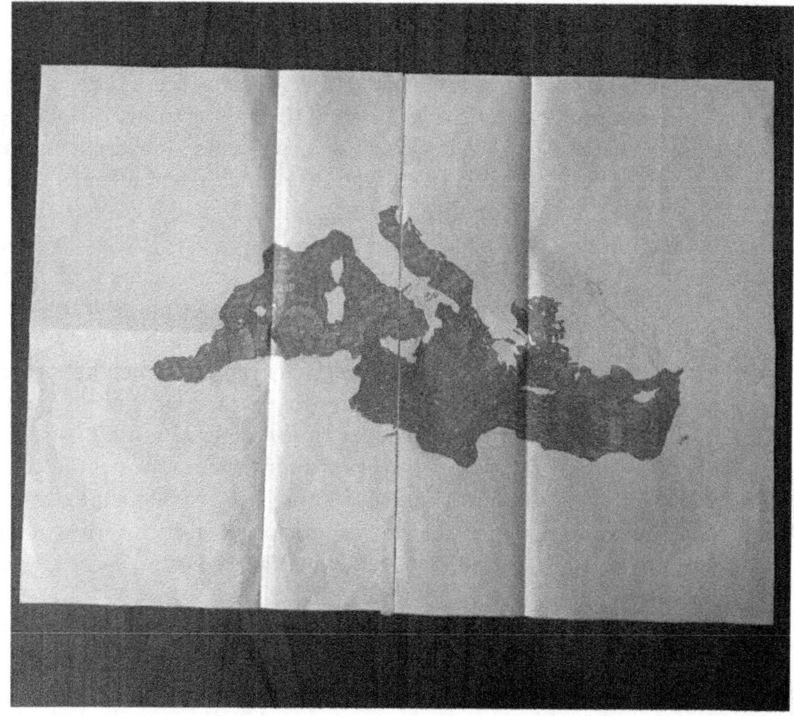

Figure 32.1 Original hand-drawn maquette for the 'Ze hayah be-leil Shabbat—The Eve of the Shabbat' creative project.

Source: Courtesy of the artists, Martina Melilli and Michela Nanut

The map&letter explores the story of Rina Messika, a Libyan Jewish lady I interviewed in Israel, and the way she performed her recollections in the interview. I had already discussed Rina's story when addressing the complexity and contradictions which characterised the historical interpretations of the mass departure of Jews from Libya between 1948 and 1951 (Rossetto, 2017).[8] In that publication, I had highlighted how Rina's story clearly demonstrated how 'not only different persons experienced different situations and therefore discuss the departure in certain terms but also how the different narratives coexist in the same person' (Rossetto, 2017: 179). In fact, although Rina left Libya for Israel illegally with the help of a Zionist Jewish youth organisation, during the interview, she had never situated her departure clearly within an ideological frame, nor had she described it as a flight from persecution or as the accomplishment of a religious aspiration. Rather, her account witnessed how, in much more universal terms, the crossing of the Mediterranean Sea had represented a tremendous change in her life as a young woman. Rina's story, I commented in this chapter, was an example of how 'a single narrative cannot achieve the goal of accurately transmitting the complexity of the process she underwent, and yet what if we could produce a form of narrative that will not seek to blur the differences and diversities of the performances' (Rossetto, 2017: 181)? Is it possible to dwell in the contradictions produced by personal experiences and individual recollections and yet create the conditions to guide

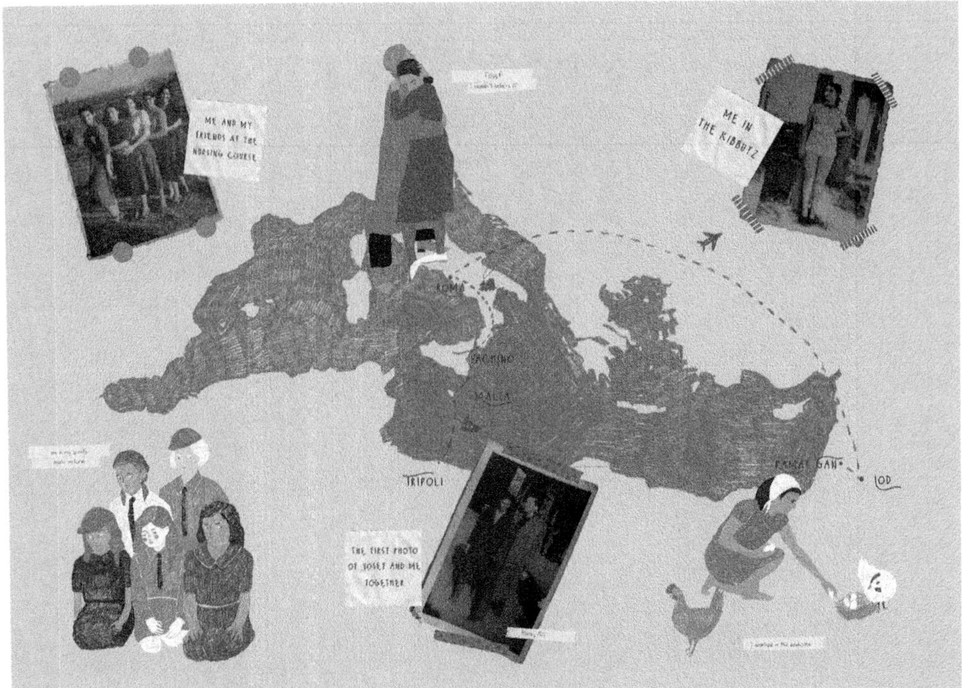

Figure 32.2 'Ze hayah be-leil Shabbat—The Eve of the Shabbat', the Mediterranean basin (internal side of the foldable printed artwork). Martina Melilli (Art direction and Texts), Michela Nanut (Graphic design and illustrations), Piera Rossetto (Scientific supervision).

us in the process of 'making sense' of those same disorienting experiences? My hypothesis was that 'deep maps'—metaphorically described by Ridge et al. 'as an experience that begins when you step off the "bus" and begin explore data more deeply' (2013: 185)—could serve this purpose, but how?

Visual artist Martina Melilli, illustrator Michela Nanut and I started working on Rina's recollections: a fascinating narrative, full of details, mostly revolving around the epic journey which took her in 1948, alone at the age of 14, to leave her family in Tripoli, cross the Mediterranean Sea and land in Sicily. After a week spent in the village of Pachino, she continued her journey to Rome where 'by chance'—in her words—she met the young man she was in love with in Tripoli, and from there, they left together for Israel, where they joined the kibbutz movement.

The materials we could rely upon were scarce: an audio interview in Hebrew I had collected and its full transcript I had prepared in Italian; very few photographs that belonged to Rina, and others I took, during the interview, of her and some objects in her flat. The result is a foldable map of the Mediterranean basin (Figures 32.2 and 32.3), on which a few of Rina's original photos are reproduced and which is accompanied by a long, imaginary letter by Rina to a friend. Text and drawings intertwine and shape a creative form of mapping, able to shed light on the more individual, micro perspective on the event of Jewish mass migration from Libya.

Figure 32.3 'Ze hayah be-leil Shabbat—The Eve of the Shabbat', the imaginary letter (external side of the foldable printed artwork). Martina Melilli (Art direction and Texts), Michela Nanut (Graphic design and illustrations), Piera Rossetto (Scientific supervision).

The main components of the map and its fundamental structure could be analysed and understood under three categories inspired by Hertha Müller's poignant quote: 'But to be certain of our own existence, we need the objects, the gestures, and the words'.[9]

The words. In the interview, Rina explains that she managed to remember the journey with such richness of detail because she was used to writing everything down in a notebook, a habit she kept for the rest of her life when she travelled. This inspired Melilli to imagine a letter that Rina would have written to a friend in Tripoli. The text, elaborated by Melilli, reproduces Rina's voice as if she were reading aloud, to her friend Nini, the notes she took more than 60 years before. This choice resonates with practices of representing ethnographic data through the epistolary form as a way 'to acknowledge and communicate the "emotional engagement" of ethnography' (Carroll, 2015: 693; Iedema and Carroll, 2015; Smith and Kleinman, 2010).

Objects and gestures. Neither Melilli nor Nanut knew the history of Jewish migrations from North Africa. They were not familiar with Jewish culture and traditions and did not know the various Hebrew terms used by Rina, such as *Shaddai*, *Mezuzah* and *Apalah*. Although they ignored their meaning, both artists perceived the centrality of objects these words referred to, such as the Shaddai. Usually translated as the Almighty—one of the names of God—the Shaddai is a protective amulet, which can take the shape of the hand, thus resembling the Hamsa, the number five, or the hand of Fatima in Islamic tradition. Among the communities of both religious traditions in the Middle East and North Africa, this hand-amulet is believed to have the power to ward off the evil eye. Rina says that she received it from her mother on the eve of her departure. The Mezuzah was another emotionally charged object linked to a powerful gesture: Rina recalls kissing it the moment she left her house for good, in what is a typical gesture in the Jewish tradition upon entering or leaving a room or place.

The art of listening

I discussed these objects with Melilli and Nanut and their significance in the Jewish tradition. More importantly, we pondered the place each object occupies in the plot and its 'resonance': the power

> to reach out beyond its formal boundaries to a larger world, to evoke in the viewer the complex, dynamic cultural forces from which it has emerged and for which—as a metaphor or, more simply as metonymy—it may be taken by a viewer to stand.
>
> *(Greenblatt, 1990: 19–20)*

In Rina's story, in fact, these objects 'resonate' with a larger world of relations and visions of the self, and with a more complex web of significance and meanings:

> I left with him [a family acquaintance who had offered to accompany her on the trip], it was the eve of the Shabbat. Everyone was there: my father, my grandmother, everyone, the table, the *kiddush* [the blessing over the wine to honour the Shabbat], and my mother, who was not a strong woman but had a kind soul and whose voice everyone listened to, said: 'Listen, tonight, Rina is leaving for Palestine'. . . . My father made the kiddush then he kissed me, I remember it well. Afterwards, my mother gave me the silver Shaddai that my grandmother had given her when my brother was born, after the death of three other children. At that point, everyone cried, but we couldn't tell why, it was a secret, then I touched the mezuzah and left.[10]

One of the peculiarities of oral sources is represented by their capacity to restore subjectivity in a privileged way. In Portelli's words, oral sources 'tell us not just what people did, but what they wanted to do, what they believed they were doing, what they now think they did' (Portelli, 1981: 99–100). More importantly than the concrete details about the circumstances of her departure, by recalling words, gestures and objects, Rina takes us closer to the familiar, religious and traditional meanings in which she frames it.

The map&letter *Ze haya be-leil Shabbat, The Eve of the Shabbat* combines an exercise of 'close listening' (Bernstein, 1998) to the personal recollections performed by Rina in the inter/view with an exercise of 'close reading' (Herrnstein Smith, 2016; Gallop, 2000) of a creative text elaborated by the artists. In this sense, words, objects and gestures act as the essential elements of the legend of the map that, drawn to the scale of the individual memory, aims at representing both a geographical trajectory and a human journey, the 'skin and the guts' (Olmedo and Caquard, 2022) of a complex migratory experience. Ultimately, the map is an invitation to refine our 'Art of listening . . .: an imaginative attention [that] takes notice of what might be at stake in the story itself and how its small details and events connect to larger sets of public issues' (Back, 2007: 7).

As a final concluding remark, let us turn for a moment to the dotted, red line that curves and twists on Rina's map (Figure 32.3). As Tim Ingold (2020) observes:

> Drawing, said the artist Paul Klee, is like taking a line for a walk. Try it for yourself. Take a pencil in your hand and let it alight on a sheet of paper. As the tip glides into contact with the paper, a line begins to appear. And it carries on until, with a slight flick of the wrist, you allow the tip to lift off again. What remains on the sheet is the trace of a manual gesture. Depending on how you moved your hand and fingers, it may curve, twist or loop, this way and that. But it will never be perfectly straight.

The line on Rina's map curves and twits following the movement of the illustrator's hand, inspired by the close reading of the artist in dialogue with the author who witnessed that unique narrative performance. However close to the listening and reading of Rina's story, this line will never perfectly retrace her whole geographical and human journey: it is only one of the many that are woven into the fabric of her life. Actually, what this line perfectly does, in its partiality and limitation, is to remind us that our ambition to make sense of both large-scale migration phenomena and the unique, individual fragments they are composed of is

> part of an embrace with and connection to the dance of life with all its heavy and cumbersome steps. It is an aspiration to hold the experience of others in your arms while recognising that what we touch is always moving, unpredictable, irreducible and mysteriously opaque.
>
> *(Back, 2007: 3)*

Notes

1 These forced migrations occurred following the establishment of the State of Israel (14 May 1948), post-colonial tensions, the rise of Arab nationalism and the consequences of the Six-Day War (5–11 June 1967), and they almost put to an end the Jewish presence in North Africa and the Middle East, the main exceptions being Turkey and Iran.

2 "An approach to research that combines creative and academic research practices, and supports the development of knowledge and innovation through artistic expression, scholarly investigation, and experimentation. The creative process is situated within the research activity and produces critically informed work in a variety of media (art forms)."
Social Sciences and Humanities Research Council, Canada, *Definitions of Terms*, www.sshrc-crsh.gc.ca/funding-financement/programs-programmes/definitions-eng.aspx#a22 (accessed 23 June 2023).

3 Sohei Nishino about the project *Diorama Map London* [2010] in Roberts, 2012: 6. Regarding the project, see http://soheinishino.net/dioramamap-london (accessed 22 June 2023) and www.michaelhoppengallery.com/exhibitions/6/overview/ (accessed 22 June 2023).

4 Interview by the author with Victor, Rome, May 2012.

5 Interview by the author with Yoram, Milan, November 2018.

6 The collaboration developed in the framework of the FWF (Austrian Science Fund) project *Europe (In)Visible Jewish Migrants*, PI Piera Rossetto, Centre for Jewish Studies—University of Graz, Grant No. T1024-G28.

7 For an overview of the maps, see www.pierarossetto.eu/eijm-creative-mapping/

8 In the late 1940s, the Jewish community of Libya was the smallest in the North African region. An estimated population of about 30,000 individuals lived in the Tripolitanian region, while Jews living in the Cyrenaica numbered about 6,000. The migration of about 30,400 Jews from Libya in a very short period—between 1949 and 1952—represents a quite unique phenomenon since

almost 90% of the Libyan Jewish community left for the state of Israel in the immediate aftermath of its establishment (May 1948). The literature on this phenomenon of mass migration tends to concentrate on the major reasons for it, and more rarely explores the personal, intimate journey of its protagonists.

9 Nobel Lecture, 7 December 2009, at the Swedish Academy, Stockholm, www.nobelprize.org/uploads/2018/06/muller-lecture_en.pdf

10 Interview by the author with Rina, Ramat Gan, 6 March 2013.

References

Back L (2007) *The Art of Listening*. London and New York: Bloomsbury Academic.

Ballas S (2009) *Be-Guf Rishon (First Person Singular)*. Tel Aviv: Hakibbutz Hameuhad. (in Hebrew)

Bernstein C (1998) *Close Listening*. New York: Oxford University Press.

Carroll K (2015) Representing ethnographic data through the epistolary form: A correspondence between a breastmilk donor and recipient. *Qualitative Inquiry* 21(8): 686–695.

Coplan A (2004) Empathic engagement with narrative fictions. *The Journal of Aesthetics and Art Criticism* 62(2): 141–152.

David E (2014). *The Daily Life of the Upper-Middle Class Jews in Tripoli, Libya (1951–1967)*. MA Thesis, Hebrew University of Jerusalem. (in Hebrew)

Gallop J (2000) The ethics of reading: Close encounters. *Journal of Curriculum Theorizing* 16(3): 7–17.

Greenblatt S (1990) Resonance and wonder. *Bulletin of the American Academy of Arts and Sciences* 43(4): 11–34.

Herrnstein Smith B (2016) What was 'close reading'? A century of method in literary studies. *The Minnesota Review* 87: 57–75.

Iedema R and Carroll K (2015) Research as affect-sphere: Towards spherogenics. *Emotion Review* 7: 67–72.

Ingold T (2020) Lines, threads & traces. *Toast Magazine*, 31 May. Available at: https://eu.toa.st/blogs/magazine/lines-threads-traces-tim-ingold?loc=IT&preselectedCountry=true (accessed 23 June 2023)

Jones B and Ficklin L (2012) To walk in their shoes: Recognising the expression of empathy as a research reality. *Emotion, Space and Society* 5(2): 103–112.

Katz N et al. (eds) (2017). *Colonialism and the Jews*. Bloomington: Indiana University Press.

Meir-Glitzenstein E (2018) Turning points in the historiography of Jewish immigration from Arab countries to Israel. *Israel Studies* 23(3): 114–122.

Miccoli D (2020). The Jews of the Middle East and North Africa: A historiographic debate. *Middle Eastern Studies* 56(3): 511–520.

Miccoli D (2022) *A Sephardi Sea*. Bloomington: Indiana University Press.

Moreno A (2020) Beyond the nation-state: A network analysis of Jewish emigration from Northern Morocco to Israel. *International Journal of Middle East Studies* 52(1): 1–21.

Olmedo É and Caquard S (2022) Mapping the skin and the guts of stories: A dialogue between geolocated and dislocated cartographies. *Cartographica* 57(2): 127–146.

Papotti D (2011) L'approccio della geografia alla letteratura dell'immigrazione: Riflessioni su alcune potenziali direzioni di ricerca. In: Pezzarossa F and Rossini I (eds) *Leggere il testo e il mondo: Vent'anni di scritture della migrazione in Italia*. Bologna: CLUEB, pp. 65–84.

Portelli A (1981) The peculiarities of oral history. *History Workshop* 12: 96–107.

Portelli A (2007) *Storie orali: Racconto, immaginazione, dialogo*. Roma: Donzelli Editore.

Ridge M, Lafreniere D and Nesbit S (2013) Creating deep maps and spatial narratives through design. *International Journal of Humanities and Arts Computing* 7(1–2): 176–189.

Roberts L (ed.) (2012) *Mapping Cultures: Place, Practice, Performance*. London: Palgrave Macmillan.

Rossetto P (2017) Dwelling in contradictions: Deep maps and the memories of Jews from Libya. *Ethnologies* 39(2): 167–187.

Rossetto P (2022) Mind the map: Charting unexplored territories of invisible migrations from North Africa and the Middle East to Italy. *Jewish Culture and History* 23(2): 172–195.

Rossetto P (2024) The AesthEt(h)ics of the fragment: Jewish memories and co-creative mapping practices. In: Kumar V, Lamprecht G, Nievoll L, Oelschlegel G and Stoff S (eds) *Erinnerungskultur und*

Holocaust Education im digitalen Wandel: Georeferenzierte Dokumentations-, Erinnerungs- und Vermittlungsprojekte. Bielefeld: Transcript, pp. 55–72.

Rossetto P and Melilli M (2021) Mapping memories, charting empathy. Framing a collaborative research-creation project. *From the European South* 8: 145–151.

Rossetto P and Spadaro B (2014) Across Europe and the Mediterranean: Exploring Jewish memories from Libya. *Annali di Ca' Foscari* 50: 37–52.

Smith L and Kleinman A (2010) Emotional engagements: Acknowledgement, advocacy and direct action. In: Davies J and Spencer D (eds) *Emotions in the Field*. Stanford, CA: Stanford University Press, pp. 171–187.

Tedlock D (1983) *The Spoken Word and the Work of Interpretation*. Philadelphia: Pennsylvania Press.

Tuan YF (1977) *Space and Place: The Perspective of Experience*. Minneapolis, MN: University of Minnesota Press.

Wood D (1993) What makes a map a map? *Cartographica* 30(2–3): 81–86.

33
DRAWING (ON) CARTOGRAPHIC INTIMACIES

Laura Lo Presti

Introduction

We usually associate intimacy with deep and private feelings, with stories hard to tell and share with others, especially with strangers. This understanding of intimacy makes it difficult and counterintuitive to imagine that a map, a medium commonly deployed to share spatial information in an objective, comprehensive way, could be used to utter and whisper the visceral experience of unease and secrecy demanded by an intimate, often difficult, story. Yet both intimacy and maps could be understood otherwise. First, intimacy has not just to do with a personal or individual experience. Berlant (1998) argues that intimacy is not a fixed state but a complex, dynamic and ever-changing process that is shaped by both personal *and* social forces. She emphasises the idea that intimacy, often tied to unrealistic expectations and aspirations, is a process that involves vulnerability, the willingness to open oneself up to others, and emotional investment.

> To intimate is to communicate with the sparest of signs and gestures, and at its root intimacy has the quality of eloquence and brevity. But intimacy also involves an aspiration for a narrative about something shared, a story about both oneself and others that will turn out in a particular way.
>
> *(Berlant, 1998: 281)*

While pertaining to an oral linguistic dimension, 'eloquence' and 'brevity' are also core principles of the cartographic semantic (Casti, 2000), and the unspoken aspiration for a public narrative betrays an overlooked expression of the map: its capacity to recover memories and make them shareable (see Olmedo, Kayiganwa and Caquard, this volume), to make visible inner worlds (della Dora, this volume) and to make explicit, ideally 'with the sparest of signs and gestures' (Berlant, 1998: 281), introspective as well as relational, individual as well as collective, spaces. Seen in this way, maps work as infrastructures of feelings; they are deep surfaces emerging out of inner emotions but with the intended purpose of being shareable with others, with the outside. In fact, maps may become the experiential sites of public intimacies at any moment. Like any image, they move opinions,

 DOI: 10.4324/9781003327578-40

discussions and feelings; they can be the vehicles whereby an idea, a sensation or a concept is made passionately visible. For these reasons, I will use the term 'cartographic intimacies' to consider the ways in which maps can reveal and shape the spatial dimension of emotions, since '[a]ffects not only are markers of space but are themselves configured as space and they have the actual texture of atmosphere' (Bruno, 2014: 19). The term can also refer to how maps retain impressions of the conflicts and power struggles embedded within societal norms and expectations, exposing the tension between personal beliefs and external influences.

Cartographic intimacies may expand the experiential frame of emotional and affective cartographies. When referring to the category of emotional mapping (Caquard and Griffin, 2019) or affective mapping (Wilmott, 2020), the idea of engaging with emotions as carnal and vivid spaces, although conceptually acknowledged, does not seem to be fully practised (with a few exceptions, e.g. Luckett and Bagelman, 2023; Sweet and Escalante, 2015). While emotional mapping deals with '(1) the emotions that we place on maps; (2) the emotions that shape the mapping process and the map; and (3) the emotions people experience in response to maps' (Caquard and Griffin, 2019: 5), affective mapping is defined as 'the representation of affects, including emotions, feelings, sensations, intensities, memories, or hauntings through maps, cartographic depictions (such as cartograms), and geographic information science' (Wilmott, 2020: 53). Both usually end up materialising through psychological responses—either topophiliac or topophobic (Relph, 1976)—that people have to or in specific locations, depending on their life stories, as well as on the cultural and social meanings that they attach to such places. Yet there are very few examples indicating how feelings, intensities or emotions can be represented *as* spaces without forcefully overlapping with real locations. Emotional and affective cartographies, articulated in several forms of creative, mental, sensible and deep mapping, have in fact been engaged differently by geographers with respect to scholars from the humanities and beyond, who are traditionally inclined to unravel, elicit and spatialise thoughts, desires and social and political constraints of individuals and communities (e.g. in pedagogy, anthropology, contemporary art and psychology).

An opportunity to muse on the idea of cartographic intimacies as an extension of the category of emotional mapping emerged in the activities of the Laboratory of Cartographic Humanities (HuMaps Lab), which I organised in 2022 and 2023 for a Geography course in the Humanities bachelor's degree programme at the University of Padova. This chapter draws on this didactic experience to show how emotions can be grasped, and spatially interpreted, through creative and critical map-centred activities. It highlights the potential of the cartographic humanities to access inner worlds differently, by enhancing people's understanding of the correspondence between the psychic, somatic and social spatial dimensions of their being through a multisensory modality; one that fosters a perception of the map that takes the body as its referent.

HuMaps as infrastructures of feelings

The Laboratory of Cartographic Humanities (HuMaps Lab) was organised in Spring 2022 and 2023 as a didactic experiment. In two years, 349 university students have been asked to map either aspects of their biographies—what I introduced to them as 'auto-cartographies', 'biomaps', 'body maps' and 'inscapes'—or social, cultural and political phenomena that were addressed during the course (e.g. migration crises, climate change issues, female and racial oppressions, geopolitical hotspots, urban challenges) and that were brought together under the category of 'social cartographies'. Even in this second case, subjectivity has always been at the centre of the mapping process: students personalised and introjected such external matters of concern, according to their individual feelings and opinions. The only request that was made to the participants of the Lab, following a series of theoretical and methodological lessons developed around the idea of 'cartographic humanities' (see Introduction, this volume) and by proposing examples of maps' eclectic reuse in literary studies, media studies and contemporary art, was to abandon the notion of the map as a mere physical orientation tool. Through a multidisciplinary approach, the notion of cartographic humanities aims to expand the understanding of maps and mapping beyond their technical and scientific aspects and to provide a more holistic and nuanced understanding of their potential in shaping the world creatively and our experiences within it.

The HuMaps Lab has been built on three main principles. The first concerns the semantics of the humanities. Humanities recall the adjectives 'human' and 'humane', which underline the added potential of this transdisciplinary field of study, that is, making maps more sensitive to human interiority. A humanistic sense of cartography has, to some extent, been practised since the Middle Ages and the Renaissance, and, generally, in all the cartographic practices and styles that, running parallel to the canon, have experienced space in a sensible way. In this respect, the Lab explored the use of cartography in any possible field of the humanities (cartography and art, cartography and literature, cartography and history, cartography and philosophy, and so on). The second feature of the HuMaps Lab relies on a cultural cartography-oriented approach (Cosgrove, 2008) which takes as a point of reference theories and methods of the new cultural geography. The new cultural geography, just like the humanistic one, brings attention to human experiences and decisions and to a subjective conception of culture, but it is also influenced by radical and social geography, addressing culture as an attribute of power (Mitchell, 2000). For a cultural geographer, communication and consumption are fundamental processes for the study of space. Therefore, a cultural approach to mapping focuses on the production, communication and consumption of the map, one of the oldest forms of 'iconotext' (Cosgrove, 2001), a representation that incorporates both text and graphic images and, potentially, any multisensory gesture that is elicited by the contact between (and with) words and images. Cultural mapping (differently from Duxbury and Garrett-Petts, this volume) is mainly interested in the ways in which maps are used in literature, film, art and other forms of media. The third principle subtending the Lab is creativity. Creativity is about breaking with norms or practices, doing something unexpected or unpredictable, but still satisfying certain constraints. Therefore, it makes explicit the desire to break with norms and conventions often associated with cartography.

What these three directions make clear is that activities framed under the label 'cartographic humanities' are always characterised by openness and unpredictability: cartography, like geography, is not an isolated and normative academic field; it does not belong only to geographers. We all experience space and can represent it in an endless variety of ways.

When reading and viewing the HuMaps Lab projects, the idea of 'cartographic intimacies' was prompted immediately in my mind to refer to the emotional, intimate and personal connections that students developed with space (in a wider, not merely geographical, sense). For this to happen, students were trained to conceive the mapping experience as an infrastructure of feeling. If we rely on the idea that the word 'infrastructure' etymologically means what stands beneath, below or within a structure, suggesting relationships of depth and layering, the concept of maps as infrastructures of feelings hints at the idea that they are powerful mediums, platforms and substrates for capturing the emotional dimensions of human experiences and connections to space. In the cartographic humanities, parallels to this infrastructural understanding of mapping emerge through the words of della Dora in her beautiful book, *The Mantle of the Earth* (2020). Considering the cartographic mantle, della Dora argues that '[i]t naturally directs the gaze to the surface, but it also implies the existence of a hidden depth' (2020: 2) and that 'cartographic mantles give tangible expression to the metaphor and to the complex workings of human society' (2020: 253).

As a tangible metaphor, the map, like infrastructure, can be reasonably conceived—to borrow Larkin's characterisation—as a 'matter that enables the movement of other matter' (2013: 329), thus eliciting movements of different kinds. In fact, post-representational map scholars have often implicitly underlined the *infrastructure-ness* of mapping in the way in which a cartographic substrate may become an enabling meshwork where people, objects, ideas and information move and connect. Addressed from a humanistic stance, maps both are and result from an expanded architecture of texts and textures and are thus affective and effective in generating and organising specific feelings.

The maps produced in the HuMaps Lab followed such a conceptual path, ranging from—but more often overlapping—the map as a figure and a ground, as a visual text and an anchoring texture. In particular, two processes can be enquired by examining the ways in which maps have been used infrastructurally by students: a *mapping out* process where they made public and shareable an intimate, hidden and often traumatic story; and a *mapping in* process in which students made intimate and subjective a topic of social and political relevance. For space reasons, I will focus on the first movement of such cartographic intimacy: the internal-external process of making an intimate space tangible through a mapping experience.

Mapping out through somatopias

As Giuliana Bruno states: 'the surface can contain even our most intimate projections' (2014: 8). This is how the cartographic surface has been interpreted by most students interested in mapping their inner thoughts, problems, emotions, trauma and desires. The *mapping out* process has been chosen to highlight the potentials, often emerged through a reflection of the same limits of the map ontology, that a creative and critical understanding of cartography may bring into the reconceptualisation of the map as a form of bodily intimacy. With topics ranging from psychological and food disorders and experiences of serious illness to more joyful expressions such as oneiric mapping and developmental and growth maps, the category of autobiographic maps has been expressed mainly through metaphorical images related to the body and its organs.

Anthropomorphic maps, also known as 'cartographic caricatures' or 'cartoon maps' (Mangani, 2006), are certainly not new to the history of cartography. Seen as creative and whimsical representations of geographic areas where features of the land, cities and landmarks are transformed into human or animal-like figures, these maps often employ exaggerated or playful illustrations to convey a sense of character, charm and storytelling. The concepts of 'maps as bodies' and 'bodies as maps' are also at the core of feminist theory. For instance, they have been discussed by Lewes (1996) when looking at the practice of depicting female bodies as utopian sexual landscapes. Lewes coined the term 'somatopia', combining two words: *soma*, which refers to the body, and *topos*, which means place or location. Somatopia can be understood as the notion of the body as a place or a spatial entity. Broadly speaking, it involves perceiving and experiencing the body not only as a physical object but also as a site where various experiences, emotions and meanings are materially embodied and expressed.

One of the students, MF, for instance, decided to map out her 'dark period', where she used to practise self-harm, and different ways of coping[1] that she put to work to respond to this stressful condition, by using a somatopia. Playing with the notions of 'geography of the bones' and 'geography of the flesh' (introduced in the first lecture of the geography course), she reflected on how scars can be mapped and can therefore fall within the so-called 'geography of the flesh', or rather, as she writes in her project '*on* the flesh' (see Figure 33.1):

> I started to map my arm, drawing its profile on a sheet where I had previously printed all the words I repeated to myself during my 'dark period'—the same words that led me to punish myself. Then, I traced with a red marker some cuts in the points where the real scars are: I made them 'live' cuts to indicate that those wounds, however closed and healed by now, sometimes still burn. Finally—even if I am not sure one can perceive it well—I wet and then dried the sheet, to symbolise all the tears shed now dried. The arm rests on a blue background, the colour of calm and peace, and [the background] is not uniform, to indicate the various intensities of this serenity. It can be understood both as a sky to which the arm reaches out, but also as a sea on which islands of daisies and words float.
>
> *(MF,* Geographies *on* the flesh, *2023, HuMaps Lab project, author's translation)*

Viewing the body as a place can thus illuminate the impact of trauma and processes of healing. Just as a political map can show areas scarred by conflicts or natural disasters, the *body-as-a-map* can bear the imprints of emotional, physical or psychological wounds. Likewise, mapping the body as a place of healing and resilience emphasises the transformative potential for growth and recovery that pertain to this activity. Many other students inquired into the fine line between the psychic, the somatic and the social by spontaneously transforming other parts of their bodies, such as hands, hearts, brains, sexual organs and faces, into maps, giving them symbolical meaning (see Figure 33.2; see also Brückner, this volume). GC, for example, decided to transform her face into an island to share the idea of isolation (see Figure 33.3). She writes: 'During the geography lesson on the cartographic humanities everyone was sitting next to their friends, and I was alone, as always'. Hence, the idea of representing her face as 'an island or better an iceberg, surrounded by a tide of people united with each other and a sea to divide me from them' (GC, *No Man is an Island, Maybe . . .*, HuMaps Lab project, 2023, author's translation). Negative adjectives referring to herself (e.g. stupid, depressed, ugly, boring) float on the sea. They stand for what she thinks others think of her, coming to the conclusion that they are just a mirror of what she thinks of herself.

Not by chance, the relationship between the map and the face is one of the most featured in the history of cartography. For instance, the Stoke's Capital Mnemomical Globe (1868) is famous for reproducing a terrestrial globe with the features of a human face to facilitate the memorisation of cities, continents and islands (Mangani, 2006). As Deleuze and Guattari remind us, 'the face is a map . . ., the face has a correlate of great importance: the landscape' (1987: 170).

Figure 33.1 MF (2023) *Geographies on the flesh*.
Source: HuMaps Lab

By depicting bodies as the spatial background of an intimate story or biography, these metaphorical contemporary mappings acknowledge the intricate layers of human existence, navigating themes of identity, power, desire, trauma and healing. They recognise the depth and complexity of individual experiences and reflect the ongoing negotiation between personal claims and societal pressures: 'I decided that I was going to shape my subconscious, my incessant anxiety, my dread of other people's opinion', writes GC (2023, *No Man is an Island, Maybe*, author's translation).

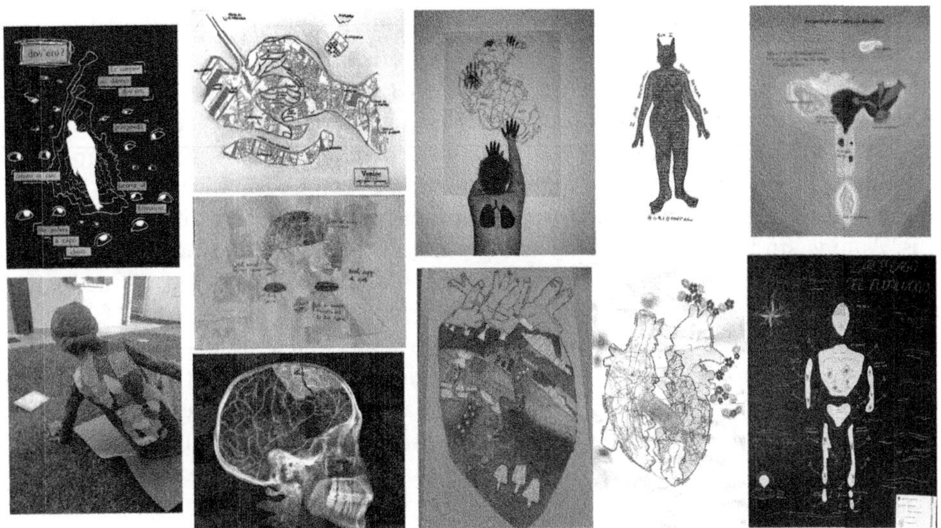

Figure 33.2 A selection of creative cartographies centred around the idea of anthropomorphic maps or somatopias.

Source: HuMaps Lab, 2022 and 2023

Figure 33.3 GC (2023) *No Man Is an Island, Maybe . . . .*

Source: HuMaps Lab

The main consideration that emerged during the mapping out process is that maps can be addressed as soul spaces, and they can be explored through all senses and languages. Cartographic intimacies foregrounded a vivid sense of the surface of emotions. Feelings are movements often ungraspable and invisible over map surfaces, yet, through a combination of words, colours, patterns and materials, they can become visible components of the mapping event. Sense of matter, embodiment, tactility—with their various manifestations in drawing, touch and manipulation—often prevailed over a digital production of maps. Among students, there was a strong opinion that a unique way to humanise—to make emotional or intimate—a map is to work with hand techniques (drawing, painting, collaging, sewing, kneading). MF (*Geographies* on *the flesh*, 2023) used, for instance, artistic techniques that she loved to practise when she was a child—spreading tempera with a sponge, tracing the profile of the arm, cutting out and pasting the various 'islands' in the blue sea—to create a bridge between the past her, still unaware and innocent, and today's her, much more entangled in social constraints. Standing on the idea that GIS, as it is currently made, cannot provide multisensory and affective experiences of inner spaces, some students decided to replace it with image processing software (e.g. Adobe Photoshop, Canvas and Procreate) to reach a similar experience of personalisation of the map.

Overall, the activities have been mainly seen as cathartic and therapeutic. MF, for instance writes: 'This map has certainly been a therapeutic way of facing my past and making it concrete through an artistic expression' (*Geographies* on *the flesh*, 2023, author's translation). ES comments:

> When I was assigned the task, I thought that this might be an opportunity to bare my feelings a little, to communicate, through a humanistic three-dimensional map, what the path I had to face was, done of a thousand obstacles, climbs and prohibitions.
>
> (Cancer: a vortex of emotions, *2023, author's translation)*

GC reflects on the point that 'this project allowed me to explore myself internally through the spatial dimension, opening up new itineraries of the geography of the soul' (*No Man is an Island, Maybe*, 2023, author's translation).

What has been peculiar of this lab is that the expanded architecture of texts, materials and images of which such maps have been made, or became part of, made more explicit the processuality of map production, the story embedded in maps and the constant elicitation of feelings occurring both during their creation and in the re-evocation of the experience.

Conclusion

Emotions are fundamental to human experience, and they play a significant role in shaping the surfaces of individual and collective bodies (Ahmed, 2014), including objects (Turtle, 2014). Likewise, cartographic surfaces can shape and by turn be shaped by subjective and social emotional forces (Rossetto, 2019), making them visible as carnal spaces. Maps affect the way we interact with space and how we perceive our surroundings but also provide an anchor to remember our past experiences, be they traumatic or joyful. As a fabric for a wide range of emotions, they can be used to creatively extend the reach of our reflections, our fears and desires, how we see relationships between ourselves and others, our paths, bringing our world into them.

By drawing or assembling pieces of literary, visual, lyric and video narratives, students have been encouraged to freely experiment with the endless potentialities provided by a cultural and humanistic approach to mapping. The use of body mapping, in particular, has allowed them a more creative and nuanced representation of their life experiences, as opposed to the more traditional, scientific geographical maps. The participants in the lab were also encouraged to think critically about the ways in which map gestures and map images can be used to convey emotional meaning, and to consider both the limits and potentialities for their thinking and feeling in space inherent in these forms of mapping. As they have engaged in this *mapping out* process, the Lab has helped them understand the power of maps as infrastructures of feelings, as emotional intermediaries. Maps, in this context, are not seen as a representation of physical spaces but as a material spatialisation of visceral feelings and stories, one which drives orientations, disorientations, agitation and care for humans, non-humans, memories and future experiences. Drawing (on) cartographic intimacies, the HuMaps Lab has provided a research method and a didactic experiment that may extend the potential reach of emotional mapping, suggesting the idea that maps and mapping can be mutually experienced as personal and social, visceral and surficial processes of healing and care, of getting to know each other profoundly while remaining respectful of a certain distance. It is as if the map becomes a letter, an intimate message which often tells of a feeling of inadequacy, in search of a reader who can identify with what she reads, touches and sees.

Note

1 In psychology, coping is conceived as a response to external stressful/negative events.

References

Ahmed S (2014) *Cultural Politics of Emotion*. Edinburgh: Edinburgh University Press.

Berlant L (1998) Intimacy: A special issue. *Critical Inquiry* 24(2): 281–288.

Bruno G (2014) *Surface. Matters of Aesthetics, Materiality, and Media*. Chicago: Chicago University Press.

Caquard S and Griffin A (2019) Mapping emotional cartography. *Cartographic Perspectives* 91: 4–16.

Casti E (2000) *Reality as Representation: The Semiotics of Cartography and the Generation of Meaning*. Bergamo: Bergamo University Press.

Cosgrove D (2001) *Apollo's Eye: A Cartographic Genealogy of the Earth in the Western Imagination*. Baltimore: Johns Hopkins University Press.

Cosgrove D (2008) Cultural cartography: Maps and mapping in cultural geography. *Annales de Géographie* 660/661(2): 159–178.

Deleuze G and Guattari F (1987) *A Thousand Plateaus: Capitalism and Schizophrenia*, Massumi B (trans.). Minneapolis: University of Minnesota Press.

della Dora V (2020) *The Mantle of the Earth: Genealogies of a Geographical Metaphor*. Chicago: University of Chicago Press.

Larkin B (2013) The politics and poetics of infrastructure. *Annual Review of Anthropology* 42(1): 327–343.

Lewes D (1996) Utopian sexual landscapes: An annotated checklist of British somatopias. *Utopian Studies* 7(2): 167–195.

Luckett T and Bagelman J (2023) Body mapping: Feminist-activist geographies in practice. *Cultural Geographies*. Epub ahead of print 14 June 2023. DOI:10.1177/147447402311794.

Mangani G (2006) *Cartografia Morale: Geografia, Persuasione, Identità*. Modena: Franco Cosimo Panini.

Mitchell D (2000) *Cultural Geography: A Critical Introduction*. Oxford: Blackwell.

Relph E (1976) *Place and Placelessness*. London: Pion.
Rossetto T (2019) *Object-Oriented Cartography: Maps as Things*. London and New York: Routledge.
Sweet E and Escalante S (2015) Bringing bodies into planning: Visceral methods, fear and gender violence. *Urban Studies* 52(10): 1826–1845.
Turtle S (2014) *Evocative Objects: Things We Think With*. Boston: MIT Press.
Wilmott C (2020) Affective mapping. In: Kobayashi A (ed.) *International Encyclopedia of Human Geography*, 2nd edition, vol. 1. Amsterdam: Elsevier, pp. 53–60.

34

AUTO-CARTOGRAPHY

(Fictional) ethnographies of the self and the map in the field

Giada Peterle

Introduction: the map and the researcher

There is a deliberate ambiguity in the title of this chapter: whereas bringing cartography and autoethnography together may inevitably evoke a reflection about the manifold interconnections between the map and the *self*, the subject to which the auto-cartographic reflection is referred to is not explicit. Indeed, the *self* that is narrated and examined in auto-cartography could be either the researcher(s) producing, analysing, using and engaging with the map, or the map itself. Even the speaking-I—or the narrator—of autoethnographic accounts may be different, from time to time, especially when we use fiction to explore non-human narration rather than our own anthropocentric points of view. Therefore, the question about what *self* we come to know better through the practice of auto-cartography and carto-fictional writing remains deliberately vague: in fact, the use of fiction may become an expedient to explore both the researcher-I, as an emerging subject that influences and is influenced by the encounter with maps in the field, and the map itself, as an emerging process that needs to be explored through stories. Letting a map speak for itself through fiction may be an option to discover its characteristics (and character!), coming to know the process of map production and reception from an original angle; it could be also an expedient to de-centralise the human gaze of researchers and read their own research practice, positionality and attitude towards maps in the field through a more-than-human lens. Following recent studies in emergent cartography, the practice of considering maps as narrators is an opportunity to explore them as (narrative) mapping processes. 'Maps have stories to tell and even autobiographical accounts to disclose' (Peterle, 2019b: 76), and attention towards the life of the map should be paid, especially by those researchers who intend the map as an emerging political, social and emotional process (Boria and Rossetto, 2017; Oliver, 2016). As Rossetto affirms (2019: 125), this idea of 'attending to the "biographical qualities of maps" or following cartographic artefacts through history' moves beyond mere archival interest in the provenience and dating of maps and helps in considering the social aspects of their lives and afterlives (see Brückner, this volume). This chapter proposes a specific form of auto-cartography as a creative research practice that enables us to understand maps as processes through stories and to explore how they impact *our self*, influencing both our research practices and our lives as researchers.

 DOI: 10.4324/9781003327578-41

Mapping the self

Any map is a potential series of stories that we actualise by telling or reading them: the 'map is a story and as a map reader we become a very important actor in the interaction with the map' (Field, 2014: 99), since we are actualising the narrative potential of a cartographic representation. According to Field (2014: 100), a narrative approach to maps considers 'the relationship between maps and the stories they tell' and, importantly, also 'the way in which we tell those stories through the map-making process'. Scholars in narrative cartography (Caquard, 2013; Engberg-Pedersen, 2017; Ryan, 2020) have extensively discussed how maps are used to inspire writers and analyse their literary works. They have also shown that there is a wide spectrum of interconnections between stories and maps that move beyond the subdiscipline of literary cartography. '"Story maps", "fictional cartography", "narrative atlas" and "geospatial storytelling" are some of the terms that characterise the growing interest in the relationship between maps and narratives' writes Caquard (2013: 135) in a literary review of narrative cartography. Whereas literary scholars interested in cartographic representations have been asking '[h]ow to map paths and routes of fictional characters through fictional space?' (Piatti and Hurni, 2011: 220), meanwhile, questions about how we can either map stories or use StoryMaps, digital cartography and cartographic design for storytelling have become crucial in the field (Caquard and Fiset, 2014; Mocnik and Fairbairn, 2018; Roth, 2021; Saxinger et al., 2021). This means that a narrative approach to maps has moved beyond the mere analysis of their interconnections with literary texts to consider a wider set of cartographic narratives and storytelling practices that can be either artistic or vernacular, individual or collective, digital or analogue.

Practices that use maps to tell stories have become popular activities, and 'individuals and communities are now using the online versions of maps to locate and trace their own stories' (Caquard and Cartwright, 2014: 102). This set of emerging vernacular practices shows that, somehow, 'the power of maps to stimulate and support narrative processes' (Caquard and Cartwright, 2014: 104) has already been used as a tool to visualise and reconstruct identities. Caquard and Cartwright further highlight that observing maps from a processual positioning 'emphasizes the importance of taking into account both the production and the consumption of the map instead of focusing on the map as a representation' (2014: 104). Whereas a 'narrative of mapping' focuses on the stories we tell about maps as processes, fewer stories have been told about the maps we encounter or produce during fieldwork and how they influence the research path and our positionality in the field. How do stories of research and maps interact? What do these carto-centred stories tell about ourselves, in the field, and about our research processes? Not only is the map an (auto)ethnographic object that has its own stories to tell; any map that is part of a research has something to say about the process of doing fieldwork and about the researcher. Therefore, any map has stories to tell that let autoethnographic reflections about how we inhabit the field emerge.

A reflection about the interconnections between cartography and autoethnography inevitably refers to the 'four senses' in which any map 'is biography' as mentioned in a famous piece by Harley (1987: 18). According to Harley, any 'map encompasses not so much a topography as an autobiography', and retracing our 'steps across the map is far more than a sentimental journey' (1987: 20). Therefore, this chapter explores auto-cartographic writing as an opportunity to promote reflexivity; it suggests focusing on how we '*commune with* the maps we collect' (Harley, 1987: 20), encounter, navigate, analyse, sketch or scratch during fieldwork. This may turn the process of research into a narrative mapping practice with accounts of how different maps became useful and meaningful to us (Besio, 2017: 192).

A carto-centred autoethnography

Autoethnography 'is becoming an increasingly common research and representational orientation in the social sciences and humanities' (Butz and Besio, 2009: 1660). This sort of self-narrative, often but not necessarily written in first person, may take a variety of forms (Adams et al., 2017; Holman Jones et al., 2016) and emerge from different speaking positions: for example, Ellis and Bochner (2000) mention 'short stories, poetry, fiction, novels, photographic essays, personal essays, fragmented and layered writing, and social science prose' (Ellis and Bochner, 2000: 739), and we could add to this long list research-comics aimed at sketching the self during the research process (Kuttner et al., 2021). Besides this wide range of genres and forms, there are some common features that allow individuating autoethnographic writing as an autobiographical research method that explores different layers of consciousness bringing the personal, the cultural and the public together (Ellis and Bochner, 2000: 739). 'The *auto* in this use of autoethnography', writes Besio (2017: 191), 'refers to the ethnographising of the researcher', who becomes both the subject and the object of ethnographic research. According to Butz and Besio (2009: 1660):

> [A]utoethnography may be understood as the practice of doing this identity work self-consciously, or deliberately, in order to understand or represent some worldly phenomenon that exceeds the self; it is 'a form of self-narrative that places the self within a social context'.

Autoethnography is never just a self-referred process; rather, it implies a storytelling practice that is also addressed to others. By sharing (our) stories, we come to know ourselves better but also give others the opportunity to do the same, knowing our positionality: this is particularly relevant when thinking of academic research as an ongoing, shared and dialogical process of knowledge production. As Besio (2017: 191) writes, autoethnography conveys something about the researcher's personality and positionality 'that often goes unsaid in academic writing, but that is important to how *we* research'. This kind of reflexivity is not just an opportunity to map our presence within the field but provides researchers with a relevant analytical tool to understand interactions between the spatial, the social, the public and the intimate when conducting carto-centred ethnographic research. Autoethnography is not just a set of manifold autobiographical research writings but rather a 'sensitivity to the autoethnographic characteristics of what we learn from research participants as well as to our own situatedness in relation to the people and worlds we are studying' (Butz, 2009: 140). This autoethnographic sensibility, Butz and Besio explain (2009: 1664), recognises that clear distinctions 'among researchers, research subjects and the objects of research are illusory, and that what we call the research field occupies a space between these overlapping categories'.

According to DeLyser (2015: 211), 'an autoethnographic sensibility encourages a research practice critically aware'. For example, Sullivan (2015: 124) describes his 'auto-geography' as a form of telling of the story of the self that is hooked onto place instead of time. Sullivan, thus, reflects on the meaning of the word 'auto-geography', affirming that he wanted 'to supply a term that other geographers could use, for geographers have a certain "unease with the term biography," perhaps because that term evades the centrality of place in the telling of our stories' (2015: 125). Feminist geographers have invited researchers to make themselves deliberately visible in their research writings, contrasting the idea of the 'view from above, from nowhere' (Haraway, 1988: 589) with a situated subjectivity, whose characteristics, fragilities, biases, body and affective relations influence the research process. Writing the self through place, as Sullivan invites us to do, may be an autoethnographic practice to train our reflexivity and understand ourselves in relation to space. What is particularly intriguing, from a cartographic humanistic perspective, is the possibility to write ourselves in relation to maps in order to train our cartographic reflexivity. An 'auto-cartography' may both trace our own biography through a map as well as narrativise our processual engagements with mapping practices, letting emotional, affective, relational and bodily aspects emerge (see Lo Presti, this volume; Olmedo et al., this volume). The result is a kind of carto-centred autoethnography, a practice of autoethnographic writing that is focused on maps as processes and on the ways in which we both influence and are influenced by the maps we create and encounter. This writing practice may be a means to explore the different stages of the mapping process, indeed, to follow the map's life and afterlives, thus placing the map at the centre of auto-cartography writing; otherwise, carto-centred autoethnography may place the researcher at the centre of the writing practice, exploring the interactions between the map, the self and the field. In this section, we have seen how maps stimulate carto-centred stories to emerge as well as how carto-centred stories, and the practices of writing and reading them, can 'have the power to make authors and readers aware of their everyday engagement with maps as embodied and affected, subjective and relational mapping experiences' (Peterle, 2019a: 1076). The next section introduces empirical examples of auto-cartographic and carto-fictional writing.

Experimenting with auto-cartography

Auto-cartographic interlude: ethnographising the researcher

It was no later than January 23 when I entered the Visit Turku's Tourist Information Point, close by the Market Square in the city centre. It was one of the first days of my two-week-long stay in the city, and precisely between day 2 and day 4. I cannot be more precise than this because my fieldwork diary is now somewhere where I cannot find it. It was January 2020. Neither myself nor the young guy behind the desk knew anything about our near dystopic pandemic future, though. I was in search of a map for my visual archive. More precisely, a map of the public transport system. I was in Turku to write a comic story about the golden age of the tramway and needed to collect as many stories and visual materials about the history of public transport in the city as possible. They would have helped me to compare the current bus routes with the former tramlines, to see where they virtually overlapped in the urbanscape of Turku. The guy provided me with a very traditional 'kartta' (map) of the region: on one side, the city centre; on the other side, the wider Turku region, with the centres of Kaarina, Raisio, Naantali, Paimio and the Turku archipelago. The map was full of ads, and the sponsors—bars, restaurants, bookshops—were accurately mapped on it. Yet there was no information about the public transport network except for a yellow rectangle in the bottom right corner, saying 'FÖLI. Menopeli kaikilla herkuilla'. I googled it: 'go play with all the goodies!' it said. The goodies were 'bussit' (buses), 'föllärit' (bikes), and 'vesibussit' (water buses). Unfortunately, there have been no tramlines available in Turku since 1972. Yet the map helped me more than I expected. According to it, the Föli office was no more than 100 metres from Visit Turku. With the kartta in my hand, I greeted the nice guy and walked out the door, pretty sure I would find the right map soon.

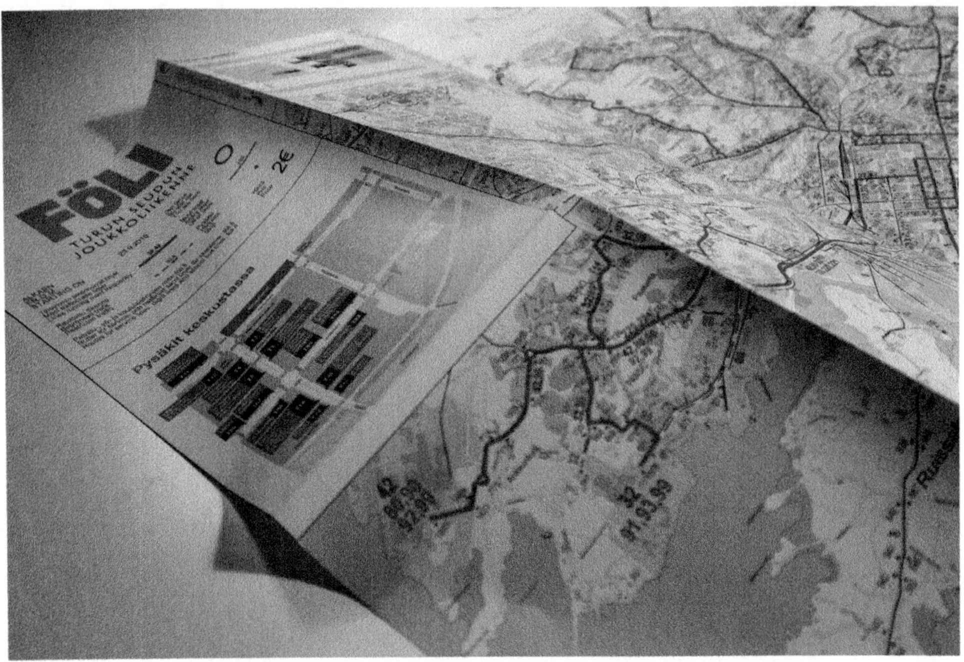

Figure 34.1 Föli map of public transport in the Turku region, Finland, January 2020.

Source: Photograph: Giada Peterle

Carto-fictional interlude: a Föli map's life and afterlife

Apparently, there's not much to say about me (Figure 34.1). I'm a banal map of public transport in the city region of Turku, in southwestern Finland. I was printed by Föli around the end of September 2019 and sold for a pretty low price. Only two euros to take me home, touch me, unfold me, use me, maybe even sketch on me, or scratch me in your pocket, forget me in your backpack. Me and my kind, we usually don't have any high expectations about our lives and afterlives. Especially since when people have been using digital maps and apps to check the timetables and catch the bus. In most cases, when you're a map like me, you're going to be in the hands of a tourist for a few days and then thrown away once the holiday is over. If you're lucky enough to be bought by a map enthusiast and collector, you're going to be stored in a dusty drawer. Maybe they'll help you to stretch your tired limbs, from time to time, unfolding you just to see the colours on your skin slowly fading away as time passes by. So, compared to these quite low expectations, mine was a kind of unexpectedly glorious destiny. I knew it was my turn, in the central Föli office in Turku, as soon as the map which had been my predecessor in the stack for almost a month was sold to an old lady from Tampere coming to visit her son. What I couldn't know, though, was that a young researcher would turn me into an elicitation tool during interviews, as a source of inspiration for her research-comic about public transport (Figure 34.2). I admit, I'm pretty photogenic, with my soft white and light blue colours on the background and the complex network of bright red lines tracing the public transport network. Yet, never have I dreamed of seeing my portrait published in academic books and papers, even in a

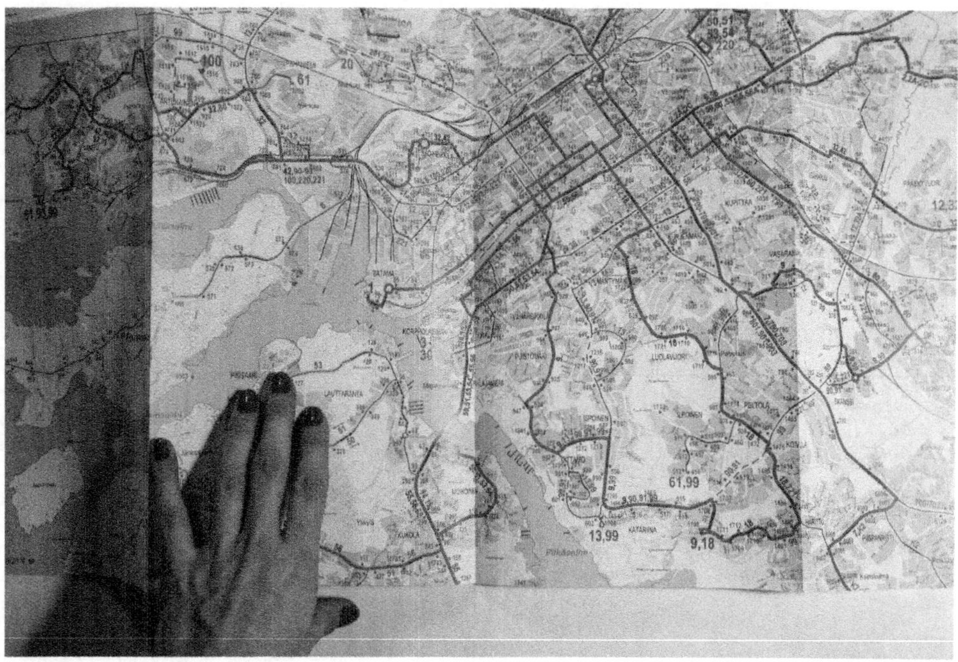

Figure 34.2 Föli map unfolded on the floor of a small, one-room flat in the city of Turku, Finland, in January 2020.

Source: Photograph: Giada Peterle

comic book, which, I am told, is now in the hands of people outside Finland. Well, even if my representations are incredibly mobile, my material present is not that glorious. It's been a while since the researcher took me out of this fieldwork diary to unfold me. Sometimes I've the impression she doesn't know where I am.

Conclusions: writing (with) maps

The interludes presented in the previous paragraphs provide examples of engaging with carto-centred, autoethnographic and fictional writing to let cartographic thinking emerge but also to narrativise our research process through maps. These auto-cartographies help to ethnographise myself and the mapping practices I engaged with while conducting ethnographic research in Turku in January 2020, with the aim of composing a comic book that was later published (in 2021) with the title *Lines. Moving with Stories of Public Transport in Turku*. Maps became, from the very beginning, an elicitation tool during interviews, a means to trace walking routes across the city as well as visual inspirations for the composition of the narrative structure of the comic book. These *interludes* show the relevance of maps but also let the emotional aspects of fieldwork research emerge through carto-centred writing. Moreover, through irony, they show limits and unexpected aspects related to ethnographic research: as, for example, the loss of the fieldwork diary, the initially unsuccessful search for the public transport map, and the unpredicted role of the Föli map.

Having recognised the intrinsic 'power of maps to stimulate and support narrative process' (Caquard and Cartwright, 2014: 104), auto-cartographies can become generative means 'of autoethnographic reflections on subjective mapping experiences and cartographic emotions' (Peterle, 2019b: 1073). This narrative approach to ethnographic research and fieldwork stems from the assumption that 'at its heart, research is storytelling' (Christensen, 2012: 232); but also that we can use stories to make research processes more accessible. Auto-cartography, in this context, is an invitation to think of research as storytelling through an autoethnographic and carto-centred perspective which promotes reflexivity and pays attention to the manifold ways in which maps and mapping practices become part of our research. Carto-fiction is a further opportunity for both mapping the self and writing (with) maps, letting emotional and subjective cartographies emerge.

In recent years, fiction has become one of the methodological tools used for qualitative research in the social sciences (Sjöberg, 2008): it is now considered 'a valid stylistic writing choice for presenting ethnographic research findings' (VanSlyke-Briggs, 2009: 344); for engaging readers more thoroughly and reaching wider audiences; for challenging the researcher's point of view through a multivocal perspective; and for avoiding the presentation of research findings as assertive truth. Carto-fiction is an epistemological tool (Peterle, 2019a: 1072) that changes the way we both present (tell) and conduct (embody, design, experience) research 'on' and 'with' maps. Indeed, as Leavy (2016: 20) affirms, 'the practice of writing and reading fiction allows us to access imaginary or possible worlds, to reexamine the worlds we live in, and to enter into the psychological processes that motivate people and the social worlds that shape them'. Fiction is an alternative mode of theorising and narrativising maps as complex processes that are related to emotions and bodily experiences, which often remain untold in more traditional research accounts. According to Caquard and Griffin (2018: 14), 'maps are somehow shy' and, for this reason, '[t]hey tend to hide their emotional side behind their clear lines, precise points, minimalistic words, numerical data and informative purpose. But when we scratch the cartographic surface, maps appear to be impregnated with all sorts of emotions'. Likewise, research accounts often tend to hide the cartographic emotions we experience during fieldwork. Auto-cartographies have the power to make us, as researchers, and our readers aware of our everyday engagements with maps as embodied, affected, subjective and relational mapping practices. Auto-cartography suggests the exploration of fictional, autoethnographic and creative writing practices as methods of research in the field of cartographic theory: this encounter between cartography, autoethnography and fictional writing could let the affective, emotional and embodied aspects of maps and mappings emerge (Peterle, 2019a: 1076), making them visible through stories.

References

Adams TE, Ellis C and Holman Jones S (2017) Autoethnography. In: Matthes J, Davis C and Potter R (eds) *The International Encyclopedia of Communication Research Methods*. Hoboken: Wiley, pp. 1–11.

Besio K (2017) Hiding in the garden: Autoethnography and intimate spaces. In: Moss P and Donovan C (eds) *Writing Intimacy into Feminist Geography*. Abingdon: Routledge, pp. 188–202.

Boria E and Rossetto T (2017) The practice of mapmaking: Bridging the gap between critical/textual and ethnographical research methods. *Cartographica* 52(1): 32–48.

Butz D (2009) Autoethnography as sensibility. In: DeLyser D, Aitken S, Herbert S, Crang M and McDowell L (eds) *Handbook of Qualitative Geography*. London: Sage, pp. 138–155.

Butz D and Besio K (2009) Autoethnography. *Geography Compass* 3(5): 1660–1674.
Caquard A and Griffin A (2018) Mapping emotional cartography. *Cartographic Perspectives* 91: 4–16.
Caquard S (2013) Cartography I: Mapping narrative cartography. *Progress in Human Geography* 37(1): 135–144.
Caquard S and Cartwright W (2014) Narrative cartography: From mapping stories to the narrative of maps and mapping. *The Cartographic Journal* 51(2): 101–106.
Caquard S and Fiset JP (2014) How can we map stories? A cybercartographic application for narrative cartography. *Journal of Maps* 10(1): 18–25.
Christensen J (2012) Telling stories: Exploring research storytelling as a meaningful approach to knowledge mobilization with Indigenous research collaborators and diverse audiences in community-based participatory research. *The Canadian Geographer/Le Géographe Canadien* 56(2): 231–242.
DeLyser D (2015) Collecting, kitsch and the intimate geographies of social memory: A story of archival autoethnography. *Transactions of the Institute of British Geographers* 40(2): 209–222.
Ellis C and Bochner AP (2000) Autoethnography, personal narrative, reflexivity: Researcher as subject. In: Denzin N and Lincoln YS (eds) *The Handbook of Qualitative Research*. Thousand Oaks, CA: Sage, pp. 733–768.
Engberg-Pedersen A (ed.) (2017) *Literature and Cartography: Theories, Histories, Genres*. Cambridge, MA: MIT Press.
Field K (2014) The stories maps tell. *The Cartographic Journal* 51(2): 99–100.
Haraway D (1988) Situated knowledges: The science question in feminism and the privilege of partial perspective. *Feminist Studies* 14(3): 575–599.
Harley JB (1987) The map as biography: Thoughts on ordnance survey, six-inch sheet Devonshire CIX, SE, Newton Abbot. *The Map Collector* 41: 18–20.
Holman Jones S, Adams TE and Ellis C (eds) (2016) *Handbook of Autoethnography*. Abingdon: Routledge.
Kuttner PJ, Weaver-Hightower MB and Sousanis N (2021) Comics-based research: The affordances of comics for research across disciplines. *Qualitative Research* 21(2): 195–214.
Leavy P (2016) *Fiction as Research Practice: Short Stories, Novellas, and Novels*. London: Routledge.
Mocnik FB and Fairbairn D (2018) Maps telling stories? *The Cartographic Journal* 55(1): 36–57.
Oliver J (2016) On mapping and its afterlife: Unfolding landscapes in Northwestern North America. *World Archaeology* 43(1): 66–85.
Peterle G (2019a) Carto-fiction: Narrativising maps through creative writing. *Social & Cultural Geography* 20(8): 1070–1093.
Peterle G (2019b) Story of a mapping process. The origin, design and afterlives of the *Street Geography* map. *J-READING Journal of Research and Didactics in Geography* 2: 73–87.
Piatti B and Hurni L (2011) Cartographies of fictional worlds. *Cartographic Journal* 48(4): 128–223.
Rossetto T (2019) *Object-Oriented Cartography: Maps as Things*. London and New York: Routledge.
Roth RE (2021) Cartographic design as visual storytelling: Synthesis and review of map-based narratives, genres, and tropes. *The Cartographic Journal 58*(1): 83–114.
Ryan ML (2020) Narrative cartography. In: Richardson D, Castree N, Goodchild MF, Kobayashi A, Liu W and Marston RA (eds) *International Encyclopedia of Geography*. Hoboken: Wylie, pp. 1–8.
Saxinger G, Reinoso AS and Wentzel SI (2021) Cartographic storytelling: Reflecting on maps through an ethnographic application in Siberia. *Fennia-International Journal of Geography* 199(2): 242–259.
Sjöberg J (2008) Ethnofiction: Drama as a creative research practice in ethnographic film. *Journal of Media Practice* 9(3): 229–242.
Sullivan R (2015) Three sections from the auto-geography of Rob Sullivan. *GeoHumanities* 1(1): 124–130.
VanSlyke-Briggs K (2009) Consider ethnofiction. *Ethnography and Education* 4(3): 335–345.

35

RE-SITUATING PARTICIPATORY CULTURAL MAPPING AS COMMUNITY-CENTRED WORK

Nancy Duxbury and W. F. Garrett-Petts

What is participatory cultural mapping?

Participatory cultural mapping is rooted in practices of community engagement and collaboration, working to make visible and co-produce knowledge that is of value for community identity formation, reflection, decision-making, advocacy and development. As historian Jo Guldi (2017: 80) records, the 'first recognizably participatory maps . . . emerged [as] embedded in global social movements where writers and activists stressed a variety of [graphic] tools that social activists could use'. By the 1970s, the emphasis, especially in North America, moved from social advocacy to a more pragmatic social and municipal planning agenda. Still, the promise of individual and collective empowerment—of giving voice to the many through mapping—persists.

Today such mapping aims to recognise and make visible the ways local stories, practices, relationships, memories, rituals and physical elements constitute places as meaningful locations. It embodies knowledge-building processes and comprises a platform for sharing and dialogue. On the one hand, cultural mapping is a highly pragmatic knowledge-building process of 'collecting, recording, analyzing and synthesizing information in order to describe the cultural resources, networks, links and patterns of usage of a given community or group' (Stewart, 2007: 8). On the other hand, it is also a humanistic conversational platform and meeting place for discussion, sharing and learning, which is intentionally used as a methodological tool to bring a diverse range of stakeholders into conversation about the cultural dimensions and potentials of a place. Furthermore, there is growing recognition of the importance of participatory cultural mapping as a platform in which knowledge is not only *shared* but also *co-created* in conversation (Duxbury, 2022). In other words, beyond the value of making visible/documenting individual experiences and knowledges, it is extending and building new collective knowledge in the process.

The phenomenon of cultural mapping has gained extraordinary international currency during the last 30 years as an instrument of communal expression, empowerment, intercultural dialogue and community building (Abrams and Hall, 2006; Bryan, 2011; Caquard, 2013; Crawhall, 2007; Gerlach, 2010, 2014; Guldi, 2017; Hunter, 2019; Kerski, 2014; Kitchin and Dodge, 2007; Roth, 2009). Participatory cultural mapping seeks to combine

 DOI: 10.4324/9781003327578-42

the tools and techniques of cartography with vernacular and participatory methods of storytelling to represent spatially, visually and textually the 'authentic' knowledge and memories of local communities. It is a social practice that invites multiple forms and modes of non-specialised vernacular discourse—from Indigenous communities, locals, those with lived/living experience, peers and those from not-for-profits and grassroots organisations representing multi-sectoral viewpoints—into the public sphere of community identity formation, political and social advocacy, local knowledge production, municipal planning, cultural sustainability planning, participatory decision-making and community engagement.

Participatory cultural maps take many forms, with choices ranging from the simple spatial arrangement of post-it notes on a flip chart, to mind map diagrams, to photo-voice exhibitions, to cultural asset maps, to group discussions and survey responses documented through graphic facilitation or web-based inventories, to detailed hand-drawn renderings of places and experiences and journeys, to multimedia compendia and even works of art (Cochrane et al., 2014; Corbett et al., 2016; Duxbury et al., 2019; Stewart, 2007). Of these choices, it is the hand-drawn sketch maps, journey maps and story maps that are becoming increasingly recognised as rich cultural texts redolent with individual experience and especially worthy of greater attention from both scholars and practitioners (e.g., see Crawhall, 2007; Duxbury et al., 2015; Pillai, 2013; Poole, 2003; Roberts, 2012; Sletto, 2009; Strang, 2010).

The potential of map creation implies possibilities for 'moving ideas into the world whether through representations of data or platforms for imagining' (Longley in Duxbury et al., 2019: 1). Cultural mapping projects create spaces and processes for collaborative research, learning, visualising, dreaming and community action. They provide opportunities to critically examine the past, assess the present, examine representations, make connections, address absences and envision continuities and change into the future. As Rike Sitas (2020: 16–17) points out, participatory cultural mapping and related initiatives can play a key role in inclusive urban planning:

> To leverage culture and heritage for more just cities, pluralistic narratives that link fundamentally to places and people's lives are critical. These stories exist and are always in the making but need avenues through which to be surfaced. . . . These narratives help shift our social imagination—the capacity to imagine alternative future worlds. . . . Liberating culture, heritage and the imagination from rigid frames also opens up ways of thinking spatially and temporally . . . and this can foster the ability to speculate for more fantastical futures.

In this way, participatory cultural mapping is much more than assembling information, and its role in creating participatory platforms for sharing, discussion, thinking together and imagining future possibilities forms an essential dimension.

Extending from this, participatory mapping can be viewed as a mechanism to foster citizen-led interventions and democratic governance, based on processes that spearhead new modes of participatory interaction with citizens (Ortega Nuere and Bayón, 2015). However, in most situations, participatory cultural mapping tends to be employed as a one-time initiative, a project rather than a long-term strategy, and thus typically remains not fully articulated or integrated within community planning and development practices (Duxbury, 2019; Garrett-Petts et al., 2021; Garrett-Petts and Gladu, 2021). If cultural mapping is to become sustainable and transformative, community engagement must be based on partnerships that are more than merely transactional—that is, focused on more than operational tasks and fulfilment of short-term expectations (Duxbury and Garrett-Petts, forthcoming).

In the face of rapidly changing societies and diversifying forms of social exclusion, new approaches to citizen empowerment, citizen participation and social inclusion require ideas, knowledge(s), experiences, resources and capacities that are (dis)located across an array of arenas and distributed among different actors. Participatory cultural mapping processes and participant-generated cultural maps assert that 'local inhabitants possess expert knowledge of their environments and can effectively represent a socially or culturally distinct understanding of the territory that includes information excluded from mainstream or official maps' (Duxbury and Garrett-Petts, forthcoming, n.p.). This approach is akin to counter-mapping, open to diverse perspectives, knowledges and ways of expressing, and to bringing these perspectives and knowledges into a public sphere. It foregrounds the importance of building cartographic literacy within communities—as is the focus of many counter-mapping and Indigenous mapping (see Pareira and Sletto, this book) initiatives in recent years. It acknowledges that the process of making implicit knowledge explicit and mobilising the symbolic forms through which local residents understand and communicate their sense of place, also have ethical and political dimensions.

Contextualising: community–academe collaboration

In contemporary academe, aspirations to 'co-create' knowledge with communities are heightening and becoming more visible, but we also observe resistances to fully embrace the challenges and implications embodied in meaningful community–academe collaboration. Participatory cultural mapping as generally practised is also a highly mediated activity, inevitably informed by learned disciplinary assumptions and practices, and scaffolded by those with expert technical knowledge about community-based mapping processes.

Cultural mapping typically involves scholars and activists, cartographers and GIS specialists, workers from development organisations, local researchers, NGO volunteers, municipal workers, hired consultants, or others working with the community members in workshop settings or in the field, gathering and organising information. While both the local participants and the researchers/facilitators share many aims, there are nonetheless differences at play. The voices and individual perspectives embedded in the maps are easily paraphrased or otherwise subsumed by a more authoritative, synthesising and official discourse—ironically, a discourse driven by the desire to faithfully celebrate, preserve and learn from local voices (Garrett-Petts, 2016; Garrett-Petts and Karsten, 2019). At the end of the day, the maps and the processes that produced them become data available for expert collation, analysis, interpretation and re-representation; and, as might be expected in any emerging field, proponents of cultural mapping defer to what they already know, invoking processes and disciplinary practices which lend themselves more to the collection and analysis of the tangible (as opposed to the intangible) elements of local culture.

Lacking an agreed-upon and informing theory of participation, cultural mapping has proven at best inconsistent in its efforts to represent local individual voices in all their dimensions. As Robin Roth notes (2009: 207), 'Community-based mapping, despite, or perhaps because of, its popularity, is . . . recognized as having unintended effects on rural communities'. Drawing on the work of Jefferson Fox, who has documented what he refers to as the 'ironic' effects of mapping, Roth cites increased conflict, increased privatisation of land, loss of Indigenous conceptions of space and increased regulation by the state as examples of 'the potential epistemic violence associated with counter-mapping and the entanglements of power that can shape mapping projects in unfortunate ways' (2009: 207). Roth notes further as follows:

> The unintended effects of mapping . . . stem from the dominant conception of space that frames and guides the cartographic representations of indigenous territories. They are . . . an outcome of rendering a complex spatiality into abstract space; allowing the 'more-than-abstract' spatial practices to go unrepresented. Community-based mapping using abstract space insists upon a singular representation of indigenous territory in a way legible to the state; insists upon fitting 'indigenous people into the spatial configurations of modern politics'.
>
> *(2009: 207)*

A chorus of similar concerns has been expressed recently and mainly by academics, in particular cultural geographers and anthropologists, who urge consideration of 'unintended' or 'unforeseen' impacts of mapping, especially mapping of Indigenous cultures, that gloss over or minimise local difference, including identity formation, histories and the lived experience of personal landscapes (Abrams and Hall, 2006; Hale, 2005; Hodgson and Schroeder, 2002; Peluso, 1995; Sletto, 2012). As Brenda Parker (2006: 470) argues, while participatory mapping aspires to the values of 'inclusion, transparency, and empowerment', questions nonetheless remain regarding the maps' composition, 'how they should be evaluated, and the relationship between community maps and power'. She concludes:

> How mapmakers think through and engage ideas of empowerment and for whom it is envisioned and occurs need to be better understood. Furthermore, these projects need to be understood in relation to extant power relations and possibilities for social change.
>
> *(2006: 479–480)*

Furthermore, among those encouraging participatory democracy, championing local self-government and the right to the city, citizen participation and the techniques of cultural and community mapping are increasingly viewed through a cautionary lens:

> *consultations with the public make a lot of sense, and this can take the form of surveys, focus groups, open houses, community mapping, community visioning, and much else. The techniques are now quite sophisticated. The key thing is that the authority to make the final decisions—and decide what sort of consultations are to occur—remains where it always was. As Leonard Cohen might have put it, 'everybody knows' that the consultations are meant to help the authorities, not displace them. So, if you don't want to be a helper—like a little child in the kitchen—the attractions of taking part are not very great. Is it surprising that people refuse to participate?*
>
> *(Magnusson, 2015: 58; italics in the original)*

All these doubts and concerns inevitably raise questions about the broader implications of initiatives aiming to democratise knowledge through meaningful community-engaged processes.

Questions about the nature of individual participation and the depth of engagement in the public sphere are not new: since the 1960s, and especially with the publication of Sherry Arnstein's 1969 manifesto, 'A Ladder of Citizen Participation', planners, community organisers, social activists, artists and others have been categorising public participation on a spectrum of involvement. Arnstein references eight degrees of participation, steps on a ladder leading from 'manipulation' to full 'citizen control'—from informing to empowering. Anticipating the current academic commitments to inclusive excellence, including meaningful collaboration and co-creation involving both expert and lay knowledge, Arnstein makes us more aware of how context, power relations, social justice and motives influence the degrees of participation possible, allowing for conditions of no citizen power, counterfeit power and actual power. How, then, might participatory cultural mapping fulfil its potential as a field of inquiry dedicated to principles of empowerment, equitable inclusion and collaboration, co-creation and validation of diverse vernacular cartographies in the public sphere?

Re-situating and positioning participatory cultural mapping

In a sense, just asking such questions seems a critical first step. There is no doubt that participatory cultural mapping remains a valuable, relevant and increasingly deployed qualitative method for cultural inquiry; and critical self-reflection evidenced among its theorists and practitioners seems poised to provide the theory, history and examples requisite for the further development of participatory cultural mapping as an evolving field of study and practice. The work of scholars like Guldi, Roth, Gerlach and others makes us more acutely aware of the field's core principles, and the importance of using those principles to inform the mapping guides, handbooks, practices and assumptions employed.

In addition, and crucially, we need to re-situate and position participatory cultural mapping within the larger field of cultural mapping generally. The contemporary roots of cultural mapping intertwine academic and artistic research with policy, planning and advocacy contexts. The field's current methodological contours have been informed by six main cultural mapping trajectories: (1) community empowerment and counter-mapping, (2) cultural policy, (3) cultural planning and municipal governance, (4) mapping as artistic practice, (5) academic inquiry and (6) literary, music and film mapping (see Figure 35.1). The complexity, strength and vitality of cultural mapping arise through interconnecting these perspectives, sources of knowledge, approaches and methods, and trajectories of work. As we explore in detail elsewhere (Duxbury and Garrett-Petts, forthcoming):

> While these trajectories can be distinguished in terms of their relative emphasis on the instrumental or the immediately pragmatic, they inevitably overlap, as suggested by the involvement of artists or social activists or academics in counter-mapping, cultural policy, and municipal cultural mapping initiatives. At the same time, each trajectory establishes a definable rhetorical purpose for mapping from the ground up. For example, the public documentation of land claims, the public representation of authentic cultural resources and traditions, the public inventorying of tangible and intangible cultural assets, the public and private deployment of cartographic techniques and sensibilities for aesthetic practices, or the public and ongoing interrogation of the visual and spatial turns in disciplinary research. The common challenge

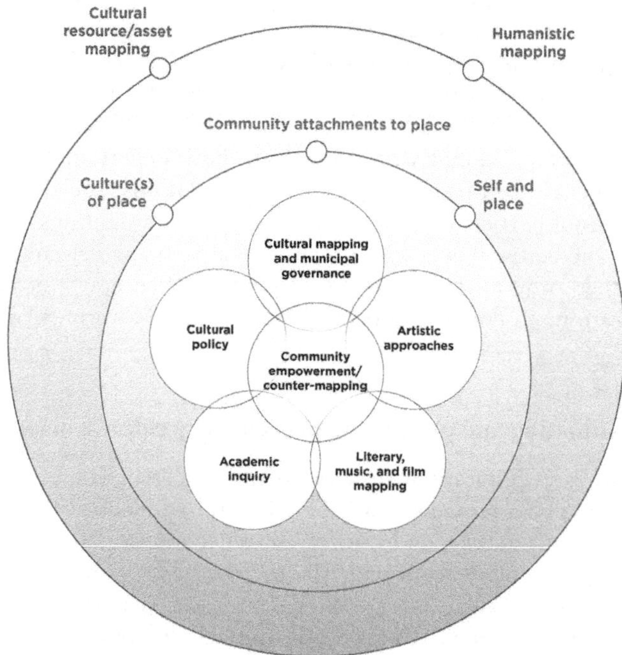

Figure 35.1 Organising the field of cultural mapping: key branches and orientations.
Source: Created by the authors

> for cultural mapping in each context is the garnering of *deep community involvement* and the *affirmation of local knowledge*.
>
> (*n.p.*)

Given this diversity of motives and approaches, it is helpful to provide a bird's eye view of the field in two organising schemas: First, evolution along *two main branches*, corresponding to 'ideal types': (1) cultural resource/asset mapping and (2) 'humanistic' mapping approaches (Freitas, 2016). Second, *three general orientations* of cultural mapping projects, according to the purpose, context and thematic focal point of a project. This general organisation, we believe, has influenced the way the field sees and defines itself, as well as the ways in which efforts are applied to advance methodological practices in each of these areas.

Branches of cultural mapping

Developing since the 1960s, *cultural resource/asset mapping* seeks to identify and document tangible and intangible assets of a place, usually in order to develop and incorporate culture and creative industries in strategies that address broader issues of a locale. In this context, a general distinction has often been made between *asset mapping* (physical or tangible cultural assets, such as cultural venues, public art works, historic sites, monuments and identifiable organisations and persons) and *identity mapping* (intangible elements—both historical and contemporary—that provide a sense of place and identity for a locale). *'Humanistic' mapping* approaches are rooted in social advocacy and community development work and tend to be associated with the rise of critical cartography (Dodge et al., 2009). This type of mapping adopts a culturally sensitive, humanistic approach to understanding specific issues of a place, creating a multivocal platform for discussion and finding community-based solutions. Focusing on the people who are resident, living and interacting within a territory, it considers their knowledges, experiences, movements and memories integral to defining the cultural assets and meanings of the territory. The topics being examined, discussed and mapped can include both tangible and intangible elements. While *cultural resource/asset mapping* tends to emphasise the documentation of 'information' and the development of 'cultural or creative sector intelligence', *humanistic mapping* approaches tend to focus more on 'participation' and 'meaning'. However, they are not mutually exclusive and are increasingly blended and mutually informing approaches.

Three orientations

Characterised by the motive and purpose of a mapping project (rather than the types of items mapped), three general orientations can be observed within cultural mapping: (1) self and place, (2) community attachments to place and (3) culture(s) of place (Duxbury et al., 2019). *Self and place* projects focus on personal attachments and connections between an individual and a place. The 'self' may be the mapper or another person. The mapping can document travel routes and sites of experiences, and may explore personal feelings, impressions, memories and other place-specific narratives and connections. The compilation of multiple individual maps can create a collective sense of a place. Informed by journey mapping approaches in health care and business, this type of cultural mapping practice has been applied to other social issues.

Community and place projects focus on relations between a collective of people, their culture(s) and the territory they inhabit, and on how places are meaningful to the communities that live there. The knowledge collected can be collective in nature (i.e. without enabling individualised extractions) or can be a pluralistic compilation of individual voices to provide collective messages and impressions, highlighting shared commonalities and differences.

Culture(s) of place projects attend to the cultural dimensions and aspects particular to a locale that make it distinct or significant. While also examining people—place entanglements, they focus more on the landscape itself, considering how a place itself is a repository of cultural information and impressions. These projects often emphasise understanding a place through multisensorial, material and experiential encounters. They may also attend to the immaterial dimensions that generate a 'sense of place'.

Moving forward

At a conceptual level, the project of re-situating the field of participatory cultural mapping still needs a more sophisticated and case-specific theory of participation specific to cultural mapping. This will help us understand when and where levels of participation are most crucial, and how they inform the integrity and impacts of cultural mapping projects. While its alignment with other types of participatory mapping work will inform this endeavour, the essential cultural dimensions of place-based meanings, memories and knowledges require additional input from allied culture-focused fields.

We also need further reflection on what we might call the rhetoric of the vernacular in cultural mapping: in practice, the inscription, validation and interpretation of individual viewpoints will remain at best inconsistent without a working theory of the vernacular. As Gerlach's (2010, 2014) groundbreaking work in this area suggests, an enhanced understanding of vernacular theory would provide a coherent body of ideas, helping ground mapping practices, particularly for participatory cultural mapping seeking to make visible personal viewpoints and insights drawn from lived experience.

Increasingly, participatory cultural mapping is recognising its obligation to make room for and embrace the different forms in which knowledge is found and the means through which it is communicated. At an operational level, developing theoretically grounded, pragmatic approaches to recognising, appreciating and bringing together different types of knowledges and perspectives is needed. This includes articulating methodologies for processing and analysing results in ways that retain the original voices and meanings of the participants, underlined by attending to the inherent plurality in ecologies of knowledge (Sousa Santos et al., 2008). Incorporating artistic approaches in participatory cultural mapping projects may help address this imperative for embracing and synthesising diverse forms of knowledge. Artistic processes often aim to engage with the 'felt sense' of community experiences, an element often missing from conventional mapping practices, and can challenge conventional asset mapping by animating and honouring the local, giving voice and definition to the vernacular, recognising *place* as inhabited by story and history, and highlighting the importance of the aesthetic as a key component of community self-expression and self-representation (Duxbury et al., 2019).

In concert, continued attention to enhancing sensitivities and knowledge concerning community–academe collaboration is needed to resist hierarchical arrangements and perspectives while retaining the application of research skills and knowledge in these horizontal relationships. The implementation of participatory cultural mapping projects can be a platform for strengthening community–academe relations that fosters cartographic literacies and capacity-building among community members. Furthermore, the question of what comes about as a result of participatory cultural mapping projects, how it might feed into other public and strategic processes, and who makes these follow-on decisions must be considered as integrated elements of participatory cultural mapping projects and initiatives.

References

Abrams J and Hall P (2006) *Else/Where: Mapping New Cartographies of Networks and Territories*. Minneapolis, MN: University of Minnesota Design Institute.

Arnstein SR (1969) A ladder of citizen participation. *Journal of the American Institute of Planners* 35(4): 216–224.

Bryan J (2011) Walking the line: Participatory mapping, indigenous rights, and neoliberalism. *Geoforum* 42: 40–50.

Caquard S (2013) Cartography II: Collective cartographies in the social media era. *Progress in Human Geography* 38(1): 141–150.

Cochrane L, Corbett J, Keller P and Canessa R (2014) *Impact of Community-based and Participatory Mapping*. Research Report, Institute for Studies and Innovation in Community-University Engagement, University of Victoria, Canada.

Corbett J, Cochrane L and Gill M (2016) Powering up: Revisiting participatory GIS and empowerment. *The Cartographic Journal* 53(4): 335–340.

Crawhall N (2007) *The Role of Participatory Cultural Mapping in Promoting Intercultural Dialogue—'We are not Hyenas'*. Concept Paper Prepared for UNESCO Division of Cultural Policies and Intercultural Dialogue.

Dodge M, Kitchin R and Perkins C (eds) (2009) *Rethinking Maps: New Frontiers in Cartographic Theory*. London: Routledge.

Duxbury N (2019) Cultural mapping: Addressing the challenge of more participative and pluralist cultural policies and planning. *Mouseion: Revista do Museu e Arquivo Histórico La Salle*, no. 33 (May–August): 17–29.

Duxbury N (2022) *Making Visible Distributed Knowledges: The Promise of Cultural Mapping*. Presentation at 'Horizons: Crisis and Social Transformation in Community-Engaged Research' Conference, Organised by Community-Engaged Research initiative, Simon Fraser University, Vancouver, Canada, 26–29 May 2022.

Duxbury N and Garrett-Petts WF (forthcoming) A guide to cultural mapping. In: Latham A (ed.) *Community Arts and Cultural Development and Engagement Open Textbook*. Edmonton: MacEwan University.

Duxbury N, Garrett-Petts WF and Longley A (2019) An introduction to the art of cultural mapping: Activating imaginaries and means of knowing. In: Duxbury N, Garrett-Petts WF and Longley A (eds) *Artistic Approaches to Cultural Mapping: Activating Imaginaries and Means of Knowing*. London: Routledge, pp. 1–21.

Duxbury N, Garrett-Petts WF and MacLennan D (eds) (2015) *Cultural Mapping as Cultural Inquiry*. New York: Routledge.

Freitas R (2016) Cultural mapping as a development tool. *City, Culture and Society* 7(1): 9–16.

Garrett-Petts WF (2016) *The Vernacular Rhetoric of Cultural Mapping: Everyday Cartography in the Public Sphere*. Keynote Address Presented at 'Where is Here? A Conference on Small Cities, Deep Mapping and Sustainable Futures', Comox Valley Art Gallery and Vancouver Island University, 21 July 2016. Available at: www.whereishereculturalmapping.com/publications/.

Garrett-Petts WF, Baker D, Gladu C, KuTschker T and Robertson T (2021) *Artful Engagement in Small Cities: Beyond the Project*. Curated Video Exhibition, Panel Presentation and Roundtable in 'Field Stories: Community-Engaged Research in Times of Crisis', CERi Visual Symposium, Simon Fraser University, 20–26 March 2021.

Garrett-Petts WF and Gladu C (2021) A researcher-in-residence initiative for cultural mapping, economic development, social inclusion, and urban planning in small cities. *Livingmaps Review* 11 (November): 1–12.

Garrett-Petts WF and Karsten S (2019) *Artist-Led Cultural Mapping: A Catalyst for the Re-Imagination, and the Re-Formation, of Municipal Power Hierarchies*. Annual Meeting of the Fachverband Kulturmanagement. Universität für Musik und darstellende Kunst Wien, Vienna, 9 January 2019.

Gerlach J (2010) Vernacular mapping and the ethics of what comes next. *Cartographica* 45(3): 165–168.

Gerlach J (2014) Lines, contours and legends: Coordinates for vernacular mapping. *Progress in Human Geography* 38(1): 22–39.

Guldi J (2017) A history of the participatory map. *Public Culture* 29(1): 79–112.

Hale CR (2005) Neoliberal multiculturalism: The remaking of cultural rights and racial dominance in Central America. *Political and Legal Anthropology Review* 28(1): 10–28.

Hodgson DL and Schroeder RA (2002) Dilemmas of counter-mapping community resources in Tanzania. *Development and Change* 33(1): 79–100.

Hunter V (2019) Vernacular mapping: Site dance and embodied urban cartographies. *Choreographic Practices* 10(1): 127–144.

Kerski J (2014) Mapping for understanding community, region, and the world: Using GIS in native education. In: Cole DG and Sutton I (eds) *Mapping Native America: Cartographic Interactions between Indigenous Peoples, Government, and Academia*, vol. 3. North Charleston, SC: CreateSpace Independent Publishing Platform.

Kitchin R and Dodge M (2007) Rethinking maps. *Progress in Human Geography* 31(3): 331–344.

Magnusson W (2015) *Local Self-Government and the Right to the City*. Montreal and Kingston: McGill-Queen's University Press.

Ortega Nuere C and Bayón F (2015) Cultural mapping and urban regeneration: Analyzing emergent narratives about Bilbao. *Culture and Local Governance* 5(1–2): 9–22.
Parker B (2006) Constructing community through maps? Power and praxis in community mapping. *The Professional Geographer* 58(4): 470–484.
Peluso NL (1995) Whose woods are those? Counter-mapping forest territories in Kalimantan, Indonesia. *Antipode* 27(4): 383–406.
Pillai J (2013) *Cultural Mapping: A Guide to Understanding Place, Community and Continuity*. Petaling Jaya, Malaysia: Strategic Information and Research Development Centre.
Poole P (2003) *Cultural Mapping and Indigenous Peoples*. Report Prepared for UNESCO. Available at: www.iapad.org/publications/ppgis/cultural_mapping.pdf.
Roberts L (2012) *Mapping Cultures: Place, Practice, Performance*. Basingstoke: Palgrave Macmillan.
Roth R (2009) The challenges of mapping complex indigenous spatiality: From abstract space to dwelling space. *Cultural Geographies* 16: 207–227.
Sitas R (2020) Cultural policy and just cities in Africa. *City* 24(3–4): 473–492.
Sletto B (2009) We drew what we imagined: Participatory mapping, performance, and the arts of landscape making. *Current Anthropology* 50(4): 443–476.
Sletto B (2012) Indigenous rights, insurgent cartographies, and the promise of participatory mapping. *Portal* 12: 12–15.
Sousa Santos B, Nunes JA and Menses MP (2008) Opening up the canon of knowledge and recognition of difference. In: Sousa Santos B (ed.) *Another Knowledge is Possible: Beyond Northern Epistemologies*. London: Verso Books, pp. ix–ixii.
Stewart S (2007) *Cultural Mapping Toolkit*. Vancouver: Creative City Network of Canada and 2010 Legacies Now.
Strang V (2010) Mapping histories: Cultural landscapes and walkabout methods. In: Vaccaro I, Smith EA and Aswani S (eds) *Environmental Social Sciences: Methods and Research Design*. Cambridge: Cambridge University Press, pp. 132–156.

36

MAPPING NARRATIVES ON HISTORICAL TOURS

Stephen P. Hanna, Amy E. Potter and Derek H. Alderman

Introduction

Each year, hundreds of thousands of people visit Monticello, the famous home of Thomas Jefferson and the hundreds of enslaved Black people held under his control (see Figure 36.1). While there, the overwhelming majority of visitors meet an interpreter on the mansion's East Portico who guides them through eight furnished rooms before concluding the tour on the North Terrace—a walkway atop the mansion's north wing. In each room, the interpreter calls forth specific portraits, pieces of furniture or some of the building's unique architectural elements as they weave a narrative featuring Jefferson's personal life, his contributions to the founding of the United States and the uncomfortable but no longer denied fact that the man had a coercive sexual relationship with the enslaved Sally Hemings.

Guided tours at Monticello are not strictly scripted. While interpreters receive the same training and oversight and follow the same tour route, they are encouraged to personalise their tours. For example, in the South Square Room where portraits of Jefferson's White family adorn the walls, we witnessed one interpreter emotionally emphasise the coercive nature of Jefferson's relationship with Hemings and his decision to never recognise her children as his. Another interpreter, however, ended their brief acknowledgement of Jefferson's relationship with Hemings by joking that 'families are complicated'.

More tour variations emerge as interpreters and visitors interact in the spaces of the house. Visitor questions, bodily reactions to the interpreter's performance or the presence of children can all alter a tour's narrative. Busy days force interpreters to fill narrative space in one room while waiting for the group ahead to clear the next room. Even rain can have an impact, causing most interpreters to end tours before the North Terrace.

Such variations mean that the narrative mapped onto the spaces of Monticello is not predetermined. Instead, these mappings of history emerge each and every time an interpreter and their tour group move from room to room, gaze on particular artefacts, nod or shake their heads, ask questions or laugh at jokes. To us, therefore, it is impossible for a researcher to summarise a tour's content after taking only one or two iterations of the tour.

We reached this conclusion while studying whether and how enslavement was incorporated into visitor experiences at other plantation museums across the Southeastern United

 DOI: 10.4324/9781003327578-43

Figure 36.1 Visitors approach Monticello's East Portico to start their tour.
Source: Photo by Stephen P. Hanna

States (Potter et al., 2022). This work, along with the experiences one of us had as a guide at a plantation museum, led our research team to develop a reiterative and mobile methodology for documenting the narratives, spaces, material culture and interactions of guided house tours. We call this methodology, 'narrative mapping' (Hanna et al., 2019; Potter, 2022).

In this chapter, we explore the inherently spatial nature of tour narratives and how the bodily movements of tour groups map and remap these narratives into the spaces of historical sites and the public memories these sites reproduce. We then describe narrative mapping as a method before using our research at Monticello in 2019 as an example. We conclude by suggesting that narrative mapping could be used to document tour narratives and experiences in other settings.

Spatial narratives as mobile mappings

Tours of historic sites are acts of *mobile storytelling*. Because narratives are inscribed within and across spaces, to apprehend them, one must move across them (Jung, 2014). The recent 'mobilities turn' in geography acknowledges that the world happens not only through inhabiting spaces in fixed ways but also through bodies navigating spaces and the narratives, objects and memories embedded within them (Cresswell, 2012). Inspired by this mobility turn, Sheehan and colleagues (2021) call on us to move beyond discussions of the fixity of commemorative spaces by examining how memory is constituted through the mobility of social actors. As they argue, the meaning of commemoration is always in an emergent state because of the networks of socio-spatial relationships which put memory on the move.

Azaryahu and Foote's (2008) work on spatial narratives provides a theoretical framework for creating methods that capture the importance of spatially guided movement at heritage sites. They explore historical space as narrative medium and contend that the spatial configurations of tours structure narratives. Building on this, Smith and Foote (2016) and Brasher (2021) emphasise that site architectures are much more than stages or containers for historical representation—they have agency because they affect visitor engagement. Chronis (2015) adds that places and the narratives giving them meaning are not separate from what guides and visitors do in those spaces.

A museum's historical message is never as tightly controlled as a researcher or tourist may assume—even if it is delivered through a closely scripted guided tour. Touring a museum is a form of *conditional storytelling* that emerges through interactions between tour guides and visitors. These interactions are always situated within, affected by and responsive to the act of moving through a site's lighting and sound, curated paths, furnished rooms and contextualised material objects (Smith and Foote, 2016; Brasher, 2021). The bodily movements of people can follow, supplement or even question established spatial narratives on tours and, thereby, repeatedly enliven what are otherwise staid depictions of the past.

Our narrative mapping method is also rooted in an 'affective turn' in heritage studies (Tolia-Kelly et al., 2016) characterised by analyses of people's embodied experiences, performances and mobilities and how they work to support, erode or revise a historical site's curated message (Alderman et al., 2020). The haptic interactions of bodies with the spatial design of museum paths, exhibited objects and narratives contribute to a tour's commemorative atmosphere—the way people feel a tour in addition to what they learn (Brasher, 2021). While spatial narratives move people to and through the places where history occurred and can be powerful sources of direct, didactic education, they can also lead to 'emotional learning', shaping how people psychologically connect (or not) with the identities, lives and struggles of people from the past (Micieli-Voutsinas, 2021: 3).

Recognising that spatial narratives are representational, inherently affective, embodied and produced through mobility, requires a rethinking of how such narratives are mapped. Del Casino and Hanna (2006: 36) argue that maps are 'not simply representations of particular contexts, places and times. They are mobile subjects, infused with meaning through contested, complex, intertextual and interrelated sets of socio-spatial practices'. Thus, the practice of mapping narrative into space is never a once-and-for-all affair. Instead, it is always in a state of becoming as the bodily movements of visitors and guides on the same tour, but at different times and under different conditions, render new iterations of spatial narratives that may alter how public memory is produced and consumed.

This attention to how spatial narratives are shaped by and in turn shape meaningful movements within historical sites is part of a processual approach to the map that recognises the full range of practices and contexts that constitute mapping—a shift 'from ontology (what things [maps] are) to ontogenetic (how things [maps] become)' (Kitchin et al., 2013: 494). Spatial narratives encountered and performed through the spaces and media of museums can be thought of as 'mobile mappings' of historical meanings, what Wilmott (2020: 11) describes as 'the practice of drawing relations together in and through movement' and moderating the tension between 'the fixity of representation and the openness of space'.

Therefore, the embodied, mobile and always emergent qualities of spatial narratives produced through guided tours can be captured by theorising such tours as mappings. While the tour itself may originate in a physical or mental map created when a site's management plans the visitor experience, the spatial narrative imagined during planning is only made real when it is practised—when guides and visitors engage in mapping by moving through and interacting with museum spaces. To understand these mappings, researchers need a methodology attuned to their mobile, iterative and emergent qualities.

Narrative mapping as a mobile methodology

Mobile methods require the researcher either to follow the subject through space or to make the subject mobile for the purposes of the research' (Ricketts-Hein et al., 2008: 1269). Examples include Herbert's (1996) ride-alongs with police and deployments of walking interviews to capture embodied experiences (Anderson, 2004; Stevenson and Farrell, 2017; Mackay et al., 2018). The central assumption motivating these studies is that meaning is spatial. Therefore, gaining a richer understanding of subjects and phenomena requires researchers to traverse space.

This relationship between space and narrative is most explicit within guided tours (Hallin and Dobers, 2012). Since tour narratives are contingent on interactions between guides, visitors and material culture, the interpretive impact or role of any particular space along a tour's route may vary from tour to tour. Larsen and Meged's (2013) ethnography of tour groups conducted through tour-alongs is the approach most similar to the narrative mapping method we present in this chapter. More directly, narrative mapping was a response to Potter's (2016) argument for a deeper understanding of the narrative and affective complexities and variations of tours as experienced.

When employed at plantation museums, our narrative mapping instrument collects information about which tour topics are dominant in different tour spaces—particularly the prevalence and placement of content concerning enslaved Black people relative to White enslavers on a tour (Hanna et al., 2019). In addition, the instrument captures visitor questions and the emotions guides employed while performing the narrative to help us track the impacts guides and visitors have on tours. The result is an eight-question form accompanied by plantation house floor and site plans that we filled out after observing a tour. While we strove to be unobtrusive tour observers, we acknowledge that our presence and identities as researchers could have affected the tour (see Table 36.1).

The form's header provided space to document the number of visitors on a tour, the tour's length and the guide's demographic characteristics. The first three questions required tour observers to note which topics were discussed on their tour, rank the topics most prevalent along the tour's route and tally the number of times enslavement was mentioned. The subsequent questions asked observers to describe how the guide discussed enslavement, note the questions visitors asked, gauge the performative/emotive qualities of the guide and provide additional details about interactions among the guide, visitors and material spaces and objects. The final question prompted observers to draw the tour's route on a map or floor plan and mark where the topics listed in the first question were discussed. This information was then aggregated and interpreted to reveal tendencies and variations among the spatial narratives emerging during multiple iterations of a tour. This method of capturing

Table 36.1 Narrative mapping instrument questions developed by authors for Monticello

Question 1: Which of these topics* were discussed on the tour? Include all topics for which the guide provided at least a few details (circle all that apply).

Lives of Women	Thomas Jefferson	Enslaved/Slavery
Grounds/Gardens	Jefferson Family History	Agriculture/Crops
Furnishings	Social life/Culture	Archaeology
National history	Architecture	Legacy of slavery
Descendant community	Other	

Question 2: Of all the topics listed above, which received the most attention?

Most

Second most

Third most

Fourth most: if Jefferson was ranked first

Question 3: Tally the number of times that enslavement was mentioned on the tour (count each occurrence of slave, slavery, enslaved, name of an enslaved person).

Question 4: How was enslavement discussed on the tour? Please note whether names and/or biographies were included; whether enslaved persons were quoted or given voice; whether enslaved only mentioned as numbers, labourers, or units of master's wealth; whether slavery's legacy into present was mentioned; whether enslaved descendants were mentioned.

Question 5: Was the Thomas Jefferson associated with slavery? If so, how? ('good' or 'bad' owner, position in debates over slavery, relationships with specific enslaved individuals, etc.).

Question 6: What geographies of slavery were discussed on the tour (limited to plantation or including local, regional, national, and/or global information)?

Question 7: Use this space for additional notes. Elaborate on any questions listed above and/or note topics of questions visitors asked and whether these changed the guide's narrative. Finally, how did the guide use emotion and/or artefacts/furnishings when talking about any of the topics you circled in question 1?

Question 8: Please use map of grounds and floor plans to note where topics you circled in Question 1 were discussed on the tour.

* Tour topics were developed by Butler (2001), refined during fieldwork at Louisiana plantation museums (Potter et al., 2022) and modified for use at Monticello (Hanna et al., 2022).

spatial narratives enabled us to track how narrative themes, material culture and people interact in the spaces of a plantation museum such as Thomas Jefferson's Monticello.

Narrative mapping Monticello's house tour

Four of the first five presidents of the United States owned Virginia plantations and enslaved Black people. These White men's homes, Mount Vernon, Monticello, Montpelier and Highland, have been preserved and turned into historical sites which collectively attract over 1.5 million visitors per year. As managers at each of these sites grapple with how to more justly represent the experiences of the Black people enslaved by America's 'Founding Fathers', they have, to varying degrees, worked with Black descendants of people forced to live, work and usually die at the four plantations.

In 2019, our research team investigated how the involvement of enslaved people's descendants in designing tours and exhibits affected visitors' engagement with narratives of enslavement at Monticello, Mount Vernon, Montpelier and Highland (Hanna et al., 2022). To do this, we interviewed museum staff and descendants, photographed exhibit spaces, surveyed visitors and used narrative mapping to document guided tours. At Monticello, the well-publicised struggle of Sally Hemings's descendants to be accepted as descendants of Thomas Jefferson, the museum's subsequent decision to acknowledge the relationship between Hemings and Jefferson, and the museum's efforts to gather oral histories from descendants of Monticello's enslaved population gave us ample reason to expect visitor experiences to be affected by relationships between Monticello and its descendant community.

Visitors who purchase the standard admission ticket to Monticello can take the standard house tour, the Gardens and Grounds tour and the Slavery at Monticello tour. From our exit survey of visitors, we learned that 87% took the standard house tour making it the centrepiece of most visitors' experiences.

In 2019, our research team mapped the standard house tour narrative 18 times. On these tours, guides were almost always White women ranging in age from 25 to 70, the number of visitors on a tour ranged from 17 to 25, and the average tour lasted 47 minutes. We ranked Thomas Jefferson as the topic receiving the most attention on 17 tours, furnishings or architecture as a top three topic on 15 tours, and the history of the United States as a top three topic on 10 tours. While enslavement or the biographies of particular enslaved people formed part of the narrative on all but one tour, the topic was only ranked among the top three twice.

This statistical summary only sets the stage for how guides and visitors mapped and remapped Jefferson's biography, his place in U.S. history, Monticello's architectural elements and furnishings and the lives of the people he enslaved along the tour's route. Narrative mapping pins tour topics to specific spaces for each iteration of the tour enabling us to note where and how such themes were embodied by different tour groups moving through the house (see Figure 36.2).

For example, discussions of enslaved people most often occurred in two rooms. In the South Square room, interpreters used portraits of Jefferson's White wife and daughter to introduce visitors to his family life. After Monticello recognised that Sally Hemings and Jefferson had a sexual relationship that produced children, interpreters have used this room to discuss the relationship and introduce the Hemings family. As Figure 36.2 shows, narratives about enslavement were intertwined with information about Jefferson's life and his White family in this space. In the Dining Room, guides typically talked about the work performed by enslaved people as they prepared and delivered food to Jefferson and his guests. Visitors often asked about the room's dumb waiter and some guides noted that Jefferson designed it so wine and food could be delivered without his guests having to see Black enslaved servants. As our mapping indicates, tour themes associated with enslavement in this room typically included furnishings and architecture.

In contrast, enslavement was seldom made present by guides or visitors in public rooms like the Entrance Hall and the Parlor where furnishing and artefacts tend to be associated with Jefferson's roles as president, statesman and author of the Declaration of Independence. Similarly, the inventions and books on view in the Library excited some visitors as guides described Jefferson's intellectual prowess, but narratives emerging in that space rarely reminded people that the Jefferson family's history as enslavers enabled him to study

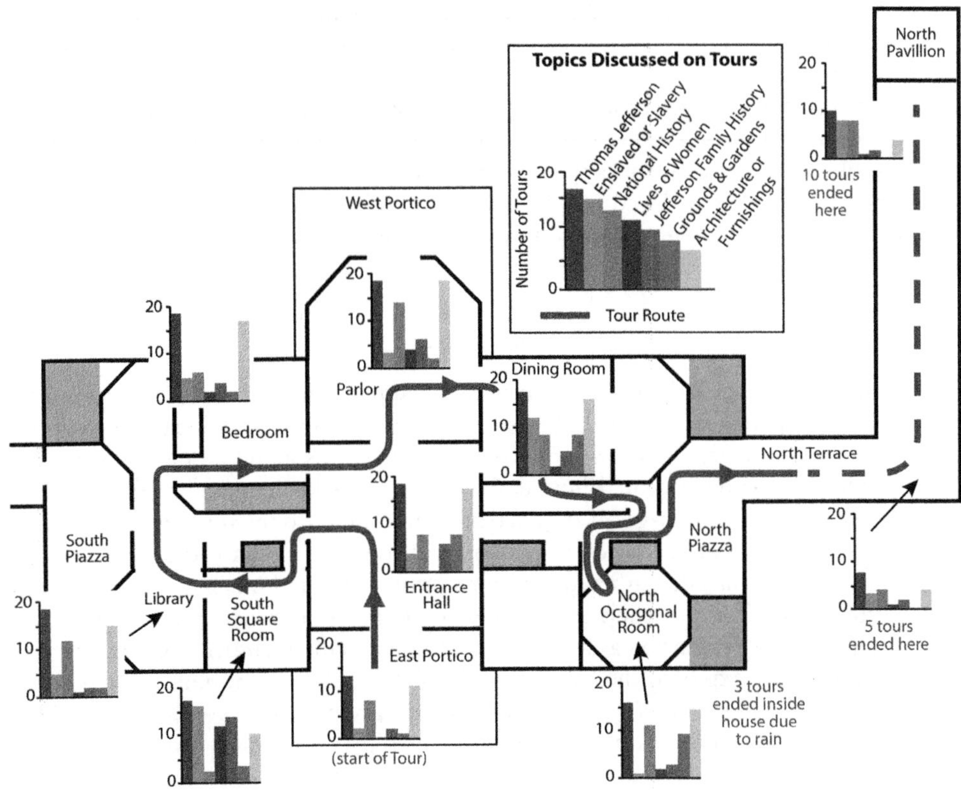

Figure 36.2 Narrative map of Monticello's standard house tour.

Source: Map by Stephen P. Hanna

political theory and architecture. The only space where Jefferson's biography, the lives of enslaved people and United States' history were all present was at the tour's end on the North Terrace. There, visitors gazed at the mansion's famous West Portico while guides described how Black families were ripped apart when, on the portico, auctioneers sold enslaved people to settle Jefferson's debts after his death. And, as interpreters summarised Jefferson's place in American history, some noted that Jefferson did nothing to end enslavement despite describing slavery as evil. Unfortunately, guides only took visitors to the North Terrace on 10 of the 18 tours.

Through our use of narrative mapping, we were also able to capture some of the emotional performances and affective responses that occurred as guides and visitors mapped stories onto Monticello. Guides often used a deep, respectful tone when describing Jefferson's contributions to the nations' founding. Some interpreters worked to excite visitor interest by gesturing to specific furnishings or architectural elements as evidence of his brilliance. When discussing enslavement or noting brief biographical facts about a few enslaved house servants, however, almost all guides reverted to more 'matter-of-fact' tones. This understated approach was less likely to elicit emotional responses among the visitors that can place a great deal of stress on interpreters (Potter, 2016).

While most visitors remained quiet on tour, they glanced about to take in different objects in each room, sometimes the ones a guide mentioned, but sometimes other portraits or curios. When visitors did ask questions, they were usually inspired by a piece of furniture or artwork thereby elevating the importance of furnishings and architecture within these tours' narratives. Questions about the women and men Jefferson enslaved were rarer and more likely to occur on tours where the guide had already talked about Sally Hemings or the lives of other enslaved people. Only once did we record that a visitor's question prompted a fuller discussion of slavery and we never witnessed a visitor openly expressing sadness or anger about the topic.

Conclusion

As illustrated by its use at Monticello, narrative mapping allowed us to understand how narratives and experiences emerged as visitors and guides moved and engaged with each other, with material culture and with affective bodily actions along a guided tour's route. More specifically, we documented how enslavement tended to emerge in places separate from the more prevalent discussions of Jefferson's importance to the founding of the United States. We also noted that, in spaces where guides introduced and named the Black women and men Jefferson enslaved, neither guides nor visitors exhibited much emotion. This contrasts with guides' efforts to elicit respect and or spark excitement among visitors when describing Jefferson's reputation as a founding father and enlightenment figure.

Thus far, we have explored the efficacy of a mobile narrative mapping methodology in the context of museums that narrate the history of enslavement, but we encourage scholars and practitioners to explore its potential in other settings. For almost any kind of guided tourist experience, there is a need to understand the embodied and emergent ways that spatial narratives are constructed and performed within geographic contexts. In particular, site managers can use the method to assess the implications of where guides and visitors engage with what is said, or not said, and how different characteristics of these locations shape tour narratives. It is also our hope that narrative mapping can create spaces of collaboration between practitioners and researchers to both theorise and practise museum narratives in more complex, spatially fluid and contingent ways.

Key to the broader cartographic humanities project and the work described here is realising that the mapping of commemorative meanings on the move is more than merely a method. Rather, the tourist destination in and of itself constitutes and operates as a mobile mapping by bringing stories, curated objects, paths and social actors together into spatially dynamic relationships. While in any one moment, such relationships may appear settled or fixed, they actually create potentially varied narrative and affective experiences actualised by the embodied mapping of historical meanings into space at that moment in time and space. In this respect, cartography is always part of the work of museums.

References

Alderman DH, Brasher JP and Dwyer III OJ (2020) Memorials and monuments. In: Kobayashi, A (ed.) *International Encyclopedia of Human Geography*, 2nd edition, vol. 9. Amsterdam: Elsevier, pp. 39–47.

Anderson J (2004) Talking whilst walking: A geographical archaeology of knowledge. *Area* 36(3): 254–261.

Azaryahu M and Foote KE (2008) Historical space as narrative medium: On the configuration of spatial narratives of time at historical sites. *GeoJournal* 73(3): 179–194.
Brasher JP (2021) Creating 'confederate pioneers': A spatial narrative analysis of race, settler colonialism, and heritage tourism at the Museu da Imigração, Santa Bárbara d'Oeste, São Paulo. *Journal of Heritage Tourism* 16(1): 20–42.
Butler DL (2001) Whitewashing plantations: The commodification of a slave-free antebellum South. *International Journal of Hospitality & Tourism Administration* 2(3–4): 163–175.
Chronis A (2015) Moving bodies and the staging of the tourist experience. *Annals of Tourism Research* 55: 124–140.
Cresswell T (2012) Mobilities II: Still. *Progress in Human Geography* 36(5): 645–653.
Del Casino Jr VJ and Hanna SP (2006) Beyond the 'binaries': A methodological intervention for interrogating maps as representational practices. *ACME: An International Journal for Critical Geographies* 4(1): 34–56.
Hallin A and Dobers P (2012) Representation of space. Uncovering the political dimension of guided tours in Stockholm. *Scandinavian Journal of Hospitality and Tourism* 12(1): 8–26.
Hanna SP, Alderman DH, Potter AE, Carter PL and Forbes Bright C (2022) A more perfect union? The place of Black lives in presidential plantation sites. *Memory Studies* 15(5): 1205–1231.
Hanna SP, Carter PL, Potter AE, Bright CF, Alderman DH, Modlin EA and Butler DL (2019) Following the story: Narrative mapping as a mobile method for tracking and interrogating spatial narratives. *Journal of Heritage Tourism* 14(1): 49–66.
Herbert S (1996) The normative ordering of police territoriality: Making and marking space with the Los Angeles Police Department. *Annals of the Association of American Geographers* 86(3): 567–582.
Jung Y (2014) Mindful walking: The serendipitous journey of community-based ethnography. *Qualitative Inquiry* 20(5): 621–627.
Kitchin R, Gleeson J and Dodge M (2013) Unfolding mapping practices: A new epistemology for cartography. *Transactions of the Institute of British Geographers* 38(3): 480–496.
Larsen J and Meged JW (2013) Tourists co-producing guided tours. *Scandinavian Journal of Hospitality and Tourism* 13(2): 88–102.
Mackay M, Nelson T and Perkins HC (2018) Interpretive walks: Advancing the use of mobile methods in the study of entrepreneurial farm tourism settings. *Geographical Research* 56(2): 167–175.
Micieli-Voutsinas J (2021) *Affective Heritage and the Politics of Memory After 9/11: Curating Trauma at the Memorial Museum*. New York: Routledge.
Potter AE (2016) 'She goes into character as the lady of the house': Tour guides, performance and the southern plantation. *Journal of Heritage Tourism* 11(3): 250–261.
Potter AE (2022) 'A pledge of allegiance to the south': Commemorating the enslaved at two historic house museums in Kansas City, Missouri. *The Public Historian* 44(3): 110–138.
Potter AE, Hanna SP, Alderman DH, Carter P, Forbes-Bright C and Butler D (2022) *Remembering Enslavement: Reassembling the Southern Plantation Museum*. Athens, GA: George University Press.
Ricketts-Hein J, Evans J and Jones P (2008). Mobile methodologies: Theory, technology and practice. *Geography Compass* 2(5): 1266–1285.
Sheehan R, Brasher J and Speights-Binet J (2021) Mobilities and regenerative memorialization: Examining the equal justice initiative and strategies for the future of the American South. *Southeastern Geographer* 61(4): 322–342.
Smith SA and Foote KE (2016) Museum/space/discourse: Analyzing discourse in three dimensions in Denver's history Colorado Center. *Cultural Geographies* 24(1): 1–18.
Stevenson N and Farrell H (2017) Taking a hike: Exploring leisure walkers embodied experiences. *Social & Cultural Geography* 19(4): 429–447.
Tolia-Kelly DP, Waterton E and Watson S (eds) (2016) *Heritage, Affect and Emotion: Politics, Practices and Infrastructures*. London: Routledge.
Wilmott C (2020) *Mobile Mapping: Space, Cartography and the Digital*. Amsterdam: Amsterdam University Press.

PART 7

Public cartographic humanities

37
THE SOCIAL LIFE OF MAPS

Martin Brückner

Introduction

In modern European cultures, as in many of those that for better or worse bear the imprimatur of western influence, the history of cartography is closely intertwined with the social life of maps. The process of mapmaking, the habits of map use and the map as artefact are steeped in social practices, involving the exchange of ideas and knowledge, craft and memory, not to mention diverse modes of communication. Yet, to truly understand the role maps have played as an expression of the cartographic humanities, we must recognise that what bridges the gap separating maps (as artificial spatial representation) from people (as active or passive actors) is the maps' materiality and status as commodity. Once maps are considered not only as an information system but as objects, they lead a fascinating double life. On the one hand, maps exist outside or next to us and are usually considered useful things, as long as their basic properties satisfy the human desire for spatial orientation. On the other hand, people's daily proximity to maps changes the maps' social character, and different sets of values arise during the process of encounter and exchange—namely, from the moment when maps are looked at, read or used in non-cartographic contexts their meaning as usable things becomes entangled with or contingent on ritual actions and social performances, on patterns of transaction, sources of motivation and a multitude of idiosyncratic personal experiences (Brückner, 2017: 1–12).

To illustrate the logic of the social life of maps, it helps to recall the critical movement from the 1980s and 1990s when cultural criticism, fuelled by discourse studies and political theory, discovered cartography and the profound relationship between modern mappings and the rise of nationalism. Scholars working across the disciplines, such as political science and sociology, historical geography and literary studies, issued path-breaking interpretations that Benedict Anderson (1991) would sum up best: in modern societies, the image and paper construct of the national map gave substance to the artificial and often elusive concept of the imagined community called the nation. Newly formed sovereign states turned the outline image of national maps into avatars of nationalism. In particular, the 'map-as-logo' emerged as a highly persuasive signifier breathing life into fledgling political unions. By removing the lines of longitude and latitude, place names and neighbourly

 DOI: 10.4324/9781003327578-45

Figure 37.1 Plate gift of Mr. and Mrs. John Mayer, bowl gift of S. Robert Teitelman, and jug bequest of Henry Francis du Pont.

Source: Courtesy of Winterthur Museum, Garden & Library

relations, logo maps not only were 'instantly recognizable, everywhere visible, [but they] penetrated deep into the popular imagination, forming a powerful emblem for the anticolonial nationalisms being born' (Anderson, 1991: 175).

That the national map emerged as a prolific logo and unifying emblem was made possible by the rapid transmission of map images across diverse social institutions and material platforms. Considering one of the first post-colonial map exchanges that took place in the United States, maps, like William McMurray's *The United States* (1784), balanced the new nation's expansive geography against the compressed shorthand of the logo map; the painter Ralph Edward snuck the nation as map logo into the portrait of *Mrs Noah Smith and Her Children* (1798); and American educators instructed schoolgirls how to stitch the nation's logo into the continental texture of North America (Brückner, 2006: 134–141). But it was the global marketplace that introduced the United States as a map logo to a broader audience via transfer printing and other material adaptations. By the 1800s, map-like replications of the nation's territorial outline adorned cream or dinnerware (Figure 37.1), book frontispieces, or fabrics like printed handkerchiefs. With the commercialisation of the nation's outline over the next decades, the logo map increased its social currency as 'cartifacts' in the form of map puzzles and map carpets; in the shape of cookie-cutters for baking enthusiasts; or as decals marking goods bearing the inscription 'Made in America'.

The social life of maps, then, comprises the activities that occur between actual maps and their many uses defined by material exchange and social interactions. This approach to studying maps, and implicitly the cartographic humanities, follows Arjun Appadurai's observation that 'focusing on the things that are exchanged, rather than simply on the forms or functions of exchange, makes it possible to argue . . . that commodities, like persons, have social lives' (1986: 3–4). Or, to connect the critical momentum of the 1990s to that of the 2020s, in particular to theories and methods proposed by material culture studies, to examine the social life of maps is to paraphrase W. J. T Mitchell's (2005) question, 'What Do Maps Want?'. Instead of thinking about maps as merely representational systems, their versatility as social things render them 'as if they were environments where images live, or personas and avatars that address us and can be addressed in turn' (Mitchell, 2005: 203).

Thinking about the early United States, the logo maps' material diversity and circulation gives more than licence to explore the meaning of maps from both within and from the outside. Rather, to make sense of the social life of maps, 'we have to follow the things themselves', because, as suggested by Appadurai (1986: 3), 'their meanings are inscribed in their forms, their uses, their trajectories'. To make the many cartographic and non-cartographic moments of map encounters manifest, we need to ask, along with Igor Kopytoff (1986: 66–67), where did maps come from, and who made them? What was their career before and during their circulation, and what did people consider to be the ideal career for maps? What were the recognised 'ages', or periods, in the 'life' of a map, be it as a manuscript, print or in other material formats? And what were some of the cultural markers defining their uses? How did map usage change with age, and what happened to map materials when they reached the end of their usefulness? In short, what were the biographical possibilities inherent to a map's status as a usable thing broadly defined during a particular period in culture?

Several studies using different methodologies and historical contexts have engaged with the social lives of maps (Mukerji, 1983; Pritchard, 2001; Jacob, 2006; Bosse, 2007; Dillon, 2007; Dunlop, 2015; Jaffee, 2010; Brückner, 2017). To illustrate some of the maps' biographical possibilities, the following three sections describe representative paths that maps have taken while becoming woven vertically and horizontally into the fabric of western everyday life. The examples concentrate on the role maps have played in the public sphere; on patterns of map socialisation in educational institutions; and on personal map encounters shaping experiences of shared intimacy and identity.

Public cartography

In the western world, the social life of maps is closely linked to the phenomenon that is today called the public sphere. A distinct cultural formation, it meted out an area of social life that was separate from the more narrowly defined public realm of the state, with its ruling classes and their association with public authority, and the private sphere comprising everyday activities, commercial exchanges and civil codes of conduct. According to the philosopher Jürgen Habermas (1989), it was as much a conceptual space (e.g. public opinion, the free press as 'fourth estate') as a pragmatic one (e.g. coffee houses, voluntary associations, theatres) in which individuals and groups could associate and discuss matters of mutual interest through the medium of speech and letters. As best-selling printed artefacts, maps accompanied the structural formation of the public sphere. From the late seventeenth century onward, state-sponsored mapmakers and commercial publishers produced a vast array of maps at affordable prices in historically unprecedented numbers. As maps became commodified articles, they entered a growing list of public goods deemed to be versatile tools essential for a modern society intent on expanding the reach of its economy, on the one hand, and the sphere of political influence, on the other hand.

Map displays were a common feature in public settings by the eighteenth century. Visitors of imperial London's Board of Trade or colonial Boston's Old State House would have encountered map giants like John Rocques' *An Exact Survey of the Citys of London* (1746; 485 by 660 cm) or Henry Popple's *A Map of the British Empire in America* (1733; 251 cm × 230 cm). The latter map was also found on display in 1776 in Pennsylvania's state house, today's Independence Hall in Philadelphia, which John Adams described in a letter to his wife, Abigail: 'It is the largest I ever saw, and the most distinct. Not very accurate. It is Eight foot square' (Brückner, 2017: 127). Specifically designed for display, wall maps like these responded to a political rhetoric of symbolic size. In their capacity as theatrical props, they provided the dual sense of empirical reality and evidentiary gravitas, serving as dramatic background intent on impressing local citizens and foreign diplomats alike. In particular, official map displays were spectacles projecting raw political power, turning even a sceptical gaze, like that of John Adams, into one of respect and admiration.

As state and commercial interests often aligned, the social life of maps fostered a new kind of 'public cartography' (McHaffie, 2011: 130) when nineteenth-century audiences saw map displays become a regular feature in spaces identified as the crucibles of public life, namely in coffee houses and taverns, postal and magistrate offices, as well as in commercial settings, from stationary shops to large-scale exhibition halls. Already in 1807, American travellers reported the display of large maps in local post offices. At the same time, postmasters across the United States documented the receipt of Abraham Bradley's *Map of the United States Exhibiting Post Roads & Distances* (1805). As the nation's Postmaster General, Bradley ostensibly used his political position to sell his map for personal gain by exploiting the nation's formal network of postal routes and office spaces. Yet it was travellers like Fortescue Cuming who reported the way maps were folded into people's public lives. Seeking to post a letter after hours in a village outside of Harrisburg, Pennsylvania, he observed how

> the postmaster very civilly invited me into his parlour, to settle for the postage, where seeing a large map of Pennsylvania, I took the opportunity of tracing my journey. . . . There were some ladies in the room, apparently on a visit, and there was an air of sociality and refinement throughout, which was very pleasing.
>
> *(Brückner, 2017: 197)*

In public spaces, then, map displays became conversation pieces, serving people's needs not only for community and sociability but also for self-orientation in both the spatial and social senses. Nineteenth-century art and science exhibitions, especially the map shows at New York's Crystal Palace (1853) or Philadelphia's Centennial Exhibition (1876), aligned the maps' visual spectacle with performance culture (Brückner, 2017: 117–124). A self-conscious audience wearing their Sunday's best meeting before a giant map would have prompted polite exchanges and perhaps friendly competition as visitors tested their geographic knowledge. They would point to borders and call out familiar places, engaging their senses and neighbours while reminiscing about past and future journeys or memories of distant places. In doing so, people transformed public maps into interactive objects through which to explore their place in the world in relation to the various spaces of map displays (e.g. public architecture), the decorative arts (e.g. prints and paintings) and the material culture of everyday life (e.g. picture frames, overmantels and furniture).

Schooling with maps

What is significant to realise is that mid-nineteenth-century audiences would have felt very much at home with interactive map displays thanks to the rise of public education and pedagogic reforms. Responding to educational goals dating back to John Locke's treatise, *Some Thoughts Concerning Education* (1693), early map publishers in different countries—including the firms of Homann and Lotter in Germany, Didier and Vaugondy in France, Van Keulen and Van der Aa in the Netherlands, Bowen and Sayers in England, and Carey and Melish in the United States—transformed cartographic works into a widely recognised educational medium. Collectively, they introduced an array of map products—and these included everything from wall maps and textbook maps, to pocket maps and folio atlases—that were directly responsible for the increase in map distribution and for the widespread growth in map literacy.

One audience targeted by commercial cartographers were educators, that is, the teachers, principals and trustees of school boards who were charged with implementing cartography into the general curriculum. Between the 1810s and 1850s, curricular reforms in England, the United States, several German states, and Austria shifted geographic education from colleges and universities to public schools offering primary as well as secondary education. France pursued a similar push for map education after the Franco-Prussian War of 1870, following a commission report by Ferdinand Buisson who had studied public schools in Europe and America. By the end of the century, educators informed by western pedagogic principles had universally adopted a slate of topographical and historical maps and atlases. By then, maps had become indispensable educational tools, influencing everything from teacher selections and their training to primary and secondary school curricula (Brückner, 2006: 240–259; Jacob, 2006:39, 180, 353–355).

While commercial textbooks popularised map encounters for pupils, throughout the nineteenth century, the social life of maps intensified through the adoption of map-centred lesson plans advocated by pedagogues such as Johann Heinrich Pestalozzi in Switzerland (universal literacy), Wilhelm von Humboldt in Germany (standardised education) and William Woodbridge and Emma Willard in the United States ('home geography'). As a result, schoolbook authors such as Mary Somerville and John Guy in England, Heinrich Berghaus and Georg Westermann in Germany, or Arnold Guyot in the United States developed curricula that had pupils step outside the classroom and prepare maps of school grounds, streets and neighbourhoods. One upshot of these curricular changes was mandated map-drawing exercises. Inexpensive school atlases paired with pre-printed outline maps promoted exercises in the art of 'mappery', that is, in the art of planning and designing maps from scratch. Students aged 6–12 years learned how to map their worlds through drawings done in ink, pencil and watercolours. In some cases, pupils accentuated paper maps with needlework embroidery, as map samplers were already a common staple in female education. Throughout the nineteenth century and into the twentieth century, projects like these were recognised by educational authorities during graduation ceremonies and put on display at local trade fairs as testaments of skill and proficiency (Brückner, 2017: 304–309; Schulten, 2012; Mayar, 2022).

Coinciding with the educational reforms that accompanied the rise of modern nation states, school curricula frequently paired maps with competence in literacy (Brückner, 2006). Basic and advanced geography books promoted map-reading lessons that closely followed the catechistic recitation method deployed by alphabet primers, 'Readers' or '*Lesebücher*'. In the United States since the 1820s, the monitorial approach to teaching, like the one proposed by Joseph Lancaster, implemented group exercises in map pointing and map reading. By 1860, American school teachers commented that students 'sing the capitals and [the] bound[aries of] the states . . ., while they point out the places on the map' (Brückner, 2017: 290). Lessons like these facilitated the internalisation of cartographic knowledge through the performance of map-based *recitatifs* or chants. Some enterprising teachers took the oral map lesson one step further. Their broadsides announced a 'geographical concert and public recitation' performed by 40 students. According to the broadsides' illustration (Figure 37.2), a giant wall map transformed the public lecture hall into a map-o-rama in

EXCURSION!
TO LEOMINSTER,
Wednesday, April 28, '52,

To enjoy the Entertainment to be given by
J. G. McKINDLEY,
AND
GEOGRAPHICAL CLASS.
CARS WILL LEAVE FITCHBURG AT 5 P. M.,

Giving about two hours of time previous to the commencement of the Concert. Arrangements have been made in order to have all enjoy themselves as they please, either by rambling about the village, visiting the Comb Shops, or have a social time at the Hotel.

Tickets, TO LEOMINSTER AND BACK, ADMITTING TO THE RECITATION, to be had at the

Post Office and Depot, for 25 cents.

The people of Fitchburg will not fail to improve the best opportunity ever offered them, to visit their worthy neighbors of Leominster.

Fitchburg, April 27, 1852.

P. S. Cars will return at about half past 9, or after the Concert.

Figure 37.2 Broadside, *Excursion! To Leominster, Wednesday, April 28, '52, to Enjoy the Entertainment to Be Given by J.G. McKindley, and Geographical Class* ([Fitchburg, MA] [1852]).

Source: Courtesy, American Antiquarian Society

which the map recitation prompts comparisons to a national spelling bee or the singing of a national anthem (Brückner, 2017: 290).

While public map encounters or recitals involved interaction or mutual supervision by peers or disciplinary figures, nineteenth-century diaries suggest that the act of map reading elicited a 'disciplinary intimacy' between map and children (Brodhead, 1993: 13–47). Diaries of students and parents frequently reflected upon the fact that map lessons were self-conducted without a monitor or prompter. The American middle-schooler, Mary Ann Bacon, proudly noted in 1802 how she got up 'at six[,] devoted the morning to studying the boundarys [sic] on the map [and] in the four noon . . . recited it'. Or, consider the young mother Ann Cary who in 1828 recorded her six-year-old son's educational progress, writing about 'his knowledge of geography—it is really curious to hear him going over all the names of places, States, lakes, rivers, etc., on his map' (Brückner, 2017: 290–191). In the wake of pedagogic reforms that eschewed corporeal punishment pupils practised their lessons with a self-discipline similar to Eastman Johnson's painting called *The Lesson* (1874). Showing two unsupervised children self-absorbed in the study of a large atlas, the object of the map substitutes institutional authority for social relationships, transferring the formal discipline of the absent teacher and schoolroom with the intimacy of a fellow child and the domestic interiors of the home.

Cartographic intimacy

With maps acting upon various cultural practices, the social life of maps involved thickly folded patterns of mediation in which map content and cartographic signification led to an increased sense of map intimacy. Especially in private settings, that is, in the physical spaces and intellectual domains used by people for work and communication without recourse to or intervention from governmental or other corporate institutions, historical records from inventories and wills to diaries and letters identify maps as prized possessions (Brückner, 2017: 161–180). Put on display in parlours and bedrooms, near hearths and windows, they were prized not in the sense of actual material value, but for personal value. Viewed and discussed in privacy, maps offered opportunities for remembrance and sentimental reflection, providing guidance as much as comfort while making people feel at home with themselves and each other.

One expression of map intimacy is revealed by the way in which the social life of maps over time proved to be a mutually constitutive force integral to self-representation. Working in painterly traditions that ranged from portraiture and architectural interiors to natural landscapes and city views, artists and engravers frequently incorporated map images to comment on the symbolic significance of their human subjects. Similarly, the constant encounter of maps during literacy instructions informed the verbal limning of the human body in cartographic terms, be it in the form of a face covered 'all over [with] lines, like a railway map' or a body looking 'seamed, scarred, and indented into as many lines and angles as a school map . . . [or] map of the world' (Brückner, 2017: 313). That the social life of maps engendered more than bodily metaphors can be deduced from the popularity of ephemeral maps (see also Lo Presti, this volume). The inclusion of cordiform 'heart maps' celebrating places as well as virtues in friendship albums, scrap books or letters illustrates the depth to which social engagements with maps had become internalised.

Educators never adapted *The New England Primer's* exhortative couplet declaring that instead of tending to books like the Bible 'This Map attend/thy Life to Mend'. But everyday encounters with maps added a certain emotional intelligence and even compassion to the cartographic representation of spaces and places. This cartographic intimacy is the message central to Edward Everett Hale's Civil War novella, *The Man Without a Country* (1891; c. 1863). In the story, an officer of the American army is court-marshalled to never again set foot on American soil or read about the nation or hear its name spoken. When a visitor meets the banished soldier on his deathbed on board of a navy ship, he discovers the ship's cabin had been transformed into 'a little shrine'. This shrine contains the icons familiar to readers of national maps: '[t]he stars and stripes', 'a picture of Washington, and . . . a majestic eagle'. But at the end of the deathbed scene, the prisoner points to 'a great map of the United States, as he had drawn from memory' followed by the exclamation, 'Here, you see, I have a country', which—similar to school exercises—he then followed up with a recitation of place names (Hale, 1891: 39). The man who had lost his place in the country had instinctively latched on to the map as the transitional object affording a sense of belonging. Of course, the problem with transitional objects is that it is their fate to become gradually detached from their users. In the case of public or school maps, they are neither forgotten, nor mourned. Instead, over time they lose their meaning for the citizenry as intimate objects and today, with the predominance of digital mappings and phone apps, we are left to speculate how the social life of maps continues to hold sway over the emotional fabric of both the individual and the collective.

Conclusion

Throughout the eighteenth and nineteenth centuries, the maps' primary function as a tool of wayfinding and spatial representation was invariably imbricated with a host of secondary uses that appropriated the map's innate visual and material adaptability to different, non-cartographic effects. Examining the discrete biographical stages in a map's life, the social life of maps comprises the labour and rituals with which the materiality of ordinary maps fostered a carto-coded culture. On the production side, artistic talent, venture capital, not to mention professional networks committed to mapmaking created far-reaching affiliations between the object of the map and the ever-expanding cohort of map draftsmen and shopkeepers, salesmen and exhibition planners. On the consumer side, a host of public and private rituals amplified the factor of intimacy, rituals that included stately displays and gimmicky salesmanship, tavern talks and ballroom dances, widely publicised school examinations and discreetly held gift ceremonies. The ultimate impact that the social life of maps had on western culture is to foster a widespread map intimacy that emerged from countless mapmaking efforts conducted by school children, who, through the work of mappery, integrated their self-made maps into much larger intellectual and ideological frameworks: following years of map education, the image and object of the map constituted a social experience for young and old, permeating basic spatial knowledge, political sentiments and personal bonds.

References

Anderson B (1991) *Imagined Communities. Reflections on the Origin and Spread of Nationalism.* London: Verso.

Appadurai A (1986) *The Social Life of Things: Commodities in Cultural Perspective.* Cambridge: Cambridge University Press.

Bosse D (2007) Maps in the marketplace. *Cartographica* 42(1): 1–51.

Bradley A (1805) *Map of the United States Exhibiting Post Roads & Distances*. Washington: Bradley. Available at: www.davidrumsey.com/luna/.
Brodhead R (1993) *Cultures of Letters: Scenes of Reading and Writing in Nineteenth-Century America*. Chicago: University of Chicago Press.
Brückner M (2006) *The Geographic Revolution in Early America*. Chapel Hill: University of North Carolina Press.
Brückner M (2017) *The Social Life of Maps in America, 1750–1860*. Chapel Hill: University of North Carolina Press.
Dillon D (2007) Consuming maps. In: Akerman J and Karrow R (eds) *Maps: Finding our Place in the World*. Chicago: University of Chicago Press, pp. 289–343.
Dunlop C (2015) *Cartophilia. Maps and the Search for Identity in the French-German Borderland*. Chicago: University of Chicago Press.
Edward R (1798). *Mrs Noah Smith and Her Children*. Available at: www.metmuseum.org/art/collection/search/10835.
Habermas J (1989) *The Structural Transformation of the Public Sphere*. Cambridge: MIT Press.
Hale E (1891; c.1863) *The Man Without a Country*. Boston: Roberts Brothers.
Jacob C (2006) *The Sovereign Map: Theoretical Approaches in Cartography Throughout History*. Chicago: University of Chicago Press.
Jaffee D (2010) *A New Nation of Goods. The Material Culture of Early America*. Philadelphia: University of Pennsylvania Press.
Johnson E (1874). *The Lesson*. Available at: www.wikiart.org/en/eastman-johnson/the-lesson-1874.
Kopytoff I (1986) The cultural biography of things: Commoditization as process. In: Appadurai A (ed.) *The Social Life of Things*. Cambridge: Cambridge University Press, pp. 64–91.
Mayar M (2022) *Citizens and Rulers of the World: The American Child and the Cartographic Pedagogies of Empire*. Chapel Hill: University of North Carolina Press.
McHaffie P (2011) Manufacturing metaphors: Public cartography, the market and democracy. In: Dodge M, Kitchin R and Perkins C (eds) *The Map Reader: Theories of Mapping Practice and Cartographic Representation*. London: Wiley, pp. 129–133.
McMurray W (1784) *The United States*. Available at: www.loc.gov/item/gm71005423/.
Mitchell WJT (2005) *What Do Pictures Want? The Lives and Loves of Images*. Chicago: University of Chicago Press.
Mukerji C (1983) *From Graven Images: Patterns of Modern Materialism*. New York: Columbia University Press.
Popple H (1733) *A Map of the British Empire in America*. London: Willm Henry Toms & RW Seale. Available at: www.davidrumsey.com/luna/.
Pritchard M (2001) Maps as object of material culture. *The Magazine Antiques* (January): 212–220.
Rocques J (1746) *An Exact Survey of the City's of London, Westminster, ye Borough of Southwark, and the Country near Ten Miles round*. London: J Pine. Available at: www.bl.uk/onlinegallery/onlineex/crace/a/007zzz000000019u00018000.html.
Schulten S (2012) *Mapping the Nation: History and Cartography in Nineteenth-Century America*. Chicago: University of Chicago Press.

38

PUBLIC MAP EXHIBITIONS

What goes in and what comes out

Tom Harper

Introduction

Unlike their ambiguous exhibits, map exhibitions seem fairly easy to define. They are in-person or online events in which maps and other related objects are selected and arranged in physical and virtual spaces to present narratives about places, events or even maps themselves. Contiguous with an increase in temporary exhibitions over recent decades (Te Heesen, 2018: 59), map exhibitions have increased in number, variety and location (see Campbell, 1980–2022[1]; Doktor, 2023). Many of these involve public map collections housed in national, regional and civic research libraries and archives. Since access to these 'special' collections tends to be restricted to reading rooms, exhibitions provide, among other things (Marini, 2019: 14–16), a crucial means of showcasing them and work involving them to a broader audience in order to demonstrate value and relevance, a commitment to public service, justifying public funding. Physical map exhibitions continue to be a major activity (Harley, 1987: 21), generally carried out using host collections and staff, forming part of wider public programmes, incorporating loans, touring multiple venues, with accompanying publications and (since the mid-1990s) online companions. Given these efforts, exhibitions have arguably become more central to their library hosts' operations than the peripheral connotations of 'outreach' suggests.

A recent history of map exhibitions

Though by no means their core activity, there was a public-facing dimension to the earliest work of public map collections created during the 19th century from older private, royal, military and other collections (Wolter, 1973). For example, in the decades after its foundation in 1828, the public were admitted to view maps in the département des Cartes et plans of the Bibliothèque nationale de France (Richard, 2014: 65). Maps were included in later-19th-century science and industry exhibitions in the United States and UK (Edney, 2022), while the British Museum (BM)'s earliest documented temporary exhibition to include maps occurred in 1880 (Harris, 1998: 319).

The themes of BM—from 1973 British Library (BL)—map exhibitions, assessed by Barber and others (Barber, 2020: 136–137; Baigent and Millea, 2020: 300–301)—were

DOI: 10.4324/9781003327578-46

Figure 38.1 The evolving design and layouts of map exhibitions. Left: James Cook bicentenary exhibition, British Museum, 1968. Right: 'Maps and the 20th Century: Drawing the Line', British Library, 2016 (© British Library Board).

broadly reflected internationally. They included histories of particular places, marking centenaries, public events and conferences, history of science landmarks, new acquisitions, and technical or artistic aspects. From the 1980s, and perhaps earlier (Edney, 2023), these were joined by an additional theme that reflected the contextual treatment maps were receiving from academics (Barber, 2020: 134). Where maps in previous exhibitions had frequently illustrated 'Anglo-Saxon achievements' (Barber, 2020: 136) the new exhibitions focused upon maps' shared and universal cultural value, language and function ('What Use is a Map?' BL, 1989; 'The Power of Maps', Cooper-Hewitt Museum, 1992; 'Talking Maps', Bodleian Libraries, 2019; 'Karten: Navigeren en Manipuleren', Leiden University Library, 2022). A number were large 'blockbuster' shows ('Cartes et figures de la Terre', Centre Pompidou, 1980; 'Segni e Sogni della Terra', Palazzo Reale Milano, 2001; 'Maps: Finding our Place in the World', Chicago Field Museum, 2007) (Figure 38.1).

Later exhibitions combined this contextual approach with a focus on host collection strengths, such as map subjects and genres ('London: A Life in Maps', 2007; 'Magnificent Maps', 2010, both BL; 'L'âge d'or des cartes marines', Bibliothèque nationale de France, 2012; 'Geo Graphic', National Library Board, Singapore, 2015; 'The World in Maps', Beineke Library, 2022). Other exhibitions revisited earlier themes with new perspectives ('1492: An Ongoing Voyage', Library of Congress (LC), 1992; 'We Are One: Mapping America's Road From Revolution to Independence', Leventhal Map Center, Boston Public Library, 2015), and traditional perspectives ('Mapping our World', National Library of Australia, 2014). In the 2020s, the range of themes in large and small public map exhibitions across the world is as diverse as at any time.

Like all exhibitions, map exhibitions are products of a range of roles and 'complex interactions of competing parties and interests' (Lavine and Karp, 1991: 2). What goes into map exhibitions can therefore be difficult to identify. An approach isolating the key elements of host institution, curator, collection and audience provides the following insights.

The host institution

The host institution, in this case a library, commissions and sponsors the exhibition, provides space to house it, funds to pay for it (or the means to fundraise), a collection to exhibit and staff to fulfil the exhibition brief. These will vary in the case of loan and touring exhibitions. The library stakes its reputation on each exhibition, and each exhibition is an expression of the library insofar as the subject is given cultural legitimacy by its sponsorship. The library context for exhibitions lacks the copious analysis available for museum exhibitions (Fouracre, 2015: 378; Ryan and Quinn, 2022: 2–3) and has particular complexities. Many, such as the relative weakness of books as exhibits, are outlined in Rogatchevskaia's honest evaluation of curating the BL's 2017 'Russian Revolution: Hope, Tragedy, Myths' exhibition (Rogatchevskaia, 2018). Unlike books, however, maps are often a strong exhibiting medium. As 'Magnificent Maps' demonstrated, many were created expressly to be displayed (Barber and Harper, 2010: 9). The BL Map Library's (perhaps more than other BL special collections') pre-1973 museum existence may also account for its solid exhibiting reputation. Nevertheless, an understanding of library exhibitions should incorporate a perceived tension between research (traditionally 'library') and culture (traditionally 'museum'). Such tension was most apparent during the BL's early years as it attempted to forge an identity, its exhibiting resolve tested by a mindset within which a sceptical director general asked 'wouldn't we be better to devote our efforts to supporting other people's exhibitions?' (BL, 1980).

A range of motives lie behind a library's commitment to hosting exhibitions and will vary over time. A 1980 perspective is contained in a draft policy paper by the BL's Exhibitions Committee. Exhibitions should link with anniversaries and occasions 'which the national library is expected to commemorate', highlight strengths and little-known parts of the collection, reflect the interests, knowledge and enthusiasm of staff, 'sum up the state of scholarship', and 'entertain and stimulate the visitor' (BL, 1980). While there is continuity in the BL's 2015 commitment to 'engage everyone with memorable cultural experiences' (BL, 2015), four decades have brought change too, for example in what qualifies for national celebration.

Today exhibitions are closely aligned with a library's strategic and corporate objectives. These may include prioritising revenue generation or cultivating particular audiences and stakeholders. All will perform as an advocacy tool for the library, which is accountable to a higher body. For example, by hosting map exhibitions alongside centenaries, libraries fulfil a role in public life (Anon, 1955). By exhibiting newly acquired collections, libraries demonstrate prudent use of public funds (Hébert, 2003: 187). There are wider strategic relationships. Exhibition loans enable libraries to perform cultural diplomacy as part of wider strategic objectives (LC, 2000). Corporate sponsors can help forge truly spectacular map exhibitions, which in turn facilitate sponsors' agendas as outlined in their forwards to exhibition catalogues (National Library of Australia, 2013: xi; Palazzo Reale, 2001: ix).

The curator

The requirements and ethos of the host institution are channelled through exhibition curators, who shape the exhibition through subject specialism and knowledge of the exhibited collection. Like all exhibitions, map exhibitions are more than 'key instruments in the diffusion of specialist knowledge to a lay audience' (Moser, 2010: 22). Every exhibition creates knowledge through the decisions made by their protagonists, drawing on their 'cultural assumptions and resources' (Lavine and Karp, 1991: 1). This knowledge, manifested in the exhibition narrative, themes, choice and presentation of exhibits, will reflect the professional and personal perspectives of the curator. The BL curator was not reciting the exploits of Sir Francis Drake in the eponymous 1977 exhibition, but creating a narrative of an episode of British imperialism for a patriotic contemporary audience. Despite overtures of a universal mapping culture, 'Segni e Sogni della Terra' created a celebratory image of western (particularly Italian) science and industry. 'Mapping our World' crafted a narrative of the emergence of Australia through maps, reinforcing the identity of the 21st-century Australian state.

Map exhibitions' knowledge creation is particularly interesting to observe through the popular post-1980s exhibition narrative of a shared, universal tradition of maps and map-making across space and time. Maps' everyday social relevance was a main theme of the BL's 'What Use is a Map?'. '[T]he motorist as much a patron as the King of Spain' noted the exhibition catalogue, juxtaposing Matthew Paris's 13th-century itinerary map with a 1980s road atlas to demonstrate unity of purpose (Campbell, 1989: 10–11). There was also plenty to distinguish a priceless medieval manuscript from a charity shop bookshelf-warmer. The curator chose to identify this unified purpose and create a narrative of ubiquitous, enduring mapping heritage. It may well be that such an exhibition narrative, which came from the academic environment the curator inhabited, served to satisfy an already heightened public enthusiasm for maps in popular culture. It may also have gone towards establishing such an interest, hosted by such an acknowledged authority as a national library and reinforced by a publication, public lecture and press release (Figure 38.2).

The universal, transcultural mapmaking exhibition narrative has proven particularly popular and enduring. For example, 'Cartes et figures de la Terre' claimed that maps 'invest almost every human activity, from the noblest to the most frivolous' (Centre Pompidou, 1980: 1, author's translation). 'Maps: Finding our Place in the World' demonstrated mapping to be 'a universal phenomenon in all human societies' (Anon, 2008: 223), while the BL's 2016 'Mapping the 20th Century' juxtaposed maps from entirely different contexts to illustrate their ubiquitous graphic language. But there is a potential problem if the theoretical basis for the narrative is flawed, or, as has been suggested by Edney (2019: 125–127), actually reinforces an idealisation of mapping that was developed as a moral, modernising tool of western culture. Following this argument, there is no platonic idea of 'the map' however much it might support exhibition curators' 'ingrained cultural commitment' (Edney, 2019: 4) or (by including a broad range of maps) help the host institution to demonstrate diversity. Bringing together disparate artefacts in an exhibition and calling them all maps creates, rather than reflects, the knowledge that they share an identity. In doing this, exhibitions follow the leads of their collections.

The collection

The knowledge created by map exhibitions will reflect the exhibited collection, interpreted through the curator, who is influenced by it in turn. Unsurprisingly, European and North American map collections defined during the 19th-century prioritised acquisition of European-western scientific mapping, which facilitated their states' domestic and imperial activities. Ironically, this tight collecting focus poses problems for the painting of 'universal' exhibition narratives. Non-European maps, forming a minority of such collections, are frequently included in exhibitions to support universal narratives and to demonstrate diversity, but their inclusion can appear tokenistic or worse (Edney, 2019: 71; National Library of Australia, 2013: xx). Individual examples are often over-exhibited and their perception as exotic and unusual—often the original reason for their acquisition—perpetuated.

Figure 38.2 The effective cover illustration of the British Library's 1989 'What Use Is a Map?' exhibition catalogue. Like the exhibition, the illustration brought together images of very different types of objects under the umbrella term of 'map' (© British Library Board).

Though not a public collection, an interesting perspective is provided by the 2018 Musée Guimet exhibition 'Le Monde vu d'Asie' which made use of that collection's maps from different Asian cultures. The exhibition's curators, external academics unaffected by institutional loyalty, created a narrative in direct opposition to European maps (of which none were included) and the colonial endeavour, urging the visitor to 'shift the gaze, reverse the perspectives' (Singaravélou and Argounès, 2018: 11, author's translation). In acknowledgement that the Guimet collection is a western appropriation, mostly acquired through the same processes the exhibition narrative was critiquing, a chapter of the catalogue investigated European collecting. The attempts of 'Le Monde vu d'Asie' to overcome the profound difficulties of escaping a western perspective demonstrates rising awareness of the issue, certainly when compared with the BL's 1974 'Chinese and Japanese Maps' exhibition which aimed to 'redress the balance of a series of exhibitions devoted primarily to the cartographic and astronomical achievements of the western world' (Nelson, 1974: 357). That is, on western terms, with no acknowledgement that some exhibits had probably been looted by invading British forces a century earlier.

Over the past decade, a change in social attitudes and the creation of policies by host institutions (often following equality legislation) has produced greater inclusion and diversity in map exhibitions and therefore an increasing commitment to social relevance. Exhibitions are becoming better at providing openness over items' acquisition, agency and, as in the Royal Geographical Society's 'Hidden Histories Made Visible' exhibition, previously obscured authorship (Driver, 2013). An improvement is detectable in interpretation. For example, comparing the BM's 1968 Cook bicentenary exhibition with the BL's 250th centenary exhibition 'Cook: The Voyages' (2018) reveals a noticeable change in the type of exhibits accompanying maps (portraits of British heroes in the former are replaced by ethnographic portraits and scenes in the latter) and the neutral tone of the latter's labels (Anon, 1968; Frame, 2018).

Yet do such interventions simply skirt the issue? For all that recent map exhibitions strive to demonstrate inclusion and diversity through narrative, interpretation and selection, there remain questions around how their collections can adequately fulfil the brief. For every Phyllis Pearsall and Marie Tharp, 'Mapping the 20th Century' displayed at least 30 maps attributed to male mapmakers. Socially inclusive map exhibitions are not created through diverse exhibits alone, or even inclusive narratives which often, as we have seen, remain couched in narrow, traditionally western perspectives. Particular effort is required, for example, to involve communities in exhibitions (Perkins, 2018).

The audience

Shared approaches and co-production blur the roles between exhibition curation and the final component, the audience. In line with recommendations for audience focus in exhibition manuals (Smithsonian, 2018: 8), and a recognition that public participation 'enhance[es] both the legitimacy and accountability of publicly funded culture', audience input into various stages of museum exhibitions' development is increasing (Davies, 2010: 305–306). A similar picture is not overly evident in map exhibitions, however. Despite the inclusion of user-generated content through interactive exhibits, community-produced exhibits (Harper, 2010a) (Figure 38.3), and maps by non-specialists (Campbell, 1989: 5), such engagement can appear superficial compared with exhibitions for which themes and narratives have been generated through external collaboration. Research suggests that

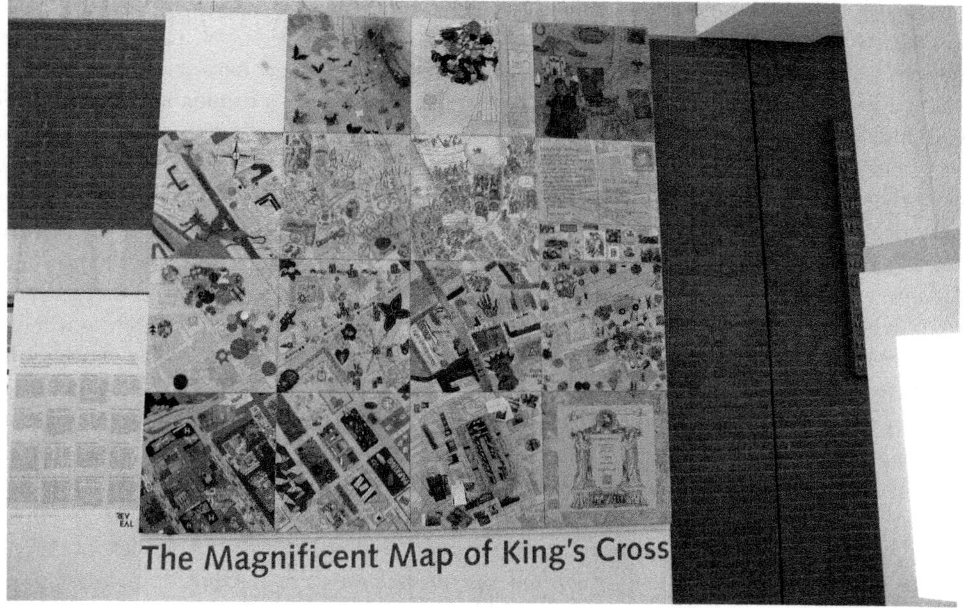

Figure 38.3 The 'Magnificent Map of King's Cross', a community-produced wall map produced as part of the King's Cross 'Reveal' festival during the British Library's 'Magnificent Maps: Power, Propaganda and Art' exhibition, on display in the lobby of the British Library in 2010.

Source: Image credit: Tania Rossetto

smaller cultural venues are better at co-producing exhibitions, and that larger institutions worry about losing authority and control of exhibition narratives (Davies, 2010: 316–318).

Limited audience input stands in contrast to the sophisticated methods by which many institutions' marketing and press operations promote exhibitions. Newsworthy stories deploying 'never-before-seen' tags provide valuable publicity (Anon, 2022). Additional media can also widen exhibitions' exposure. The loan of the 1450 Fra Mauro map formed a key plank of 'Mapping our World's' media (Rudra, 2013), a live satellite broadcast was staged in 'Cartes et figures de la Terre' (Centre Pompidou, 1980: 1), while a TV documentary 'The Beauty of Maps' aired in the UK two weeks before 2010s' 'Magnificent Maps' opened (Harper, 2010b).

Measures of success

The documentary undoubtedly contributed to 'Magnificent Maps' impressive 226,000 individual visits. Other 'blockbuster' map exhibitions also performed well (Anon, 2008, 2014), though map exhibition figures are dwarfed by those of major art shows. Touring exhibitions can further extend reach, generating social impact. For example, various BL exhibitions travelled to North American venues in the 1970s and 1980s. The University of Wisconsin's 'Maps and the Columbian Encounter' visited 15 venues in 12 states (Campbell, 1993: 140). National Library of Scotland's facsimile 'You Are Here' toured local Scottish libraries (National Library of Scotland, 2023). But attendance figures are just one measure of success. As a free admission show, 'Magnificent Maps' did not perform equally well financially. Its admission-charging successor 'Mapping the 20th Century' performed comparatively better while receiving only a quarter of the visitors.

Free entry to cultural attractions increases footfall but does not necessarily diversify the profile of visitors (Martin, 2007: 409–411). 'Assessment beyond numbers' (Marini, 2019: 18), the extent that map exhibitions are judged on their ability to draw broad audiences, is difficult to ascertain since targets and closing reports tend not to be publicly available. Nevertheless, research into visitor types has been carried out for decades. 'Casual visitors, especially tourists', 'regular non-specialist visitors', 'educational groups, sometimes ethnic or religious' and 'specialists, experts and scholars' constituted four attendee categories identified by the BL Exhibitions Committee in 1980 (which concluded 'that it is at the casual visitor that possible improvements should be aimed') (BL, 1980). Though this may demonstrate a commitment to public engagement, a lack of proper evaluation processes has led to ongoing scepticism of library exhibitions' effectiveness as outreach exercises (Fouracre, 2015: 382).

With notable exceptions (Le Blanc, 2016), specific audience engagement tends to be conducted by libraries' community and learning teams once map exhibitions are open. As occurred during 'Maps: Finding Our Place in the World', 'twelve year old children from a public school in Chicago's inner city deeply engaged in examining the great Leardo mappamundi' (Anon, 2008: 224) seems to describe the perfect circumstances for map exhibitions to 'question the relationship between looking, knowledge and power' (Hooper-Greenhill, 2020: 76). Another measure of success, how well map exhibitions' narratives resonate with audiences, is similarly difficult to gauge due the paucity of available reports. A rare publicly available report, into the 1992 'The Power of Maps' exhibition, demonstrated that visitors' appreciation of maps as subjective interpretations rather than mirrors of the world did improve between the exhibition's beginning and end (Doering et al., 1999: 20–31). But regardless of how digestible the narrative may be, it is unwise to assume that audiences cannot decide themselves what is at stake (Karp, 1991: 15), or are as concerned with sociocultural map narratives as simply wanting to look at maps. That map exhibition reviews have continued to appear in newspapers' travel rather than culture sections (e.g. Calder, 2016) suggests that efforts to imbue maps with a higher cultural status have not been entirely successful.

Conclusion: the future of map exhibitions

Despite the considerable digital focus of library map collection outputs, often with accompanying agonies (Piekielek and Bidney, 2020), and notwithstanding successful switches to online exhibitions during the 2020–2022 COVID pandemic (Nelson, 2020), it seems doubtful that exclusively online exhibitions will adequately supersede in-person map exhibitions soon. However, an increasing cache of younger visitors for whom paper maps mean increasingly little, and the participatory nature of much contemporary mapping, suggests that future map exhibitions might incorporate new mapping modes to provide relevance. It will also be interesting to see whether socially engaged, post-representational mapping by artists, activists and theorists (Rossetto, 2015; Cohen and Duggan, 2021) will find a meaningful way into public map exhibitions. Such maps tend to sit outside of library collecting remits and are consequently untainted by map collections' often problematic histories. Perhaps salvation lies in a stronger dialogue between the two.

For the majority of people, maps will continue to function as conventional objects of navigation, education and enjoyment regardless of public map collections' attempts to mediate through exhibitions. However, it seems probable that their exhibitions have in some way broadened the public perception of maps. A desire for social relevance underpins much of this effort. Such endeavours are assisted by the dynamism in map studies exemplified by early map research and the cartohumanities, emergent mapping practices, and maps' growing prominence in a diverse range of exhibitions across the world. Together these may help fulfil their advocates' hopes for maps to operate as forces for change in societies.

Note

1 Tony Campbell's 'Chronicle' has been included in successive issues of Imago Mundi since 1980 and documents the increase in number, variety and location of map exhibitions.

References

Anon (1955) The celebrations of the 700th anniversary of Marco Polo's birth at Venice. *Imago Mundi* 12(1): 139–140.

Anon (1968) *An Exhibition to Commemorate the Bicentenary of Captain Cook's First Voyage Round the World. The Kings Library, the British Museum 19 July to 27 October 1968*. London: The British Museum.

Anon (2008) Conference reports. *Imago Mundi* 60(2): 221–226.

Anon (2014) Maps break national library attendance record. *Canberra City News*, 9 March. Available at: https://citynews.com.au/2014/maps-break-national-library-attendance-record/ (accessed 19 January 2023).

Anon (2022) NLB's past and ongoing exhibitions. *The Straits Times*, 11 May. Available at: https://www.straitstimes.com/askst/nlb%E2%80%99s-past-and-ongoing-exhibitions (accessed 1 January 2023).

Baigent E and Millea N (2020) 'Intelligent strangers as well as members': Enlightening maps and social and political spaces for cartographic conversations. *The Cartographic Journal* 57(4): 294–311.

Barber P (2020) 'Context is everything . . .': Ruminations on developments in the history of cartography since the 1970s and their consequences. *Imago Mundi* 72(2): 131–147.

Barber P and Harper T (2010) *Magnificent Maps: Power, Propaganda and Art*. London: The British Library.

British Library (1980) *Map Departmental Archives, PB/MAPS/AA2/HW/DL*.

British Library (2015) *Living Knowledge: The British Library 2015–2023*. Available at: www.bl.uk/about-us/our-vision#:~:text=Living%20Knowledge%20(PDF%20format)%20explains,for%20research%2C%20inspiration%20and%20enjoyment (accessed 1 January 2023).

Calder S (2016) Back on the map: Why paper sometimes isn't enough to help you find your way. *The Independent*, 28 October. Available at: www.independent.co.uk/travel/news-and-advice/british-library-maps-exhibition-maps-and-the-20th-century-apps-ordnance-survey-a7383611.html (accessed 1 January 2023).

Campbell T (1980–2022) Chronicle. *Imago Mundi*, 32–74.

Campbell T (1989) *What Use Is a Map?* London: British Library.

Campbell T (1993) Chronicle. *Imago Mundi*, 45: 137–148.

Centre Georges Pompidou (1980) *Cartes et figures de la Terre: Information Presse CCI*. Available at: www.centrepompidou.fr/media/document/14/de/14de72ff434d630eeeab4c2ded1d94ad/normal.pdf (accessed 15 December 2022).

Cohen P and Duggan M (eds) (2021) *New Directions in Radical Cartography: Why the Map Is Never the Territory*. Lanham, MA: Rowman and Littlefield.

Davies SM (2010) The co-production of temporary museum exhibitions. *Museum Management and Curatorship* 25(3): 305–321.

Doering D, Bickford A, Karns, DA and Kindlon AE (1999) Communication and persuasion in a didactic exhibition: The Power of Maps study. *Curator: The Museum Journal* 42(2): 88–107.
Doktor JW (2023) *Cartography: Archive of Past Events and Exhibitions*. Available at: www.docktor.com/archive.htm (accessed 20 February 2023).
Driver F (2013) Hidden histories made visible? Reflections on a geographical exhibition. *Transactions of the Institute of British Geographers* 38(3): 420–435.
Edney M (2019) *Cartography: The Ideal and Its History*. Chicago: University of Chicago Press.
Edney M (2022) The history of map exhibitions. In *Mapping as a Process*. Available at: www.mappingasprocess.net/blog/2022/11/27/the-history-of-map-exhibitions (accessed 2 January 2023).
Edney M (2023) The history of cartography at MOMA, 1943. In *Mapping as a Process*. Available at: www.mappingasprocess.net/blog/2023/2/17/the-history-of-cartography-at-moma-1943 (accessed 28 February 2023).
Fouracre, D (2015) Making an exhibition of ourselves? Academic libraries and exhibitions today. *The Journal of Academic Librarianship* 41(4): 377–385.
Frame W (2018) *James Cook: The Voyages*. London: British Library.
Harley JB (1987) The map and the development of the history of cartography. In: Harley JB and Woodward D (eds) *The History of Cartography Volume One: Cartography in Prehistoric, Ancient and Medieval Europe and the Mediterranean*. Chicago: University of Chicago Press, pp. 1–42.
Harper T (2010a) Magnificent map of King's Cross. In *Maps and Views Blog*. Available at: https://blogs.bl.uk/magnificentmaps/2010/05/magnificent-map-of-kings-cross.html (accessed 9 November 2022).
Harper T (2010b) The beauty of maps #1. In *Maps and Views Blog*. Available at: https://blogs.bl.uk/magnificentmaps/2010/04/the-beauty-of-maps.html (accessed 10 March 2023).
Harris PR (1998) *A History of the British Museum Library 1753–1973*. London: British Library.
Hébert JR (2003) The map that named America: Library acquires 1507 Waldseemüller map of the world. *Library of Congress Information Bulletin* 62(9): 187–193.
Hooper-Greenhill E (2020) *Museums and the Interpretation of Visual Culture*. London: Routledge.
Karp I (1991) Culture and representation. In: Lavine SD and Karp I (eds) *Exhibiting Cultures: The Poetics and Politics of Museum Display*. Washington and London: Smithsonian Institution Press, pp. 11–24.
Lavine SD and Karp I (1991) Introduction: Museums of multiculturalism. In: Lavine SD and Karp I (eds) *Exhibiting Cultures: The Poetics and Politics of Museum Display*. Washington and London: Smithsonian Institution Press, pp. 1–9.
LeBlanc M (2016) American Revolution maps in the classroom: K-12 education at the Norman B. Leventhal map center. *Journal of Map & Geography Libraries* 12(3): 281–294.
Library of Congress (2000) *John Bull & Uncle Sam: Four Centuries of British American Relations*. Washington, DC: Library of Congress.
Marini F (2019) Exhibitions in special collections, rare book libraries and archives: Questions to ask ourselves. *Alexandria* 29(1–2): 8–29.
Martin A (2007) The impact of free entry to museums. In: Sandell R and Janes RR (eds) *Museum Management and Marketing*. Abingdon: Routledge, pp. 406–415.
Moser S (2010) The devil is in the detail: Museum displays and the creation of knowledge. *Museum Anthropology* 33(1): 22–32.
National Library of Australia (2013) *Mapping Our World: Terra Incognita to Australia*. Canberra, A.C.T.: National Library of Australia.
National Library of Scotland (2023) *On tour: You are Here*. Available at: www.nls.uk/exhibitions/touring-displays/you-are-here/ (accessed 5 March 2023).
Nelson GD (2020) Bending lines: Maps and data from distortion to deception. *Cartographic Perspectives* 96: 51–60.
Nelson H (1974) Chinese maps: An exhibition at the British Library. *The China Quarterly* 58: 357–362.
Palazzo Reale (2001) *Segni e sogni della Terra: il disegno del mondo dal mito di Atlante alla geografia delle reti*. Novara: De Agostini.
Perkins C (2018) Community mapping. In *Oxford Bibliographies Online in Geography*. Available at: www.oxfordbibliographies.com/view/document/obo-9780199874002/obo-9780199874002-0184.xml (accessed 14 April 2023).

Piekielek N and Bidney M (2020) What is everyone supposed to be doing? *Journal of Map & Geography Libraries* 16(1): 1–6.
Richard H (2014) Jomard et la diffusion des sciences géographiques. *Bulletin de la Sabix* 54: 61–66.
Rogatchevskaia E (2018) Revolution collected and curated Russian Revolution: Hope, tragedy, myths at the British Library. *Slavic & East European Information Resources* 19(3–4): 175–200.
Rossetto T (2015) Semantic ruminations on 'post-representational cartography'. *International Journal of Cartography* 1(2): 151–167.
Rudra N (2013) National Library of Australia's Mapping Our World exhibition reveals much about our past. *The Sydney Morning Herald*, 9 November. Available at: www.smh.com.au/entertainment/national-library-of-australias-mapping-our-world-exibition-reveals-much-about-our-past-20131107-2x47y.html (accessed 9 November 2022).
Ryan T and Quinn B (2022) Understanding the library as a commemorative exhibition space. *Public Library Quarterly*. Available at: Citation Manager | Taylor & Francis Online (tandfonline.com) [0].
Singaravélou P and Argounès F (2018) *Le Monde vu d'Asie: Une Histoire Cartographique*. Paris: Musée national des arts asiatiques-Guimet.
Smithsonian Institution (2018). *Guide to Exhibit Development*. Available at: https://exhibits.si.edu/wp-content/uploads/2018/04/Guide-to-Exhibit-Development.pdf (accessed 18 December 2022).
Te Heesen A (2018) On the history of the exhibition. *Representations* 141: 59–66.
Wolter J (1973) Geographical libraries and map collections. *Encyclopedia of Library and Information Science* 9: 236–266.

39

PARTICIPATORY NETWORK MAPPING FOR PUBLIC ACTION

Barbara Brayshay and Aldo de Moor

Introduction

In this chapter, we demonstrate the role of participatory community network mapping as a tool for addressing the societal problems of social and economic inequality that persist for people in marginalised communities across a wide range of spatial and temporal scales and settings. Frequently embedded in the post-industrial and post-colonial geographies and histories of communities, such deeply entrenched problems are a complex interlinked set of issues that perpetuate cultural and social disadvantage and the power relations that sustain them and as such they can appear to be intractable. Like communities everywhere, they are made up of complex social networks of relationships, interactions and connections embedded in the mesh of wider structures of the communities and agencies they interconnect with. These networks include the micro-scale informal networks of community members and local groups, meso-scale networks of community support agencies and NGOs and macro-scale networks of institutional support services and the wider societal context. In attempting to unpick these complexities, there is a need for tools and processes that are attuned to the specific needs of the community that can identify the networks of the community support ecosystem and the agencies and stakeholders that operate in the wider social and public sector domains to help build collaborative solutions. In the midst of this complexity people seeking solutions to their problems can find themselves lost in the bewildering landscape of fragmented service provision and access to information, help and advice. Equally service providers struggle to connect with the communities they are supposed to serve. These boundaries between community members, community support services and the wider networks of external institutions and agencies can be hard to bridge. What they need is a map!

We introduce the case study which was commissioned to create a systems map of the community support ecosystem available to unemployed people in the Black Caribbean and other minority communities in the London Borough of Lambeth and identify the barriers and leverage points to their economic engagement. Next, we outline our design philosophy to enable the community to find its voice using participatory mapping and storytelling. We show how we applied this philosophy in the Lambeth case study in which the community takes stock of its issues and the available support services through storytelling, using this

 DOI: 10.4324/9781003327578-47

as an input for mapmaking; creating a community support network map; and initial steps for redesigning community support together with the stakeholders. We present a participatory mapping meta-model of bridging the community–institution support divide which outlines dimensions to further explore using mapping and storytelling for stronger community support.

The Lambeth case study

To illustrate the potential of participatory community network mapping as an approach for addressing these challenges, we present the findings from an urban case study from research undertaken with The Ubele Initiative,[1] a Black diaspora organisation who commissioned a social systems' map as part of the 'Black on Track' initiative, a project that aimed to address issues of unemployment and underemployment and the deeper systemic changes needed to remedy the barriers that are limiting the opportunities, prosperity and well-being of the community (Brayshay and Mackie, 2023).

Today Lambeth's Black communities are struggling to overcome a legacy of economic marginalisation that can be traced back to the early days of the 1940–1950s Windrush,[2] when people arrived from the Caribbean Islands and West Africa in response to the British government's call for workers to come to help rebuild post-war Britain. Decades of social and racial inequality followed their arrival. Located in the complex entanglements of post-colonial history, Lambeth's Black communities have relied on local councils and central government to deliver public services and provide a safety net for support in times of hardship. However, as central government austerity policies have continued to reduce state support, many communities have found themselves falling into a gap, having neither the top-down protection of state provision nor the bottom-up grassroots resource networks within the community (Mould, 2022). The impact of the COVID-19 pandemic showed all too clearly that Black, Asian and minority ethnic people have been acutely affected by pre-existing inequalities across a range of areas, including health, employment, accessing Universal Credit, housing, and the no recourse to public funds policy (House of Commons Women and Equalities Committee, 2020). In response to this crisis, community-led initiatives are working to self-organise and build community resourcefulness that recognises the agency needed to be present within a community to resist oppression and marginalisation (MacKinnon and Derickson, 2013). This has resulted in new approaches to co-production and collaborative forms of co-working and co-design to social problem-solving and building a more equitable civil society (Chatterton, 2019, Sendra, 2023). A major challenge is one of building bridges across the gaps between an isolated community, and its community organisations and external institutions to maximise the social capital available to them.

A further barrier facing both service providers and service users in their capacity-building efforts is the fragmented nature of the intra-community and inter-institutional/public support ecosystem. Although partially embedded in each other's networks, significant gaps emerge as the community struggles to find the available resources to help meet its needs, and organisations and institutions in the support ecosystem also struggle to reach those most in need. Previous research undertaken in Lambeth's Black communities with a specific focus on evaluating dissatisfaction with council services (Equinox Consulting, 2013) and building capacity in the Community and Voluntary Sector (Equinox Consulting, 2017) addressed issues of major concern and the availability of support services in the Borough. Unemployed people who took part in the research reported that a lack of ready access to the job market was not only affecting them financially but also impacting on their pride, dignity, and mental and physical health.

It is into this relatively uncharted territory of Lambeth's Black community support networks that we developed a mapmaking and map-reading methodology to begin to visualise its social support ecosystem and identify pathways to bridge the gaps between the community and institutional support networks. It comprised of two participatory mapping workshops with members of the Black on Track project. These comprised of a storytelling workshop to discover participants' experiences and perceptions of the community and the support network which formed the basis of issue and support network maps. Additional desktop internet searches for service providers supplemented the support network data set. This was followed by a 'sensemaking' workshop to explore pathways to the support network identified in the mapping.

Finding the community voice: participatory community network mapping and storytelling

The map design process is inspired by the CommunitySensor methodology for participatory community network mapping of de Moor (2017, 2018) who describes it as a core communal sensemaking activity, a participatory process of capturing, visualising and analysing community network relationships and interactions. A process in which a community maps its objectives, participants and resources to give meaning to their collective experiences and to gain an understanding of who they are and what they aspire to. An essential part of the process includes mapping aspirations as well as identifying community issues, assets and resources so that the mapping becomes a collective re-imagining not only to make sense of the community but also to build a collective vision for the future. Key to this mapping approach is that the community does not just map itself but also the wider context outside of its boundaries to enable the community to collaboratively design solution directions for the problems it encounters. By explicating and jointly making sense of not just community needs but also the collaboration ecosystem in which it lives and works, collective, scalable and impactful solutions can be woven together with stakeholders both inside and outside the community, including the institutions it needs to engage with. However, empowering the community remains of the essence. The starting point for the participatory mapping and collaborative sensemaking of the CommunitySensor methodology is for the community to define its needs on and in its own terms, including problems and capabilities, resources and any solutions they already have. In other words, the community needs to find its own voice first.

The first step was to identify the communities' needs and capabilities. However, just mapping the resource base would do little to incorporate the subtle interrelationships between the community's needs, help identify leverage points to economic engagement nor address the many sociocultural barriers to realise community support.

Undertaking work in communities with people who are ethnically, culturally or economically different from ourselves provokes us to think deeply about issues of positionality and situated knowledge production (Pratt, 2009; Kinkaid, 2022). Recent research by Black geographers has focused attention on the complex spatialities of Black life, those of decolonisation, racism, marginalisation, justice and representation, evidenced in the continued economic and social marginalisation of communities such as Lambeth. Our research practice is guided by the work of Fricker (2009) and Hawthorne (2019). Kidd et al. (2017)

and particularly Walker and Boni (2020), who highlight issues of epistemic in/justice as foundational to a reflexive, inclusive and decolonial approach to knowledge production and its importance in participatory research practice.

To this end, we step back so that the community can speak for itself through the medium of storytelling to discover its own concepts, language and proto issues as well as strengths and capabilities that can then be further situated, discussed, validated and co-owned by the community. Our storytelling approach is based on the principles of Participatory Narrative Inquiry (PNI) which uses storytelling as a tool for communities to make sense of complex situations and work towards socially innovative solutions for problem-solving, focusing on weaving together individual perspectives through the recounting and interpretation of lived experiences (Kurtz, 2014; Copeland and de Moor, 2018).

Step 1—community stock-taking: telling the stories

The first participatory workshop aimed to explore with participants the barriers and challenges they had faced that had brought them to their current situation and their needs and aspirations to plan for a better future for themselves and the community.

The research tools used in the storytelling were derived from persona creation and user journey mapping, frequently used in the design of web-based products and services (Følstad and Kvale, 2018). Personas are fictional characters created by participants based on real life user experience. The creation of fictionalised characters as a vehicle for storytelling allows participants to transfer their very personal experiences to the character that can be less revealing for those who may not want to disclose sensitive information publicly. The workshop took place during the COVID-19 pandemic and therefore had to be held online. A group of 20 participants were divided into smaller breakout groups of five or six participants. Each group was asked to create a persona that was a typical character from their community challenged by unemployment. Participants then took their personas on a user journey drawn from the experiences that had brought them to their current situation, the obstacles they face going forward as well as their aspirations and goals.

Two persona examples distilled from the storytelling workshop will be used to illustrate the methodology.

Persona women's voices 'Lydia'

Lydia's story speaks to issues of racial and gender discrimination experienced by women in the Black community. Her key issues are single parenting, gang culture, police surveillance of Black youth, affordable childcare, and balancing work and childcare commitments. Her aspirations are to find a pathway out of zero-hour, low-paid work into self-employment and achieve a better work–life balance for herself and her children.

Persona older unemployed men 'Sam'

Sam has been made redundant and is now the primary carer for his wife. Redundancy and his domestic situation have cut him off from both the financial benefits and social networks of work. Sam's key issues are depression, poor mental and physical health, isolation,

Table 39.1 Issues identified by the community in the storytelling workshop

Community	*Thematic Cluster*	*Issues in Community Terms*
	Health issues	Alcoholism, autism, addiction, drugs, fitness for work, mental health, obesity, physical health
	Social/cultural issues	Antisocial behaviour, dysfunctional families, food poverty, gang culture, gender discrimination, isolation, knife crime, lack of parenting skills, loneliness, low expectations in education, low expectations in family, no entrepreneurial culture, racial discrimination, racism, unsafe streets, single parents
	Personal issues	Lack of ambition, lack of confidence, lack of education,
Institutional	Service barriers	Access to housing, access to local jobs, access to business advice, gentrification, local poverty, lack of tech/digital skills, policing of Black youth, poor local economy, poor local environment

loneliness, lack of ambition and financial stress. Sam's story illustrates the complexity of his interlinking health and support needs to help him into active and productive employment.

These examples of personas and user journeys illustrate the complexity of unemployment-related needs and ways to address them, which are much more intricate than simply matching an unemployed person up to the local Job Centre as would often be seen as the quick fix, whereas the community can and should be a catalyst in finding support. The storytelling workshop generated a range of issues and barriers experienced by participants summarised in broad thematic clusters in Table 39.1.

Lydia and Sam each have support needs and issues some of which they can find within the Lambeth Black community, but they also need institutional support from public services such as healthcare, employment, education, training and social services, and they need to know where to find them.

Significant findings from the workshop were: participants were not aware of the range of support available to them and that that a great deal of service provision is invisible to them; conversely, the council know that they offer so much and that people can't find them. As a result, institutional services may not be in line with the complex needs identified by the community; all of which demonstrates the need for service redesign.

Step 2—mapping the community support networks

Moving on to visualise and collectively redesign the community support network began with creating a composite map linking the resource base mapping of community and institutional support networks together with the issues mapping from the storytelling workshop (Figure 39.1). It shows which community-based organisations provide community support around which issues. It also includes the same for the institutional services available, although that is still only an incomplete and partial view through the eyes of the community, as the institutions themselves were not yet included in the process. The composite map showed that the Lambeth support ecosystem is a loosely coordinated set of organisations; some of which collaborate with each other and whose members have different

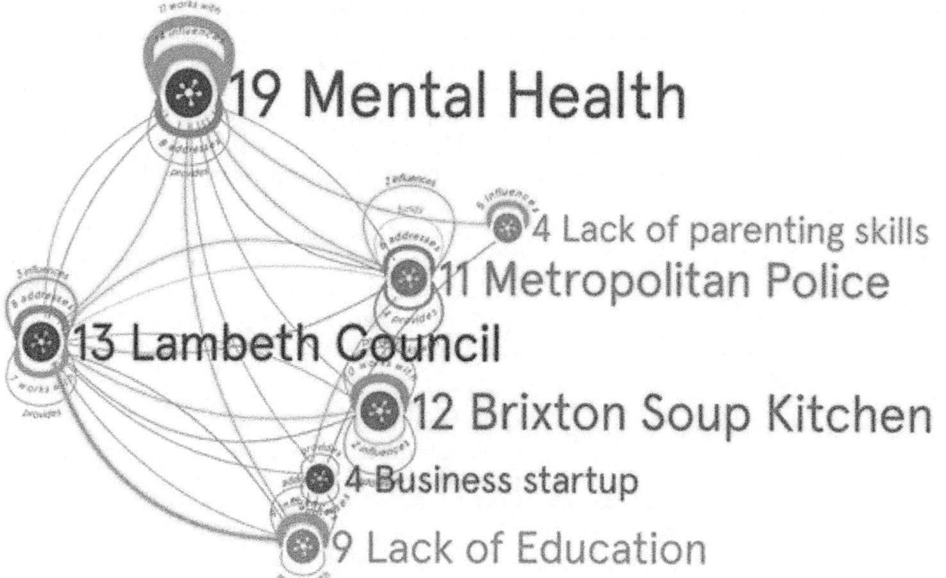

Figure 39.1 A snapshot from the composite map of community issues and support services. See Brayshay and Mackie (2023) for the complete network maps.

goals, command different resources and follow different processes. It includes community organisations and groups, such as community centres, local voluntary groups and community resources such as schools, nurseries, libraries and NGOs and mutual aid groups such as Age UK and Alcoholics Anonymous who provide services in the borough. Institutional-state services that relate to the participant's issues are also included in the resource base mapping such as Lambeth Borough Council, Unemployment Services, the NHS and Metropolitan Police (note the map is dynamic and content is increasing as the project continues).

The composite map thus visualises the overall structure of the network and can be used to trace development pathways for both individuals and organisations by connecting personal needs to support organisations and vice versa. As such, apart from taking stock of support provided by the community itself, it represents a first step towards cross-boundary knowledge sharing between the 'life world' of the community and the 'systems world' of service providers. Future work with representatives of these institutions themselves will explore how they contribute to building service bridges into the community from their point of view.

Step 3—redesigning community support

An essential next step in the process was to make the outcomes from the mapmaking accessible and usable for collaborative 'sensemaking' with the community itself. To this end, data cards with details of the issues and organisations on the map were generated by the Graph Commons mapping software.[3] The cards formed the basis of an iterative Domino-style game as a collaborative exercise in which participants used the cards to explore the

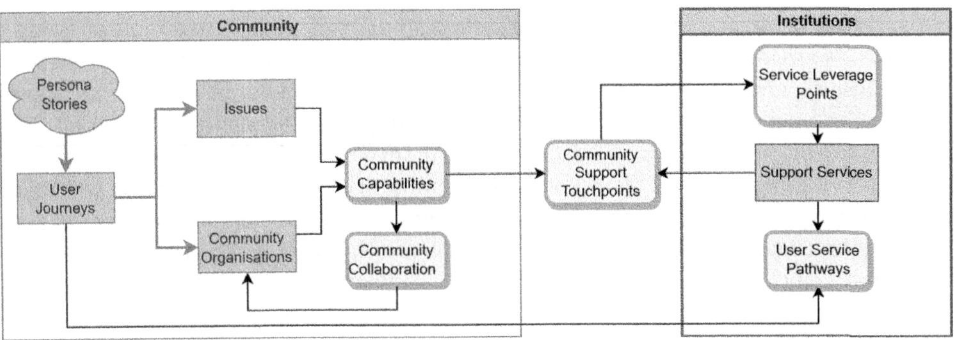

Figure 39.2 A participatory mapping meta-model of bridging the community-institution support divide.

community support ecosystem by creating existing and potential combinations of issues and organisations and discussing them in the group, jointly coming up with suggestions for redesigning community support provided by both community organisations and institutional service providers.

COVID-19-pandemic restrictions resulted in opportunities to play the beta version of the game being limited to an online workshop. Future envisioned in-person sessions with the community and different stakeholder situations will allow for more direct forms of engagement with the cards, which is especially important in marginalised community settings for building trust and a sense of community.

Now that the community is finding its voice and its resources how do we continue?

Still missing is how to build bridges into the world of the institutions that are or should be servicing the community, and vice versa. Our current composite map focuses on mapping issues with community capabilities and institutional support services as identified. However, it does not say anything about how to redesign community support that aligns community capabilities with additional institutional services. In Figure 39.2, we present a participatory-mapping meta-model that we designed to capture and extend the community support design dimensions to be put on the map. These dimensions contain ideas about providing and improving community support consisting of community capabilities, institutional services and their amalgams.

The five main design dimensions of the model are as follows:

1. Community Capabilities: Community organisations provide more or less informal capabilities/opportunities for meeting citizens' needs.
2. Community Collaboration: Community organisations working together more effectively and efficiently to provide better community-owned support.
3. User Service Pathways: How citizens find and interact with institutional community support services.
4. Community Support Touchpoints; Support interfaces where the community and institutional worlds meet, matching community-owned support capabilities with institutional services.
5. Service Leverage Points: Institutional service (re)design for providing more integrated service palettes that better meet community support needs.

A narrative using the persona of Sam illustrates how the model could hypothetically be applied as a source of inspiration for future mapping and sensemaking exercises that would help to inspire ideas for community support redesign, amplifying existing community resources, and better aligning community-owned support with institutional services. The persona narrative begins with an initial starting point within the Black community's capabilities, Sam joining the local Domino Club [Community Capability]. There is potential to build powerful collaborations that weave community organisations into intra-community networks. An example could be that the Brixton Soup Kitchen setting up in the Lyon Lloyd Community Centre has the potential to reach more clients who come there for other activities [Community Collaboration]. Every month in the Community Centre, Age UK holds an advice session just prior to the Soup Kitchen starting to deliver food [Community Support Touchpoint]. While Sam is waiting for his soup, he gets to talk to the Age UK representative who introduces him to their Lambeth My Social programme which offers a range of social and advice services for elderly people [User Service Pathway]. The Age UK representative also directs Sam to Mind, a U.K. mental health charity, who can give Sam more personal advice and support to help him with his depression. By piggybacking on the institutional capacity of Age UK, in this way, Mind can reach many more clients than they otherwise could [Service Leverage Point].

Such network weaving, which the mapping project aims to address, will help build Black equity and result in a much more effective support system within the community. Reaching beyond the community network to bridge the gap into the institutional-state sector, service touchpoints draw institutional services into the ecosystem creating possibilities for inter-community and intra-community collaborations that are needed to strengthen the community support ecosystem. The service leverage points are directed towards possibilities for service redesign in which institutional-state sector provision is also more integrated and connected.

Further mapping of the networks of organisations and individuals in Lambeth, including research organisations, community organisations and government services, aims to increase connectivity and share knowledge that can impact systems change at community, institutional and combined levels. It will have the ability to encourage rich webs of connections so that there are always alternative support pathways available.

Conclusion

Marginalised communities face many issues, but they also have many capabilities and strengths to address them. Communities are not islands, but part of a much larger societal ecosystem. Many institutional services exist that could better benefit communities, if only the divide between the community and institutional service provision could be bridged. Better community support is thus a resultant of the community being better able to identify its needs and capabilities and more effectively engage with relevant institutional service providers for additional support. In this, it is of the essence that the community can find its own voice to engage with its surrounding institutions on an equal footing.

We have presented a participatory community mapping approach consisting of a community first identifying its needs and existing support through storytelling, then creating a composite map matching issues with potential community and institutional resources, and then engaging with the community and ideally institutional representatives to redesign community support. We have introduced a case study in which we used this approach to help the Black community in Lambeth to co-create better community support for its unemployed people. We outlined the next steps that could and will be taken in future work to further extend the approach, in particular for aligning the (re)design of community capabilities with institutional service mixes.

All too often, communities and institutions see each other as 'the other side'; however, they need and can work together to create joint, more sustainable, impactful solutions for many of the societal problems communities face. We have begun to map the uncharted 'territory' of Lambeth's Black community networks, and we are just at the beginning of our exploration of how participatory mapping can help bridge current community–institution divides also characteristic of other societal problem domains like climate action. We hope that our tale from the lived experience of the Lambeth community inspires some to join us in our quest.[4]

Notes

1 www.ubele.org/
2 The MV Empire Windrush arrived at Tilbury Docks on 22 June 1948.
3 All maps were created using Graph Commons software (https://graphcommons.com/).
4 Drew Mackie at Drew Mackie Associates designed and produced the network maps and Domino game. We would also like to thank The Ubele initiative for giving us the opportunity to develop this project which was funded by Black Thrive Lambeth's Employment Project, part of Impact on Urban Health's Multiple Long-Term Conditions Programme.

References

Brayshay B, Mackie D (2023) *New Maps: Social Systems Mapping in the London Borough of Lambeth*. Community Informatics Research Network (CIRN) Conference Proceedings. Monash University Centre, Prato, Italy, pp. 66–81. Available at: www.monash.edu/__data/assets/pdf_file/0006/3282018/CIRN_Proceedings_final.pdf.

Chatterton P (2019) *Unlocking Sustainable Cities: A Manifesto for Real Change*. London: Pluto Press.

Copeland S and de Moor A (2018) Community digital storytelling for collective intelligence: Towards a storytelling cycle of trust. *AI & Society* 33(1): 101–111.

de Moor A (2017) CommunitySensor: Towards a participatory community network mapping methodology. *The Journal of Community Informatics* 13(2): 35–58.

de Moor A (2018) A community network ontology for participatory collaboration mapping: Towards collective impact. *Information* 9(151): 1–37.

Equinox Consulting (2013) *An Insight into the Black Caribbean Community in Lambeth*. London: Equinox Consulting.

Equinox Consulting (2017) *Voluntary and Community Sector Strategy Beyond Survival: A Thriving VCS in Lambeth*. London: Equinox Consulting.

Følstad A and Kvale K (2018) Customer journeys: A systematic literature review. *Journal of Service Theory and Practice* 28(2): 196–227.

Fricker M (2009) *Epistemic Injustice*. Oxford: Oxford University Press.

Hawthorne C (2019) Black matters are spatial matters: Black geographies for the twenty-first century. *Geography Compass* 13(11): e12468.

House of Commons Women and Equalities Committee (2020) *Unequal Impact? Coronavirus and BAME People*. Third Report of Session 2019–2021.

Kidd I, Medina J and Pohlhaus G Jr (2017) Introduction. In: Kidd I, Medina J and Pohlhaus Jr. G (eds) *The Routledge Handbook of Epistemic Injustice*. London and New York: Routledge, pp. x–xx.

Kinkaid E (2022) Positionality, post-phenomenology, and the politics of theory. *Gender, Place & Culture* 29(7): 923–945.

Kurtz C (2014) *Working with Stories in Your Community or Organization: Participatory Narrative Inquiry*. Kurtz-Fernhout Publishing.

MacKinnon D and Derickson KD (2013) From resilience to resourcefulness: A critique of resilience policy and activism. *Progress in Human Geography* 37(2): 253–270.

Mould O, Cole J, Badger A and Brown P (2022) Solidarity, not charity: Learning the lessons of the COVID-19 pandemic to reconceptualise the radicality of mutual aid. *Transactions of the Institute of British Geographers* 47(4): 866–879.

Pratt G (2009) Positionality. In: Derek G, Johnston R, Pratt G, Watts M, and Whatmore S (eds) *The Dictionary of Human Geography*. West Sussex: John Wiley and Sons, pp. 556–557.

Sendra P (2023) The ethics of co-design. *Journal of Urban Design*. Epub ahead of print 8 February 2023. DOI:10.1080/13574809.2023.2171856.

Walker M and Boni A (2020) Epistemic justice, participatory research and valuable capabilities. In: Walker M and Boni A (eds) *Participatory Research, Capabilities and Epistemic Justice: A Transformative Agenda for Higher Education*. Cham: Palgrave Macmillan, pp. 1–25.

40

THE PUBLIC OUTREACH OF THE ICA COMMISSION ON ART AND CARTOGRAPHY

Taien Ng-Chan

Introduction

Art and cartography—as a topic of investigation and subject of many books, papers, conferences, workshops and exhibitions—underwent an explosive period of growth in the early 2000s, as seen in many publications such as *You Are Here* (Harmon, 2003), *The Map as Art* (Harmon, 2009), *Else/where: Mapping New Cartographies of Networks and Territories* (Abrams and Hall, 2006), *Infinite City* and *Unfathomable City*, Rebecca Solnit's poetic and fantastic atlases about San Francisco (2010) and New Orleans (with Rebecca Snedeker, 2013). In cinema studies, a cartographic turn occurred, as indicated in such books as *Cartographic Cinema* (Conley, 2007), *Atlas of Emotion: Journeys in Art, Architecture, and Film* (Bruno, 2002), and *Locating the Moving Image* (Roberts and Hallam, 2013); literary cartographies mapped out locations of settings (Piatti and Hurni, 2011). Many works of artistic cartography are also counter-maps that protest the status quo, make critiques, reframe public discourse, preserve 'un-official' memories or focus on embodied spatial practice, taking inspiration from the Situationists (Debord, 1956), critical cartographers (Kitchin and Dodge, 2007; Crampton, 2010), landscape architects (Corner, 1999) and spatial theorists (Massey, 2005). The Situationist dérive is often seen as a precursor to many of the practices central to alternative or artistic walking and mapping (Careri, 2002; O'Rourke, 2013). Although various spatial turns in the latter half of the 20th century had already brought spatial theory to the forefront of many fields in the arts and humanities (Bruno, 2006), the Internet and the introduction of the iPhone in 2007 were certainly landmark developments that contributed to this period of growth (Lemos, 2010; Gordon and de Souza e Silva, 2011; Farman, 2012). Location and mobile media then became typical entry points to constant networked communication, leading to the creation of digital hybrid spaces and locative media art as new fields of investigation and critique (Pinder, 2013; Southern, 2015). Around this period, practices such as 'research-creation' and 'arts-led research' became more common as well (Chapman and Sawchuk, 2012), making inter- and transdisciplinary methodologies integral to the cartographic humanities.

The formation and development of the Commission on Art & Cartography (ArtCarto), one of the newest official commissions of the International Cartographic Association (ICA), mirror these larger developments in the field, as can be seen in the themes of conference

 DOI: 10.4324/9781003327578-48

papers that have been given over the years (archived on the ICA website back to 1993), as well as within its own Terms of Reference (a document outlining the group's mandate and goals). The meaning of 'art' in relation to the discipline of cartography has shifted from its designation as an element of design in mapmaking to include engagement with contemporary ideas in the arts and humanities, from increased focus on process and practice, to the social, relational, participatory and co-creative aspects of mapmaking, as well as increased awareness of bias in representation and data visualisation. The definitions and practices of art and cartography are expanding constantly. The availability of networks and spaces for gathering and knowledge exchange is important for a field to develop further and gain acceptance both within its own discipline and in the wider world. This chapter traces the development of the Commission on Art & Cartography and the role it has played in providing an international space of connection, exchange and public outreach for those interested in art and cartography in its myriad forms.

History and mandate of the ICA

The ICA is the world's pre-eminent international organisation on cartography, currently with 73 member countries and 37 affiliate member organisations, including commercial map publishers, GIS software developers and archival data initiatives. The ICA's mission, according to its website, is simple: 'To promote the disciplines and professions of cartography and GIScience in an international context' (https://icaci.org/). The ICA was founded in Bern, Switzerland, on 9 June 1956, where it remains to this day 'officially registered as an idealistic non-profit organization'. It was built on discussions held over three years of conferences between 1956 and 1959, and its first General Assembly was in Paris in 1961. Its research programme addresses a wide range of areas, from Atlases to Visual Analytics, conducted through 28 current Commissions and five current Working Groups, each of which are responsible for delineating and upholding a specific mandate to advance their field. The Commissions and Working Groups have changed over time, naturally following the interests of the researchers involved.

Conferences and publications are among the foremost methods through which the ICA conducts knowledge exchange and dissemination. Since its inception, its main event has been the International Cartographic Conference (ICC), held every two or three years in different cities, beginning in Frankfurt, Germany (1962), and moving to such locales as Delhi, India (1968), Madrid, Spain (1974), Warszawa, Poland (1982), Ottawa, Canada (1999), Beijing, China (2001), Moscow, Russia (2007), and Cape Town, South Africa (2023). A truly international gathering, ICA has provided a place and networks for all who have a passion for maps, from government mapping agencies, earth and environmental scientists, cartographic publishers, researchers of all kinds concerned with the representation of space, including, fairly recently, artists as well.

The first Symposium on Art & Cartography took place in Vienna in 2008, organised by William Cartwright (then President of the ICA and initiator of the Art & Cartography Working Group), Georg Gartner (Vice-president of the ICA, co-initiator of the working group and host of the Symposium) and Antje Lehn (professor in Fine Art and co-host of the Symposium), bringing together 'interested scientists and artists who attempt to communicate space and place by means of modern media or experimental artefacts. . . . The only limitation was that they needed to keep to the "spatial context"' (Cartwright et al., 2009: 4). Attendees were able to present and reflect on each other's works, and to discuss the potential for future collaborations. The official Art & Cartography website (https://artcarto.wordpress.com/) was launched that year as well, and has documented the group's activities, as well as general items of interest, since its founding.

Between the Commission's website and the archives of the ICA's international conferences, subfields and threads can be traced, such as embodied, ephemeral, critical, cinematic and literary cartographies, which represent larger interdisciplinary streams of study. As in the wider field of the Fine Arts, the question of 'what is art' has been challenged again and again, from the modernists who broke with traditional forms of representation to current post-modernist appropriation and remediation techniques. In a similar fashion, the question of 'what is art and cartography' has been challenged and changed even in the short decade and a half of ArtCarto's existence, as the web archives reveal.

Tracing ArtCarto through the archives

Each Commission of the ICA sets its own Terms of Reference as to its areas of focus and main concerns, and generally runs a pre-conference workshop and a business meeting during each ICC. When the Commission on Art & Cartography first started, it seems that part of its tasks was to educate the rest of the organisation on the designation of art as being more than aesthetics in design. Hence the first ArtCarto Terms of Reference (2011–2015) begins by establishing this fact immediately with this line in the first Term: 'Explore the art element of cartography—Art in cartography means much more than designing aesthetically pleasing maps' (https://icaci.org/commissions-archive/ under the years 2011–2015).

Expressing the difference between art and design was necessary at the time, as art was not a consideration in the ICA. The ArtCarto founders write that prior to the 1950s, 'cartographic artefacts were built under the theoretical and practical "umbrella" of this partnership of art, science and technology' (Cartwright et al., 2009: 3). They argue that cartography then became focused on science and technology, disavowing art on 'a "quest" to gain "scientific legitimacy" by using scientific visualisation as a lodestone for gauging the "quality" of theories and applications' (Cartwright et al., 2009: 3). The soon-to-be Commission on Art & Cartography was founded as a way to 'bridge the gap between science and art in cartography initiatives' (Cartwright et al., 2009: 3).

In a search of the ICC digital archives on the organisation's website, which go back only to 1993's meeting in Cologne, Germany, the word 'art' does not appear in the conference themes at all before 2007. There are presentations that could fall under media and communications studies, such as a session called 'Mass Media Cartography' including 'Weather maps on television in the USA' (J.R. Carter) and 'Newspaper maps in Germany' (W. Scharfe). There is a presentation on 'Graph Theory and Network Generalization in Map Design' (K. Beard and W. Mackaness). In subsequent years, 'design' grew to become a regular topic within the ICC conference proceedings. At ICC1995 in Barcelona, Spain, sessions included 'Multimedia and Hypermapping', 'Cartographic Design with Digital Techniques' and 'Desktop Technologies', pointing to the revolution in desktop publishing that occurred in the late 1980s and 1990s. By the ICC2007 in Moscow, the last ICC before the founding of ArtCarto, while 'art' is still not mentioned, Map Design and Production is a prominent theme, and other topics such as 'Maps in the Media' and 'Space and Time in GIS' lean towards the cartographic humanities.

The founding of the Working Group on Art & Cartography had an immediate impact on the ICA, as can be seen at the following ICC2009 in Santiago, Chile, where for the first time, 'art' is listed as a theme in the programme and as an organising principle. The topics at the ICC2009 are more overtly arts-and-humanities-based, with presentations such as 'The Art of Being Lost: An Alternate Approach to Mapping' (L. Vaughan), 'Maps as Comics, Comics as Maps' (A. Moore), 'Literary Geography or How Cartographers Open Up a New Dimension for Literary Studies' (B. Piatti), and 'Reframing the Digital Cartographic Frame: Examples from The Cybercartographic Atlas of Canadian Cinema' (S. Caquard). In addition, ArtCarto presents the Surrealist short film 'Zig-Zag, Snakes and Ladders (a didactical fiction about cartography)' by Raúl Ruiz, explicitly bringing in an example of cartography in art as separate from design, and challenging traditional cartographic ideas in the manner of critical cartographers but through cinema.

At the ICC2011 in Paris, both 'Art & Cartography' and 'Map Design' became official Commissions of the ICA. Sébastien Caquard and Barbara Piatti became the first Chair and Vice-Chair of ArtCarto. Caquard and Piatti's work in cinematic and literary cartographies helped to bring together a network of scholars interested in mapping places in movies, novels, short stories, plays and other works of fiction, as well as ephemeral cartographies such as dreams, emotions and senses. The influence of the Internet and location-based smartphone apps as well as performance theory could be seen at the ICC2013 in Dresden in panels such as 'Social Mapping', 'NeoCartography' (the social web and collaborative mapping) and 'Playing with Maps' with presentations on location-based smartphone apps, navigation and play.

In addition to conference presentations, workshops, publications and collective art/mapping events have also helped establish the Commission on Art & Cartography. That first year at the ICC2011, the first ArtCarto pre-conference workshop was held, titled 'Mapping Processes and Practices: Arts, Maps and Society' organised jointly with the Commission on Maps and Society, as well as an experimental walking tour titled 'Exploring visible and invisible borders in Paris—a dialogic walk with cartographers and artists'. Subsequent pre-conference workshops include 'Maps and Games' at ICC2013 in Dresden, and 'Mapping Ephemeralities/Ephemeral Cartographies' at ICC2015 in Rio de Janeiro. ArtCarto members published several special journal issues, and there were also a few international collective cartographic filmmaking experiments during this time.

In only a few years, it is evident that the overall diversity has grown in the represented themes of the ICA conferences as well as associated workshops and publications, which have become much more arts-based and interdisciplinary. At the ICC2015 in Rio de Janeiro, 'Art, Culture and Cartography' was an official theme, and the Commission for Art & Cartography's Terms of Reference were renewed for the years 2015–2019. This time, they no longer expressed the need to disclaim that art and cartography 'means much more than designing aesthetically pleasing maps'.

Artistic practices of embodied and ephemeral mapping

As the current Chair of the Commission on Art & Cartography, I oversaw the development of the latest Terms of Reference (2019–2023), which now include a more descriptive list of practices associated with art and cartography, 'including but not limited to such subfields as narrative cartography, cinematic cartography, sensory and phenomenological approaches to mapping, locative media, performative and performance-based cartographies, and media archaeological and other research-creation or practice-led processes' (https://artcarto.wordpress.com/terms-of-reference/). At the ICC2019 in Tokyo, I took over as Chair, and since

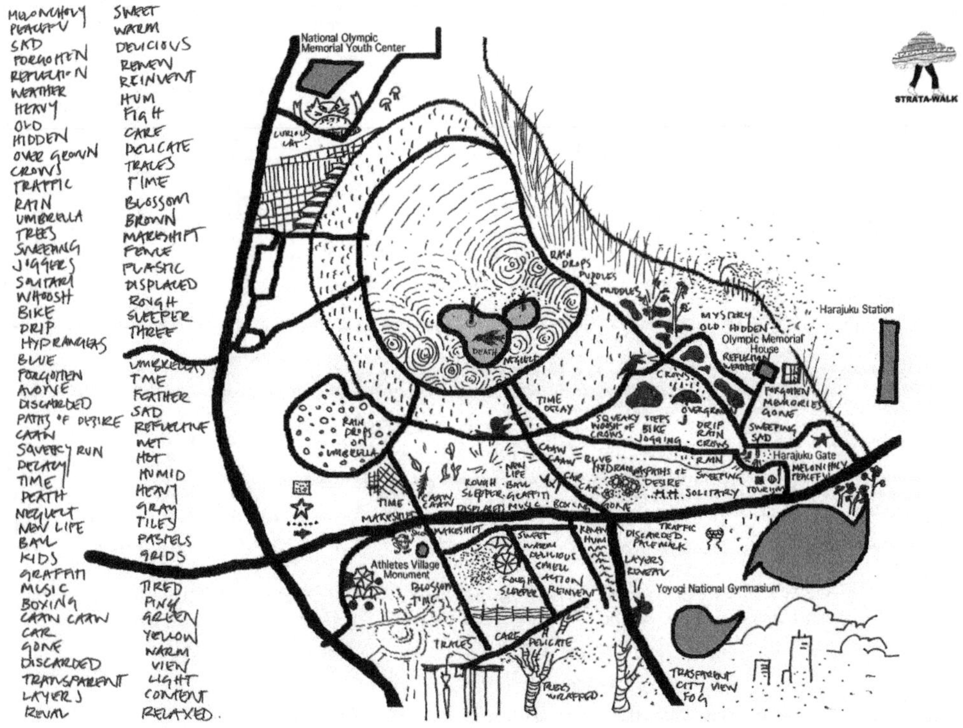

Figure 40.1 Emotional map of Tokyo's Yoyogi Park by Joanna Gardener, published with permission.

then, my own interests in embodied and digital hybrid mapping processes and experimental cartographic cinema have contributed to the direction of the public workshops. Sharon Hayashi and Joanna Gardener, as Vice Chairs, have also influenced the direction of ArtCarto, bringing their expertise in digital mapping, visual culture and history, and sensual and emotional cartographies.

As part of the pre-conference workshop in Tokyo for ICC2019, ArtCarto collaborated with my artist collective, the Hamilton Perambulatory Unit or HPU (Donna Akrey and myself), to present our combined methodologies of 'strata-mapping' and historical research in 'Reclaiming Through Mapping: The Olympic Sites of Tokyo'. The HPU has been holding performative walking and mapping events since 2014, and the methods that we have developed in our public 'strata-walks' are geared towards a wide variety of audiences. Our 'strata-mapping' framework aims to highlight the multiplicity of sensory, social and historical layers that make up place, from the ephemeral to digital to the built and historical, always with acknowledgement that the layers together comprise the whole (Ng-Chan, 2020). Our Tokyo workshop was the first of many to explore strata-mapping practices in the context of the ICA (Figure 40.1).

The following year, ArtCarto had planned a workshop in Portugal as part of a walking arts festival, again in collaboration with the HPU, which had to pivot online, as did much of the rest of the world. The COVID-19 pandemic shut everything down and the practice of lockdown began. Our online workshop had the tongue-in-cheek title of 'A Sense of

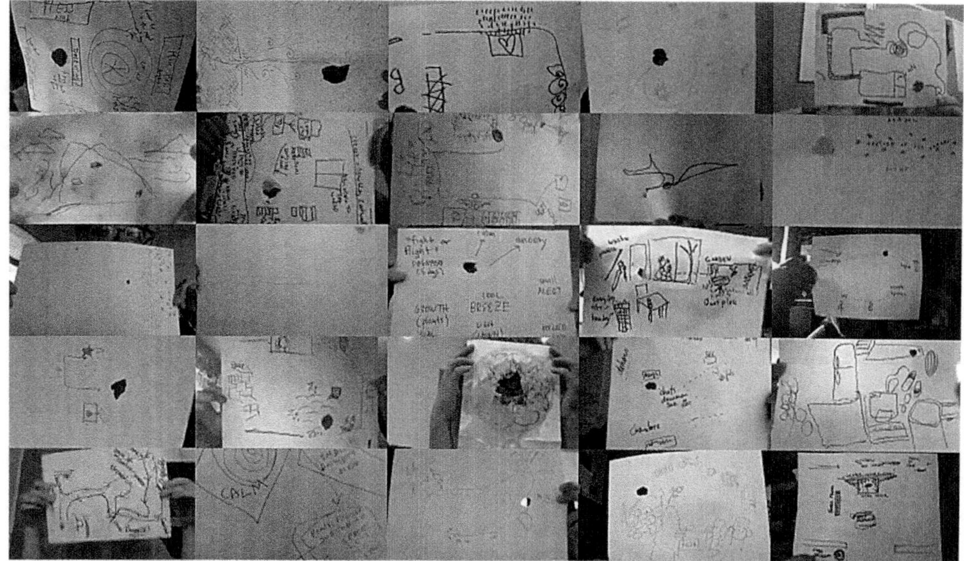

Figure 40.2 A screenshot of the Zoom meeting taken by the author (with permission), showing the hand-drawn maps of lockdown spaces by ArtCarto/HPU workshop participants.

Impending Doom: a strata-walk for turbulent times'. Inspired in part by Xavier de Maistre's parody travelogue *Voyage Around My Room*, we devised a series of prompts designed to tune one's body into one's emotional landscape, to map our lockdown spaces with the eye of a traveller, to see anew. Our group of participants, individually located in rooms all over the world, were able to connect and sense each other and collectively find some catharsis in the virtual space of Zoom.

Although these activities are valuable in providing space for exchange and networking, they can also be seen as a form of art in themselves, which focuses on producing and exploring human relations through collective experience, rather than aesthetic objects for consumption (Bourriaud, 1998). This is also a practice of participatory art that 'directly engages the audience in the creative process so that they become participants in the event' (www.tate.org.uk/art/art-terms/p/participatory-art). Strata-walks—whether in a city or around one's room—function as modes of public pedagogy, research-creation and participatory/relational art. The emphasis is on the interrelationships between people and environments; the creative element does not lie in the making of an object but a social event or a situation. The relations forged during the event also become time-space strata as memory and new knowledge of place (Figure 40.2).

Conclusion

Not all of ArtCarto's events are works of art, of course, but these do illustrate the distance that the concept of art and cartography has come within the ICA, from being considered as an element in the design of maps to current thinking around ephemeral processes of participatory mapping where there might not even be a tangible map produced. In the essay *Rethinking Maps*, Rob Kitchin and Martin Dodge (2007: 331) argue that cartography should be thought of as 'a processual, rather than representational, science', and this is clearly one of the major underlying motivations in such 'post-representational' mapping practices (Caquard, 2014) that have informed the development of the Commission in recent years.

The ICA has always done outreach to the public, for instance through the Barbara Petchenik Children's World Map Competition, run by the Commission on Cartography & Children, and open to children under 15 years of age from ICA member countries. However, since its founding as a working group, ArtCarto has played a central role within the ICA in expanding the definition of art and the humanities in relation to cartography, and because of this, is also able to better engage with the wider public beyond the ICA. For instance, some of the maps from our 'Impending Doom' workshop made their way into *The Quarantine Atlas* (Bliss, 2022), a book featuring COVID-19 lockdown space maps submitted to Bloomberg CityLab's MapLab. ArtCarto is also engaged with public forms of 'walkshops' and exhibitions that invite a wide range of participants, whether in partnership with artist-research collectives such as the HPU, or at arts festivals and conferences, or on social media.

As technologies develop, maps will become even more ubiquitous and accessible. It is important to keep pushing towards new forms of spatial experimentation with new kinds of emerging media such as augmented, virtual and extended realities. Publications, exhibitions and events such as strata-walks and workshops can provide a way for practitioners of art and cartography to expand their methodological toolboxes, exchange knowledge, invite public connection and grow their networks in an international context. As participants in an art event, we can also experience ways of honing our senses and attuning to our environments, to co-create representations and processes of the time and space shared together. Hopefully, the Commission on Art & Cartography will continue for a long time to provide such opportunities for sharing spatial language, new ways of looking and sensing the world, new ways to create, document and pass on spatial knowledge.

References

Abrams J and Hall P (eds) (2006) *Else/Where: Mapping New Cartographies of Networks and Territories*. Minneapolis: University of Minnesota Design Institute.

Bliss L (ed.) (2022). *The Quarantine Atlas: Mapping Global Life Under COVID-19*. New York: Black Dog & Leventhal.

Bourriaud N (2020 [1998]) *Relational Aesthetics*. Dijon: Les presses du reel.

Bruno G (2002) *Atlas of Emotion: Journeys in Art, Architecture, and Film*. New York: Verso.

Bruno G (2006) Visual studies: Four takes on spatial turns. *Journal of the Society of Architectural Historians* 65(1): 23–24.

Caquard S (2014) Cartography III: A post-representational perspective on cognitive cartography. *Progress in Human Geography* 39(2): 225–235.

Careri F (2002) *Walkscapes: Walking as an Aesthetic Practice*. Barcelona: Editorial Gustavo Gili.

Cartwright W, Gardner G and Lehn A (eds) (2009) *Cartography & Art*. Berlin: Springer-Verlag.

Chapman O and Sawchuk K (2012) Research-creation: Intervention, analysis and 'family resemblances'. *Canadian Journal of Communication* 37(1): 5–26.

Conley T (2007) *Cartographic Cinema*. Minneapolis: University of Minnesota Press.

Corner J (1999) The agency of mapping: Speculation, critique and invention. In: Cosgrove D (ed.) *Mappings*. London: Reaktion Books, pp. 213–252.

Crampton J (2010) *Mapping: A Critical Introduction to Cartography and GIS*. Chichester: Wiley-Blackwell.

Debord G (1996 [1956]) Theory of the dérive. In: Andreotti L and Costa X (eds) *Theory of the Dérive and Other Situationist Writings on the City*. Barcelona: Museu d'Art Contemporani di Barcelona, pp. 22–27.

Farman J (2012) *Mobile Interface Theory*. London: Routledge.
Gordon E and de Souza e Silva A (2011) *Net Locality: Why Location Matters in a Networked World*. West Sussex: Wiley-Blackwell.
Harmon K (2003) *You Are Here: Personal Geographies and Other Maps of the Imagination*. Princeton: Princeton Architectural Press.
Harmon K (2009) *The Map as Art: Contemporary Artists Explore Cartography*. Princeton: Princeton Architectural Press.
Kitchin R and Dodge M (2007) Rethinking maps. *Progress in Human Geography* 31(3): 331–344.
Lemos A (2010) Post-mass media functions, locative media, and informational territories: New ways of thinking about territory, place, and mobility in contemporary society. *Space and Culture* 13(4): 403–420.
Massey D (2005) *For Space*. London: Sage Publications.
Ng-Chan T (2020) Strata-mapping the Detroit river border with the Hamilton Perambulatory Unit. *Intermedialities* 34: n.p. DOI:10.7202/1070880a.
O'Rourke K (2013) *Walking and Mapping: Artists as Cartographers*. Cambridge: MIT Press.
Piatti B and Hurni L (2011) Cartographies of fictional worlds. *The Cartographic Journal* 48(4): 218–223.
Pinder D (2013) Dis-locative arts: Mobile media and the politics of global positioning. *Continuum: Journal of Media & Cultural Studies* 27(4): 523–541.
Roberts L and Hallam J (eds) (2013) *Locating the Moving Image: New Approaches to Film and Place*. Bloomington: Indiana University Press.
Solnit R (2010) *Infinite City: A San Francisco Atlas*. Berkeley: University of California Press.
Solnit R and Snedeker R (2013) *Unfathomable City: A New Orleans Atlas*. Berkeley: University of California Press.
Southern J (2015) Locative awareness: A mobilities approach to locative art. *Leonardo Electronic Almanac* 21(1): 178–195.

41

THE (AESTH)ETHICS OF PUBLISHING GEOPOLITICAL MAPS

Laura Lo Presti and Tania Rossetto

Introduction

In their *Manifesto for Map Studies*, which has tremendously inspired cultural cartographic research in the last 15 years, Dodge et al. (2009) proposed a set of routes, articulated through a number of 'modes', 'methods' and 'moments', for the development of the study of both philosophies and practices of mapping. The focus on the domain of *authorship*, with an emphasis on new 'prosumer' or 'counter-mapping' practices, is one of the modes highlighted. Among the methods, *ethnography* features as an innovative crucial way to capture how maps come into being, with particular reference to everyday performances of map use. Within the moments listed in the manifesto, then, we find *moments of change and decision-making*. On this point, Dodge et al. (2009: 236) state that 'the role mapping plays in the construction and maintenance of different global world orders, and its contributions to moments of change such as revolutions, boundary disputes or regime change is seriously under-researched'.

Taking these three routes in a kind of triangulation, in the last ten years, we, the handbook's editors Tania Rossetto and Laura Lo Presti, have engaged in an inspiring conversation with Laura Canali, designer of geopolitical maps and map artist based in Rome. Through regular moments of encounter in different settings and situations (Figure 41.1), in fact, we have delved deeply into her experience as an author of maps, exploring such particular authorship through ethnographic methodologies that have seldom been adopted in the field of map studies and especially in the study of living mapmakers. The fact that Canali is specialised in the design of geopolitical maps, then, put us in front of a peculiar cartographic genre that is imbued with public responsibility and ethical issues. While geopolitical maps have the potential to influence visions of the world, changes and decisions, it is also the geopolitical mapmaker that continuously faces important moments of decision on how to envision such world(s). The following sections include a selection of excerpts from research encounters between the three of us that occurred beginning in 2015 which reveal different aspects of Canali's authorship and relationship with her audiences. Her mapmaking job, in fact, happens in an inextricable relation to such audiences, which gives her job a strong public dimension.

 DOI: 10.4324/9781003327578-49

Figure 41.1 Encountering the mapmaker. A map-elicitation session during fieldwork in Rome with Laura Canali in 2017.

Source: Photograph by Tania Rossetto

We will hear the mapmaker's voice sharing her ideas, perceptions, beliefs and feelings regarding such a public dimension.

Making maps for the public

Laura Canali is a graphic designer, cartographer and artist born in Rome and responsible for the production and design of geopolitical maps at *Limes: Rivista Italiana di Geopolitica [Limes: Italian Review of Geopolitics]* since its foundation. *Limes* is a popular Italian geopolitical journal and eminent opinion maker established in 1993 by political scientist Lucio Caracciolo. It is the fruit of a collaboration between scholars (historians, geographers, sociologists, political scientists, lawyers and anthropologists) but also journalists and decision-makers (politicians, diplomats, military, entrepreneurs and managers). With 200 pages per issue, *Limes* is published monthly by a major Italian publishing group (Gruppo Editoriale GEDI). It has around 25,000 paper copies sold each month in addition to its e-book and iPad versions and thousands of daily hits on its website (for both free and paid content). Recently, the magazine has encountered particular fortune, as the geopolitical situation, particularly in Europe, has been shaken by the Russia—Ukraine war. Every issue of

Limes is dedicated to a topic of geopolitical interest and is replete with original maps, which constitute the most recognisable mark of the journal. To date, Canali has made about 3,500 geopolitical maps. In such variegated mixture, where policy makers 'meet' critical scholars, Canali's maps have to mediate, translate and communicate the many reflections and perspectives on the geopolitical space made by the various analysts and held in the different sections of the online site as well as of the printed journal.

Canali became a self-educated graphic artist while working at her father's screen printing company and attending evening classes in the applied arts. She arrived at *Limes* in 1993 at the age of 25, when the graphic designer employed at the periodical, Roberto Steve Gobesso, decided to devote himself exclusively to the cover design and to delegate the making of the internal maps, considered a less creative task, to the new employee. Later, starting from 2009, Canali also devoted herself to the cover design, which gave her a new space and freedom of cartographic expression. The map artworks of the covers for *Limes* always deal with the geopolitical theme of the magazine's issue, yet they could also be considered part of a parallel 'map art' work that Canali initiated in 2008.

By means of fieldwork in various settings of the mapmaker's everyday life, mobile interviewing, participant observation at meetings and exhibitions, map-elicitations, narrative and object-based interviews, online and email conversations, in subsequent steps, since 2015 we have explored the complex personal and professional world of Laura Canali. One of the main features that constantly emerged during our conversations was the attention given to the audience. Laura considers her making of geopolitical maps a practice of 'translation' for an audience. In fact, by translating words into maps, she provides a graphical form to the articles hosted within the magazine. The aim is that of complementing the verbal content of the articles to increase their effectiveness. Within the process of map production at *Limes*, the article that offers the more illuminating view on a given topic is often the best candidate for obtaining a coloured map (colour was introduced only 15 years ago and, due to printing costs, coloured maps remain a small part of the total amount of the maps included in printed issues). For the other ones, 'a miraculous balance of greys', as Canali puts it, is needed. Colour, as we will see in the next section, is one of the main distinctive features of Canali's geopolitical maps and artwork, as well as one of the main ethical preoccupations in terms of the emotional effects provoked among the public.

As Boria (2008: 280) put it, 'geopolitical maps take into consideration opposing political and economic interests and spell them out to the reader through highly simplified, stylised cartographic drawings aimed at conveying a specific point of view with regard to the represented situation or phenomenon'. The mapmaker translating one of these views is directly and practically involved in such 'spelling out'. Actually, Canali's maps are the result of a labour chain and of a whole set of practices and processes accompanying their creation (see Boria and Rossetto, 2017: 43–45), yet she strongly feels her protagonism and authorship, as when she states that 'I am the medium between the reader and the map' (Laura Canali, interview excerpt, 2016). Seeing herself as a guide to the reader, she recognises the cartographer and her maps as having a fundamentally pedagogical and not merely informative role.

Indeed, while authorship is shared with the article's authors and other staff members of the magazine, the ultimate decisions on the cartographic renderings published in *Limes* pend on the responsibility of the map designer.

Asked about her idea of the readership of *Limes*, Canali responded:

> There is an attached audience that has followed the magazine since 1993. After the attack on the Twin Towers, the readers increased, and they have become more differentiated. At the beginning they were people specialised in geopolitics, and therefore the maps were more specific and well understood. Then I had to take a step back, to be clearer and to take nothing for granted. Readers today are university students, diplomats, military and many curious people who want to get a more precise idea of the world in which we live.
>
> *(Laura Canali, interview excerpt, 2023)*

Significantly, at some point of her career, Canali has felt the necessity to explain her work to readers, curating two personal columns, *Ricamando il Mondo* [Embroidering the World], since 2009, and *Cartografie dell'Immaginario* [Imaginary Cartographies], since 2011. The first, in particular, was born from the need to explain her work process and the visual choices and ethical claims underpinning her mapping production.

As in more recent times her work has also become subject to comments on social media, Canali also expresses the need to sometimes 'defend' her work.

> Normally I expose myself as little as possible. Yet I always defend my maps from those who interpret them in a wrong way or invent non-existing meanings. This occurs frequently on social media by people who do this as a job. I think that in such cases it's necessary to come out. . . . Yet you need the right argument and to have your back covered by the sources. If your work is contested, it's necessary to show the reasoning at the base of your work.
>
> *(Laura Canali, interview excerpt, 2023)*

The chromatic ethics of geopolitical mapmaking

In a recent interview, while asked about what it means to communicate cartographically to the public, Laura responded by prioritising the role played by colours in her mapmaking practice.

> Being aware of designing for the public gives a lot of responsibility. I can't say that the public has an influence in the choices I have to make, as in the choice of colours, but I am very careful in providing all the information in the map legend. I care that texts are clear and that the map, more in general, proves to be legible. I am used to thinking that colours are tools that can generate feelings in the people who look at them, so I very much reflect before crafting a certain assemblage of colours. I seek to find a balance between tones and words.
>
> *(Laura Canali, interview excerpts, 2023)*

Indeed, while colours are certainly a distinctive formal feature of Canali's work, they also entail a deep ethical dimension. There are many different perspectives that could be applied to the understanding of colours on maps, from the empirical to the cognitive, from the semiotic to the aesthetic, etc. (see Rossetto, 2018 for a review). At the beginning of the twenty-first century, Johnson (2001) lamented the lack of studies concerning the connotations and meanings attached to colours, beyond the merely physiological perceptual aspect, thus calling for an updated critical, cultural, contextual and (de)constructivist approach to map colour choices and the particular stories they tell. Yet, within this variegated panorama, the ethnographic approach provides the opportunity to explore further this aspect from the perspective of cartographic authorship. During a session on map-elicitation held in 2017 and specifically devoted to the 'chromocartographies' generated by her geopolitical maps, Laura extensively considered her use of colour in relation to the audience.

> When an article of *Limes* involves very complex issues and several concepts, it's selected for including a coloured map. Colour is like a multiplier.
>
> First there is a main concept, a nucleus, or two. Then there are the following steps. You need colour to put emphasis on this nucleus and the concepts around which the map develops.
>
> I am careful with colours, because with colours you may give sensations . . . colours are linked to the soul. If I put in some red bordeaux, it becomes wicked. I'm careful with colour, I try to use it without misinterpreting the message expressed by the author [of the article]. I try not to add my impressions. I make so many decisions . . . but I feel myself as a means. Here I am at the service of the reader; I am not an artist.
>
> *(Laura Canali, interview excerpts, 2017)*

As Bláha and Štěrba (2014: 203) contend, the emotional potential is 'considerably larger in colour than in other cartographic means of expression'. During fieldwork, Canali spontaneously expressed her emotional engagement with colours and the need to control this emotional dimension for ethical purposes. Colour is a matter of emotion, and the geopolitical mapmaker, for Canali, has the responsibility to dose the emotional impact of colours. We could read in such discussion of the ethics of colour something of an 'emotional geopolitics' (Pain, 2009) in action:

> Colours help in changing the perspective from which the world is viewed.
> When I use black, I always say that it's a defeat, but in other cases black is clear-cut. If you need to point out something precisely, then black helps. Black is not for states; it must be dosed, and neutrally used. Black holds power, it attracts the eye; it's like a gunsight.
>
> Like USA blue, Russia violet, Egypt yellow . . . well, I re-shuffle the chromatic cards, I don't put labels. If you always use the same colour for a nation-state, you give a fixed idea of that state, but states change their roles in different scenarios. Within every map the scenario changes, and thus the colours change.
>
> But knowing too much about colour risks influencing me. Sometimes I feel too aware, and I want to preserve my naivety, my enchantment. I don't want to be overwhelmed by colour codes, I want to remain creative, to escape. Pale colours make me feel uncomfortable. I need strong colours to manage complexity. . . . Geologists use colours on secure landforms. Those 'technical' coded colours don't perform the task of representing the world of today, this changing world.
>
> The world is complex and fearsome. When I make geopolitical maps, I try not to be alarmist. Other magazines use fear to treat geopolitical themes.
>
> *(Laura Canali, interview excerpts, 2017)*

Elsewhere, in a written meditation (see Rossetto, 2019: 107–109), Canali indulged in an intimate expression of her struggles with colours.

> My map will be full of bright and bold colours—pop colours, one could say.
> I love red, it's always there tempting me. I would like to put it on the map in all its shades. No, no . . . I'd rather not. Red has to just pop up here and there. It can't spread, because it hurts the eyes. You immediately seek refuge. It alarms and then . . . it's conceited, it's the first to catch the eye. You can't blend it with other colours. It comes out as a cricket, it can't resist. I have to trace a special border for it . . . an arrow, a circle, a sign of explosion. It will be the first to be noted. It will be there, above the others, to point out the newest and most dynamic facts.
>
> *(Laura Canali, interview excerpts, 2018)*

The use of—and rumination on—colour is what makes Laura Canali simultaneously a cartographer and an artist. The following section deals with her parallel work as an art practitioner and the related additional ways in which she expresses publicly her mapmaking practice.

Being public otherwise: exhibiting geopoetical/political art

Even though Canali started painting abstract canvases in 2008, it is only in more recent years that she initiated a more regular activity as an art practitioner and in particular as a *map* artist. While initially abstract painting was somehow a way to escape her main job, gradually she has found a way to combine the two sides of her creativity. Saliently, on the home page of her personal website, ideally devoted to her 'separate' map artist career (www.lauracanali.com, Figure 41.2), Canali now displays a gallery which includes, in scattered order, geopolitical maps from *Limes*, creative artworks for the covers of the magazine, abstract canvases and pieces of map art.

As she said:

> My artistic production is a way to deviate the feelings emerging when I design geopolitics. In the maps I craft, I am free to express myself and 'use' geography as a tool for imaginary trips. My personal exhibitions are often a mix of geopoetics and geopolitics, because the one is linked to the other. Geopoetics gets ideas from the facts I analyse in geopolitics, makes me learn about poets and writers who work on themes related to many geographical areas. Geopoetics is a way to look at the world with feelings, to link the past with the present; it's a journey that sometimes proceeds through the terrestrial depth to emerge on the surface; it sometimes pushes you to ascend to the universe, and then to return down to Earth. Geopolitics runs across the terrestrial surface; it's made of strategy, tactic, powerful forces behind handshakes and fake smiles.
>
> *(Laura Canali, interview excerpts, 2023)*

Canali now regularly exposes both her creative and her geopolitical production in personal art exhibitions (see Marocchi, 2022), or during public events and festivals regularly organised by *Limes*.

Figure 41.2 Laura Canali's website.
Source: Screengrab by Tania Rossetto, 2023

In the development of her artistic activity, there has been a particular event that signified a crucial passage. In 2016, at the Rotonda della Besana, a public space in Milan, she exhibited for the first time a canvas with a cartographic subject at an open-air exhibition hosting large-sized copies of maps produced for *Limes* as well as for the World Bank. This canvas, titled *The Mediterranean Is Stone* (Figure 41.3), was derived from a cover of *Limes* first published on the online platform of the magazine on 17 April 2015. It shows the shape of the Mediterranean Sea in the guise of dried land, with a striated, rocky surface in order to express the idea that the Mediterranean, transformed into a space of tragedy by the frequent shipwrecks of migrants, is 'no more liquid, and no more poetic', as Canali explained in her personal column *Ricamando il Mondo*.

> I made the cover design of Limes' new issue to paint a picture dedicated to the concept of limit/border. I chose our sea because today it is one of the most problematic frontiers of the world. . . . When it came to use colour to fill the sea, I realised that the land colours were more suitable than marine colours because, in truth, this sea was no more fluid in my eyes, but harsh and rocky like a chain mountain.
>
> *(Canali, 2015)*

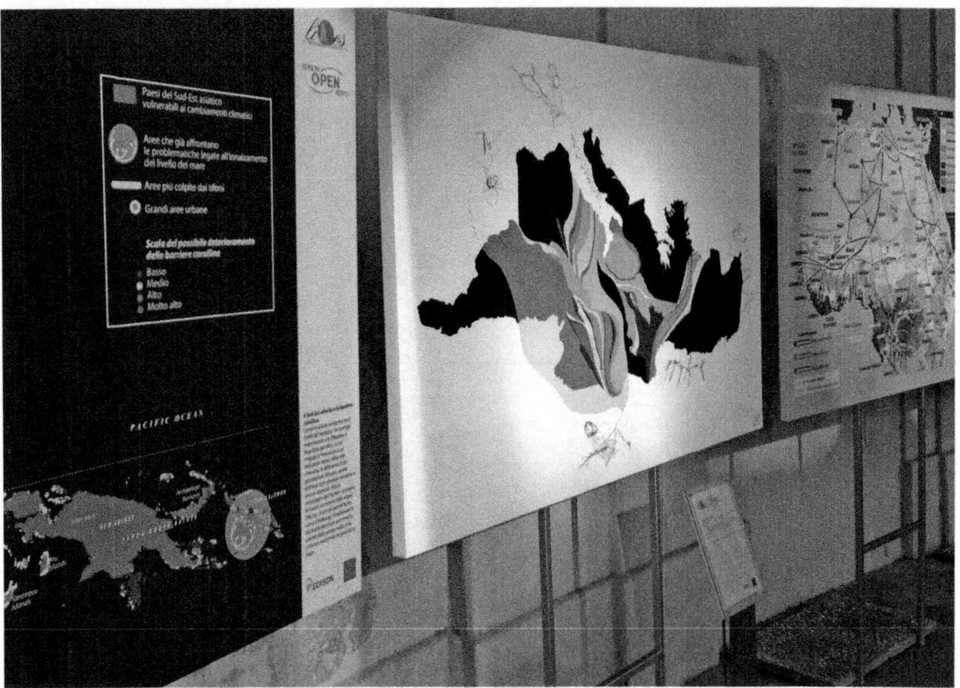

Figure 41.3 *The Mediterranean Is Stone* exhibited at the Rotonda della Besana in Milan between two geopolitical maps, all by Laura Canali (2015).

Source: Photograph by Tania Rossetto

The public exhibition of the derived painting, held in 2016, was a fundamental point in her career since, as Canali remembers in a dense ethnographic encounter held in that year (see Lo Presti, 2017):

> Now everything is mixed, even painting. That was a thing apart, and now I paint geography even there! Because maps saturated me before, at some point there was too much geography, I had enough. So I made many geometric drawings—the leftovers, as I call them—which I slammed on the canvas. Since I have painted *Il Mediterraneo è pietra* [The Mediterranean Is Stone], geography has entered on the canvas as well.
>
> *(Laura Canali, interview excerpts, 2016)*

The Mediterranean Is Stone is also revealing of an additional layer of Canali's relationship with the public. This artwork has a dense (geo)political meaning and constitutes a strong statement about a particularly debated issue in Italy, namely the national policy towards the shipwrecks of migrants navigating towards our coasts (see Lo Presti, 2019). In a striking and dramatic coincidence, the cover for *Limes* designed by Canali was released one day before the terrible night of 18 April 2015, when more than 800 people disappeared 73 km from the Libyan coast, shipwrecking on a fishing vessel headed to Lampedusa. The decision of transforming the cover into a work of art, in fact, clearly reinforced the performance of despair and the act of accusation launched by the cartographer.

We could also see at work here the ethical dilemma underlying the aesthetic choices of the *Limes* cartographer, who considers it frustrating and painful to bring out the migratory 'thingness' and 'obstacles' using lines and ridges but at the same time cannot help but denounce the necropolitic regime on which the Mediterranean is built today (see Lo Presti, 2019). Importantly, she treats the map as a powerful image, as she believes that it can help to influence the perception of events, in this case geopolitical ones. The solidity and mournfulness of the seascape are mostly addressed as the fatal consequences of international and European policies.

> I felt that the concept of hardness was necessary to describe the world around us today. All the wars, suffering, injustice, death, pain that are shaking the countries on the other coast of the sea took me away from the idea that the sea might be liquid.
>
> *(Laura Canali, interview excerpts, 2016)*

Visually, such hardness is translated as a removal of water—but especially of those little islands that are used as handholds by people escaping their former countries. This is a deliberate choice made by the cartographer, even if she suffered in making it.

> There are no more islands on the map. I removed all the islands in that sea. First, I thought a bit about it, then I decided that the islands should be taken away because I wanted to give the idea of the hard path, while the island is still a landing place, a place where one can hang on. You land there, it's a place where you can rest and be welcomed. Removing them was like taking the stairs away from the pool.
>
> *(Laura Canali, interview excerpts, 2016)*

Conclusion: displaying precarious worlds and maps

As a geopolitical cartographer, Canali is constantly dragged into the latest international crises, political conflicts, war battlefields and border disputes. Despite dealing with very stressful topics such as migrations, famines, injustices, economic divides, abuses of power or terrorism, she recognises her position as that of a 'guide' who cannot shirk from revealing her point of view on events, expressing it through the choices of colours, shapes, lines or symbols of her maps. Yet this 'guiding' act is not made with the ambitious aim of showing the 'evidence' of such harsh geopolitical topics, processes and events. Rather, the aim is that of infusing interest and awareness for the complexities of the world.

> In my opinion, thanks to the use of colour with harmonic forms, with graceful signs, the reader slowly finds the courage to look in depth. You somehow cuddle him, you help him to look, because understanding the world is difficult and could be distressing. However, you need to know it to live it better. It's like a disease, isn't it? When a disease affects you, there is fear—the initial shock—but then, the more you know it, the more you can deal with it. After all [with geopolitical maps], it is the same mechanism. You should understand what's going on in the world to have less fear. So, I think it's important to ensure that the reader reads, that he comes to be informed without running away. Overreaching with colour means getting involved. Those neutral things, which remain in the balance, you can't say that they are bad or good, you can't make a choice on them.
>
> *(Laura Canali, interview excerpts, 2016)*

It is interesting that when Canali tries to explain one of her maps, she always ends up with an animated description of a concrete situation on the terrain, verbally visualising refugees, soldiers, tracks, raids, polluted wells or arm traders on her map. She cannot abstain from animating her maps. Canali carries the burden of dealing with power conflicts and violence in her mapmaking. Indeed, she shares this pain and the related responsibilities with the staff of *Limes* when she uses the plural pronoun, saying, 'We do it without scaring people, without scaremongering'. Maps are emotional entities, and to control and direct this emotional content is a crucial task for the map designer. In performing this task, Canali continually makes and re-makes both her maps and the ethical foundation of her work, which she approaches through a workable and practical manner, rather than critically or cynically.

> Colours are there to provide something that invites people to watch and that guides the gazes through the stories the map aims to describe. A coloured path on an ever-changing Earth.
>
> *(Laura Canali, interview excerpts, 2018)*

> I work day by day. My maps are not static maps, and I want to educate the reader [about] this continuous change. . . . Maps are always in movement, they are of the moment; I never think that my maps will last for a long time.
>
> *(Laura Canali, interview excerpts, 2017)*

The fragility and precariousness of the world are thus transferred to the maps, which are fragile and precarious entities, too. The ethnographic encounter with the mapmaker helped us in understanding what is going on behind the pages and the paintings, the vibrant colours and the neat lines, the analyses and the imaginations of the ever-changing world she publicly displays.

References

Bláha JD and Štěrba Z (2014) Colour contrast in cartographic works using the principles of Johannes Itten. *The Cartographic Journal* 51(3): 203–213.

Boria E (2008) Geopolitical maps: A sketch history of a neglected trend in cartography. *Geopolitics* 13(2): 278–308.

Boria E and Rossetto T (2017) The practice of mapmaking: Bridging the gap between critical/textual and ethnographic research methods. *Cartographica* 52(1): 32–48.

Canali L (2015). Il Mediterraneo non è più liquido. E non è più poetico. *Limesonline*. Available at: www.limesonline.com/rubrica/il-mediterraneo-non-e-piu-liquido-e-non-e-piu-poetico (accessed 24 August 2023).

Dodge M, Kitchin R and Perkins C (eds) (2009) *Rethinking Maps: New Frontiers in Cartographic Theory*. Abingdon: Routledge.

Johnson JM (2001) Mapping ethnicity: Color use in depicting ethnic distribution. *Cartographic Perspectives* 40: 12–22.

Lo Presti L (2017) *(Un)Exhausted Cartographies: Re-Living the Visuality, Aesthetics and Politics in Contemporary Mapping Theories and Practices*. PhD Thesis, Università degli Studi di Palermo.

Lo Presti L (2019). Terraqueous necropolitics: Unfolding the low-operational, forensic, and evocative mapping of Mediterranean sea crossings in the age of lethal borders. *ACME: An International Journal for Critical Geographies* 18(6): 1347–1367.

Marocchi S (ed.) (2022) *Laura Canali: Pietre e Miraggi*. Exhibition Catalogue. Rome: Fondazione Besso.

Pain R (2009) Globalized fear? Towards an emotional geopolitics. *Progress in Human Geography* 33(4): 466–486.
Rossetto T (2018) Chromocartographies: An ethnographic approach to colours in Laura Canali's geopolitical maps. *Livingmaps Review* 4: 1–20.
Rossetto T (2019) *Object-Oriented Cartography: Maps as Things*. London and New York: Routledge.

42

MAPLAB

A Bloomberg newsletter connecting maps and the news

Laura Bliss and Marie Patino

Introduction

MapLab, a newsletter published by Bloomberg CityLab, was launched in 2017. It explores the world of mapping, and how it intersects with the news for an audience of designers, architects, mapmakers and other people interested in cartography and datavisual graphics. It is unique within the U.S. news media landscape, covering mapping in several ways: as an industry, as a powerful shaper of news consumption and geographic understanding, and as an artform that illuminates human experience. In this interview, Bloomberg journalists Laura Bliss and Marie Patino discuss the newsletter, its objectives, offshoot projects and the readers it has brought together.

How would you define/describe Bloomberg CityLab's MapLab? What are its major objectives?

MapLab[1] is a newsletter primarily about how maps and cartography intersect with the news. It was born out of Bloomberg CityLab journalist Laura Bliss's brain back in 2017, when CityLab was part of the Atlantic magazine (CityLab was acquired by Bloomberg from the Atlantic in 2020). Originally Laura wrote and published MapLab as a biweekly newsletter, and we turned it into a monthly product in 2022. Marie Patino, a graphics reporter at Bloomberg, has been a regular author since 2022; Laura now edits it.

We approach our work in a few different ways: for one, we highlight mapping projects by cartographers and researchers that provide important information and perspectives on key news events, covering topics including climate change (Patino, 2023b), elections (Bliss, 2020b), geopolitics (Bliss, 2020a), and urban planning (Dudley, 2021). We also cover cartography as an industry, comprising well-established GIS software companies such as ESRI, satellite imagery providers like MAXAR and PlanetLabs, and governmental agencies like NASA and Copernicus (Patino, 2021a).

We also write about how maps shape our perception of the world, for better or worse. Each and every one of us interacts with maps on a daily basis, so it's important to understand on a basic level how they're made, and what's at stake in some of the maps we're fed on social media, or that we see popping up in the news (Patino, 2022a). A map is not

DOI: 10.4324/9781003327578-50

neutral! It's the result of a series of choices taken by the cartographer(s) working on it. That's one of the objectives of MapLab in a way: sharpening readers' visual literacy, even if it's just a tiny bit. Maybe next time you'll see a map you'll ask yourself one more question, like, 'Why did the mapmaker use dots instead of a solid line to show the borderline between these two regions? Whose priorities does that reflect?'. And we think that's a huge win.

It's also really fun to look at the history and aesthetic dimensions of mapping. Cartography is a precise science with a long history, full of interesting people and endeavours. One of Marie's favourite examples is probably the Franco-Italian Cassini family who made the first map of France ever. It was done manually, was incredibly detailed and took . . . 70 years, and multiple generations of Cassinis (Baena, 2022)!

Marie is particularly interested in the design aspect of mapping, which is really an art form: some maps are just beautiful, and you can stare at them for hours. For example maps of ski resorts (Hirschfeld, 2022)—they have to thread a thin line between geographical accuracy and completeness, and they are so pretty and rich in detail: you see the slopes, you can distinguish little pine trees, the details of the mountains, and you can use these maps for orientation as well.

Laura also finds the evolution of map design and how maps have been used to affect social change to be captivating. When she was starting MapLab, she was especially inspired by the muckraking work of the Hull House Maps and Papers,[2] which used hand-collected survey data—and hand-painted maps—to show how poverty and overcrowding affected immigrants in the 1890s. They were made by women working at Jane Addams' Chicago settlement house in the 19th century. Completely untrained in cartography, these mapmakers were essentially proto-social workers and journalists, and they produced some of the first-ever maps that incorporated that kind of urban demographic data, which is now so common to see.

Cartography is such a rich field, and for us as journalists, that makes it fascinating to cover.

When and in which context was it established?

The idea for MapLab grew out of writing about maps and creating map-driven journalism for CityLab, which has a large audience of city planners, architects and designers, and other kinds of visually oriented people. Laura, who was then a staff writer at CityLab, found that stories that featured maps tended to draw significant interest and traffic; just like her, she realised, people love to stare at maps! There's something very transporting about them. Thinking there was great audience potential, she pitched MapLab as a newsletter. The format and frequency have changed a bit since its launch in November 2017 (Bliss, 2017), but the newsletter still always includes one central story (Laura likes to call it a 'featurette') that highlights what we think is the most interesting/important thing in maps that week/month, plus a round-up of links to articles that we think our thousands of map-hungry readers will enjoy.

How does MapLab fit in with the news coverage of CityLab and Bloomberg News more generally?

At CityLab and across the Bloomberg News newsroom more generally, we publish articles about maps, and articles that use maps, outside of the newsletter. Much like with MapLab, this includes a large range of topics and formats, from covering industry news to running long-form features about mapping projects to undertaking ambitious datavisual projects with interactive maps. We've found MapLab to be a powerful channel for connecting with an audience that loves maps and responds well to our stories and reporting projects.

One example of this is Laura's pandemic mapping project. She and Jessica Martin, CityLab's audience editor, used Bloomberg's social media channels and newsletters, including MapLab, to ask readers to make maps of what their lives looked like during the COVID lockdowns of 2020.[3] We received hundreds of responses from around the world, and published them throughout the pandemic. In 2022, Laura turned that project into a book called *The Quarantine Atlas* (Bliss, 2022) which collected some of those maps and entwined them with essays about how COVID reshaped cities and neighbourhoods and the writers' relationships to those places (Figure 42.1). Promoting the book at events over the past year has been really fun because it's been a chance for Laura to connect with MapLab subscribers in person for the first time.

Not only do we write about maps, but we also write with maps. We use them as a storytelling tool to help us explain things in a visually straightforward way. We think a map can hold 1,000 times more power than a paragraph of text when it's well crafted. Maps have this power to make issues much more tangible and catch the eye of readers, along with enabling them to geolocate the very topic of the story they're reading.

Bloomberg graphics team is lucky to have several very talented map-practitioners in-house, who help with this on a daily basis. Hayley Warren, for example, covers a lot of climate-change-related issues (Roston et al., 2022) and knows Copernicus databases inside and out. Jeremy CF Lin turns any map into a true eye-candy design (Stock et al., 2023). And most of the mapmakers at Bloomberg are good designers who also have a lot of expertise when it comes to manipulating any weird data format that's thrown their way. They know the limitations of the data they work with, and are able to contextualise it and question its sources. This knowledge is crucial especially when making geopolitical maps of current events, and in the newsroom we do this a lot. We have made a lot of maps of the war in Ukraine,[4] including offensives, counter-offensives and Russian attacks on the country. Getting this information and representing it accurately is very important and can be hard when you don't know where to look, and when the situation on the ground is evolving fast (Patino, 2022d).

Marie loves to work on maps that don't necessarily look like maps (Patino, 2022b), but contain enough geographical information for readers to understand what is being shown to them. For example, she worked on a story about people relocating from and to Manhattan during the pandemic, and represented the five boroughs with blob-like shapes (Holder and Patino, 2022), rather than the regular New York City map most people are used to seeing. That's fun and different, and that can help get people's attention. Looking for different ways to represent geographical information has helped her find cartographers who are attracted to maps that look odd to feature in MapLab, such as the artist Peter Gorman, who makes very minimalist maps (Patino, 2023a).

What does MapLab mean for the wider activity of the Bloomberg company?

MapLab is one of many newsletters that Bloomberg News publishes.[5] Virtually every news vertical—Businessweek, Technology, Green, CityLab, you name it—has at least one newsletter running at different frequencies. It's really nice to have MapLab, which is such a unique product within the news landscape, at Bloomberg. In addition to covering the world of maps in all the ways we do, we are also able to highlight the great maps and graphics stories that our journalists create.

Figure 42.1 The cover of *The Quarantine Atlas*.

Source: Credit: Black Dog & Leventhal. Courtesy Black Dog & Leventhal Publishers

What were and what are the intended audiences of this Lab? How do you reach such audiences? How do they interact with or comment on such materials?

Our intended audience is really all map 'nerds' out there, but also people interested in urbanism, architecture, transportation, environment and design, because cartography touches all of these things. If you think of public transportation, for example, maps of subway systems and public transit networks are a whole topic area that we've written about many times in MapLab—for example, about the potential need for a redesign of the London Underground map after the Elizabeth line was inaugurated in 2022 (Patino, 2022c). If you're a city dweller, you see these maps every day and might not even notice them anymore, or might have never questioned why they look the way they look. With MapLab, we try to approach these maps from a fresh angle, or introduce you to someone whose original work touches upon them, and hopefully help you notice them again.

We also get quite a few pitches from people who are active in the field of cartography and want to send us their projects, and these are always so interesting to look at. We can't cover them all unfortunately, but we always feel so privileged when people take the time to share their personal cartography projects with us.

Professionals and hobbyists in the field make up an incredibly creative and generous community, and every time we feature someone's work in MapLab, the conversations we have with them are fascinating. One example is the 30DayMapChallenge, which was launched in 2019 by a mapmaker named Topi Tjukanov. The challenge encourages people to make a map a day around a specific prompt for a month, and post their creations on social media. We covered the 2021 edition in MapLab (Patino, 2021b).

During the pandemic you launched the 'How 2020 remapped your worlds' campaign. Tell us about this initiative and its different stages. What can the success of this initiative tell us about the role of maps in people's life?

The idea for the COVID mapping project grew out of a very normal journalistic question—how is this global event affecting peoples' everyday lives?—combined with the special attention we give to maps at CityLab, thanks in part to MapLab. We know that people love maps and that maps can be an incredibly powerful medium for expressing things that are harder to put into words. We were so blown away by the level of response we got to those original call-outs in April 2020. In that moment where there was so much uncertainty and fear around the novel coronavirus, we suspect it was cathartic for a lot of people to sit down and make a map, using whatever materials they had on hand.

But we think the response also reflects the fact that the changes that COVID wrought on peoples' lives were geographic and spatial in nature. Border shutdowns, one-kilometre lockdown zones, keeping six-feet apart; the way downtowns and suburbs were inverted in terms of daily activity; the way kitchens and living rooms became offices, classrooms, etc. All of this could be mapped and people found many, many beautiful, creative ways to do so.

Are there any other particular initiatives you proposed in the past years?—'The maps that make us' campaign. Such example regards forms of private, existential, visceral mappings, as in the case of the 'How 2020 remapped your worlds' campaign. How do you see such

more 'humanistic' forms of mapping in comparison with other kinds of map projects featured in the MapLab?

'The Maps That Make Us' (Bliss, 2019) was a series of personal essays that Laura commissioned and edited for CityLab in 2019. Writers included professional cartographers, artists, a map librarian and some wonderful journalists, all of whom reflected on the role a particular map had played in their lives. The essays were quite heartfelt and beautiful, and looking back, the project probably helped set the stage for the COVID mapping project in terms of its focus on the more humanistic dimensions of mapping. Personal maps and essays are certainly quite different from, say, covering partisan gerrymandering or the science behind mapping ocean heatwaves. But in the world of MapLab (and CityLab), all of it is fair game. You might even say that we're all over the map, no pun intended. But in our world, that's a good thing.

Notes

1 See www.bloomberg.com/citylab/maplab
2 See https://florencekelley.northwestern.edu/historical/hullhouse/
3 See www.bloomberg.com/features/2020-coronavirus-lockdown-neighborhood-maps/
4 See www.bloomberg.com/graphics/2022-ukraine-russia-us-nato-conflict/
5 See www.bloomberg.com/account/newsletters

References

Baena V (2022) Revolutionary cartography and the Cassini map of France. *New York Public Library*, 18 October. Available at: www.nypl.org/blog/2022/10/18/revolutionary-cartography-and-cassini-map-france.

Bliss L (2017) Introducing MapLab: A Biweekly tour of the ever-expanding cartographic landscape. *Bloomberg City Lab—MapLab Newsletter*, 14 November. Available at: www.bloomberg.com/news/articles/2017-11-14/introducing-maplab-the-newsletter-for-map-lovers.

Bliss L (2019) What are the maps that make us? *Bloomberg City Lab—MapLab Newsletter*, 9 August. Available at: www.bloomberg.com/news/articles/2019-08-09/exploring-the-maps-that-make-us.

Bliss L (2020a) MapLab: Contexted territories. *Bloomberg City Lab—MapLab Newsletter*, 6 February. Available at: www.bloomberg.com/news/newsletters/2020-02-06/maplab-contested-territories.

Bliss L (2020b) MapLab: The rise of micro-politics. *Bloomberg City Lab—MapLab Newsletter*, 21 October. Available at: www.bloomberg.com/news/newsletters/2020-10-21/maplab-the-rise-of-micro-politics.

Bliss L (2022) *The Quarantine Atlas: Mapping Global Life Under COVID-19—A Bloomberg City-Lab Project*. New York: Black Dog & Leventhal Publishers.

Dudley D (2021) MapLab: New York City's enduring subway map debate. *Bloomberg City Lab—MapLab Newsletter*, 15 December. Available at: www.bloomberg.com/news/newsletters/2021-12-15/maplab-new-york-city-s-enduring-subway-map-debate.

Hirschfeld C (2022) Is this the end of the trail map? *The New York Times*, 8 March. Available at: www.nytimes.com/2022/03/08/travel/ski-trail-maps-disappear.html.

Holder S and Patino M (2022) More people are moving to Manhattan than before the pandemic. *Bloomberg City Lab*, 8 June. Available at: www.bloomberg.com/graphics/2022-manhattan-real-estate-moving-data/.

Patino M (2021a) MapLab: What it takes to map mars. *Bloomberg City Lab—MapLab Newsletter*, 24 March. Available at: www.bloomberg.com/news/newsletters/2021-03-24/maplab-what-it-takes-to-map-mars.

Patino M (2021b) MapLab: Thirty days of maps. *Bloomberg City Lab—MapLab Newsletter*, 29 December. Available at: www.bloomberg.com/news/newsletters/2021-12-29/maplab-thirty-days-of-maps.

Patino M (2022a) MapLab: The case for a cartographer's code of ethics. *Bloomberg City Lab—MapLab Newsletter*, 10 March. Available at: www.bloomberg.com/news/newsletters/2022-03-09/maplab-the-case-for-a-cartographer-s-code-of-ethics.

Patino M (2022b) MapLab: (Almost) 101 ways to map a dataset. *Bloomberg City Lab—MapLab Newsletter*, 6 April. Available at: www.bloomberg.com/news/newsletters/2022-04-06/maplab-almost-101-ways-to-map-a-dataset.

Patino M (2022c) MapLab: Is it time for an overhaul of the London tube map? *Bloomberg City Lab—MapLab Newsletter*, 2 June. Available at: www.bloomberg.com/news/newsletters/2022-06-01/maplab-is-it-time-for-an-overhaul-of-the-london-tube-map.

Patino M (2022d) MapLab: Monitoring the invasion of Ukraine from outer space. *Bloomberg City Lab—MapLab Newsletter*, 19 October. Available at: www.bloomberg.com/news/newsletters/2022-10-19/maplab-monitoring-the-invasion-of-ukraine-from-outer-space.

Patino M (2023a) MapLab: The power of minimalist maps. *Bloomberg City Lab—MapLab Newsletter*, 22 February. Available at: www.bloomberg.com/news/newsletters/2023-02-22/maplab-an-interview-with-artist-and-mapmaker-peter-gorman.

Patino M (2023b) MapLab: Tracking marine heatwaves. *Bloomberg City Lab—MapLab Newsletter*, 27 July. Available at: www.bloomberg.com/news/newsletters/2023-07-26/maplab-tracking-marine-heatwaves.

Roston E, Kaufman L and Warren H (2022). How the world's richest people are driving global warming. *Bloomberg News*, 24 March. Available at: www.bloomberg.com/graphics/2022-wealth-carbon-emissions-inequality-powers-world-climate/.

Stock K, Lin JCF and Ellis R (2023) America's loneliest road is finally EV-ready. *Bloomberg Green*, 23 August. Available at: www.bloomberg.com/graphics/2023-ev-charging-map-road-trip/.

INDEX

Page numbers in *italics* indicate figures; page numbers in **bold** indicate tables.

For Product Safety Concerns and Information please contact our EU representative GPSR@taylorandfrancis.com Taylor & Francis Verlag GmbH, Kaufingerstraße 24, 80331 München, Germany

Batch number: 10399615

Printed by Printforce, the Netherlands